THE GLOBAL CITIES READER

Since the mid-1990s, research on global cities has exploded throughout the social sciences. It has now become one of the most exciting, if controversial, approaches to the study of urban life today.

Fifty generous selections, including contributions from John Friedmann, Michael Peter Smith, Saskia Sassen, Peter J. Taylor, Manuel Castells and Anthony D. King, explore the interrelationships between cities and globalization. The seven parts with accompanying editorial introductions guide the student through the key theoretical, methodological and empirical debates.

The Global Cities Reader explores the major foundations of research on globalized urbanization. Classic and contemporary case studies of globalizing cities from Europe, North America and East Asia as well as from emerging world city regions of the global South are presented. The political and cultural dimensions of global city formation are examined in separate parts. The Reader concludes by examining the refinement and critique of global cities research since the 1990s.

Neil Brenner is Associate Professor in the Department of Sociology and Metropolitan Studies Program, New York University.

Roger Keil is Professor in the Faculty of Environmental Studies, York University, Toronto.

THE ROUTLEDGE URBAN READER SERIES

Series editors

Richard T. LeGates
Professor of Urban Studies, San Francisco State University

Frederic Stout
Lecturer in Urban Studies, Stanford University

The Routledge Urban Reader Series responds to the need for comprehensive coverage of the classic and essential texts that form the basis of intellectual work in the various academic disciplines and professional fields concerned with cities.

The readers focus on the key topics encountered by undergraduates, graduates and scholars in urban studies and allied fields. They discuss the contributions of major theoreticians and practitioners and other individuals, groups, and organizations that study the city or practice in a field that directly affects the city.

As well as drawing together the best of classic and contemporary writings on the city, each reader features extensive general, section and selection introductions prepared by the volume editors to place the selections in context, illustrate relations among topics, provide information on the author and point readers toward additional related bibliographic material.

Each reader will contain:

- Approximately thirty-six *selections* divided into approximately six sections. Almost all of the selections will be previously published works that have appeared as journal articles or portions of books.
- A *general introduction* describing the nature and purpose of the reader.
- Two- to three-page *section introductions* for each section of the reader to place the readings in context.
- A one-page *selection introduction* for each selection describing the author, the intellectual background of the selection, competing views of the subject matter of the selection and bibliographic references to other readings by the same author and other readings related to the topic.
- A plate section with twelve to fifteen plates and illustrations at the beginning of each section.
- An index.

The types of readers and forthcoming titles are as follows:

THE CITY READER

The City Reader: third edition – an interdisciplinary urban reader aimed at urban studies, urban planning, urban geography and urban sociology courses – will be the *anchor urban reader*. Routledge published a first edition of *The City Reader* in 1996 and a second edition in 2000. *The City Reader* has become one of the most widely used anthologies in urban studies, urban geography, urban sociology and urban planning courses in the world.

URBAN DISCIPLINARY READERS

The series will contain *urban disciplinary readers* organized around social science disciplines. The urban disciplinary readers will include both classic writings and recent, cutting-edge contributions to the respective disciplines. They will be lively, high-quality, competitively priced readers which faculty can adopt as course texts and which will also appeal to a wide audience.

TOPICAL URBAN ANTHOLOGIES

The urban series will also include *topical urban readers* intended both as primary and supplemental course texts and for the trade and professional market.

INTERDISCIPLINARY ANCHOR TITLE

The City Reader: third edition
Richard T. LeGates and Frederic Stout (eds)

URBAN DISCIPLINARY READERS

The Urban Geography Reader
Nicholas R. Fyfe and Judith T. Kenny (eds)

The Urban Sociology Reader
Jan Lin and Christopher Mele (eds)

Forthcoming:

The Urban Politics Reader
Elizabeth Strom and John Mollenkopf (eds)

The Urban and Regional Planning Reader
Eugenie Birch (ed.)

The Urban Design Reader
Michael Larice and Elizabeth Macdonald (eds)

TOPICAL URBAN READERS

The City Cultures Reader: second edition
Malcolm Miles and Tim Hall, with Iain Borden (eds)

The Cybercities Reader
Stephen Graham (ed.)

The Sustainable Urban Development Reader
Stephen M. Wheeler and Timothy Beatley (eds)

The Global Cities Reader
Neil Brenner and Roger Keil (eds)

■ ■ ■ ■ ■ ■

For further information on The Routledge Urban Reader Series
please visit our website:

www.geographyarena.com/geographyarena/urbanreaderseries

or contact:

Andrew Mould
Routledge
Haines House
21 John St
London WC1N 2BP
United Kingdom
andrew.mould@tandf.co.uk

Richard T. LeGates
Urban Studies Program
San Francisco State University
1600 Holloway Avenue
San Francisco, California 94132
(415) 338-2875
dlegates@sfsu.edu

Frederic Stout
Urban Studies Program
Stanford University
Stanford, California 94305-6050
(650) 725-6321
fstout@stanford.edu

The Global Cities Reader

Edited by

Neil Brenner

and

Roger Keil

Routledge
Taylor & Francis Group

LONDON AND NEW YORK

First published 2006
by Routledge
2 Park Square, Milton Park, Abingdon, Oxon OX14 4RN

Simultaneously published in the USA and Canada
by Routledge
711 Third Avenue, New York, NY 10017

Routledge is an imprint of the Taylor & Francis Group, an informa business

Typeset in Amasis and Akzidenz Grotesk by
Graphicraft Limited, Hong Kong

British Library Cataloguing in Publication Data
A catalogue record for this book is available from the British
Library

Library of Congress Cataloging in Publication Data
The global cities reader / edited by Neil Brenner and Roger Keil.
    p. cm. — (Urban reader series)
  Includes bibliographical references and index.
  1. Cities and towns. 2. Urbanization. 3. Globalization.
I. Brenner, Neil. II. Keil, Roger, 1957– III. Series: Routledge urban
reader series.
  HT119.G64 2005
  307.76—dc22   2005009601

ISBN10: 0–415–32344–4 (hbk)
ISBN10: 0–415–32345–2 (pbk)
ISBN13: 9–78–0–415–32344–4 (hbk)
ISBN13: 9–78–0–415–32345–1 (pbk)

We dedicate this book to John Friedmann,
friend and mentor

Contents

Plates

Contributors

Janet Abu-Lughod is Professor Emerita of Sociology at the New School for Social Research in New York City, USA.

Jonathan V. Beaverstock is Professor of Economic Geography at Loughborough University, UK.

Fernand Braudel (1902–1985) was President of Ecole Pratique des Hautes Etudes and a Professor at the Collège de France, Paris, France.

Neil Brenner works in the Department of Sociology and Metropolitan Studies Program at New York University, USA.

Simone Buechler is Assistant Professor of Metropolitan Studies at New York University, USA.

Manuel Castells holds the Wallis Annenberg Chair in Communication, Technology and Society at the University of Southern California, USA

Robert B. Cohen is a Fellow with the Economic Strategy Institute in Washington, DC and President of Cohen Communications Group in New York City, USA.

Mike Douglass is Director of the Globalization Research Center and a Professor in the Department of Urban and Regional Planning at University of Hawaii Manoa, USA.

Susan S. Fainstein is Professor of Urban Planning in the Graduate School of Architecture at Columbia University, New York City, USA.

Joe R. Feagin is Ella McFadden Professor of Liberal Arts at Texas A&M University, USA.

Steven Flusty teaches in the Department of Geography at York University, Toronto, Canada.

John Friedmann is Professor Emeritus of Urban Planning in the Graduate School of Architecture and Planning at UCLA, USA and Honorary Professor in the School of Community and Regional Planning at the University of British Columbia, Canada.

Stephen Graham is Professor of Human Geography at Durham University, UK.

Richard Grant teaches in the Department of Geography at the University of Miami, Florida, USA.

Anne Haila is Professor of Urban Studies in the Department of Social Policy at the University of Helsinki, Finland.

Sir Peter Hall is Bartlett Professor of Planning at University College London, UK.

Ulf Hannerz is Professor of Social Anthropology at Stockholm University, Sweden.

Richard Child Hill is Professor Emeritus of Sociology at Michigan State University, USA.

Roger Keil is Professor of Environmental Studies at York University, Toronto, Canada.

June Woo Kim teaches in the Centre for Advanced Studies in the Faculty of Arts and Social Sciences at the National University of Singapore.

Anthony D. King is Bartle Professor of Art History at the State University of New York at Binghamton, USA.

Stefan Krätke is Professor of Economic and Social Geography at the Europa-Universität Viadrina in Frankfurt (Oder), Germany.

Henri Lefebvre (1901–1990), a French urbanist and philosopher, was one of the most influential urban theorists of the twentieth century.

Ute Lehrer teaches Environmental Studies at York University, Toronto, Canada.

Timothy Luke is University Distinguished Professor in Political Science at the Virginia Polytechnic Institute and State University, Blacksburg, Virginia, USA.

Takashi Machimura is Professor in the Graduate School of Social Sciences at Hitotsubashi University, Tokyo, Japan.

Warren Magnusson is Professor of Political Studies at the University of Victoria, Canada.

Peter Marcuse is Professor of Urban Planning in the Graduate School of Architecture at Columbia University in New York City, USA.

Margit Mayer is Professor of Political Science in the John F. Kennedy Institute of the Free University Berlin, Germany.

Jan Nijman is Professor of Geography at the University of Miami, Florida, USA.

Kris Olds is Associate Professor of Geography at the University of Wisconsin, Madison, USA.

Nihal Perera is Associate Professor of Urban Planning and Director of Asian Studies at Ball State University, Indiana, USA.

Riccardo Petrella is Professor of Political and Social Sciences at the University of Louvain, Belgium.

Jennifer Robinson teaches in Geography at The Open University, UK.

Nestor Rodriguez is Professor and Chair of Sociology and Co-Director of the Center for Immigration Research at the University of Houston, Texas, USA.

Klaus Ronneberger is an independent scholar and author based in Frankfurt am Main, Germany.

Robert Ross is Professor in the Department of Sociology at Clark University, Massachusetts, USA.

Michael Samers is Senior Lecturer in Geography at the University of Nottingham, UK.

Leonie Sandercock is Professor in the School of Community and Regional Planning, University of British Columbia, Vancouver, Canada.

Saskia Sassen is Ralph Lewis Professor of Sociology at the University of Chicago and Centennial Visiting Professor in the Department of Sociology at the London School of Economics, UK.

Christian Schmid is an urban geographer who works in the Institute of Geography at the University of Berne and in the Department of Architecture at the Swiss Federal Institute of Technology (ETH) in Zurich, Switzerland.

Allen J. Scott is Distinguished Professor of Public Policy and Geography at the University of California, Los Angeles, USA.

Gavin Shatkin teaches Urban Planning in the Taubman College of Architecture and Urban Planning at the University of Michigan, USA.

David Simon is Professor in Development Geography at Royal Holloway, University of London, UK.

Michael Peter Smith is Professor of Community Studies and Development at the University of California, Davis, USA.

Richard G. Smith teaches in the Geography Department at Leicester University, UK.

Edward W. Soja is Professor in the Department of Urban Planning at the University of California, Los Angeles, USA.

Deyan Sudjic is architecture critic for the *Guardian* and a journalist and editor based in London, UK.

Peter J. Taylor is Professor of Geography at Loughborough University, UK.

Kent Trachte is the Dean of the College at Franklin & Marshall College, Pennsylvania, USA.

Goetz Wolff is an independent researcher, teacher and consultant in Los Angeles and serves as Lecturer in the Urban Planning Department at the University of California, Los Angeles, USA.

Henry Wai-Chung Yeung is Associate Professor of Geography at the National University of Singapore.

Sharon Zukin is the Broeklundian Professor of Sociology at Brooklyn College and the Graduate Center of the City University of New York in Manhattan, USA.

Acknowledgments

We have accumulated many debts in completing this Reader. We are extremely grateful to Ahmed Allahwala, who helped us with various editorial tasks at an early stage of our work on the manuscript. Sincerest thanks also go to Cornelia Sussmann, who tirelessly pursued permissions for most of the contributions we have assembled in this book. In the process, she learned much more about the thorny thicket of publishing than she had ever hoped. We are deeply indebted to Matthew J. Murphy in New York City, who scanned the texts and images with professional skill and rapid turnover. New York University's Digital Studio provided a superb computer infrastructure for this work; thanks go to Richard Malenitza for cheerfully coordinating our work there. In Toronto, we must express our deepest gratitude to Jakub Lisowski, who worked closely with us around the clock in preparing the final text and tables; we could not have done it without him.

We are grateful to the editors of the Routledge Urban Reader Series, Richard T. LeGates and Frederic Stout, for their unfailing support for our project from day one. Their advice and feedback were always appreciated. We would like to convey our warm thanks to Andrew Mould of Routledge, who kept us on target. We are grateful as well to our friendly yet determined editors, Melanie Attridge and Zoe Kruze, who indulged our repeated requests for deadline extensions. Our production editors at Routledge, Jocelyn Lotstrom and Tristan Lee, did a superb job with preparation of the final text.

Credit goes to our referees as well as to colleagues John Friedmann, Saskia Sassen and Kris Olds for their good advice. Steve Graham's *The Cybercities Reader* has been a hard act to follow and a superb model to try to emulate. Thanks, Steve, for your continued support.

Some of the readings in this volume are original contributions. Our appreciation goes out to our friends and colleagues who succumbed to our prodding to write up a fresh piece for this project: Simone Buechler, Steven Flusty, Tony King, Stefan Krätke, Ute Lehrer, Peter Marcuse and Christian Schmid. We thank all of the other contributors for their extreme helpfulness in commenting on the painfully abbreviated versions of their work which we have presented in this volume.

Special recognition goes to several friends who generously provided us with photographs to illustrate their own and other readings in the book: Anne Haila, Ute Lehrer (who also contributed the frontispiece), Takashi Machimura, Kris Olds and Henry Wai-Chung Yeung.

Neil Brenner thanks his friends in the Central Park Track Club for setting a fast pace, and for many lessons learned.

Roger Keil thanks his family for giving him some space to work on this book while they went skiing during this cold Toronto winter which has just ended. There will be more slopes for me to join you on next time I am sure.

Neil Brenner and Roger Keil
New York and Toronto, March 2005

EDITORS' INTRODUCTION

Global city theory in retrospect and prospect

Plate 1 Toronto (Roger Keil)

It is possible you have been to Troy without recognising the city. The road from the airport is like many others in the world. It has a superhighway and is often blocked. You leave the airport buildings which are like space vessels never finished, you pass the packed carparks, the international hotels, a mile or two of barbed wire, broken fields, the last stray cattle, billboards that advertise cars and Coca-Cola, storage tanks, a cement plant, the first shanty town, several giant depots for big stores, ring-road flyovers, working class flats, a part of an ancient city wall, the old boroughs with trees, crammed shopping streets, new golden office blocks, a number of ancient domes and spires, and finally you arrive at the acropolis of wealth.

(Berger 1990: 170)

A global city – the global city

At some point, most travelers have been on some version of the road into the imagined city of Troy, which British novelist John Berger describes in this opening quotation. We all seem to know this kind of metropolitan region, which appears to exist in one form or another around the globe. This metro-politanized environment is now the place where most of us live. It is an internationalized network of local places that are now increasingly bound together and interdependent, an "archipelago of enclaves" (Hajer and Reijndorp 2001: 53) which most people in the world – including the editors of this Reader – now call their "home."

Our initial approach to the theme of this book is necessarily grounded in our own everyday lives. Both of this book's editors have lived in several major North American and western European global cities – including Amsterdam, Chicago, Frankfurt, Los Angeles, New York and Toronto – and these experiences provide us with a useful starting point for exploring the globalized character of urban life today. Since New York City is featured prominently elsewhere in this book (see Reading 16 by Zukin), let us begin our journey by traveling briefly to Toronto, where the globalization of urbanization has been manifesting itself in striking, if place-specific, ways since the mid-1970s.

Toronto is the city in Canada that most closely resembles the fictional Troy described by John Berger in the passage quoted. The municipality of Toronto is a dense core city of 2.5 million people surrounded by another 2.5 million in sprawling suburbs, exurbs and edge cities. While Toronto is not situated within the top tier of the global urban hierarchy, it does serve as the urban core of a second-tier global financial, cultural and manufacturing region. It thus exemplifies many of the features of global city formation that are examined at length in the contributions to this Reader. Toronto is the economic hub of the province of Ontario, Canada's industrial heartland; it is a basing point for significant transnational corporations and for Canada's financial industries; it hosts the country's largest and busiest airport; and it is a major hub within the highway networks of Canada and the northeastern United States. Additionally, Toronto has also recently become a major site in which a neoliberalizing, market-oriented restructuring of Canadian capitalism has been attempted (Keil 2002; Kipfer and Keil 2002). As Todd (1995) and Sassen (2000) have pointed out, the case of Toronto is particularly interesting because its transformation into a global metropolitan center has occurred relatively recently. Additionally, Toronto is a noteworthy example of a global city because it serves as the port of entry, and often as the final destination, for a large segment of Canada's extremely diverse immigrant population. Indeed, some observers have suggested that it contains the most multicultural urban population in the world.

Toronto's character as a "global city" has been forged through the interaction of two basic dimensions of contemporary urban life. On the one hand, processes of economic globalization are expressed in the *verticality* of the downtown central business district, densely packed with highrise office towers, where capitalist command and control functions and global financial industries are centralized. On the other hand, the materialization of these processes in people's everyday lives is expressed in the *horizontality* of the sprawling urban region, an extraordinarily diverse social, cultural and political space in which class, cultural and gender differences are continuously produced, contested and reworked at a neighborhood scale. It is through the collision of these opposed social dynamics – globalizing and localizing; homogenizing and differentiating – that Toronto has been transformed into a globalized city-region, a site of apparently seamless connectivity to world markets, the global urban system, global diasporic networks and global cultural flows.

While Toronto's global connectivity is viewed in an unambiguously positive light by those who are concerned to market the city as a site for transnational capital investment, it also entails certain dangers and vulnerabilities. The latter are illustrated starkly in the outbreak and management of the SARS (Severe Acute Respiratory Syndrome) virus in Toronto during the Spring of 2003. The SARS epidemic eventually killed 44 individuals, made hundreds sick and led to the quarantine of thousands over a four-month period in the Toronto metropolitan region. The spread of the disease was immediately linked to Toronto's role as a hub in international air traffic, in particular due to its web of connections to cities in East Asia, where the virus originated. Indeed, the same channels of transnational connectivity (in this case, transnational air routes) that are intended to facilitate business and financial transactions between Hong Kong and Toronto now became the basis for the transmission of a highly infectious disease. Interestingly, even as the city recovered from the actual SARS epidemic, its status as a globally connected urban center became a topic of intensive debate and public inquiry. When the World Health Organization (WHO) issued an advisory warning against traveling to Toronto, the city's international business relationships (including tourism, conferences and cultural exchanges) were seriously disrupted. Subsequently, the multicultural fabric of this particular "cosmopolis" (Sandercock 1998) was also seriously strained as the disease was characterized, in some quarters, in racist language – for instance, as a "Chinese disease" (Ali and Keil 2006).

As this example illustrates, cities such as Toronto are today tightly embedded within a broad range of global networks – demographic, cultural, economic, ecological, epidemiological – that have major implications for the everyday lives of their inhabitants. These worldwide urban networks are the subject of this volume. In particular, through a broad range of interdisciplinary investigations, the contributions to this Reader are intended to explore the origins, characteristics and consequences of this heightened global connectivity among contemporary urban centers.

Globalized urbanization and the formation of global cities

In 1900, only about 10 per cent of total world population lived in cities. At some point around the turn of the millennium, more than half of the world's population was located in urban settings (see Reading 32 by Luke). Urbanization processes are rapidly accelerating, and extending ever more densely across the earth's surface. Indeed, the combined demographic, economic and sociocultural trend of urbanization has resulted in the formation of a globalized "skein of the urban" in a transnational system of spatially concentrated human settlements (Friedmann 2002: 6). This pattern of increasingly globalized urbanization contradicts earlier predictions, in the waning decades of the twentieth century, that the era of urbanization was nearing its end. As Manuel Castells (2004) has observed, many writers during this period predicted that, due to the rise of new informational technologies since the 1980s, cities would disappear, remote work would become more pervasive, activities and people would be dispersed, and a demographic move back to rural areas would ensue. However, as Castells and several other prominent urbanists (e.g., Sassen 2000) have demonstrated, such predictions have proven to be gravely mistaken. Instead, in most countries, urbanization levels are now higher than ever, with some world regions, such as the Americas and Europe, now becoming more than 80 per cent urban (Castells 2004). The urbanization process is thus being consolidated, intensified and accelerated under contemporary conditions of globalization.

Just as importantly, as many of the contributors to this Reader have observed, the globalization of urbanization during the late twentieth and early twenty-first centuries has entailed the emergence of a new type of city, the *global* or *world* city. In the first instance, these terms refer to a set of global command and control centers that are connected in transnationally networked hierarchies of economic, demographic and sociocultural relationships. To be sure, some participants in debates on contemporary urbanization have questioned the existence of a specific class of global cities that can be positively identified through empirical data (see, for instance, Reading 46 by M.P. Smith and Reading 49 by R.G. Smith). Others have argued for the investigation of "globalizing cities" across the world economy rather than focusing solely on a limited set of leading global centers (see Reading 25 by Shatkin, Reading 26 by Robinson, Reading 27 by Grant and Nijman and Reading 44 by Marcuse). It is evident, however, that a number of major urban centers have emerged in recent decades that increasingly transcend their respective national city systems and have come to articulate localized economic, demographic and sociocultural processes to a broader, globalized configuration of capitalism. Hierarchized, networked or otherwise tightly interconnected, these "world-city nodes" (Friedmann 2002: 9) arguably constitute an important part of the global economic architecture that has been emerging since the worldwide economic recession of the 1970s. They also have a major impact upon people's everyday lives, both within and beyond cities' formal juridical boundaries. These globalizing cities, and the broader processes of globalized urbanization that have produced them, will be examined at length in the contributions to this Reader.

Aims of the Global Cities Reader

It is only relatively recently, since the early 1980s, that urban scholars have begun to explore the question of how global forces and dynamics impact local and regional social spaces. While a number of pioneering scholars, such as Fernand Braudel and Janet Abu-Lughod (see Readings 2 and 4), had previously considered the role of transnational linkages in their historical investigations of urban development, this issue only became a major preoccupation for urban scholars across the social sciences as of the 1980s. During that decade, the contributions of urbanists such as Manuel Castells, John Friedmann, Anthony D. King, Saskia Sassen, Michael Peter Smith, Edward Soja and Michael Timberlake, among others, helped generate widespread interest in the interplay between globalization and urban development under contemporary conditions (for overviews, see Alger 1990; Yeoh 1999; D.A. Smith 2000; Scott 2001; Taylor 2003; Gerhard 2004). The publication of influential volumes such as Michael Timberlake's *Urbanization in the World-Economy* (1985), Michael Peter Smith and Joe R. Feagin's *The Capitalist City* (1987), Saskia Sassen's *The Global City* (1991) and Anthony D. King's *Urbanism, Colonialism and the World-Economy* (1991) consolidated this developing research agenda within urban studies.

During the course of the 1990s, the sustained attention to the "impassable dialectic of local and global" (Lipietz 1993: 16) among world cities researchers generated an extraordinarily creative outpouring of research on cities throughout the world economy. One of the major contributions of world cities research has been to relate the dominant socioeconomic trends within these cities – for instance, industrial restructuring, changing patterns of capital investment, the expansion and spatial concentration of the financial and producer services industries, labor-market segmentation, sociospatial polarization and class and ethnic conflict – to the emergence of a worldwide urban hierarchy and the global economic forces that underlie it. Additionally, scholars of global city formation have introduced innovative studies of, among other topics, urban governance restructuring, the transformation of urban sociospatial form, the emergence of new forms of urban sociopolitical contestation and the reorganization of global urban hierarchies. In the wake of these extensive, wide-ranging research forays, the concept of the global city is now used throughout the field of urban studies and has animated intense debates on the changing character of urban life in late twentieth and early twenty-first century capitalism. More recently, the publication of major books such as Paul Knox and Peter Taylor's *World Cities in a World-System* (1995), Peter Marcuse and Ronald van Kempen's *Globalizing Cities* (2000), Allen J. Scott's *Global City-Regions: Trends, Theory, Policy* (2001), a fully revised second edition of Sassen's *The Global City* (2002) and Peter Taylor's synthesis, *World City Network* (2003), arguably signals that the "world city hypothesis" initially proposed in the mid-1980s by John Friedmann (1986) has now become one of the most exciting, if also controversial, approaches to the study of the contemporary urban condition.

Against the background of these multifaceted investigations and debates, the present volume is intended to accomplish a number of intertwined goals. First, the Reader provides an overview of the major theoretical and methodological foundations for interdisciplinary research on cities and globalization. Second, the readings present a general survey of the key substantive areas of global cities research and explore some of the core debates and controversies that have emerged from such investigations. Third, although the contributions to the book can necessarily focus upon only a limited selection of empirical case studies, those which have been included are intended to illuminate the divergent national and local pathways through which global city formation is occurring in different zones of the world economy. Fourth, many of the readings explore the multifaceted problems of urban policy, planning and design that are specific to large globalizing city regions, and thereby underscore the manifold ways in which contemporary patterns of urban restructuring are mediated through state institutions, policies and regulatory strategies. Finally, this volume is intended to advance recent approaches to urban studies that emphasize the role of local and transnational agents in the production of urban economic, political and cultural spaces (Keil 1998; M.P. Smith 2001). Thus, while much of global cities research has, quite appropriately, examined the role of large-scale organizations such as transnational corporations (TNCs) and state institutions, many of the contributors to this Reader argue that sociopolitical struggles and grassroots mobilizations likewise figure centrally in the production and transformation of globalized urban spaces.

Given the impressive breadth and depth of this massive, constantly evolving research field, it would be impossible in a single volume to provide a comprehensive survey of all of the relevant perspectives that may be said to fall under the broad rubric of global cities research. Therefore, we have selected readings that, in our judgment, introduce some of the most essential theoretical categories, elaborate new modes of conceptualization, illuminate important empirical trends and/or present particularly provocative political perspectives relating to the process of global city formation. In 1982, in one of the foundational statements of global cities research, John Friedmann and Goetz Wolff (1982: 320; see Reading 6) suggested that "[t]he world city 'approach' is, in the first instance, a methodology, a point of departure, an initial hypothesis. It is a way of asking questions and of bringing footloose facts into relation." While research on cities and globalization has progressed considerably since Friedmann and Wolff's initial intervention, we believe that their methodological orientation remains as salient as ever today. Accordingly, the contributions to this Reader have been chosen not least because they pose extremely incisive, provocative questions regarding the interplay between urban development and global restructuring. As we shall see, their answers to these questions, and their methods for generating such answers, remain quite contentious, and continue to provoke intense debate.

Capitalism, urbanization and the world system

Until the late twentieth century, the dominant Anglo-American approaches to urban studies tended to presuppose that cities were neatly enclosed within national territories and nationalized central place hierarchies (Taylor 2003). As Michael Timberlake (1985) notes, urbanization processes have typically been studied by social scientists as if they were isolated in time and explicable only in terms of other processes and structures of narrow scope, limited to the boundaries of such areas as nations or regions within nations.

(Timberlake 1985: 3)

Thus, for example, postwar regional development theorists such as Myrdal (1957) and Hirschman (1958) viewed the national economy as the basic container of spatial polarization between core urban growth centers and internal peripheral zones. Concomitantly, urban geographers such as Berry (1961) and Pred (1977) assumed that the national territory was the primary scale upon which rank-size urban hierarchies and city-systems were organized. Indeed, even early uses of the term "world city" by writers such as Patrick Geddes (1924) and Peter Hall (1966) likewise expressed this set of assumptions: the cosmopolitan character of world cities was interpreted as an expression of their host states' geopolitical power. The possibility that urban development or the formation of urban hierarchies might be conditioned by supranational or global forces was not systematically explored.

This nationalized vision of the urban process was destabilized as of the late 1960s and early 1970s, with the rise of radical approaches to urban political economy. The seminal contributions of neo-Marxist urbanists such as Henri Lefebvre (1968), Manuel Castells (1972), David Harvey (1973, 1982) and others generated a wealth of new categories and methods through which to analyze the specifically capitalist character of modern urbanization processes. From this perspective, contemporary cities were viewed as spatial materializations of the core social processes associated with the capitalist mode of production, including, in particular, capital accumulation and class struggle. While these new approaches did not explicitly investigate the global parameters for contemporary urbanization (but see Reading 50 by Lefebvre), they did embed cities within a macrogeographical context defined by the ongoing development and restless spatial expansion of the capitalist world economy. In particular, capitalist urbanization was now conceived as an arena, expression and outcome of the fundamental processes of uneven geographical development that rippled continuously across the landscape of state territoriality at all spatial scales, from the global and the national to the regional and the local (Harvey 1982; Massey 1985; N. Smith 1990). In this manner, the so-called "new urban sociology" elaborated an explicitly spatialized and reflexively multiscalar understanding of capitalist urbanization. Within this new conceptual framework, the spatial and scalar parameters for urban development could no longer be taken for granted, as if they were pregiven features of the social world. Instead, urbanization was now increasingly viewed as an active moment within the ongoing production and transformation of capitalist sociospatial configurations at multiple geographical scales (Harvey 1989).

Theoretical debates on the interplay between capitalism, urbanization and the production of socio-spatial configurations intensified during the course of the 1970s and early 1980s (Gottdiener 1985; Saunders 1985; Katznelson 1993). Crucially, the new urban sociology was consolidated during a period in which, throughout the older industrialized world, cities and regions were undergoing any number of highly disruptive sociospatial transformations associated with the crisis of North Atlantic Fordism and the consolidation of a new international division of labor dominated by apparently footloose transnational corporations (A. Amin 1994; Dicken 1998). Accordingly, critical urbanists began to deploy many of the categories and methods associated with this newly emergent intellectual tradition in order to decipher the tumultuous processes of urban restructuring that were unfolding around them, particularly in the cities and regions of the industrialized North. Extensive, sophisticated literatures rapidly emerged on key topics such as deindustrialization, the urban land nexus, North/South divides and territorial polarization, regionalism, collective consumption, post-Fordism, local state intervention, the politics of place and urban social movements (for useful overviews, see Dear and Scott 1981; Soja 1989; Dear and Wolch 1991). These multifaceted research initiatives indicated that the sources of contemporary urban transformations could not be understood in purely local, regional or national terms. Rather, the post-1970s restructuring of cities and regions was increasingly interpreted as an expression and outcome of ongoing worldwide economic, political and sociospatial transformations. Thus, for instance, plant closings and workers' struggles in older industrial cities such as Chicago, Detroit, Liverpool, Dortmund or Milan could not be explained simply in terms of local, regional or even national decisions or developments, but had to be analyzed in relation to broader secular trends within the world economy that were fundamentally reworking the conditions for profitable capital accumulation and redistributing the geographies of industrial production not only in the older capitalist world, but in many formerly peripheral, newly industrializing countries (NICs) as well (Bluestone and Harrison 1982).

In opening up their analyses to the global dimensions of urban restructuring, contributors to the new urban sociology also began to draw upon several new approaches to the political economy of capitalism that likewise underscored its intrinsically globalizing dimensions. Foremost among these was the model of "world system analysis" developed by Immanuel Wallerstein (1974, 1980), Samir Amin (1978) and Giovanni Arrighi (1979), which explored the worldwide polarization of economic development, relations of production and living conditions under capitalism among distinct core, semi-peripheral and peripheral zones. Like Braudel before them, who had similarly emphasized large-scale and long-term processes of economic and societal change, world system theorists insisted that capitalism could be understood only on the largest possible spatial scale, that of the world economy, and over a very long temporal period spanning many centuries. World system theorists thus sharply criticized the state-centric and methodologically nationalist assumptions of mainstream social science, arguing instead for an explicitly globalist and *longue durée* understanding of modern capitalism (for an excellent overview, see Wallerstein 1991). The rise of world system theory during the 1970s resonated with a more general resurgence of neo-Marxian approaches to political economy during this period (Lipietz 1987). In the context of diverse studies of transnational corporations, underdevelopment, dependency, class formation, crisis theory and the internationalization of capital, these new approaches to political economy likewise argued for more systematic, reflexive analyses of the global parameters of capitalism both in historical and contemporary contexts (see, for instance, the contributions to radical journals such as *New Left Review* and *Review of Radical Political Economics* during the 1970s and early 1980s; for overviews, see T. Hall 2000; D.A. Smith 2000). Taken together, these diverse literatures on the political economy of global capitalism provided crucial analytical reference points for urbanists concerned, during the late 1970s and early 1980s, to decipher the variegated impacts of large-scale processes of capitalist restructuring upon cities and regions. It was now increasingly recognized, as Timberlake (1985) explains, that

> processes such as urbanization can be more fully understood by beginning to examine the many ways in which they articulate with the broader currents of the world-economy that penetrate spatial barriers, transcend limited time boundaries and influence social relations at many different levels.
>
> (Timberlake 1985: 3)

It is against this background that the emergence of global cities research during the 1980s must be understood. For, like the other critical analyses of post-1970s urban restructuring that were being pioneered during this period, world cities theorists built extensively upon the analytical foundations that had been established by neo-Marxist urban political economists, world system theorists and other radical analysts of global capitalism. Their analytical starting point, as David A. Smith (2000: 147–148) notes, was that "cities are located and articulated in a hierarchical global system dominated by the logic of competitive capitalism." Within these broad intellectual parameters, world city theorists developed a distinctive set of methodological strategies, research agendas and interpretive claims, which we are now in a position to consider more closely.

Global city theory and global capitalist restructuring

Although the notion of a world city has a longer historical legacy, it was consolidated as a core concept for urban studies in the early 1980s in the context of attempts to decipher the ongoing, crisis-induced restructuring of global capitalism. As Peter Taylor (2003: 21) argues, "The world city literature as a cumulative and collective enterprise begins only when the economic restructuring of the world-economy makes the idea of a mosaic of separate urban systems appear anachronistic and frankly irrelevant." Numerous scholars contributed key insights to this emergent research agenda, but the most influential, foundational statements were presented by John Friedmann (see Readings 6 and 7) and Saskia Sassen (see Readings 9 and 10). To date, the work of these authors is associated most closely with the global city concept, and is routinely cited in studies of the interplay between globalization and urban development. However, for present purposes, rather than attempting to summarize the contributions of these authors, and those whom they subsequently influenced, we instead present a general interpretation of the major substantive claims that are associated with what has now come to be known as world city theory.

World city theory has been employed extensively in studies of the role of major cities as global financial centers, as headquarters locations for transnational corporations and as agglomerations for advanced producer and financial services industries. While the theory's usefulness in such research has been convincingly demonstrated, we would argue that the central agenda of world city theory is best conceived more broadly, as an attempt to analyze the changing worldwide geographies of capitalism in the late twentieth and early twenty-first centuries. From this point of view, the central project of world cities research is not simply to classify cities according to their role as command and control centers in global capitalism, but rather, as Friedmann (1986: 69; see also Reading 7) has proposed, to analyze the "spatial organization of the new international division of labor." The key feature of this newly emergent configuration of world capitalism is that cities − or, more precisely, large-scale urbanized regions − rather than the territorial economies of national states are its most fundamental geographical units (see also Scott 2001). These urban regions are said to be arranged hierarchically on a global scale according to their differential modes of integration into the world economy (see Reading 5 by Cohen, Reading 6 by Friedmann and Wolff and Reading 7 by Friedmann). Thus, as Friedmann (1995: 21−26) has argued, contemporary cities operate as the "organizing nodes" of world capitalism, as "articulations" of regional, national and global commodity flows, and as "basing points" in the "space of global capital accumulation."

World city theorists have analyzed this shift towards a city-centered configuration of global capitalism with reference to two intertwined politico-economic transformations since the mid-1970s − first, the emergence of a new international division of labor dominated by transnational corporations, and second, the crisis of the Fordist-Keynesian technological-institutional system that prevailed in the older industrialized world throughout the postwar period.

The new international division of labor

The emergence of a new international division of labor (NIDL) since the late 1960s resulted in large measure from the massive expansion in the role of transnational corporations in the production and exchange of commodities on a world scale (Fröbel et al. 1980; Dicken 1998). Whereas the old international division of labor was based upon raw materials production in the periphery and industrial manufacturing in the core, the NIDL has entailed the relocation of manufacturing industries to semi-peripheral and peripheral states in search of inexpensive sources of labor-power. In addition to the deindustrialization of many core industrial cities, this global market for production sites has also entailed an increasing spatial concentration of business services and other administrative-coordination functions within the predominant urban centers of the core. According to world city theorists, these upper-tier cities have become major nodes of decision-making, financial planning and control within globally dispersed commodity chains, and therefore, the central basing points for the worldwide activities of TNCs (see Reading 5 by Cohen; see also Feagin and Smith 1987). This intensified urban concentration of global capital flows has been further enabled through the development of new informational technologies, closely tied to the agglomeration economies of cities, that accelerate communication and coordination on a global scale (Castells 1989).

Thus, if the latest round of capitalist globalization has enhanced capital's ability to coordinate flows of value through global *space*, it has also been premised upon the construction of specific urban *places* within and through which the territorialized technological, institutional and social infrastructure of globalization is secured (see Readings 9 and 10 by Sassen). Therefore, even as the costs of overcoming the friction of distance in the global transfer of information and commodities are pushed ever closer to zero, cities have remained fundamental locational nodes through which global systems of capitalist production and exchange are organized. In this sense, as Sassen (1999, 2000) and Scott (2001) have emphasized, global city theory entails an explicit critique of mainstream conceptions of globalization which presume that territoriality, borders and places are becoming increasingly irrelevant. Against such views, global city theorists have emphasized the enhanced strategic role of urban regions in providing a fixed, place-specific and relatively non-substitutable sociotechnological infrastructure for globalized forms of capital accumulation (see also Storper 1996).

The crisis of North Atlantic Fordism

Contemporary processes of world city formation have also been closely related to the growing obsolescence of the technological, institutional and social foundations of the Fordist regime of accumulation, which is generally said to have been grounded upon mass production, mass consumption, nationally configured Keynesian demand-management arrangements, nationalized frameworks of collective bargaining and redistributive social welfare policies (Aglietta 1979; Lipietz 1987; Jessop 1992). The crisis of the Fordist-Keynesian technological-institutional system in the older industrial cities of North America and western Europe during the early 1970s was paralleled by unexpectedly dynamic growth in various so-called new industrial spaces such as Silicon Valley, Los Angeles/Orange County, Baden-Württemberg and the Third Italy, grounded upon decentralized, vertically disintegrated forms of industrial organization embedded within dense transactional networks of subcontracting arrangements and other non-market forms of inter-firm coordination (Scott 1988; Storper 1996). According to Storper and Scott (1989: 24–27), the major sectors associated with these emergent flexible production systems are to be classified in three broad categories: revitalized craft production, high-technology industries, and advanced producer and financial services. The locations and spatial structures of these industries vary extensively, but most are agglomerated within major urban manufacturing regions and – in the case of the advanced financial and producer services – within global cities such as London, New York, Tokyo, Paris, Frankfurt and Los Angeles in which large numbers of TNCs are based.

More recent contributors to the debate on post-Fordist industrial geography have advised a somewhat more cautious analytical perspective that acknowledges the impressive dynamism of flexible production systems while situating them within a global context characterized by continued geoeconomic and geopolitical disorder, pervasive uneven geographical development and neoliberal ideological hegemony (Amin and Thrift 1992; Peck and Tickell 1994). Nevertheless, the rise of global cities in recent decades as key geographical sites for the coordination of global commodity chains is broadly consistent with the proposition that flexible production systems are among the new leading edges of contemporary capitalist development. As Sassen's (1991) research has demonstrated (see also Readings 9 and 10), the propulsive growth industries of global cities are, above all, the producer and financial services sectors that serve the command and control requirements of transnational capital – for instance, banking, accounting, advertising, financial management and consulting, business law, insurance and the like. It is in this sense that capital's attempt to enhance its command and control over space on a global scale hinges upon the place-specific production complexes, technological-institutional systems, agglomeration economies and other externalities that are increasingly localized within global cities.

Taken together, the aforementioned arguments have provided world cities researchers with a methodological basis for analyzing the role of major urban regions in the currently unfolding geographical transformation of world capitalism. As Scott (2001: 4) suggests, global city-regions serve "as territorial platforms for much of the post-Fordist economy that constitutes the dominant leading edge of contemporary capitalist development, and as important staging posts for the operations of multinational corporations." In sum, global cities are:

- basing points for the global operations of TNCs
- production sites and markets for producer and financial services
- articulating nodes within a broader hierarchy of cities stratified according to their differential modes of integration into the world economy
- dominant locational centers within large-scale regional economies or urban fields.

Of course, as several contributors to this Reader explore at length, global cities also perform other functions in various globalized social, political and cultural networks. For instance, as the attack on the World Trade Center on September 11, 2001 showed, the symbolic power of finance capital was only one in a number of globalized vectors of signification that were connected to the ill-fated complex – global cultural flows and diasporic networks among them. Additionally, as Stefan Krätke (Reading 39) has shown with reference to the rise of "media cities," worldwide inter-urban networks have been consolidated in spheres of economic activity that are not directly linked to global financial transactions. Just as crucially, in the cases of Washington DC, Geneva, Brussels, Nairobi and other bureaucratic headquarters of the global diplomatic and non-governmental organization (NGO) communities, we can identify yet another network of global political centers (Gerhard 2003; see also Reading 24 by Simon). Religious centers such as Mecca, Rome and Jerusalem, among many others, constitute yet another such network (Flusty 2004). In some cases, places that ostensibly lack strategic economic assets nonetheless acquire global significance through their role in the worldwide networks of social movement activists. Porto Alegre, Brazil, where the World Social Forum has been based, and Davos, Switzerland, where the World Economic Forum takes place every January, are cases in point (Allahwala and Keil 2005).

 And yet, even as we expand and differentiate our mapping of the global urban system through such considerations, it is essential to emphasize that global urban networks represent only one dimension within capitalism's rapidly changing geographical organization. The consolidation of a world urban hierarchy dominated by an archipelago of upper-tier global cities has also produced new geographies of exclusion stretching from the economic "deadlands" of the older industrial core states into the marginalized zones of the developing world and the global periphery that contain almost seven-eighths of world population (Agnew and Corbridge 1994; Friedmann 1995). Thus, even as world city-regions increasingly supersede the territorial economies of national states as the basic geographical building blocks of global capitalism, new patterns of uneven spatial development are proliferating on all spatial scales. The question of how cities and city-regions located outside the heartlands of world capitalism have been affected by contemporary globalization processes has only recently begun to be explored by world city theorists – as well as by critics of this theoretical approach (see the contributions to Part Four of this Reader).

World city formation and the question of agency

Early contributions to global city theory adopted a largely structuralist point of view which emphasized the role of large-scale political-economic transformations and the consolidation of worldwide inter-urban networks. In this context, many scholars argued that globalized urbanization was reducing the capacity of local political actors and state institutions to influence socioeconomic life. For instance, Robert Ross and Kent Trachte (1990; see also Reading 12) identified global city formation with a "relative decline of the relative autonomy of the state" vis-à-vis capital. For them, the apparent hypermobility of "footloose capital" was subjecting local planners to irresistible "global constraints" and contributing to a systemic decline of urban working-class power (see also Ross et al. 1980).

More recent work, however, has rejected these arguments and explored more systematically the role of local social forces and political struggles in the production of globalized urban spaces. From this point of view, global city formation is indeed a political project of transnational capital, but the global city itself represents an arena and outcome of intense sociopolitical struggles among a broad range of social forces and political-territorial alliances. Diverse local agents, social forces and institutions – including fragments of the internationalized working class; various cultural and territorial communities; and state institutions – actively participate in and shape the process of world city formation (see the contributions to Part Five). Thus, as Smith and Feagin (1987: 17) explain, globalized forms of urbanization must be "understood as the long-term outcomes of actions taken by economic and political *actors* operating within a complex and changing matrix of global and national economic and political *forces*." In short, local agents act and react to pressures of global restructuring, but they are also active producers of globalization processes (see Reading 46 by M.P. Smith). They are the builders of the global city.

How to use the Global Cities Reader

As the preceding discussion reveals, the parameters of global city theory have been expanded considerably since the consolidation of this approach to urban studies in the 1980s. On the one hand, a broader range of cases of globalized urban development is now being explored (see Part Four). At the same time, new theoretical categories and methodological strategies have been introduced in order to examine the interplay between global, supranational, national, regional and local dynamics within rapidly changing urban spaces. Studies of global cities have also been extended beyond economic issues (such as industrial agglomeration, labor market restructuring and so forth) to investigate a broader range of substantive themes, including sociospatial polarization, institutional restructuring, governance, political ecology, identity formation, sociopolitical mobilization and representation.

This Reader contains a large number of selections that are intended to survey this vast, multifaceted and still evolving literature. It would be impossible in a single volume to provide a full, systematic overview of this sprawling research field. We have chosen the selections from a much larger set of relevant texts, and there are no doubt some omissions and lacunae in our presentation. Our goal here is less to represent all of the "greatest hits" within this research field than to introduce readers to its broad contours and to inspire them to explore its nuances more comprehensively. To facilitate this, we have included reference lists at the end of the editorial introductions to each part, which are intended to guide readers to further literature on the topic at hand.

Additionally, readers are strongly encouraged to consult the outstanding website of the GaWC (Globalization and World Cities) research group at Loughborough University, brilliantly organized by a team of urban scholars led by geographers Jonathan Beaverstock and Peter J. Taylor in the UK: http:www.lboro.ac.uk/gawc/. This website contains an invaluable selection of bibliographies, research bulletins, project descriptions, data sets and web links related to research on cities and globalization. The GaWC website is updated regularly and will be immensely useful to anyone who is engaged in reading or research on these matters.

We must emphasize that, with the exception of six original contributions (Reading 19 by Schmid, Reading 28 by Buechler, Reading 38 by King, Reading 40 by Lehrer, Reading 42 by Flusty and Reading 44 by Marcuse), *all of the other selections to this Reader have been edited quite significantly*. This procedure was necessary in order to conform to the rather strict length requirements that inevitably accompany a large, synthetic book of this nature. In constructing abridged versions of these materials, we have made painstaking efforts to preserve the integrity of key arguments and, wherever possible, to include at least some of the rich contextual details and empirical data that are contained in the case studies.

This task proved extremely difficult, however, as *we were generally able to include only about one-fourth of the original material* for each of the readings in this book. Thus, we were obliged to cut out significant sections of each reading, including discussions of ongoing debates in world city theory and other relevant academic literatures, various types of historical, contextual and empirical background material, most footnotes and endnotes, a variety of maps, data tables and diagrams, and significant numbers of bibliographic references. In short, this is a book composed not of fully complete articles and chapters, but of carefully edited sections of those texts that are intended to convey the main insights and arguments contained in the original versions. Readers of this book are advised to keep these severe length constraints in mind as they navigate its contents.

Wherever possible, we have consulted the authors to request permission to include their work in this volume, and to ask for their critical feedback on our edited and abridged versions of their texts. We would like to express our deepest gratitude to the contributors for generously permitting us to republish such radically scaled-down versions of their work in this Reader. We are also extremely grateful to them for taking time from their busy schedules to examine our editorial work. We have made every effort to address their concerns and to integrate their suggestions into the final version of each reading. Readers who are inspired by their writings are strongly encouraged to consult the original, full-length publications. We feel reasonably confident, however, that the edited versions published here convey the most central arguments and insights contained in the original texts.

We have organized this book into seven thematic parts, each of which is intended to survey a key set of issues in the literature on globalization and urban development.

- *Part One* introduces the themes of the volume and includes a number of classic texts on global city formation.
- *Part Two* contains a number of classic studies of global city formation, focusing on major themes within this emergent literature.
- *Part Three* examines the applications of global city theory through classic and contemporary case studies of globalizing cities in each of the three major super-regions of the world economy – Europe, North America and East Asia.
- *Part Four* examines global city formation from the perspective of research on cities located outside of the three global "triads", within the peripheralized zones of the global South.
- *Part Five* explores some of the political and institutional dimensions of global city formation, as well as various forms of sociopolitical mobilization and contestation that are emerging in globalizing cities.
- *Part Six* examines global cities as sites of processes of cultural globalization, focusing in particular upon the interplay between globalized urbanization and the formation of global and local cultures, identities, architectural forms and imaginaries.
- Finally, *Part Seven* examines the refinement and critique of global cities research since the 1990s and surveys some of the major scholarly controversies that have been generated within this field.

Research on the dual problematic of global cities and globalized urbanization continues to proliferate rapidly. Even as this book begins the production process (April 2005), recent issues of major urban studies journals such as *International Journal of Urban and Regional Research* and *Urban Studies* contain new articles that engage with emergent theoretical, methodological and empirical issues in this vibrant research field. And yet, increasingly, research on the cities/globalization problematic is no longer grounded upon a single intellectual paradigm. Instead, such research now encompasses a broad range of analytical perspectives and methodological strategies through which the contemporary global urban condition is being investigated. Meanwhile, worldwide urbanization patterns continue to evolve through the contested, contradictory interaction of diverse global, supranational, national and local political, economic, cultural and environmental forces. We hope that this Reader will inspire others to join ongoing efforts to decipher these evolutionary tendencies, their contradictions, and their manifold implications for cities and urban dwellers around the world. This Reader is intended to provide an intellectual starting point – no more, but also no less – for such inquiries. In critically evaluating the strengths and limitations of the readings included herein, we hope that readers will develop their own perspectives on globalizing cities and city-regions, whether for purposes of critical social-scientific analysis, to facilitate progressive policy development or to promote grassroots political mobilization.

References

Aglietta, M. (1979) *A Theory of Capitalist Regulation*. New York: Verso.

Agnew, J. and Corbridge, S. (1994) *Mastering Space*. New York: Routledge.

Alger, C. (1990) The world relations of cities, *Ekistics*, 57, April, 99–115.

Ali, H. and Keil, R. (2006) Global cities and the spread of infectious disease: the case of Severe Acute Respiratory Syndrome (SARS) in Toronto, Canada, *Urban Studies* (forthcoming).

Allahwala, A. and Keil, R. (2005) Introduction to a Debate on the World Social Forum, *International Journal of Urban and Regional Research*, June, 29, 2, 409–416.

Amin, A. (ed.) (1994) *Post-Fordism: A Reader*. Cambridge, MA: Blackwell.

Amin, A. and Thrift, N. (1992) Neo-Marshallian nodes in global networks, *International Journal of Urban and Regional Research*, 16, 4, 571–587.

Amin, S. (1978) *Accumulation on a World Scale*. London: Harvester.

Arrighi, G. (1979) *The Geometry of Imperialism*. London: Verso.

Berger, J. (1990) *Lilac and Flag*. New York: Vintage.

Berry, B. (1961) City size distributions and economic development, *Economic Development and Cultural Change*, 9, 573–578.

Bluestone, B. and Harrison, B. (1982) *The Deindustrialization of America*. New York: Basic.

Castells, M. (1972) *La Question urbaine*. Paris: Maspero.

Castells, M. (1989) *The Informational City*. Cambridge, MA: Blackwell.

Castells, M. (2004) Lecture: Cities in the Information Age, University of California, Berkeley, October 28.

Dear, M. and Scott, A.J. (eds) (1981) *Urbanization and Urban Planning in Capitalist Society*. London: Methuen.

Dear, M. and Wolch, J. (eds) (1991) *The Power of Geography*. Boston, MA: Unwin Hyman.

Dicken, P. (1998) *Global Shift: The Internationalization of Economic Activity*, 3rd edition. New York: Guilford.

Feagin, J.R. and Smith, M.P. (1987) Cities and the new international divison of labor: an overview. In M.P. Smith and J.R. Feagin (eds) *The Capitalist City*. Cambridge, MA: Blackwell, 3–36.

Flusty, S. (2004) *De-Coca-Colonization*. New York: Routledge.

Friedmann, J. (1986) The world city hypothesis, *Development and Change*, 17, 69–83.

Friedmann, J. (1995) Where we stand: a decade of world city research. In P.L. Knox and P.J. Taylor (eds) *World Cities in a World-System*. New York: Cambridge University Press, 21–47.

Friedmann, J. (2002) *The Prospect of Cities*. Minneapolis, MN: University of Minnesota Press.

Friedmann, J. and Wolff, G. (1982) World city formation: an agenda for research and action, *International Journal of Urban and Regional Research*, 6, 309–344.

Fröbel, F., Heinrichs, J. and Kreye, O. (1980) *The New International Division of Labor*. New York: Cambridge University Press.

Geddes, P. (1924) A world league of cities, *Sociological Review*, 26, 166–167.

Gerhard, U. (2003) Local activity patterns in a global city – analysing the political sector in Washington, DC. In Globalization and World Cities Study Group and Network (ed.) Available at: http://www.lboro.ac.uk/gawc/rb99.html, accessed July 4, 2005. *Research Bulletin 99*.

Gerhard, U. (2004) Global cities: Anmerkungen zu einem aktuellen Forschungsfeld, *Geographische Rundschau*, 56, 4, 4–10.

Gottdiener, M. (1985) *The Social Production of Urban Space*. Austin, TX: University of Texas Press.

Hajer, M. and Reijndorp, A. (2001) *In Search of New Public Domain*. Rotterdam: Nai.

Hall, P.G. (1966) *The World Cities*. New York: McGraw-Hill.

Hall, T. (ed.) (2000) *A World-systems Reader*. Lanham, MD: Rowman & Littlefield.

Harvey, D. (1973) *Social Justice and the City*. Baltimore, MD: Johns Hopkins University Press.

Harvey, D. (1982) *The Limits to Capital*. Chicago: University of Chicago Press.

Harvey, D. (1989) *The Urban Experience*. Baltimore, MD: Johns Hopkins University Press.

Hirschman, A. (1958) *The Strategy of Economic Development*. New Haven, CT: Yale University Press.

Jessop, B. (1992) Fordism and post-Fordism: a critical reformulation. In M. Storper and A.J. Scott (eds) *Pathways to Industrialization and Regional Development*. New York: Routledge, 46–69.

Katznelson, I. (1993) *Marxism and the City*. New York: Oxford University Press.

Keil, R. (1998) *Los Angeles: Globalization, Urbanization and Social Struggles*. Chichester: John Wiley & Sons.

Keil, R. (2002) "Common Sense" neoliberalism: progressive conservative urbanism in Toronto. In N. Brenner and N. Theodore (eds) *Spaces of Neoliberalism*. Oxford: Blackwell, 230–253.

King, A. (1991) *Urbanism, Colonialism and the World-economy*. New York: Routledge.

Kipfer, S. and Keil, R. (2002) Toronto, Inc.? Planning the competitive city in the New Toronto, *Antipode*, 34, 2, 227–264.

Knox, P.L. and Taylor, P.J. (eds) (1995) *World Cities in a World-system*. New York: Cambridge University Press.

Lefebvre, H. (1968) *Le Droit à la ville*. Paris: Anthropos.

Lipietz, A. (1987) *Mirages and Miracles*. New York: Verso.

Lipietz, A. (1993) The local and the global: regional individuality or interregionalism?, *Transactions of the Institute of British Geographers*, 18, 8–18.

Marcuse, P. and van Kempen, R. (eds) (2000) *Globalizing Cities: A New Spatial Order?* Oxford: Blackwell.

Massey, D. (1985) *Spatial Divisions of Labour*. London: Macmillan.

Myrdal, G. (1957) *Economic Theory and Underdeveloped Regions*. London: Gerald Duckworth.

Peck, J. and Tickell, A. (1994) Searching for a new institutional fix. In A. Amin (ed.) *Post-Fordism: A Reader*. Cambridge, MA: Blackwell, 280–315.

Pred, A. (1977) *City-systems in Advanced Economies*. London: Hutchinson.

Ross, R. and Trachte, K. (1990) *Global Capitalism: The New Leviathan*. Albany, NY: State University of New York Press.

Ross, R., Shakow, D. and Susman, P. (1980) Local planners, global constraints, *Policy Sciences*, 12, 1–25.

Sandercock, L. (1998) *Cosmopolis*. New York: John Wiley & Sons.

Sassen, S. (1991) *The Global City*. Princeton, NJ: Princeton University Press.

Sassen, S. (1999) *Globalization and its Discontents*. New York: Free Press.

Sassen, S. (2000) *Cities in the World Economy*, 2nd edition. Thousand Oaks, CA: Sage.

Sassen, S. (2002) *The Global City: New York, London, Tokyo*, 2nd edition. Princeton, NJ: Princeton University Press.

Saunders, P. (1985) *Social Theory and the Urban Question*, 2nd edition. New York: Routledge.

Scott, A.J. (1988) *New Industrial Spaces*. Berkeley and Los Angeles: University of California Press.

Scott, A.J. (ed.) (2001) *Global City-regions: Trends, Theory, Policy*. New York: Oxford University Press.

Smith, D.A. (2000) Urbanization in the world-system: a retrospective and prospective. In T. Hall (ed.) *A World-systems Reader*. Lanham, MD: Rowman & Littlefield, 143–168.

Smith, M.P. (2001) *Transnational Urbanism*. Cambridge, MA: Blackwell.

Smith, M.P. and Feagin, J.R. (eds) (1987) *The Capitalist City*. Cambridge, MA: Blackwell.

Smith, N. (1990) *Uneven Development*. Cambridge, MA: Blackwell.

Soja, E.W. (1989) *Postmodern Geographies*. London: Verso.

Soja, E.W. (2000) *Postmetropolis: Critical Studies of Cities and Regions*. Oxford and Malden, MA: Blackwell.

Storper, M. (1996) *The Regional World*. New York: Guilford.

Storper, M. and Scott, A.J. (1989) The geographical foundations and social regulation of flexible production complexes. In J. Wolch and M. Dear (eds) *The Power of Geography*. Boston, MA: Unwin Hyman, 19–40.

Taylor, P.J. (2003) *World City Network: A Global Urban Analysis*. London and New York: Routledge.

Timberlake, M. (ed.) (1985) *Urbanization in the World-Economy*. Orlando, FL: Academic Press.

Todd, G. (1995) "Going global" in the semi-periphery: world cities as political projects. The case of Toronto. In P.L. Knox and P.J. Taylor (eds) *World Cities in a World-System*. New York: Cambridge University Press, 192–214.

Wallerstein, I. (1974) *The Modern World-System I*. New York: Academic Press.

Wallerstein, I. (1980) *The Modern World-System II*. New York: Academic Press.

Wallerstein, I. (1991) A call for a debate about the paradigm. In I. Wallerstein, *Unthinking Social Science*. Cambridge, MA: Polity Press, 237–256.

Yeoh, B. (1999) Global/globalizing cities, *Progress in Human Geography*, 23, 4, 607–616.

PART ONE

Global city formation: emergence of a concept and research agenda

Plate 2 La Défense, Paris (Roger Keil)

INTRODUCTION TO PART ONE

[Paris] is grand, at this moment in time, to set well-guarded Babylon waltzing in the arms of Memphis, and to set London dancing in the embrace of Peking. . . . One of these fine mornings, France will have a rude awakening when it realizes it is confined within the walls of Lutetia, of which she forms but a crossroads. . . . The next day, Italy, Spain, Denmark, and Russia will be incorporated by decree into the Parisian municipality; three days later, the city gates will be pushed back to Novaya Zemlya and to the Land of the Papuans. *Paris will be the world, and the universe will be Paris.* The savannahs and the pampas and the Black Forest will compose the public gardens of this greater Lutetia; the Alps, the Pyrenees, the Andes, the Himalayas will be the Aventine and the scenic hills for this incomparable city – knolls of pleasure, study of solitude. But all this is still nothing: Paris will mount to the skies and scale the firmament of firmaments; it will annex, as suburbs, the planets and the stars.

(Paul-Ernest de Rattier, *Paris n'existe pas*, Paris 1857;
quoted in Benjamin 1999: 137, emphasis added)

This prescient depiction of Paris in the middle of the nineteenth century demonstrates clearly that two major aspects of contemporary discussions of global cities have much longer histories. First, there is the notion that a major city may, at least metaphorically, encompass the entire world. Second, the quotation also suggests that, even within the dominant global political order of nation-states, the role of urban systems as an organizing principle of socioeconomic life was recognized long before contemporary debates on globalization.

While the notion of *global* or *world* cities in its contemporary usage may be new, the idea that cities are of world-historical importance – economically, militarily, politically and culturally – has been around for some time. Cities played fundamental geostrategic roles and had long-distance, networked relationships prior to the consolidation of the modern interstate system. Whether in Mesopotamia or Egypt, the Indus Valley or in the Far East, wherever the first urban cultures appeared, their settlements were the core of territorial or maritime empires (Soja 2000). Athens and Rome are perhaps the most pervasively cited urban cores of two major world empires.

In the Middle Ages, through trade networks such as the German Hansa, cities once again came to serve as the spatial infrastructure of emerging continental and, eventually, during the early modern period, global economies (see Reading 2 by Braudel). Byzantium, which took the mantle from Rome and remained the important buckle in a belt that tied together Occident and Orient in the Middle Ages, was certainly a type of global city, even by today's standards. But there are also many examples of smaller, less-well-known cities, which fulfilled global city functions, in particular financial control, and which are now little more than regional centers and tourist destinations. One such place is Augsburg, in southern Germany, a city of impressive wealth in the late Middle Ages, when the Fugger family financed the global enterprises of the Hansa and other commercial, mining and manufacturing projects. Located at the northern foothills of the Alps, it was the ideal connector of the Mediterranean, eastern European, North Sea and western European economies. Another such place was the legendary Cahors in the French southwest, which is cited, alongside Sodom and Gomorrah, as a model for Hell in Dante's *Divine Comedy* (1321). At the time of Cahors' greatest power during its Golden Age in the thirteenth century, local and Lombard bankers transformed the city into the chief banking center of Europe and earned a reputation as usurers – a characteristic of the town that was subsequently noted by the great Italian poet.

The developmental trajectories of major cities and inter-city networks have been linked to the emergence and decline of precapitalist imperial systems and, subsequently, to the expansion of capitalism on a world scale (Abu-Lughod 1989; Chase-Dunn et al. 2000). Thus, according to Chase-Dunn and Jorgenson (2002):

> The long rise of capitalism was promoted by semiperipheral capitalist city-states, usually maritime coordinators of trade protected by naval power. The Italian city-states of Venice and Genoa are perhaps the most famous of these, but the Phoenician city-states of the Mediterranean exploited a similar interstitial niche within a larger system dominated in the guise of core capitalist nation-states in a series of transformations from Venice and Genoa to the Dutch Republic (led by Amsterdam) and eventually the *Pax Britannica* coordinated by the great world city of the nineteenth century, London. Within London the functions mentioned above were spatially separated: empire in Westminster and money in the City. In the twentieth century hegemony of the United States these global functions became located in separate cities (Washington, DC and New York).
>
> (Chase-Dunn and Jorgenson 2002: 2)

In each period, the identity and character of a global city is tied into the dominant mode of production: globality is defined by the scale of the world system – the large-scale framework of material, political and cultural life – in which that city is embedded. Cities are in turn connected in diverse, long-distance relationships that are designed to maintain the world system as a whole. However, as we will see in Reading 5 by Cohen, Reading 6 by Friedmann and Wolff and Reading 7 by Friedmann, since the mid-1970s we have witnessed the consolidation of a truly worldwide urban hierarchy that has significantly expanded the scale of major cities' command and control functions within the capitalist world system as a whole.

Peter Hall (1966: 7) attributed the the term "world city" to a book by Scottish urbanist Patrick Geddes, *Cities in Evolution* (1915), which emphasized world cities' centralized economic functions. In Hall's view, world cities were sites of intensive population growth, centralized political power and major commercial, financial and transportation functions. Crucially, however, Hall conceived world cities primarily as *national* centers that channeled international forces and influences towards national interests. Hall's conception of a "world city" is thus arguably a product of a period in which cities operated primarily as nodes within national urban systems. By contrast, contemporary notions of the world city emphasize the embeddedness of urban centers within an emergent system of global capitalism; this may entail their partial "delinking" from the territorialized economic spaces regulated by national state institutions (Ross and Trachte 1990; Sassen 1991). Not surprisingly, then, the current wave of world cities research began to emerge in the late 1970s and early 1980s, when the first effects of the contemporary round of global restructuring began to be articulated in cities throughout the older industrialized world.

The world city literature can be viewed as an implicit critique of what Ben Derudder has called "the zonal implementation of core-periphery models" (2003: 100, n. 30), which were developed by world system theorists such as Wallerstein (1974) to characterize the polarization of global capitalism among core, semi-peripheral and peripheral zones. These models are generally grounded upon territorialist assumptions in which economic space is conceived as being composed of clearly delineated, bounded geographical containers. However, by directing attention to inter-urban connections and interdependencies, which generally crosscut territorial borders in complex, networked relationships, world city theorists have suggested an alternative conceptualization of capitalism's underlying economic geographies, in which economic territoriality represents only one among various possible forms of sociospatial organization (for a related perspective, see Arrighi 1995). The question of how best to map the emergent global urban system in relation to the landscape of global capitalism remains one of the most controversial, and fascinating, within the entire field of global cities research, and it is likely to continue to stimulate energetic theoretical debate and empirical analysis in the years to come.

In terms of periodizing the emergence and development of global cities, there are basically two approaches – one which emphasizes the long-term role of cities as basing points for global economic flows, and one which emphasizes the historical specificity of contemporary patterns of global city formation. While we believe these approaches can be compatible, they have, in fact, led to quite divergent research agendas.

On the one hand, some urbanists have insisted that global cities are an age-old phenomenon. This position is strongly articulated in work by world city researchers such as Janet Abu-Lughod (Reading 4), Anthony D. King (Readings 23 and 38), Michael Timberlake (1985) and Christopher Chase-Dunn (1985). These researchers have systematically examined the long-term structural and historical background for processes of world city formation, and have argued that cities have long served as nodal points within large-scale economic systems both prior to and throughout the history of capitalist industrialization.

Some of the most influential work in this research tradition was produced by scholars who combined the methodology of studying long-term, macrogeographical change in the tradition of the French *Annales* school (see Reading 2 by Braudel) with that of world systems theorists such as Immanuel Wallerstein (1974). For instance, Chase-Dunn (1985) argues that the boundaries of the world system and the system of world cities have been closely interconnected since at least AD 800 (see also Timberlake 1985). An important aspect of this historically based work has been an emphasis on "urban specialization," a notion that was reformulated in the 1980s to describe the development of urban centers in global systems of cities. Reading 3 by Rodriguez and Feagin illustrates the powerful explanatory capacity of such an approach when it is applied to the dynamics of global city formation in successive stages of capitalist development.

The second, alternative approach to the periodization of global city formation emphasizes the uniqueness of contemporary global cities due to their role as basing points for a qualitatively new formation of globalizing capitalism. Scholars who have worked in this research tradition have linked the emergence of a globalized city system to the specific forms of worldwide capitalist restructuring that began to unfold as of the 1970s. This strand of research emerged in response to two intertwined transformations – first, the end of American–Fordist–Keynesian hegemony, which entailed the crisis of the postwar framework of accumulation and state regulation across the North Atlantic zone, and second, the development of a New International Division of Labour in the 1970s, which entailed the increasing industrialization of formerly peripheralized states and, concomitantly, intensive processes of industrial restructuring in the former heartlands of global capitalism (Fröbel et al. 1980).

Accordingly, this second approach to the analysis of world city formation interprets this development as a key spatial expression of the new forms of capital accumulation that have been consolidated since the 1970s (Keil 1993). This means that, while specific global cities emerge as the command and control centers of the newly restructured world economy (Sassen 1991), other cities are likewise subject to closely analogous, globally induced forms of political-economic and spatial restructuring. Thus, as several authors have emphasized (see, for instance, the contributions to Part Four of this Reader, as well as Reading 44 by Marcuse), cities that are not global command and control centers may nonetheless be transformed through, for example, globalized patterns of consumption, cultural politics and economic restructuring. Such spaces may be most appropriately characterized, according to Marcuse and van Kempen (2000), as "globalizing cities."

Part One of the Reader provides a broad survey of key contributions to both of the aforementioned strands of global cities research. It begins with a selection from Peter Hall's classic work on world cities, and then presents three contributions that examine the role of global cities in various historical phases of capitalist development (Readings 2, 3 and 4 by Braudel, Rodriguez and Feagin, and Abu-Lughod, respectively). Each of these readings emphasizes the continuities between contemporary global cities and various types of global urban centers during the long-run history of capitalism. We then turn to more recent contributions, which underscore the historical specificity of contemporary patterns of global city formation. Building on the pioneering work of economist Stephen Hymer (1979), Cohen argues in Reading 5 that a new hierarchy of global command and control centers is emerging due to the expanding reach of transnational corporations in globalized production circuits. This analysis sets the stage for the two foundational texts by John Friedmann (Readings 6 and 7). They had the effect, at the time, of synthesizing several new trends in development studies, world systems theory and urban studies into a new and powerful agenda for research and action.

References and suggestions for further reading

Abu-Lughod, J. (1989) *Before European Hegemony: The World System AD 1250–1350*. New York: Oxford University Press.

Arrighi, G. (1995) *The Long Twentieth Century*. London: Verso.

Benevolo, L. (1984) *Die Geschichte der Stadt*. Frankfurt and New York: Campus.

Benjamin, W. (1999) *The Arcades Project*. Cambridge, MA and London: Belknap Harvard.

Chase-Dunn, C. (1985) The system of cities, AD 800–1975. In M. Timberlake (ed.) *Urbanization in the World-Economy*. New York: Academic Press, 269–292.

Chase-Dunn, C. and Jorgenson, A. (2002) Settlement systems: past and present, *Research Bulletin*, 73, Globalization and World Cities Study Group and Network, Available online: http://www.lboro.ac.uk/gawc/rb/rb73.html.

Chase-Dunn, C., Manning, S. and Hall, T.D. (2000) Rise and fall: East–West synchronicity and Indic exceptionalism reexamined, *Social Science History*, 24, 4, 727–754.

Derudder, B. (2003) Beyond the state: mapping the semi-periphery through urban networks, *Capitalism, Nature, Socialism*, 14, 4, 91–120.

Fröbel, F., Heinrichs, J. and Kreye, O. (1980) *The New International Division of Labor*. New York: Cambridge University Press.

Geddes, P. (1915) *Cities in Evolution: An Introduction to the Town Planning Movement and the Study of Civics*. London: Ernest Benn.

Geddes, P. (1924) A world league of cities, *Sociological Review*, 26, 166–167.

Hall, P.G. (1966) *The World Cities*. New York: McGraw-Hill.

Hymer, S. (1979) *The Multinational Corporation: A Radical Approach*, R. Cohen (ed.). Cambridge and New York: Cambridge University Press.

Keil, R. (1993) *Weltstadt – Stadt der Welt*. Münster: Westfalisches Dampfboot.

Keil, R. (1998) *Los Angeles: Globalization, Urbanization and Social Struggles*. Chichester: John Wiley & Sons.

Marcuse, P. and van Kempen, R. (eds) (2000) *Globalizing Cities: A New Spatial Order?* Oxford: Blackwell.

Ross, R. and Trachte, K. (1990) *Global Capitalism: The New Leviathan*. Albany, NY: State University of New York Press.

Sassen, S. (1991) *The Global City*. Princeton, NJ: Princeton University Press.

Soja, E.W. (1989) *Postmodern Geographies*. London: Verso.

Soja, E.W. (2000) *Postmetropolis: Critical Studies of Cities and Regions*. Oxford and Malden, MA: Blackwell.

Taylor, P.J. (2003) *World City Network: A Global Urban Analysis*. London and New York: Routledge.

Timberlake, M. (ed.) (1985) *Urbanization in the World Economy*. New York: Academic Press.

Wallerstein, I. (1974) *The Modern World-System I*. New York: Academic Press.

Prologue

"The Metropolitan Explosion"

from *The World Cities* (1966)

Peter Hall

There are certain great cities in which a quite disproportionate part of the world's most important business is conducted. In 1915 the pioneer thinker and writer on city and regional planning, Patrick Geddes, christened them "the world cities." *The World Cities* is about their growth and problems. By what characteristics do we distinguish the world cities from other great centers of population and wealth? In the first place, they are usually major centers of political power. They are the seats of the most powerful national governments and sometimes of international authorities too; of government agencies of all kinds. Round these gather a host of institutions, whose main business is with government; the big professional organizations, the trade unions, the employers' federations, the headquarters of major industrial concerns.

These cities are the national centers not merely of government but also of trade. Characteristically they are the great ports, which distribute imported goods to all parts of their countries, and in return receive goods for export to the other nations of the world. Within each country, roads and railways focus on the metropolitan city. The world cities are the sites of the great international airports: Heathrow, Charles de Gaulle, Schiphol, Sheremetyevo, Kennedy, Benito Juarez, Kai Tak. Traditionally, the world cities are the leading banking and finance centers of the countries in which they stand. Here are housed the central banks, the headquarters of the trading banks, the offices of the big insurance

organizations and a whole series of specialized financial and insurance agencies.

Government and trade were invariably the original raisons d'être of the world cities. But these places early became the centers where professional talents of all kinds congregated. Each of the world cities has its great hospitals, its distinct medical quarter, its legal profession gathered around the national courts of justice. Students and teachers are drawn to the world cities: they commonly contain great universities, as well as a host of specialized institutions for teaching and research in the sciences, the technologies and the arts. The great national libraries and museums are here. Inevitably, the world cities have become the places where information is gathered and disseminated: the book publishers are found here; so are the publishers of newspapers and periodicals, and with them their journalists and regular contributors. In this century also the world cities have naturally become headquarters of the great national radio and television networks.

But not only are the world cities great centers of population: their populations, as a rule, contain a significant proportion of the richest members of the community. That early led to the development of luxury industries and shops; and in a more affluent age these have been joined by new types of more democratic trading: by the great department stores and the host of specialized shops which cater for every demand. Around them, too, the range of industry has widened: for the products of the traditional luxury trades, forged by craftsmen in the world cities of old, have become articles of popular consumption, and their manufacture now takes place on the assembly lines of vast factories in the suburbs of the world cities.

As manufacture and trade have come to cater for a wider market so has another of the staple businesses of the world cities – the provision of entertainment. The traditional opera houses and theatres and concert halls and luxurious restaurants, once the preserve of the aristocracy and the great merchant, are now open to a wider audience, who can increasingly pay their price. They have been joined by new and more popular forms of entertainment – the variety theatre and revue, the cinema, the night club and a whole gamut of eating and drinking places.

The staple trades of the world cities go, with few exceptions, from strength to strength. Here and there, a trade may wither and decay: thus shoemaking in nineteenth-century London, diamond-cutting in twentieth-century Amsterdam, shirt-making in twentieth-century New York. In the long view, even the world cities may themselves decline. Where now is Bruges – a world city of late medieval Europe? But so far in history, such cases are conspicuous by their rarity. Nothing is more notable about the world cities, taking the long historic view, than their continued economic strength. Not for them the fate of depressed regions which see their staple products decline: regions like the coalfields of Northumberland–Durham in Great Britain or Pennsylvania–West Virginia in the United States, or remote rural regions like the Massif Central of France or the south-east uplands of the Federal Republic of Germany. True, one disquieting note is that, during the 1970s, some great world city regions – London, New York – for the first time recorded declines in population, while in others – Paris, Tokyo – the rate of growth notably slowed. But this should be seen, in large measure, as the continuation of a long process of economic adaptation and of outward deconcentration; the statistical trends suggest that the official definitions of these city regions, big as they are, are no longer big enough.

As the economies of the advanced nations become steadily more sophisticated, and as those of the newly industrializing nations strain to catch them up, so in all world cities does the economic emphasis shift to those industries and trades most aptly carried on in the metropolis: industries and trades dependent on skill, on design, on fashion, on conduct with the specialized needs of the buyer. Associated with these trends, white-collar jobs grow faster than blue-collar ones; for every producer of factory goods, more and more people are needed at office desks to achieve good design, to finance and plan production, to sell the goods, to promote efficient nation-wide and worldwide distribution. So it is not surprising that, as they gain such new jobs, the world cities shed those activities that can be as readily performed elsewhere – mass production of standardized goods, space-consuming docking and warehousing, routine paper-processing in factory – like offices: such processes of economic invasion and succession are no novel event for the world cities.

"Divisions of Space and Time in Europe"

from *The Perspective of the World* (1984)

Fernand Braudel

Editors' introduction

In this chapter, Fernand Braudel (1902–1985), one of the most celebrated French historians of the postwar years, depicts the large-scale processes that constitute the modern capitalist urban system. Braudel integrated economics and geography into his studies of global history and changed the way history has subsequently been written. As is apparent in his great works, *The Mediterranean and Mediterranean World in the Age of Phillip II* (1972), *Capitalism and Material Life, 1400–1800* (1973) and the three-volume *Civilization and Capitalism, 15th–18th Century* (1979–1984), Braudel focused on the social and economic agency of people in relation to large-scale, long-term socioeconomic trends. While Braudel was most explicitly concerned to grasp the geographical dimensions of large-scale economic systems – as evidenced, for instance, in his key distinction between *world economy* and *world-economy* and in his emphasis on the polarization of economic development under capitalism – his analysis in this reading provides a prescient account of the interplay between major cities, economic power and systems of imperial rule. Braudel emphasizes that relatively autonomous economic systems may coexist and interact at a planetary scale, without actually covering the entire surface of the globe. In addition, Braudel attempts to decode the key elements of large-scale economic systems with reference to boundaries, patterns of urban centrality and forms of spatial hierarchy. Through his argument that a major city exists at the center of each world-economy, Braudel provides an early blueprint for the notion of the global city itself. Just as importantly, Braudel's conceptualization of spatial hierarchy anticipates the diverse types of world city models which we will encounter in subsequent readings in this volume.

WORLD ECONOMIES

To open the discussion, I should elucidate two expressions which might lead to confusion: *the world economy* and *a world-economy*. *The world economy* is an expression applied to the whole world. It corresponds, as Sismondi (1991 [1951]) puts it, to "the market of the universe," to "the human race, or that part of the human race which is engaged in trade, and which today in a sense makes up a single market." *A world-economy* (an expression which I have used in the past as a particular meaning of the German term *Weltwirtschaft*) only concerns a fragment of the world, an economically autonomous section of the planet able to provide for most of its own needs, a section to which its internal links and exchanges give a certain organic unity. There have been world-economies if not always, at least for a very long time – just as there have been societies, civilizations, states and even empires. If we take giant steps back through history, we could say of ancient Phoenicia that it was an early version of a world-economy, surrounded by great empires. So too was Carthage in its heyday; or the Hellenic world; or even Rome; so too was Islam after its lightning triumphs. In the ninth century, the Norman venture on the outer margins of western Europe laid down the lines of a short-lived and fragile world-economy which others would inherit. From the eleventh century, Europe began developing what was to be its first world-economy, afterwards succeeded by others down to the present day. Muscovy, connected to the East, India, China, Central Asia and Siberia, was another self-contained world-economy, at least until the eighteenth century. So was China, which from earliest times took over and harnessed to her own destiny such neighboring areas as Korea, Japan, the East Indies, Vietnam, Yunan, Tibet and Mongolia – a garland of dependent countries. Even before this, India had turned the Indian Ocean into a sort of private sea, from the east coast of Africa to the islands of the East Indies.

Might it not in short be said that here was a process of constant renewal as each configuration gave way almost spontaneously to another, leaving plentiful traces behind – even in a case, at first sight unpromising, like the Roman Empire? The Roman economy did in fact extend beyond the imperial frontier running along the prosperous line between Rhine and Danube, or eastwards to the Red Sea and the Indian Ocean. According to Pliny the Elder, Rome had a deficit of 100 million sesterces in its trade with the Far East every year. And ancient Roman coins are still being dug up in India today.

SOME GROUND RULES

The past offers us a series of examples of world-economies then – not very many but enough to make some comparisons possible. Moreover since each world-economy lasted a very long time, it changed and developed within its own boundaries, so that its successive ages and different states also suggest some comparisons. The data available is thus sufficiently plentiful to allow us to construct a *typology* of world-economies and at the very least to formulate a set of rules or tendencies which will clarify and even define their relations with geographical space.

Our first concern, in seeking to explain any world-economy, is to identify the area it occupies. Its boundaries are usually easy to discover since they are slow to change. The zone it covers is effectively the first condition of its existence. There is no such thing as a world-economy without its own area, one that is significant in several respects:

- it has boundaries, and the line that defines it gives it an identity, just as coastlines do a sea;
- it invariably has a center, with a city and an already-dominant type of *capitalism*, whatever form this takes. A profusion of such centers represents either immaturity or on the contrary some kind of decline or mutation. In the face of pressures both internal and external, there may be shifts of the center of gravity: cities with international destinies;
- world-cities – are in perpetual rivalry with one another and may take each other's place;
- it is marked by a hierarchy: the area is always a sum of individual economies, some poor, some modest, with a comparatively rich one in the center. As a result, there are inequalities, differences of voltage which make possible the functioning of the whole. Hence that "international division of labor," of which as P.M. Sweezy (1974) points out, Marx did not foresee that it "might harden into a pattern of development and under-development which would split mankind into haves and have-nots on a scale far wider and deeper than the bourgeois-proletarian split in the advanced countries themselves." All the same, this is not in fact a "new" division, but an ancient and no doubt an incurable divide, one that existed long before Marx's time. So there are three sets of conditions, each with general implications.

RULE ONE: THE BOUNDARIES CHANGE ONLY SLOWLY

The limits on one world-economy can be thought of as lying where those of another similar one begin: they mark a line, or rather a zone which it is only worth crossing, economically speaking, *in exceptional circumstances*. For the bulk of traffic in either direction, "the loss in exchange would outweigh the gain." So *as a general rule*, the frontiers of a world-economy are quiet zones, the scene of little activity. They are like thick shells, hard to penetrate; they are often natural barriers, no-man's-lands – or no-man's-seas. The Sahara, despite its caravans, would have been one such, separating Black Africa from White Africa. The Atlantic was another, an empty expanse to the south and west of Africa, and for long centuries a barrier compared to the Indian Ocean, which was from early days the scene of much trade, at least in the north. Equally formidable was the Pacific, which European explorers had only half-opened to traffic: Magellan's voyage only unlocked one way into the southern seas, not a gateway for return journeys. To get back to Europe, the expedition had to take the Portuguese route round the Cape of Good Hope. Even the first voyages of the Manila galleon in 1572 did not really overcome the awe-inspiring obstacle posed by the South Sea.

RULE TWO: A DOMINANT CAPITALIST CITY ALWAYS LIES AT THE CENTER

A world-economy always has an urban center of gravity, a city, as the logistic heart of its activity. News, merchandise, capital, credit, people, instructions, correspondence all flow into and out of the city. Its powerful merchants lay down the law, sometimes becoming extraordinarily wealthy. At varying and respectful distances around the center, will be found other towns, sometimes playing the role of associate or accomplice, but more usually resigned to their second-class role. Their activities are governed by those of the metropolis: they stand guard around it, direct the flow of business toward it, redistribute or pass on the goods it sends them, live off its credit or suffer its rule. Venice was never isolated; nor was Antwerp; nor, later, was Amsterdam. These metropolises came accompanied by a train of subordinates.

Any town of any importance, particularly if it was a seaport, was a "Noah's Ark," "a fair of masks," a "Tower of Babel," as President de Brosses described Livorno. How much more so were the real metropolises! They were the scene of fantas-

tic mixtures, whether London or Istanbul, Isfahan or Malacca, Surat or Calcutta (the latter from the time of its very earliest successes). Under the pillars of the Amsterdam Bourse – which was a microcosm of the world of trade – one could hear every dialect in the world. In Venice, "if you are curious to see men from every part of the earth, each dressed in his own different way, go to St Mark's Square or the Rialto and you will find all manner of persons."

This colorful cosmopolitan population had to coexist and work in peace. The rule in Noah's Ark was live and let live. Of the Venetian state, Villamont thought in 1590 "that there is nowhere in all Italy where one may live in greater liberty . . . firstly because the Signoria rarely condemns a man to death, secondly arms are not forbidden, thirdly there is no inquisition in matters of faith, lastly everyone lives as he pleases in freedom of conscience, which is the reason why several libertine Frenchmen reside there so as not to be pursued and controlled and so as to live wholly without constraint." I imagine that Venice's innate toleration helps to explain her "notorious anticlericalism" or as I would prefer to call it her vigilant opposition to Roman intransigence. But the miracle of toleration was to be found wherever the community of trade convened. Amsterdam kept open house, not without some merit after the religious violence between the Arminians and the Gomarists (1619–1620). In London, every religion under the sun was practiced. "There are," said a French visitor in 1725, "Jews, Protestants from Germany, Holland, Sweden, Denmark and France; Lutherans, Anabaptists, millenarians, Brownists, independents or Puritans; and Tremblers or Quakers." To these might be added the Anglicans, Presbyterians and the Catholics who, whether English or not, were in the habit of attending mass in the chapels of the French, Spanish or Portuguese embassies. Each sect or faith had its own churches and meeting-places. And each one was identifiable to the outside world. "The Quakers can be recognized a mile off by their dress: a flat hat, a small cravat, a coat buttoned up to the neck and their eyes shut most of the time."

Perhaps the most distinctive characteristic of all of these super-cities was their precocious and pronounced social diversification. They all had a proletariat, a bourgeoisie, and a patriciate, the latter controlling all wealth and power and so self-confident that before long it did not even bother, as it had in Venice or Genoa in the old days, to take the title of *nobili*. Patriciate and proletariat indeed grew further apart, as the rich became richer and the poor even poorer, since the besetting sin of these pulsating capitalist cities was their high cost of living, not to mention the constant inflation resulting from the intrinsic nature of the higher urban functions whose destiny it was to dominate adjacent economies.

Dominant cities did *not* dominate for ever; they replaced each other. This was as true at the summit as it was at every level of the urban hierarchy. Such shifts, wherever they occurred (at the top or half-way down), whatever their causes (economic or otherwise), are always significant; they interrupt the calm flow of history and open up perspectives that are the more precious for being so rare. When Amsterdam replaced Antwerp, when London took over from Amsterdam, or when in about 1929, New York overtook London, it always meant a massive historical shift of forces, revealing the precariousness of the previous equilibrium and the strengths of the one which was replacing it. The whole circle of the world-economy was affected by such changes and the repercussions were never exclusively economic.

The reference to dominant cities should not lead us to think that the successes and strengths of these urban centers were always of the same type: in the course of their history, these cities were sometimes better or worse equipped for their task, and their differences or comparative failings, when looked at closely, oblige one to make some fairly fine distinctions of interpretation.

If we take the classic sequence of dominant cities of western Europe – Venice, Antwerp, Genoa, Amsterdam, London – which we shall presently be considering at length – it will be observed that the three first-named did not possess the complete arsenal of economic domination. Venice at the end of the fourteenth century was a booming merchant city; but possessed no more than the beginnings of an industrial sector; and while she did have financial and banking institutions, this credit system operated inside the Venetian economy, as an internal mechanism only. Antwerp, which possessed very little shipping of her own, provided a haven for Europe's merchant capitalism: operating

as a sort of bring and buy center for trade and business, to which everything came from outside. When Genoa's turn came, it was really only because of her banking supremacy, similar to that of Florence in the thirteenth and fourteenth centuries; if she played a leading role, it was firstly because her chief customer was the king of Spain, controller of the flow of bullion, and secondly because no one was quite sure where the center of gravity really lay between the sixteenth and seventeenth centuries: Antwerp fulfilled this role no longer and Amsterdam was not yet ready: the Genoese supremacy was no more than an interlude. By the time Amsterdam and London took the stage, the world-cities possessed the whole panoply of means of economic power: they controlled everything, from shipping to commercial and industrial expansion, as well as the whole range of credit.

Another factor which could vary from one dominant city to another was the machinery of political power. From this point of view, Venice had been a strong and independent state: early in the fifteenth century, she had taken over the *Terraferma*, a large protective zone close at hand; since 1204 she had possessed a colonial empire. Antwerp by contrast had virtually no political power at her disposal. Genoa was a mere territorial skeleton: she had given up all claim to political independence, staking everything on that alternative form of domination, money. Amsterdam laid claim in some sense to the United Provinces, whether they agreed or not. But her "kingdom" represented little more than the *Terraferma* of Venice. With London, we move into a completely different context: this great city had at its command the English national market and later that of the entire British Isles, until the day when the world changed and this mighty combination dwindled to the dimensions of a minor power when compared to a giant like the United States (see Figures 1 and 2).

In short, the outline of the history of these successive dominant cities in Europe since the fourteenth century provides the clue to the development of their underlying world-economies: these might be more or less firmly controlled, as they oscillated between strong and weak centers of gravity. This sequence also incidentally tells us something about the variable value of the

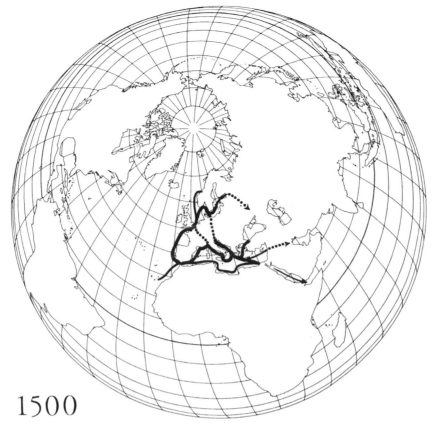

1500

Figure 1 European world-economies on a global scale, 1500
The expanding European economy, represented by its major commodity trades on a world scale. In 1500, the world-economy with Venice at its center was directly operating in the Mediterranean and western Europe; by way of intermediaries, the network reached the Baltic, Norway and, through the Levant ports, the Indian Ocean.

weapons of domination: shipping, trade, industry, credit, and political power or violence.

RULE THREE: THERE IS ALWAYS A HIERARCHY OF ZONES WITHIN A WORLD-ECONOMY

The different zones within a world-economy all face towards one point in the center: thus "polarized," they combine to form a whole with many relationships. And once such connections were established, they lasted. At ground level and sea level so to speak, the networks of local and regional markets were built up over century after century. It was the destiny of this local economy, with its self-contained routines, to be from time to time absorbed and made part of a "rational" order in the interest of a dominant city or zone, for perhaps one or two centuries, until another "organizing center" emerged; as if the centralization and concentration of wealth and resources necessarily favored certain chosen sites of accumulation.

Every world-economy is a sort of jigsaw puzzle, a juxtaposition of zones interconnected, but *at different levels*. On the ground, *at least* three different areas or categories can be distinguished: a narrow *core*, a fairly developed middle zone and a vast *periphery*. The qualities and characteristics of the type of society, economy, technology, culture and political order necessarily alter as one moves from one zone to another. This is an explanation of

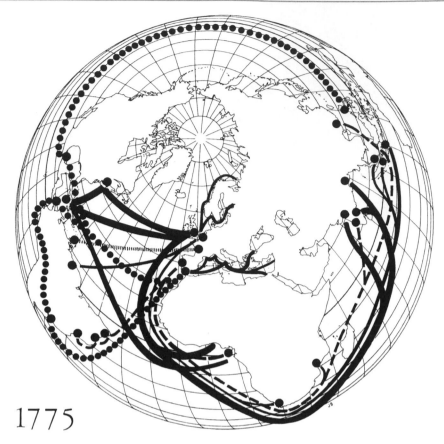

1775

Figure 2 European world-economies on a global scale, 1775
In 1775, the octopus grip of European trade had extended to cover the whole world: this map shows
English, Dutch, Spanish, Portuguese and French trade networks, identifiable by their point of origin.
(The last-named must be imagined as operating in combination with other European trades in Africa and
Asia.) The important point is the predominance of the British trade network which is difficult to represent.
London had become the center of the world. The routes shown in the Mediterranean and the Baltic simply
indicate the major itineraries taken by all the ships of the various trading nations.

very wide application, one on which Immanuel
Wallerstein (1974) has based his book *The Modern
World-System*.

The center or *core* contains everything that is
most advanced and diversified. The next zone
possesses only some of these benefits, although it
has some share in them: it is the "runner-up"
zone. The huge periphery, with its scattered
population, represents on the contrary backwardness, archaism, and exploitation by others. This
discriminatory geography is even today both
an explanation and a pitfall in the writing of world
history – although the latter often creates the
pitfalls itself by its connivance.

The central zone holds no mysteries: when
Amsterdam was the "warehouse of the world," the

United Provinces (or at any rate the most active
among them) formed the central zone; when
London imposed its supremacy, England (if not the
whole of the British Isles) formed the surrounding
area. When Antwerp found itself in the sixteenth
century the center of European trade, the
Netherlands, as Henri Pirenne said, became "the
suburb of Antwerp" and the rest of the world its
periphery. The "suction and force of attraction of
these poles of growth" were clear to see. Detailed
identification is more difficult though when it
comes to the regions outside this central zone,
which may border on it, are inferior to it but perhaps only slightly so: seeking to join it, they put
pressure on it from all directions, and there is
more movement here than anywhere else. Any

uncertainty evaporates on the other hand as soon as one enters the regions of the periphery. Here no confusion is possible: these are poor, backward countries where the predominant social status is often serfdom or even slavery (the only free or quasi-free peasants were to be found in the heart of the West); countries barely touched by the money economy; countries where the division of labor has hardly begun; where the peasant has to be a jack of all trades; where money prices, if they exist at all, are laughable. A low cost of living is indeed in itself a sign of under-development.

The backward regions on the fringes of Europe afford many examples of these marginal economies: "feudal" Sicily in the eighteenth century; Sardinia, in any period at all; the Turkish Balkans; Mecklenburg, Poland, Lithuania – huge expanses drained for the benefit of the western markets, doomed to adapt their production less to local needs than to the demands of foreign markets; Siberia, exploited by the Russian world-economy; but equally, the Venetian islands in the Levant, where external demand for raisins and strong wines, to be consumed as far away as England, had already by the fifteenth century imposed an intrusive monoculture, destructive of local balance.

There were *peripheries* in every quarter of the world of course. Both before and after Vasco da Gama, the black gold-diggers and hunters of the primitive countries of Monomotapa, on the east coast of Africa, were exchanging their gold and ivory for Indian cottons. China was always extending her frontiers and trespassing on to the "barbaric" lands as the Chinese texts call them – for the Chinese view of these peoples was the same as that of the classical Greeks of non-Greek-speaking populations: the inhabitants of Vietnam and the East Indies were "barbarians." In Vietnam however, the Chinese made a distinction between those barbarians who had been touched by Chinese civilization and those who had not. According to a Chinese historian of the sixteenth century, his compatriots called "those who maintained their independence and their primitive customs '*raw*' barbarians, and those who had more or less accepted Chinese ways and submitted to the empire '*cooked*' barbarians." Here politics, culture, economy and social model contributed jointly to the distinction. The raw and the cooked in this semantic code, explains Jacques Dourbes, also signifies the contrast between culture and nature: rawness is exemplified above all by nakedness. "When the Potao ["kings" of the mountains] come to pay tribute to the Annamite court [which was Chinesified] it will cover them with clothes."

RULE THREE (CONTINUED): DO NEUTRAL ZONES EXIST?

However, the backward zones are not to be found exclusively in the really peripheral areas. They punctuate the central regions too, with local pockets of backwardness, a district or "pays," an isolated mountain valley or an area cut off from the main communication routes. *All* advanced economies have their "black holes" outside *world time*.

A world-economy is like an enormous envelope. One would expect a priori, that given the poor communications of the past, it would have to unite considerable resources in order to function properly. And yet the world-economies of the past did incontestably function, although the necessary density, concentration, strength and accompaniments only effectively existed in the core region and the area immediately surrounding it; and even the latter, whether one looks at the hinterland of Venice, Amsterdam or London, might include areas of reduced economic activity, only poorly linked to the centers of decision. Even today, the United States has pockets of under-development within its own frontiers.

REFERENCES FROM THE READING

Braudel, F. (1972) *The Mediterranean and Mediterranean World in the Age of Phillip II*, trans S. Reynolds. New York: Harper & Row.

Braudel, F. (1973) *Capitalism and Material Life, 1400–1800*. New York: Harper & Row.

Braudel, F. (1979–1984) *Civilization and Capitalism, 15th–18th Century,* 3 volumes. Berkeley, CA: University of California Press.

Sismondi, S. (1991 [1951]) *New Principles of Political Economy: Of Wealth in its Relation to Population*, trans. R. Hyse. New Brunswick, NJ: Transaction.

Sweezy, P.M. (1974) *Modern Capitalism*. New York: Monthly Review Press.

Wallerstein, I. (1974) *The Modern World-System I.* New York: Academic Press.

3

"Urban Specialization in the World System: An Investigation of Historical Cases"

from *Urban Affairs Quarterly* (1986)

Nestor Rodriguez and Joe R. Feagin

Editors' introduction

Along with Christopher Chase-Dunn (1985), Michael Timberlake (1985), Fernand Braudel (Reading 2) and Janet Abu-Lughod (Reading, 4), sociologists Joe R. Feagin (Ella McFadden Professor of Liberal Arts at Texas A&M University) and Nestor Rodriguez (Professor and Chair of Sociology at the University of Houston and Co-Director of the Center for Immigration Research) were among the first scholars to examine systematically the process of world city formation in comparative-historical perspective. Their classic study of urban specialization in world cities was published during the mid-1980s, just as a large number of North American and western European urbanists were beginning to grapple more explicitly with the problematic of globalization. Both authors have continued subsequently to make important contributions to the study of globalized urbanization.

Rodriguez and Feagin cast their net widely in this reading to examine three sets of cities, in each case one financial center and one industrial center, situated within the globally hegemonic national state during three different phases of world capitalist development. Through their comparison of Amsterdam–Leiden (during the period of Dutch hegemony), London–Manchester (during the period of British hegemony) and New York–Houston (during the period of US hegemony), the authors explore the key role of "urban specialization" – the spatial concentration of particular types of economic activities within cities – in the dynamics of world city formation. Their major insight is that urban specialization assumes different forms not only within globally hegemonic national economies but also during successive phases of capitalist development. In each case, Feagin and Rodriguez show how advanced economic capacities in a particular sector, be it financial or industrial, serve as an essential foundation for the development and consolidation of global cities. More generally, in investigating the political economies of urbanization from early to late capitalism, they show that urban development patterns cannot be explained adequately in regional and national terms, but must be understood, instead, with reference to cities' positions in worldwide spatial divisions of labor. In this manner, Rodriguez and Feagin criticize mainstream US urban studies, including writers in the Chicago School tradition, for an excessively localist, culturalist orientation that brackets the large-scale political-economic contexts in which urban development unfolds. In addition, through their expert use of thick historical description, the authors also demonstrate how locally powerful corporate elites have managed to transform their private economic assets not only into a basis for local political influence, but also into a means for consolidating and extending their host city's global reach.

INTRODUCTION

In the history of urban sociology in the United States, much of urban-growth theory has been limited to a national level of analysis. From classical-location to uneven-development theories, conceptualization of urban growth has been contained mainly in regional or national frameworks. But in many cases, involvement in the capitalist world-economy has been a source of major urban development. Many major cities grow during certain periods mainly because of a specific function they play in the capital-accumulation circuits of the world-system. That is, some cities grow as the world-system grows because they have specialized in some function of capital accumulation, for example, as producers or as financial markets, and thus fit a specific niche in the world-economy. Taking a world-system perspective as a necessary level of analysis for understanding major urban growth, the discussion in this article elaborates on the concept of urban specialization, offers case studies of urban growth through a specialization niche, and compares the development of specialization from the standpoint of the three stages of capitalist development (mercantile, industrial, and monopoly) in which the cases are situated.

SPECIALIZATION IN THE WORLD-SYSTEM: A THEORETICAL INTRODUCTION

What is urban specialization?

It is perhaps an empirical commonplace to observe that cities tend to specialize in certain types of economic activities. London, for example is known as a financial center, and Detroit is identified with automobile production. In the present discussion, taking a world-system perspective of urban growth, the concept of specialization is developed and advanced from the standpoint of the development of specific cities in the world-economy. Structural dimensions of world capitalism are emphasized: (a) the circuits of capital, (b) the international division of labor, and (c) the spatial organization. These three dimensions are believed to be the basic matrices of socioeco-nomic development in the world capitalist system. In contrast to the conventional view, the relational aspect of specialization in a world context is emphasized. A city develops specialized economic activity and a corresponding social structure as a consequence of its relationship to other cities and regions in the capitalist world-system. This development may or may not involve the subdivision of extant industries. It may involve subdivision of an old industry, but specialization may also occur through the development of a new industry.

THREE SETS OF HISTORICAL CASES

In this section, historical cases of urban specialization in the capitalist world-system are described. The cities are financial and production centers in three historical periods of capitalist development: the mercantile, industrial, and advanced-monopoly periods. The specialized cities are situated in countries that dominated the world-economy at certain periods during its development. The core countries that acquired hegemony in the world-economy were settings where urban specialization intensified, especially since superior productive efficiency substantially undergirded hegemonic status. For the period of Dutch hegemony (1625–1671), the cases of Amsterdam and Leiden are examined; for the period of British hegemony (1815–1873), London and Manchester; and for the period of U.S. hegemony (1945–1967), New York and Houston. For each period, the dominant financial/administrative center was chosen. Leiden and Manchester were dominant industrial cities in the Dutch and British periods of hegemony, respectively. Houston was the dominant industrial-technological center of the developing petro-chemical industry during the U.S. period of hegemony.

In the development of the world-system, new stages of capitalism brought new structural conditions of urban specialization. The transitions from mercantile to advanced-monopoly capitalism contained changes in the systemic relations between financial and industrial specialized centers, in the relations between specialized industrial cities and the periphery of the world-economy, and, related to this, in the working-class development of the specialized industrial cities.

THE ERA OF DUTCH HEGEMONY

Amsterdam's financial specialization

Amsterdam developed into an international financial center during the period of Dutch hegemony in the seventeenth century. In the division of labor of the then young capitalist world-system, Amsterdam was a center of manufacturing, ship-building, and trading, a place where commodity capital was converted into money capital. But it was foremost a center of finance for business projects throughout the world. Amsterdam's Exchange Bank (established in 1609), Loan Bank (1614), and Burse (1611) made the city the focal point where money capital was converted into commodity capital and where monies were enlarged through loans and investments abroad. From a town with no special status in the mid-sixteenth century, Amsterdam grew into the "Wall Street" of the world-system in the seventeenth century.

During this growth, the population of this world city quadrupled, from about 50,000 in 1600 to about 200,000 by the end of the century. Amsterdam's specialization in finance, which stimulated the growth of its manufacturing and trading industries, involved the in-migration of impoverished peasants from northern Holland and craft workers from southern Holland.

From the perspective of the spatial organization of intra-European trade, Amsterdam was ideal. It was situated between the Baltic area countries of Russia, Poland, Sweden, and Denmark and western and southern European areas that depended on the former countries for grain imports. Amsterdam became a capital-circuit microcosm of mercantile capitalism, with "an abundance of ever-ready goods and a great mass of money in constant circulation" (Braudel 1984: 236). This position in the circuit of capital was significant for the financial rise of Amsterdam because the city's bank was originally developed to enable merchants to settle

Plate 3 Amsterdam (Roger Keil)

mutual debts. Trade and finance are thus intimately linked.

The bank of Amsterdam provided security rare in seventeenth-century banking because of the interventionist role of the state. Instead of adhering to mercantilist policies, at an earlier point in time the Dutch state (funded by Amsterdam) created conditions that facilitated private enterprise, such as the free movement of bullion, a system of bills of exchange, and credit functions. In some cases, the Dutch government guaranteed loans made by private financiers. Amsterdam's government enhanced the city's capitalistic environment by promoting the immigration of business entrepreneurs. The municipal council kept the cost of citizenship, required in some businesses, low and found housing and offered other inducements for immigrant master craft workers who agreed to start industries in silk finishing, cloth making, leather gilding, glassblowing, mirror manufacturing, salt refining, or shipbuilding.

Relations between employers and workers in Amsterdam were relatively stable, in part because of social welfare measures (e.g., lodging, hospitalization, and monetary payments for some) funded by private sources and by revenues derived from the sale of confiscated properties of the Roman Catholic church. Finance capitalists in Amsterdam invested in profit-motivated projects that greatly enhanced the development of the capitalist world-system in the seventeenth century. Examples included mining, timber exporting, and the development of sawmills in Sweden; mining, timber exporting, and salt production in Denmark; copper mining in Norway; sulfur exporting and fishing in Iceland; exporting caviar, tea, oil, and wool in Russia; grain exporting in Poland; and lake dredging and canal construction in Italy, France, and England. Amsterdam capitalists invested heavily in the Dutch East and West India companies, which through colonization helped expand the periphery of the world-economy to the West Indies, the Indian Ocean, and the Far East.

Textile specialization in Leiden

The world-trading networks that opened up with the assistance of Amsterdam finance capital stimulated the development of specialized industries in many areas in Holland. In the period of Dutch hegemony, textile production centered in Leiden. Leiden's cloth production was central to the Dutch economic ascension in the young capitalist world-system. Cloth and draperies became an important resource for the Dutch trade in the Baltic, the Mediterranean, the Levant, Africa, and the West Indies. Textile industrialization made Leiden second only to Amsterdam in population size in the United Provinces.

Within the circuits of capital of seventeenth-century Europe, Leiden was the largest single-industry setting where money capital was converted into commodity capital of greater value. In addition to large-scale manufacturing, its labor structure involved traditional craft (*fabrieken*) production that used capital mainly for the wages of the skilled workers. Many of the *fabrieken* industries produced for a consumer-market niche of new lighter-weight cloth. Textile specialization made Leiden the largest manufacturing center of the "new draperies" in seventeenth-century Europe. Leiden's textile development benefited from its geographical position at the center of an area of cloth commerce, a confined northern European area where semifinished cloths imported from England were dyed and dressed in Amsterdam, and German linens were bleached and finished in Haarlem.

Leiden's growth as a center of textile specialization involved political and class developments. The urban centers of Holland such as Leiden prospered because they dominated the countryside economically and politically. With large-scale immigration, the Leiden area became a labor-rich environment. But immigration was not the only source of labor. Leiden's labor force was supplemented with child workers, many of whom were imported. Through the use of child and immigrant labor, large manufacturers circumvented the older artisan (guild) mode of production, in which craft workers controlled the scale of operation.

In contrast to the later industrialization, in which factories operated with unorganized work forces with little power, Leiden's industrialism involved *fabrieken* industries, in which artisan guilds still had some control. The cloth guilds in Leiden and in other towns in Holland tried, albeit unsuccessfully, to resist changes in their regulation of work hours, wages, and sizes of work forces. The guilds also refused to accept alien workers.

The second half of the seventeenth century witnessed a sharp decline in Leiden's textile industry. Several factors precipitated this decline: the high wage levels of skilled cloth workers, the loss of markets in Spanish-controlled territories, English mercantile policies that restrained the export of English and Scottish wool, and the relocation of textile production by Dutch capitalists to the low-wage countryside outside Holland, and the growing popularity in Europe of the cheaper calicoes imported from India.

Leiden's textiles helped put the United Provinces (especially Holland) at the top of the spatial hierarchy of the emerging world-system in the seventeenth century. With its relatively advanced industrialization, Leiden became the leading textile area in the spatial organization of the evolving world-economy. The technical superiority of Leiden's textile specialization gave Dutch merchant capitalists an important trading advantage in other core countries and in the semiperipheral and peripheral areas of the Mediterranean, the Levant, Africa, and the West Indies. For a few places in the expanding world-economy, Leiden also provided textile technology and skilled labor.

THE ERA OF BRITISH HEGEMONY

London's financial specialization

In the period of British hegemony in the world-system (1815–1873), world trade and finance centered on London. The city was a major source of finance capital for the vast British trading network that, in addition to western Europe, included cities in the Baltic area, the West Indies, India, the Far East, and the Americas. Other major cities (e.g., Manchester) were also important markets for major primary commodities, but no city in England or abroad matched London's financial function in the world-economy. Capitalists from across Europe and a few more from the United States traveled to London to obtain funds for a diverse range of enterprises. Foreign national and state government officials from places such as Russia, Holland, Egypt, Massachusetts, Alabama,

Plate 4 London (Ute Lehrer)

and Bolivia journeyed to the banking houses of London for loans, especially for the development of transportation systems.

London bankers funded operations that enhanced transportation in many regions of the world-system. A prime example of this was the financing of railroad construction in western Europe, India, and the Americas. Railroads added to economic growth by cheapening the price of transport, and consequently the cost of producing commodities. Railroads stimulated the development of supporting industries (coal mining, locomotion production); passenger traffic became a business in itself. Investment in railroad construction was an initial event that linked London's financial houses directly to industrialization. Prior to this investment, London's financiers were involved in the growth of industrialism mainly by providing funds for the buying or transportation of industrial commodities.

London developed as a world financial center even before it developed as a national financial center; this speaks for the complexity of regional and world capital circuits that underlie the growth of urban specialization. From the chartering of the Bank of England in 1695 to the end of England's industrial revolution, merchant-bankers were the core of London's financial sector. Throughout the first half of the nineteenth century, English industrial capitalists were not prominent in London's financial establishment.

London's rise to national financial dominance grew out of the problems of small private banks. While in a few cases the new industrial capitalists obtained funding directly from London bankers, in the early years of the Industrial Revolution, the source of money capital for the development of industrial capital was the small private bank. However, an increasing number of country bank failures (311 in the period 1809–1830) prompted Parliament to pass legislation that limited the note-issuing power of these banks, allowed large joint-stock banks in and around London, and permitted the Bank of England to set up branches throughout the country. The Bank Charter Act of 1844 gave exclusive note-issue rights to the Bank of England, giving London the supreme financial position it long enjoyed in the world-economy.

State intervention was critical to the expansion of the capital circuits. The British government's designation of the Bank of England as its financial source provided an important vitality to London's finance sector. London's growth as the nineteenth-century world financial center was based on the immigration of foreign capitalists who set up offices in the city. The growth of London's merchant-banker sector led to the development of many supportive structures that contributed to making London a large metropolis. Retail stores were set up in London to cater to the upper classes.

But the majority of London's growing population (just under 5 million in 1881) consisted of working-class people. These worked in a variety of service jobs critical to the functioning of the world's financial center and in industries such as flour milling, breweries, iron-muggery, and dock loading. At least 40 per cent of the city's working class lived in poverty. Many of the poor were migrants, mainly laborers from southern England, and Irish and Jewish immigrants.

Cotton specialization in Manchester

To a considerable extent, the rise of industrial capitalism in Britain was facilitated by London's financing of industrial and trade projects in other core countries and in peripheral areas across the world-economy. Areas that used financial capital from London also depended on British manufactured products for their economic development. Thus with the extension of capital circuits in the world-system during the late eighteenth and nineteenth centuries, British industrial cities became "the workshops of the world." Because of world demand, cotton-cloth production led the way in the development of British industrial capitalism. The cotton cities experienced the most dynamic mechanization and urban growth; of these, Manchester was the most developed. Specializing in the production segment of the circulation of capital in the world-system, Manchester underwent a rapid concentration of labor and of buildings, machines, and raw materials that reached the unprecedented scale of a massive factory system.

Manchester's industrial specialization was an important factor in Britain's development as a supplier of half of the world's cotton goods. Cotton products manufactured in Manchester were exported through Liverpool to core countries

Plate 5 Manchester (Roger Keil)

such as France and Germany, and to areas in the periphery such as Africa and Asia, and especially to Latin America and India.

From a spatial perspective, Manchester specialized as a critical manufacturing node in an urban network – the means by which Britain, as a core area, was related to other core, peripheral, and semiperipheral areas in the world-system. Without these sources of raw cotton and these consumer markets, Manchester could not have achieved its remarkable industrial specialization.

Various political, spatial, environmental, and class factors were related to Manchester's cotton specialization. Manchester's spatial situation was enhanced through the development of transport systems. Two elements in the natural environment of Manchester were resources for the area's industrialization: swift streams and coal. Before steam technology, streams powered the waterwheels of the textile industry. The use of steam engines in the late eighteenth century made coal an important fuel. By the end of the Industrial Revolution in the 1840s, the geographical shape of the Manchester

textile-manufacturing district was substantially determined by the location of adjacent coalfields.

Manchester's specialization involved labor migration as well. While Manchester actually exported labor for rural industrial growth in the early development of industrialization, by the second and third decades of the nineteenth century the city was a destination for many who abandoned the handlooms in small communities and headed for the factories.

The central factor in Manchester's development of textile specialization was a radical change from artisan workers to wage-earning factory workers. Whereas at the beginning of industrial capitalism many British workers owned their tools and controlled their own work, by the mid-nineteenth century many worked for wages in factories. This transformation was not peaceful. Different artisan groups revolted as their trades became deskilled through factory organization. In the new conditions of class relations, women and children were used by British factory owners to displace the skilled workers.

THE ERA OF U.S. HEGEMONY

New York's financial specialization

New York's emergence as the world financial center in the period of U.S. political-economic hegemony (1945–1967) occurred in a setting of dramatic American economic expansion into the world-system. Rapid postwar development, the opening of market outlets in reconstructed western Europe, and investment opportunities in Latin America without competition from European or Japanese investors all helped thrust U.S. capital into a dominant position in the international circuits of capital. In the United States, the centralization of industrial capital was a salient feature throughout the period of hegemony.

Historically, the financial district of New York was "the child of the port." International trade stimulated the financial growth of New York. With the construction of the Erie Canal and other canal systems in Ohio and Pennsylvania, the port of New York took a dramatic leap in international trade. With this dynamic commercial setting, and with the immigration of impoverished European labor, a large number of industries (garment, chemical, tobacco) located in the city.

By 1900 the New York financial district had evolved into a complex formation of commercial and investment banks, insurance companies, stock markets, and the central offices of large and small firms. Of the 125 insurance firms in Manhattan, foreign-owned fire and marine insurance companies outnumbered any single type of domestic insurance company. With the decline of European capital by the end of World War I, New York financial institutions emerged as one of the world's specialized banking centers. During the 1945–1967 period of U.S. hegemony, the expansion of industrial capital stimulated the growth of financial capital. New York financial capitalists increased in wealth as they provided much external funding to nonfinancial corporations.

The foreign growth rate of U.S. banks has far exceeded their U.S. growth rate. New York banks have even expanded into nonfinancial operations in peripheral countries. New York's financial specialization has been particularly important for

Plate 6 New York City (Roger Keil)

production capital in the Latin American periphery. The city has served as headquarters for many large corporations that do business in the area.

The growth of New York's financial specialization in the world-system during and after the U.S. period of hegemony involved a restructuring of the city's labor demand. Since the 1950s, the city has lost more than 400,000 manufacturing jobs in industries such as clothing, publishing, and electronics. In the 1970 census, workers in service industries outnumbered workers in manufacturing activities for the first time. But the growth of the service sector linked to the financial and corporate headquarters complexes has involved polarized concentrations of highly paid and poorly paid jobs. High-income jobs are associated with the growth of the advanced services (e.g., financial, managerial) and of international corporate headquarters. The expansion of low-income service jobs is associated with the growth of low-skill, dead-end jobs in major service industries and with the increase of business services that cater to the top-level work forces of these industries. The city's 1.5 million Hispanics and large concentrations of Asians and West Indian immigrants – symbolic again of migration in a world-economy – have been an important supply of labor for the low-wage service jobs, as well as for the resurgent sweatshops and industrial homework.

New York's financial capitalists have played a critical role in remaking the face of the city itself. For example, they have sought state aid to provide critical infrastructure in the form of highways, bridges, and even office buildings. The two 110-story World Trade Center towers in lower Manhattan were the brainchild of the Downtown-Lower Manhattan Association, chaired by David Rockefeller, then head of Chase Manhattan Bank. Designed to house international banks, import and export companies, and world trade organizations, the twin towers involved the razing of older structures and the construction of these megastructures at local government expense (about $1 billion). Government aid in the form of special utility and street projects has also facilitated numerous private office complexes, such as the Rockefeller Center, with its 21 office towers.

New York's financial capitalists have even become the overseers of the city government itself. The very institutions that had encouraged the city to overextend itself financially, the banks holding New York municipal securities, had used the fiscal crisis in the 1970s to force the workers of New York to accept a fiscal austerity plan. Finance capital was, with the aid of corporate executives from utility and airline companies, running the city.

Oil-industrial specialization in Houston

The beginning of the U.S. period of hegemony in the 1940s coincided with an oil-industrial boom in Houston that clearly established the city as the "oil capital of the world." Post-World War II oil specialization in Houston consisted of three parts: the production and refinement of crude oil from nearby oil fields and from peripheral countries, the emergence of a major petrochemical industry, and, accelerating in the 1960s, the distribution of oil technology for the world's oil fields and industries. Since the 1910s this industrial specialization was shaped in part by New York area corporate headquarters and bank complexes; in turn, this specialization stimulated the growth of many support industries in the Houston area, from steel companies to downtown office developers. Because of its development of oil refining, petrochemical, and oil tools production, the Houston area achieved an elevated standing in many respects. By the 1960s and 1970s Houston oil-industrial specialization was so connected to the world-economy that it was not substantially affected by the decline of U.S. economic hegemony in the 1970s.

From the perspective of circuits of capital, Houston's oil-industrial specialization involves different forms and levels of capitalistic interests. Finance capital has been involved. East Coast finance capital periodically has been a critical source of loans for some of the companies that have petrochemical, oil technology, and real estate projects in the area. In addition, British, Canadian, German, Iranian, Mexican, and Saudi Arabian capital invested heavily in the city's real estate sector in the 1970s and 1980s. As a center of oil-industrial specialization in the world-economy, Houston has been the focal point for innovation in the world's oil industry. Specialization has, indeed, characterized this world city.

Consider Houston in the spatial structure of the world-system. Houston's industrial specialization has served as a means by which the United States, as a core area, developed enduring relationships with cities in outlying peripheral areas. In the post-World War II era, the growing Houston economy enhanced the incorporation of oil-rich peripheral regions into a capitalist world-system dominated by the United States. Oil-industrial specialization developed in Houston, and not in some other part of Texas because of nearby oil fields. The proximity of natural resources was very important at the start of Houston's specialization, because a pipeline infrastructure did not exist.

Yet the actions of Houston's business elite to advance the area's oil industry demonstrated that the source of Houston's specialization was shaped by political and boosterism factors as well. The local business elite promoted the economic growth of the Houston area by seeking spatial and governmental financial advantages. In spite of their widely heralded "free enterprise" philosophy, members of Houston's business elite were masters at using governmental aid to accelerate the development of local infrastructure and enterprise.

Workers from rural areas in Texas and surrounding states, from other urban areas, and from peripheral areas in the world-system have supplied Houston with labor. Migrating mainly from the peripheral countries of Mexico, El Salvador, and Guatemala, the area's undocumented Hispanic population numbers about 200,000. During Houston's economic boom in the 1970s and early 1980s, these workers played a critical labor role in construction and real estate development.

THE POLITICAL ECONOMY OF DEVELOPMENT

From the standpoint of the impact on the world-system, the development of urban specialization represents the development of a global city. This is easier to recognize in cities with financial specialization (e.g., London and New York), where multinational corporations and international banking institutions made decisions concerning economic development throughout the world. But cities with production specialization may also be considered global cities. As the cases of Leiden, Manchester, and Houston demonstrate, such specialized cities have substantial world-level impact, for example, in attracting foreign migrant labor, extending international commerce, and spatially enlarging the periphery in search of raw materials and commodity markets. Thus, for major cities with specialized production, the sources of supply and demand are global.

Though widely used in urban theory since the writing of R.D. McKenzie in 1926, the concept of "specialization" has not received the analysis it requires. Instead of exploring the dimensions and development phases of urban specialization, many analysts have been content to see specialization as an outgrowth of natural-ecological processes that proceed through an invisible-hand type of logic. The analysis presented in this article offers a different view of how specialized activity develops in cities. Far from being a "natural" process, specialization is grounded in the political economy of development. For each case of specialization the intentional business (class) and political actions that undergirded its development can be identified.

REFERENCES FROM THE READING

Braudel, F. (1984) *Civilization and Capitalism, 15th–18th Century: Volume 3, The Perspective of the World*. New York: William Collins and Harper & Row.

McKenzie, R.D. (1926) The scope of human ecology, *American Journal of Sociology*, 32, July, 141–154.

4

"Global City Formation in New York, Chicago and Los Angeles: An Historical Perspective"

from *New York, Chicago, Los Angeles: America's Global Cities* (1999)

Janet Abu-Lughod

Editors' introduction

Janet Abu-Lughod, Professor Emerita of Sociology at the New School for Social Research in New York City, has been making seminal contributions to the study of urbanization in global perspective since the early 1970s, primarily through her studies of *Third World Urbanization* (Abu-Lughod and Hay 1977) and through a series of classic monographs on urban development in Cairo, Egypt (1971) and Rabat, Morocco (1980). In the late 1980s, Abu-Lughod (1989) completed a landmark historical study that underscored the extent and dynamism of transnational urban systems, based upon dense commercial interdependencies, well before the consolidation of mercantile capitalism in sixteenth-century western Europe. Abu-Lughod's methodology resonates very closely with that of Braudel (see Reading 2) due to its emphasis on the centrality of urban systems to economic life, its adoption of a macrospatial perspective for the analysis of such systems and its focus on the long-term trajectory of historical-geographical change.

Reading 4 is excerpted from the introductory and concluding chapters of Abu-Lughod's most recent book (1999), which examines the divergent patterns of restructuring that have emerged in New York, Chicago and Los Angeles since their origins. Two key points should be kept in mind. First, Abu-Lughod insists on the importance of a long-term historical perspective to the understanding of contemporary patterns of urban development. For Abu-Lughod, therefore, the current round of globalization is not as new and unique as some scholars have claimed. Rather, it represents the latest rupture within the long history of global capitalist urbanization. Second, Abu-Lughod provides a broad geohistorical sketch of urban development in the US, suggesting that the history of each of her three cities can be understood adequately only in relation to broader, national and transnational political-economic trends. Thus, despite her criticisms of certain strands of global cities research, Abu-Lughod's study of "America's global cities" forcefully demonstrates one of its key propositions – namely, that a global political-economic and spatial perspective is required in order to decipher local developmental outcomes.

The theme of "global cities" has recently captured the imagination of urbanists, but as I shall argue, much of this exciting literature has been remarkably ahistorical, as if contemporary trends represent a sharp break from the past, if not an entirely new phenomenon. Furthermore, both the general descriptions of "world cities" and the accompanying causal analyses that attribute their commonalities to general forces residing at the highest level of the international economy neglect variations in global cities' responses to these new forces.

Contemporary scholars, trying to define the "global city," imply that it is a relatively new phenomenon that has been generated *de novo* in the present period by the development of an all-encompassing world system – variously termed late capitalism, postindustrialism, the informational age, and so on. Among the hallmarks of this new global city are presumed to be an expansion of the market via the internationalization of commerce, a revolution in the technologies of transport and communications, the extensive transnational movement of capital and labor, a paradoxical decentralization of production to peripheral regions accompanied by a centralization in the core of control over economic activities, and hence the increased importance of business services, particularly evident in the growth of the so-called FIRE economic sector – finance, insurance, and real estate. Accompanying these changes, and often thought to result from them, is a presumed new bifurcation of the class structure within the global city and increased segregation of the poor from the rich.

The value of such insights cannot be denied, but it is questionable whether these phenomena are as recent as is claimed. For example, all of these characteristics, at least in embryo form, had already made their appearance in New York City before the last quarter of the nineteenth century, when that city was clearly recognizable as a "modern" global city. And even though the pace and scale of today's globalized economy – and thus of the global cities that serve as its "command posts" – are faster and vaster, and the mechanisms of integration more thoroughgoing and quickly executed, the seeds from which the present "global city" grew were firmly planted in Manhattan during the middle decades of the nineteenth century. Chicago and Los Angeles eventually (and sequen-

Plate 7 New York City (Roger Keil)

tially) followed that model with time lags of thirty and sixty years, respectively, although they naturally did so in a changed world context and under revised regimes of production, circulation, and consumption, as well as politics.

The relatively recent appearance of these global cities in the United States alerts us to proceed cautiously in comparing global cities without taking into account variations in the depths of their historical heritages. Most megalopolitan agglomerations that today serve as "world" or "global" cities – for example, London, Paris, Amsterdam, Tokyo – developed over many centuries. They thus contain accretions of successive types of settlements that have layered, one upon the other, vastly different patterns of development and reconstruction, until the composite whole becomes difficult to grasp. Not only are their landscapes difficult to "read," but they are not easily compared with one another, because the national political and cultural contexts in which they developed are so different [...] Even to compare the differential impact of global forces on America's three largest metropolitan centers requires much closer attention to the specific historical and geographic contexts in which they developed over the last century. Such an approach needs to take into consideration: the changing shape of the world system that constitutes the largest context for developments within them; the history of the expansion of the United States over the course of the nineteenth and twentieth centuries, within which the national urban hierarchy developed; and the more detailed histories of these individual urbanized regions that, over time, have generated the physical and social "terrain" onto which the newer global forces are now being inscribed and with which they interact.

Cities as nodes in networks are not a new phenomenon. Indeed, the fact that cities lie at the center of complex networks constitutes their *essential* feature. Throughout world history, certain cities – some of them imperial capitals remarkably large for their times, but a few relatively tiny "city-states" – have served as key nodes through which wider circuits of production, exchange, and culture have been coordinated, at least minimally. But in these earlier manifestations of integration, the territorial reach of even the most extensive "transnational/transimperial" systems was limited to only small fractions of the globe. Entire continents were excluded or were in touch at their peripheries only with the outer fringes of core regions. Nevertheless, urbanization per se was, in fact, both a symptom and a consequence of the construction of such regional systems, whose cores exerted dominance over their agricultural hinterlands and/or, via rivers or even the edges of the sea, increased the surplus available to the cities through conquest and/or tribute or through favorable terms of trade with distant points.

The first of these mini-world-systems climaxed toward the beginning of the second millennium B.C.E. when the three river-valley cradles of urbanism – along the Nile, the Tigris–Euphrates, and the Indus Rivers – came in more intimate contact with one another by multiple networks of trade that threaded through deserts, skirted the shores of the eastern Mediterranean and the Arabian Sea, tran-

sited the Red Sea and Arabo-Persian gulfs, and sent out probes to more distant areas in Anatolia, the Iranian plateau, and the zones beyond the Indus River Valley. (Almost contemporaneously, another minisystem was developing in the Yellow River region of China, one that would eventually form linkages with regions south and west of it.)

A second surge in integration began during the Hellenic Age, when Alexander's conquests briefly unified the eastern Mediterranean and reached beyond it – as far as India. This system climaxed during Roman imperial hegemony, when the entire littoral of the Mediterranean Sea became part or a central core that eventually stretched into western Europe as far north as England and reached, via trading circuits, not only the eastern coast of the Indian subcontinent but, indirectly, even China.

Despite a brief hiatus caused by the fragmentation of western Europe (glossed somewhat inaccurately as the "fall of Rome" but more accurately described as the devolution of the so-called Western Roman Empire), the persistence of Eastern Christianity and an Islamic expansion throughout the Mediterranean and Eastern worlds led to the emergence of a third partial world system that extended over an even larger area. This system climaxed in the thirteenth century when very large portions of Europe, Eurasia, the Middle East and North Africa, coastal zones of east Africa, India, Malaysia, and Indonesia, and even China were becoming more interactive – in both commercial and cultural contacts and through military conflicts. The unprecedented unification of Central Asia and China under the Yuan dynasty intensified such interactions. Needless to say, northwestern Europe was then still at the periphery of this system and the New World was not yet connected to it.

Perhaps shaken by a series of pandemics that culminated in the Black Death, whose highest mortalities occurred in zones most tightly integrated into the ongoing world system, there was another hiatus. Within this breathing space, a fourth world system began to be reorganized, admittedly on the basis of the old but expanding rapidly through the so-called Age of Discovery to encompass parts of the New World and eventually other "terra incognita" in the South Pacific and southern Africa. This was the early phase of what

Immanuel Wallerstein has called "the modern world-system," and it constituted the context within which New York first developed, albeit as a subordinate node.

During this period of early modern restructuring, the "balance of power" began to shift away from the Mediterranean and Asia to the increasingly powerful Atlantic sea powers, first Portugal and Spain and then England and the Low Countries (including Holland and Spain, which were then in a common "nation"). In the process, the formerly forbidding Atlantic was added as the third central sea (albeit more treacherous) of the evolving system, joining the Mediterranean and the Indian Ocean–South China Sea, which continued to serve as major pathways of trade, commerce, conquest, and the movement of peoples. But minor European incursions were also being made into the Pacific as well. By the end of the nineteenth century, this system climaxed in classical European colonialism, achieved through the conquest of Africa and portions of Asia. By then, most countries in the Americas had been liberated.

Because throughout these earlier eras transport by water remained considerably cheaper than transport over land, the key points of exchange were, almost without exception, river or sea ports (the exceptions, of course, were oases along desert routes). It is in the context of the modern world-system, then, that the ports along the northeastern seaboard of North America became linked to a European core, and that, eventually, New York solidified its lead in the U.S. subsystem, a lead that, although later challenged by inland and Pacific coast cities, has never really been superseded.

For much of the first centuries of its existence, then, New York remained a key American link into a world system that focused increasingly on the Atlantic. Throughout the nineteenth century and into the early years of the twentieth, American history reads as the integration and eventual consolidation of a transcontinental subsystem, spreading from east to west. Even when the midcontinent was settled up to the Mississippi, and St. Louis (soon to be overtaken by Chicago) became the hinge for the drive to Manifest Destiny, New York retained and

Plate 8 Chicago (Roger Keil)

indeed strengthened its dominance as a core in its own right. It was, almost from its start, a "global city." Chicago could never have achieved the eminence it did without its prime outlet to the sea, New York.

It is important to recall that the integration of Chicago with the nascent U.S. system to its east and south was initially by water, the historically preferred transportation pathway. In the first quarter of the nineteenth century, decades before the railroad terminals consolidated Chicago's lead as midcontinental nexus, the outlets to the Atlantic coast via the Erie Canal–Great Lakes system and to the Caribbean Sea via the internal thoroughfare of the Mississippi River were already in place. What the rails did that waterways could not do, however, was link the zones west of the Mississippi to Chicago and from there on to New York. Without these rail linkages, Los Angeles's later growth (at least in the form it took) would have been inconceivable.

It was not until the tiny Mexican settlement of Los Angeles, conquered a bare quarter of a century before, was finally connected to the U.S. network via railroads – at first indirectly through San Francisco in 1875 and then via a direct route a decade later – that its modern growth spurt began. And it was not until the twentieth century, after the formation of America's first "overseas empire" (thanks to territories ceded in the 1898 Spanish–American War), that the Pacific became a true, albeit still a secondary, focus of American geopolitics. Heightened by these strategic interests, the sea circuit from the Pacific to the Caribbean was significantly shortened a dozen or so years later by the construction of the Panama Canal. Thus New York was the point of departure for Manifest Destiny, Chicago was its midwestern switching yard, and Los Angeles ultimately became its terminus.

It would be an error, however, to think in such geographically determined ways. Although an advantageously located site is a sine qua non for urban development, the agency of "men" (and they *were* mostly men in those days), acting politically and economically, has always intervened to favor certain of several otherwise equiprobable locations and to mobilize private and public financing to exaggerate the potential of such favored sites. And changes in technological capac-

Plate 9 Los Angeles (Roger Keil)

ity often have served to reduce or increase the viability of any natural setting.

Thus New York's port, so favorably endowed by nature, did not expand dramatically until the commercial invention of direct port auctions gave the city's brokers a monopoly over foreign trade, and until the engineering achievement of a through waterway to the Great Lakes made New York the dominant break-in-bulk point in internationally linked trade. And it was the capital accumulation facilitated by sophisticated institutions of insurance, banking, and credit that consolidated New York's lucrative role as broker for the slave-produced cotton crop, in preference to any southern port.

Similarly, both drainage of Chicago's waterlogged site and the clever machinations of land-speculating politicians in attracting rail termini and "hub" functions were essential in consolidating that city's lead over potential competitors, just as the later engineered reversal of the flow of

Chicago's river reduced the need for portage to the Mississippi. "Nature's metropolis" may have drawn upon a rich agrarian and mineral hinterland, but it was, in the last analysis, the city's skill at centralizing the processing of these raw riches by means of machines and accounting inventions that made it "the metropolis of midcontinent."

The case of Los Angeles is even clearer, because initially the region had neither a water supply sufficient to support a major city nor a natural harbor able to compete with the better-endowed ports at San Francisco to the north or San Diego to the south. Only the political clout of local businessmen, exploiting access to both local and national public funds, enticed a continental rail terminus to the area, secured distant water for the municipality's monopoly (assisted by the engineering skills of the compulsively driven genius immigrant Mulholland), and gained the enormous federal subsidies necessary to construct an expensive, artificially enhanced massive port complex.

Unhappily, wars also play their part in creating locational advantages out of potentials. Just as Los Angeles's modern history was born in the 1847 conquest, expanded in the 1898 Spanish-American war, and further consolidated with the construction of the Panama Canal just before World War I, so the city was not decisively catapulted into the first ranks of the American urban system until World War II, when the Pacific arena drew the United States into an irreversible involvement with the "East" (to its west). The Second World War also boosted the economies of New York and Chicago: the former primarily through its ports, from which lend-lease shipments were funneled to Europe, and its expanding shipbuilding and airplane manufacture directed to the European theater; the latter through the burgeoning demand for war *matériel* produced by its heavy Fordist industries.

By then, the world system was moving into the culminating phase of late-modern globalization. The evidence is obvious. One has only to contrast the First World War with the Second. The first had really encompassed only a portion of the European-Atlantic "world." The second signaled that the world system had incorporated the countries of Asia and the Pacific Rim as well. To this day, the postwar period has seen the "reach" of this sys-

tem extend to virtually all parts of the globe, including most of Central and South America. Only a few mountainous redoubts, some interior deserts in Africa, Asia, and Australia, and a handful of off-course islands lie temporarily beyond global reach, and their days are numbered.

Weapons of war, produced first in the United States for its own defense, have fueled the remarkable economic prosperity of the Southwest, including Los Angeles; have partially infused the economies of the Northeast and Mid-Atlantic states, including New York's extended region; and have, by their absence, further undermined the economies of the Midwest, including Chicago's. But weapons produced for export have also enhanced the hegemonic position of the United States in the world economy and, through sales to Third World countries and the deployment of forces in subregional conflicts, have reconfigured the shape of the entire world system.

History does not end with globalization. The present fates of the urbanized regions of New York, Chicago, and Los Angeles are linked to a changing geography of power, and thus, ultimately, to the shape of the larger system. Reflecting the Janus-like position of the contemporary United States as both an Atlantic and a Pacific power, and the increased integration of North America with the Caribbean and the Latin American continent, the three seacoasts of the United States have become even more important magnets for people, both through internal migration and external immigration. In recent decades the population of the United States has continued to decant toward those coasts, not only in the conventional directions of east and west but southward as well. The rapid rise of gateway Miami almost to world-class status is certainly linked to the growing importance of the Caribbean and "our neighbors to the south," as that zone of influence is increasingly integrated with the American core, if only, it sometimes seems, by illegal traffic in drugs. Chicago's tragedy is that it is not in these growth zones.

Technological advances have continued the age-old process of disengaging decisions from actions on the ground, with the ironic effect of facilitating the dispersal of production and people while increasingly centralizing what many analysts now refer to as "command functions." We saw this at earlier moments: the substitution of the commerce in "chits" in New York for the "real" midwestern wheat that remained in place in Chicago's silos; the removal of factories to the outskirts of cities at the same time company headquarters expanded in city centers, where telephones and later computers could monitor production farther afield and even abroad; the diffusion of stock ownership at the same time professional managers concentrated their hold over important decision making.

To some extent, these processes continue, but the scales at which they now operate often disengage or camouflage any clear lines between causes and effects, between those who command and those who labor, as capital and labor move with increasing freedom beyond not only the metropolitan boundaries but national borders as well. This disengagement means that healthy growth in command functions is not incompatible with dire destitution in those parts of the system (whether highly localized, at the national level, or at the global level) that are "out of the loop." Such marginalized zones can now be found in Manchester and Sheffield, England, in downtown Detroit, in the South and West Side ghettos of Chicago, in South-Central Los Angeles, in Bangladesh, and in many parts of the African continent.

History does not end, nor do changes in the world system cease. As sites to satisfy the world's burgeoning demand for consumer goods and services relocate to Asia, the hierarchy of global cities is being reconfigured and many new centers are being added. China, the location of some of the world's oldest and largest cities, may be reclaiming her status as one of the cores of the evolving "fifth iteration" of this world system. This changing context will inevitably have an impact on the global cities of the West, but we can only guess at the role political and technological developments may play in this process.

REFERENCES FROM THE READING

Abu-Lughod, J. (1971) *Cairo: 1001 Years of the City Victorious*. Princeton, NJ: Princeton University Press.

Abu-Lughod, J. (1980) *Rabat: Urban Apartheid in Morocco*. Princeton, NJ: Princeton University Press.

Abu-Lughod, J. (1989) *Before European Hegemony: The World System, A.D. 1250–1350*. New York: Oxford University Press.

Abu-Lughod, J. and Hay, R. (eds) (1977) *Third World Urbanization*. Chicago: Maaroufa Press.

"The New International Division of Labor, Multinational Corporations and Urban Hierarchy"

from M. Dear and A.J. Scott (eds), *Urbanization and Urban Planning in Capitalist Societies* (1981)

Robert B. Cohen

Editors' introduction

Robert B. Cohen is currently a Fellow with the Economic Strategy Institute in Washington, DC and President of Cohen Communications Group in New York City, a consulting firm. Based on Cohen's dissertation research in the Department of Economics at the New School for Social Research, this reading was originally published in an influential collection of radical urban planning texts in 1981; it was subsequently cited quite widely in the burgeoning field of global cities research. Even though many of its methodological strategies and empirical data have now been superseded (see, for instance, Reading 11 by Beaverstock, Smith and Taylor), Cohen's text provided an innovative, insightful starting-point for subsequent work in this new field of inquiry. It is thus widely recognized as a foundational contribution to the study of the global urban hierarchy.

Drawing upon Stephen Hymer's (1979) pioneering work on the global relationships of multinational corporations, Cohen's original insight was to recognize the linkages between the organizational structures of major capitalist firms and the changing configuration of the global urban system. On this basis, Cohen proposed a radically new understanding of city systems, grounded in extensive data on the changing locational patterns of multinational corporations, as being global rather than national webs of articulation. Cohen's discovery of this new, truly *global* dimension of urbanization under the new international division of labor (NIDL) represented a genuine intellectual revolution for urban scholars, who now had at their disposal a number of new methodological tools and data sources through which to investigate the political economy of urban development on a world scale.

Cohen's text quickly became a standard intellectual reference point within the nascent field of global urban studies. Not all of the author's detailed predictions regarding the future development of the global urban system came to pass. For instance, there is little concern today with the Eurodollar market. Also, transnational corporate expansion did not occur in a linear fashion since the mid-1980s, as Enron-scale meltdowns, the dotcom disaster and the Asian crisis have shown. Nonetheless, Cohen's efforts to map the new global urban order remain highly salient today, not least because he presciently anticipated the large- and small-scale contradictions that have been associated with contemporary patterns of global urban restructuring. On one hand, Cohen observed that the new urban hierarchy was entrenching rather than loosening existing global power relations and North–South inequalities. On the other hand, Cohen usefully emphasized that, even within the core globalizing cities of the United States, class-based social polarization was intensifying as the process of globalization unfolded. In this manner, Cohen anticipated some of the major debates regarding social inequality in global cities that were subsequently to unfold (see, for instance, Reading 12 by Ross and Trachte, Reading 13 by Fainstein and Reading 47 by Samers).

THE EMERGENCE OF GLOBAL CITIES

Changes in the corporation and in the structure of the advanced corporate services have led to the emergence of a series of global cities which serve as international centers for business decision-making and corporate strategy formulation. In a broader sense, these places have emerged as cities for the coordination and control of the new international division of labor (NIDL). If one examines available data on international and domestic activities for the Fortune 500 firms and for a number of advanced services, it is apparent that only a few cities in the United States are vastly more important as centers of international business and corporate services than as centers of national business. Firms headquartered in New York account for 40 per cent of foreign sales of Fortune 500 companies, compared to 30 per cent of all sales of Fortune 500 firms. Banks in New York and San Francisco have 54 and 18 per cent of foreign bank deposits in US banks, compared to 31 and 13 per cent of total US bank deposits. This concentration of international corporate decision-making and corporate services in a few US cities is tied to the emergence of the NIDL, the ensuing need for changes in corporate strategy and structure, and the transformation of the advanced corporate services. These changes and the increased centralization of corporate-linked functions have changed the dimensions of urban hierarchy in the United States.

The movement of US corporations overseas had several important impacts on the institutional structure of large US cities. International decision-making by major firms was largely concentrated in two cities, which were major centers of corporate headquarters and finance: New York and San Francisco. This trend was supported by major banks from these cities which developed an elaborate network of subsidiaries to service the foreign operations of major corporations. As a result, cities which had been important centers of business in an earlier, more national-oriented phase of the economy began to lose economic stature to these global cities because their firms were not as internationally oriented as others. Jobs related to international operations did not develop as extensively in places like Cleveland, St. Louis and Boston, as they did in New York and San Francisco. While the centralization of international activities in a few cities may be reversed somewhat in the future, as more US firms develop foreign operations, this trend will probably continue. This is especially true, since once firms lose time to their competitors on the international scene, it is often regained only with great difficulty.

In conjunction with the growth of international corporate activity, advanced corporate services, including banks, law firms, accounting firms and management consulting firms, expanded their international skills and their overseas operations. Yet, these, too, grew in but a few urban centers and thus drew even those firms with international operations which were headquartered in regional or national centers of business to more international centers. Thus, just as few banks with international expertise were located outside of New York, San Francisco or Chicago, few law firms with international competence were centered outside of Washington, Los Angeles or New York. Washington became especially significant as a center for international law because of the contacts which firms there had developed with the US State Department and with foreign governments. New York emerged as perhaps the single most important center of international accounting expertise in the nation, since even some of the largest accounting firms headquartered outside of the city do much of their international work there.

To quantify the present status of different US cities as centers of international business, I have

Table 1 The urban distribution of *Fortune* 400 firms' total sales and foreign sales in 1974 – Selected SMSAs

SMSA	Fortune firms (1)	Percentage of total sales (2)	Percentage of foreign sales[a] (3)	Multinational index (3) ÷ (2)
New York	107	30.3	40.5	1.34
Los Angeles	21	4.6	3.8	0.83
Chicago	48	7.3	4.6	0.83
Philadelphia	15	1.6	1.0	0.77
Detroit	12	9.1	8.8	0.97
San Francisco	12	3.2	5.4	1.69

Sources: *Fortune Magazine* (May 1975); corporate annual reports, *Wall Street Transcript*, and Securities and Exchange Commission prospectuses.

Note: [a]The foreign sales data were compiled for 321 of the largest firms. Data for the other, mostly smaller, firms were not available.

constructed a "multinational index" for cities (see Table 1). This index compares the percentage share of a city's Fortune 500 firms in total foreign sales to their percentage share of total sales. For instance, in the New York Standard Metropolitan Statistical Area (SMSA), Fortune 500 firms account for 40.5 per cent of all foreign sales by Fortune firms, and 30.3 per cent of total sales by Fortune firms. Dividing the first percentage by the second one gives us a multinational index of 1.34. Where most of the corporations domiciled in a city have extensive international operations, the city's score on the multinational index will be almost 1.0 or higher. This defines international cities. Those metropolitan areas with some firms having international subsidiaries score between 0.7 and 0.9, a lower score, and I have defined these as national cities. Finally, where there is little evidence that the firms in a city have any share of foreign corporate sales relative to their share of total sales, I define them as regional cities.

This index permits the assessment of a city's relative strength as a center of international business in relation to its strength as a center of national business. The index defines "business strength" at three different levels based not on absolute share of foreign sales or total sales of the Fortune 500 firms, or on the size and industrial structure of a city's firms, but on the relative share of international business compared to all business of those firms headquartered in a city.

What is remarkable about the results obtained by using this index is that several large cities which we usually think of as "national" centers turn out to be very weak as centers of international business. San Francisco, New York, Houston, Boston, Pittsburgh, Seattle, Rochester, Akron and Detroit are strong centers of international business based upon this index. But Chicago, St. Louis, Cleveland, Dallas and Milwaukee are merely "regional" centers. They are cities with few businesses that have extensive international operations. Most other large cities are "national," with corporations having somewhat smaller relative shares of foreign sales compared to total sales for the Fortune 500. I would argue that merely examining the foreign and total sales of corporations headquartered in a city is not sufficient to gauge its importance as a center of international business. One has to know if the city is a strong center of international banking and strategic corporate services. Only a place with a wide range of international business institutions can be truly called a world city.

Thus, I devised a similar "multinational banking index" which compares the share of all foreign bank deposits of the top 300 banks held by banks in one city to their share of all domestic deposits held by the same banks. Using this index, several of the strong "international" corporate centers drop to a "national" or "regional" classification. Pittsburgh and Boston only rank as "national" banking centers, and Houston, Seattle and Detroit are "regional" banking centers. Rochester and Akron do not even rank as regional centers. On the other hand, Chicago and Dallas move up to positions as global cities with international centers of banking, along with New York and San Francisco. Most of the other large cities – Los Angeles, Philadelphia, St. Louis, Atlanta, Cleveland, Minneapolis – are only "regional" banking centers.

From these two indices it is clear that only two places achieve classification as global cities for both corporate and banking business: New York and San Francisco. Chicago and Houston may be moving to "international" status, but the former has relatively few internationally oriented corporations while the latter has few banks with sizeable international operations.

THE EMERGING WORLD URBAN HIERARCHY

Industrial restructuring and urban hierarchy

Changes in the international "competitiveness" of a number of world industries have been a major factor in the reshaping of urban hierarchy throughout the world. The restructuring of industry on an international scale (movements of plants from developed to "developing" areas both within and between nations, closing of "noncompetitive" plants in older, industrialized centers, and the technological improvement of industry to increase productivity) have all contributed to major shifts in employment and trade, with the greatest impacts being felt in both the urban centers of developed nations and the larger cities of developing countries.

What forces are behind this restructuring? The desire by multinational corporations to seek new markets and more profitable ways to organize production on a world scale (Hymer 1979), national policies by developed nations to strengthen the future international competitive position of selected industries, and national policies on the part of developing nations to stimulate the growth of export sectors, largely by attracting subsidiaries of multinational corporations. These strategies of multinationals and policies of governments have resulted in a situation where the profitability of

productive facilities in certain parts of the developed world relative to those in other parts of the world, particularly the "newly industrializing countries (NICs)," has declined considerably. This has resulted in the inability of many older plants to obtain adequate profits, compared to those obtainable elsewhere.

In response to this situation, corporations in many of the developed nations have begun to restructure their operations. This process has included: relocating firms even more extensively to areas where less expensive conditions for production are available, changing the process of production by introducing new technology or altering the labor process, or by closing down existing plants. Both corporate restructuring and national policies, which have been focused upon a few specific industries, have resulted in dramatic changes in the structure of employment in both developed and developing nations, and have altered the pattern of trade between nations. Between 1970 and 1976, just fifty of 422 individual SITC commodity classifications accounted for 80 per cent of the fourfold, $23,000,000,000, increase in the total value of exports by NICs to developed market economies, with the top ten items accounting for 44 per cent of the increase. While the direct effect of imports on employment is usually seen as less important than that due to technical progress in productivity, changes in the international competitive position of specific industries may have very deleterious effects on employment.

Such effects and international changes have had an important impact on corporations and governments. They have made it increasingly apparent that successful multinational companies will, in the future, not be those who exploit product life cycles, as much as they will be those who master the process of "global scanning" and are able to integrate their operations on a world scale. These changes have also resulted in increased intervention by governments in the economies of developed nations, responding both to the need to adjust to job losses in declining sectors and to support corporate restructuring strategies in more dynamic sectors. The result has been greater conflict and instability in the international economy. Governments and firms in certain nations have devised "national industrial strategies" to enable the future expansion and adjustment of their economies to progress more successfully than their rivals, but how well the private goals of companies and the public goals of governments will be integrated remains to be seen.

THE UNITED STATES' URBAN HIERARCHY

Centers of corporate headquarters and corporate services

As illustrated in the previous section, only a few centers now stand out as places where key international functions are agglomerated in the United States. This contrasts sharply with the more regional hierarchy of national corporate functions, which is readily apparent from an examination of the location of Fortune 500 corporate headquarters. In the latter list, a number of regional centers stand out as important centers of headquarters.

In the 1950s, the "metropolis and region" structure of the United States, which was fairly well established by 1913, was a rather accurate reflection of urban hierarchy. Each regional metropolis was an important center of corporate services. Cleveland, for example, had a large number of corporate law firms to complement its important group of corporations. Major accounting firms were headquartered in New York, Chicago and Cleveland. Important regional banks could be found in nearly all of the centers of corporate head offices. By 1974, although the "metropolis and region" network of corporate head offices and corporate services had expanded (see Table 2), international business functions had become much more significant than before.

Thus, an analysis of the urban hierarchy in the US must take these new and often more complex functions into account. If this is done, we find that an extremely limited number of cities act as world centers of business and corporate services. It should also be noted that as the top level cities, primarily New York and San Francisco, emerged as key international centers, even the international activities of firms headquartered outside these cities were increasingly linked to financial institutions and corporate services located within them.

Table 2 Urban centers with more than ten *Fortune* 500 corporate headquarters

Fortune 500 Headquarters 1957		*Fortune* 500 Headquarters 1974	
New York	144	New York	107
Chicago	54	Chicago	48
Pittsburgh	24	Los Angeles	21
Philadelphia	21	Cleveland	17
Detroit	17	Pittsburgh	15
Cleveland	16	Philadelphia	15
Los Angeles	14	Detroit	12
San Francisco	13	San Francisco	12
St. Louis	13	Minneapolis	10
		St. Louis	10
		Milwaukee	10

The emerging division of labor in the US and contradictions in the urban hierarchy

The new structure of urban hierarchy in the US is one dominated by two main international centers for corporate control and coordination. While nine or more regional corporate metropolises remain important as corporate centers (see Table 2), they appear to have decreased in importance if international corporate services are taken into account. In addition, with the evident centralization of the more sophisticated corporate services in just a few regional centers, places like Cleveland and St. Louis which have seen numerous blue-collar jobs leave the metropolitan area have not had a growing service sector to cushion part of the economic decline.

Indeed, the growing importance of some cities, even regional cities, as corporate centers reflects a number of contradictory forces in US urban development. First, certain places, by becoming centers of corporate operations and services, present less opportunities for blue-collar jobs and job mobility than they have traditionally done, and risk not only losing their middle class, but also "marginalizing" the lower class which has traditionally found job mobility extremely difficult. Second, as corporations and the rest of the business community realize the significant role played by certain metropolises, there is more likelihood that they will shift from a laissez-faire attitude to urban economic development to more active support of planned urban development. But if such planning focuses primarily upon corporate-linked activities it also risks alienating the middle and lower classes. Third, the relative loss of power by national and regional cities may lead them to become the hinterland of new world metropolises. This would lead such cities to follow a pattern of development similar to the dependency relationship which has occurred between Canada and the US or what Kari Levitt has called "branch plant development" (Levitt 1971).

To examine uneven development within the labor forces of different types of cities, I have, following Averitt, quantified the differences between employment in the core and peripheral industries of these cities (Averitt 1970). Core industries were defined by factor analysis and were characterized by high concentration ratios, large establishment size, high rates of unionization, capital intensity, few black workers and high male educational attainment. My analysis showed that the more important a city was as a national or regional center, the greater the share of its employment in dynamic, core industries. But once a city became a global city with a multinational index score of 1.0 or more, it lost employment in the core industries. These results indicate that international administrative centers, like New York and San Francisco, may be qualitatively different from more national centers of corporate activities. In part, the difference may reflect higher prices for rents and services which result from the pronounced agglomeration of corporate and corporate-related activities in these centers. On the other hand, it may also reflect the fact that since the Second World War, such centers have grown largely by attracting a specialized range of activities, following a pattern that may have begun, as in the case of New York, as early as the 1900s.

THE NEW HIERARCHY OF WORLD CITIES

The new world cities

The new hierarchy of world cities has a number of parallels with the changing hierarchy of US cities. For both groups, the rise of international corporate activities and of international corporate services has made a significant difference in the pattern of urban hierarchy. To examine the world hierarchy of cities of the 1950s and early 1960s, we can rely upon the world cities selected by Peter Hall. These included the traditional large national centers of business and government: London, Paris, Amsterdam–Rotterdam, the Rhine–Ruhr complex, Tokyo and Moscow. These places were certainly important for their concentrations of finance and corporate services, for as Hall pointed out, they attracted corporate-related jobs, "for every producer of goods, more and more people are needed at office desks to achieve good design, to finance and plan production, to sell the goods, to promote efficient nationwide and worldwide distribution" (Hall 1971: 9).

As the 1970s began, multinational business and international finance began to play a more dominant role in Europe and Asia. Traditional national centers which lacked concentrations of international corporations and banks declined in importance. If one had to assess the structure of urban hierarchy outside the US using the same approach which has been employed in studying the US urban hierarchy of corporate functions, a different group of cities might be called "world cities" than those selected by Hall. According to Table 3, corporate head offices of non-US multinational corporations are most highly concentrated in London and Tokyo. Only London, Frankfurt and Zurich have gained enhanced stature because they are centers of the Eurodollar market.

Table 3 Main locations of corporate headquarters, 198 largest non-US corporations, 1978

Metropolitan area	Number of headquarters
Tokyo	30
London	28
Osaka	13
Paris	12
Rhine-Ruhr	10.5[a]
Randstad Holland	3.5[a]

Source: United Nations *Transnational Corporations in World Development: A Re-examination*, Appendix IV, and *Jane's Major Companies of Europe* (1976).
Note: [a]When a firm has headquarters in two places, each locality is given 0.5 offices.

Thus, New York, Tokyo and London are predominant as world centers of corporations and finance. Osaka, the Rhine–Ruhr, Chicago, Paris, Frankfurt and Zurich are second-level world cities as far as international corporate activities are concerned. There is at least anecdotal evidence that with the emergence of Frankfurt as a major international financial center, corporate service activities began to concentrate in the city.

The world hierarchy of cities and its contradictions

The new urban hierarchy of corporate head office activities outside the US also reflects a new division of labor between centers of corporate control and coordination and more "national-oriented" urban places (Palloix 1975). The Amsterdam–Rotterdam and Rhine–Ruhr agglomerations and Paris, cited by Hall as world cities, are much less a part of the international system of corporate and financial activities than are the international financial centers in Europe. It is difficult to assess how the more "national-oriented" European centers should be compared to financial centers, such as Singapore, Panama, Hong Kong and the Bahamas. But if we examine these latter centers as places for corporate decision-making, it is apparent that they serve primarily as centers for moving and mobilizing financial resources, rather than as centers of control.

As international boundaries become blurred by the increasingly global nature of corporations, numerous contradictions will arise within the world hierarchy of cities. First, there will be contradictions that will arise because private institutions, particularly large multinational corporations and banks, are able to undermine or contravene established government policy. This contributes to the erosion of the position of certain traditional centers of government policy where corporate head offices or major financial institutions are not present in large numbers. Second, shifts in the importance of the Eurodollar market will probably create conflicts between new centers of finance and older, more national ones.

The NIDL will have a particularly strong impact on cities in the developing nations, particularly on those in the newly-industrializing coun-tries, or NICs. Because multinational corporations, aided by international banks, will probably accelerate the establishment of foreign subsidiaries in these nations over the next decade with the assistance and support of governments that view indu-strialization as a "redemptive mystique" (Johnson 1970), the perpetual crisis of cities in developing nations will almost certainly be exacerbated. In sum, the future of the city in developing nations must also be analyzed in relation to international and national development. Certain forces will certainly continue to exacerbate the flow of people into the cities, such as the international sourcing of raw materials and the use of land for more export-related (largely agricultural) activities. In addition, the need for entrepreneurial and technical skills, information from personal contacts, and the resolution of problems through extralegal procedures will contribute to the continued agglomeration of corporate manufacturing activities in the major cities of developing nations. Yet what results from the pattern of development described here is an increasingly dualistic society, with the cities of developing nations being characterized by extremely high unemployment among relatively more educated, largely male, urban immigrants, who are usually sixteen to forty-five years of age. The social and political turmoil which can result from the pursuit of such policies, even if they now result in profits for multinational companies and higher GNP per capita in developing nations, will certainly have a severe impact upon both the developed and developing nations in the 1980s and 1990s.

FUTURE TRENDS IN THE NIDL AND URBAN HIERARCHY

A number of changes will probably occur over the next decade which will restructure the NIDL. These changes may have significant impact on both corporations and nation-states. They will both create more formidable problems for corporate managers by making the international economy more complex and more competitive, by eroding the traditional industrial structure of developed nations further, and by causing sweeping changes in the employment structure of developing nations. These changes may be categorized into two groups: changes in corporate and financial institutions, and changes in developing nations. In the first group, the probable changes include: (i) a rapid expansion of the international activities of both US and non-US medium-sized corporations, including corporations from developing nations; (ii) a shift of the center of the Eurodollar market from London to New York; (iii) a continued growth in the dependency of US and foreign corporations on advanced corporate services; and (iv) a rapid growth of foreign investment in the United States. These changes would place further pressures upon management to become global in perspective, to increase its control over operations, to plan over longer time horizons, and to develop more flexible strategies.

The second group of changes would also increase pressures on corporations and financial institutions. They would probably include: (i) the rapid emergence of newly-industrializing nations as centers of production and as domiciles for new competitors for multinational firms; (ii) the assertion of power by OPEC-like commodity cartels for copper, aluminum, certain foodstuffs and other basic commodities; (iii) the erosion or enhancement of the power of OPEC by the development of current non-OPEC centers of oil reserves, Mexico being the prime example; (iv) a rapid growth of the debt burden of poorer developing nations, making them more vulnerable to potentially-disruptive economic transformations due to foreign investment and less able to purchase foreign goods; and (v) a growing political instability among the more developed of the developing nations.

These trends, since they place more pressure on corporations and financial institutions in developed nations, will probably tend to further concentrate corporate and financial decision-making in present world centers, drawing decision-making activities away from national or regional centers. This would exacerbate uneven urban development in developed nations, particularly since the continued movement of productive activities overseas would further erode urban blue-collar jobs.

REFERENCES FROM THE READING

Averitt, R. (1970) *The Dual Economy.* New York: Norton.

Hall, P.G. (1971) *The World Cities.* New York and Toronto: McGraw-Hill.

Hymer, S.H. (1979) *The Multinational Corporation: A Radical Approach.* Cambridge: Cambridge University Press.

Johnson, E.A.J. (1970) *The Organization of Space in Developing Countries.* Cambridge, MA: Harvard University Press.

Levitt, K. (1971) *Silent Surrender: The American Economic Empire in Canada.* New York: Liveright.

Palloix, C. (1975) *L'Internationalisation du Capital.* Paris: Maspero.

"World City Formation: An Agenda for Research and Action"

from *International Journal of Urban and Regional Research* (1982)

John Friedmann and Goetz Wolff

Editors' introduction

John Friedmann is one of the pioneering urbanists of the late twentieth century. Many of Friedmann's most important contributions to urban and regional studies were produced during his nearly three decades on the faculty of the Program for Urban Planning in the Graduate School of Architecture and Planning at UCLA, which he helped found in the late 1960s and where he is now Professor Emeritus. Friedmann is also currently Honorary Professor in the School of Community and Regional Planning at the University of British Columbia. International recognition for his scholarship includes honorary doctorates from the Catholic University of Chile and the University of Dortmund. Friedmann's publication record includes 14 individually authored books, 11 co-edited books, and 150 chapters, articles, and reviews. Friedmann is currently involved in research on urbanization processes with special reference to China.

Goetz Wolff, who was a PhD student at UCLA when this article was written, is an independent researcher, teacher and consultant in Los Angeles. He also serves as Lecturer in UCLA's Urban Planning Department. His research and teaching interests center on industrial change and regional economic development issues, with particular reference to the Southern California region.

Friedmann and Wolff's study of world city formation was published in 1982 in a new radical journal of urban sociology, the *International Journal of Urban and Regional Research*, and became an instant classic: it kickstarted an impressive outburst of research on this topic, including many of the contributions to this Reader. Building upon a broad range of intellectual sources – including radical international political economy, world systems theory, Marxian urban studies, urban systems theory and radical community studies – Friedmann and Wolff's "agenda for research and action" represented a genuinely original synthesis. Reminiscent of previous arguments by Jane Jacobs (1984), Friedmann and Wolff viewed cities rather than national economies as the motors of contemporary capitalist development. World cities, they argued, represented a new breed of global command and control centers within the new international division of labor (NIDL) associated with post-1970s capitalism. These cities, moreover, concentrated many of the contradictions and inequalities of the NIDL inside their own boundaries: they had to be viewed, simultaneously, as spaces of hope and as spaces of gloom, that is, as sites in which "citadel" and "ghetto" existed in uneasy proximity.

It was by no means accidental that Friedmann and Wolff wrote this article during a period in which Los Angeles was becoming what some would hyperbolically describe as the "capital of the 21st century" (Scott and Soja 1996). Indeed, Los Angeles in the 1980s provided an ideal backdrop for the development of global city theory. For, as Goetz Wolff showed in another influential paper written in the early 1980s with Edward Soja and Rebecca Morales (Soja et al. 1983), Los Angeles exemplified many of the dramatic economic changes that were unfolding in the United States during that decade, as the Fordist industrial structure of the post-World War II years was increasingly replaced by an internationalized, more flexible and less stable regime of accumulation. More generally, many patterns of urban restructuring that would subsequently unfold in cities throughout the older industrialized world were already strikingly evident in the Southern California metropolis during the late 1970s and early 1980s (see Keil 1998; see also Reading 21 by Soja).

I

Our paper concerns the spatial articulation of the emerging world system of production and markets through a global network of cities. Specifically, it is about the principal urban regions in this network, dominant in the hierarchy, in which most of the world's active capital is concentrated. As cities go, they are large in size, typically ranging from five to fifteen million inhabitants, and they are expanding rapidly. In space, they may extend outward by as much as 60 miles from an original center. These vast, highly urbanized – and urbanizing – regions play a vital part in the great capitalist undertaking to organize the world for the efficient extraction of surplus. Our basic argument is that the character of the urbanizing processes – economic, social, and spatial – which define life in these "cities" reflect, to a considerable extent, the mode of their integration into the world economy.

We propose, then, a new look at cities from the perspective of the world economic system-information. The processes we will describe lead to new problem configurations. The central issue is the control of urban life. Whose interests will be served: those of the resident populations or of transnational corporations, or of the nation states that provide the political setting for world urbanization? Planners are directly engaged on this contested terrain. They are called upon to clarify the issues and to help in searching for solutions. Obviously, they will have to gain a solid, comprehensive understanding of the forces at work. And they will have to rethink their basic practices, since what is happening in world cities is in large measure brought about by forces that lie beyond the normal range of political – and policy – control. How can planners and, indeed, how can the people themselves, living in world cities, gain ascendancy over these forces? That is the basic question. Towards the end of this paper we shall venture a few observations about the tasks we face and their implications for planning.

II

Our argument is a relatively simple one. Since the Second World War, the processes by which capitalist institutions have freed themselves from national constraints and have proceeded to organize global production and markets for their own intrinsic purposes have greatly accelerated. The actors principally responsible for reorganizing the economic map of the world are the transnational corporations, themselves in bitter and cannibalistic conflict for the control of economic space. The emerging global system of economic relations assumes its material form in particular, typically urban, localities that are enmeshed with the global system in a variety of ways.

The specific mode of their integration with this system gives rise to an urban hierarchy of influence and control. At the apex of this hierarchy are found a small number of massive urban regions that we shall call world cities. Tightly interconnected with each other through decision-making and finance, they constitute a worldwide system of control over production and market expansion. Examples of world-cities-in-the-making include such metropolises as Tokyo, Los Angeles, San Francisco, Miami, New York, London, Paris, Randstadt, Frankfurt, Zurich, Cairo, Bangkok, Singapore, Hong Kong, Mexico City and São Paulo.

To label them world cities is a matter of convenience. In each and every instance, their specific role must be determined through empirical research. Only this much we can say: their determining characteristic is not their size of population. This is more properly regarded as a consequence of their economic and political role. A more fundamental question is in what specific ways these urban regions are becoming integrated with the global system of economic relations. Two aspects need to be considered:

1. The form and strength of the city's integration (e.g. to what extent it serves as a headquarters location for transnational corporations; the extent to which it has become a safe place for the investment of "surplus" capital, as in real estate; its importance as a producer of commodities for the world market; its role as an ideological center; or its relative strength as a world market).
2. The spatial dominance assigned by capital to the city (e.g. whether its financial and/or market control is primarily global in scope, or whether it is less than global, extending over a multinational region of the world, or articulating a national economy with the world system). These criteria of world system integration must be viewed in a dynamic, historical perspective. Urban roles in the world system are not permanently fixed. Functions change; the strength of the relationship changes; spatial dominance changes. Indeed, the very concept of a world economy articulated through urban structures is as old as the ancient empires. Rome was perhaps the first great imperial city. One may think of Venice in its Golden Age, or of nineteenth-century London. While recognizing this historical continuity, we would still argue that the present situation is substantially different. What then is new?

First, we must consider the truly global nature of the world economy. Even imperial London, ruling over an empire "where the sun never sets," controlled only portions of the world. The present transnational system of space economy, on the other hand, is in principle unlimited. It is best understood as a spatial system which has its own internal structure of dominance/subdominance.

Following Immanuel Wallerstein, we may label its three major regional components as core, semi-periphery, and periphery. *Core areas* include those older, already industrialized and possibly "postindustrial" regions that contain the vast majority of corporate headquarters and continue to be the major markets for world production (northwest Europe, North America, Australia, Japan). The *semi-periphery* includes rapidly industrializing areas whose economies are still dependent on core-region capital and technical knowledge. They play a significant role in extending markets into the world periphery. Mexico, Brazil, Spain, Egypt, Singapore, Taiwan, and the Republic of Korea would be examples of semi-peripheral regions. And the *world periphery* comprises what is left of market economies. Predominantly agrarian, the people of the world periphery are poor, technologically backward, and politically weak.

This analytical scheme must be deftly handled. It is a first approximation to a deeper understanding of world city structure. Above all, it is an historical classification. Over the span of one or two generations, a country may change its position as it moves from periphery to semi-peripheral status (ROK, Spain, Brazil), from semi-periphery to core (Japan), and even perhaps back from core region status to the semi-periphery (Great Britain), or the ultimate decline into peripheral obscurity (Lebanon, Iran). What makes this typology attractive is the assumption that cities, situated in any of the three world regions, will tend to have significant features in common. As the movement of particular countries through the three-level hierarchy suggests, these features do not in any sense determine economic and other outcomes. They do, however, point to conditions that significantly influence a city's growth and the quality of urban life.

The world economy is thus no longer defined by the imperial reach of a Rome, a Venice, or even a London, but by a linked set of markets and production units organized and controlled by transnational capital. World cities are a material manifestation of this control, and they occur exclusively in core and semi-peripheral regions where they serve as banking and financial centers, administrative headquarters, centers of ideological control, and so forth. Without them, the world-spanning system of economic relations would be unthinkable.

This conception of the world city as an instrument for the control of production and market organization implies that the world economy, spatially articulated through world cities, is dialectically related to the national economies of the countries in which these cities are situated. *It posits an inherent contradiction between the interests of transnational capital and those of particular nation states that have their own historical trajectory.* World cities are asked to play a dual role. Essential to making the world safe for capital, they also articulate given national economies, with the world system. As such, they have considerable salience for national policy makers who must respond to political imperatives that are only coincidentally convergent with the interests of the transnationals. World cities lie at the junction between the world economy and the territorial nation state.

Finally, the global economy is superimposed upon an international system of states. Nation states have their own political fears and ambitions. They form alliances, and they exact tribute. They must protect their frontiers against actual and potential enemies. Wishing to ensure their continuing power in the assembly of nations, or even to enlarge their power, they must provide for a continuing flow of raw materials and food supplies.

On the other hand, although transnational capital desires maximum freedom from state intervention in the movements of finance capital, information, and commodities, it is vitally interested in having the state assume as large a part as possible of the costs of production, including the reproduction of the labor force and the maintenance of "law and order." It is clear, therefore, that they would benefit from a strategy to prevent a possible collusion among nation states directed against themselves. Being essential to both transnational capital and national political interests, world cities may become bargaining counters in the ensuing struggles.

They are therefore also major arenas for political conflict. How these conflicts are resolved will shape the future of the world economy. Because many diverging interests are involved, it is a multifaceted struggle. There is, of course, the classical instance of the struggle between capital and labor. This remains. In addition, there is now a struggle between transnational capital and the national bourgeoisie; between politically organized nation states and transnational capital; and between the people of a given city and the national polity, though this may be the weakest part.

There is, then, nothing inevitable about either the world economy or its concrete materialization in world cities. Capital is in conflict with itself and with the political territorial entities where it must come to rest. There is no manifest destiny. Yet the emerging world economy is an historical event and this allows us to formulate our central hypothesis: *the mode of world system integration (form and strength of integration; spatial dominance) will affect in determinate ways the economic, social, spatial and political structure of world cities and the urbanizing processes to which they are subject.*

What we describe is not a Weberian "ideal type" of a fully formed "world city." All that we intend is to point to certain structural tendencies in the formation of those cities that appear to play a major role in the organization of world markets and production. We have in mind a heuristic for the empirical study of world city formation.

III

In making the internationalization of capital central to our analysis, we focus upon a combination of complex processes that are indeterminate, contradictory, and irregular. There is little dispute about the fact of the worldwide expansion of market relations. But the cause – the driving force – of the internationalization of capital is debated: Is internationalization merely a working out of the internal logic of capitalism? Or has labor in the industrialized countries created a situation where capital now finds it more profitable to locate in the periphery?

We proceed under the assumption that both "windows" – to use David Harvey's felicitous image – contribute to the needed understanding of the global order. The contradictions inherent in the capitalist economy and the basic struggle which results from the domination of labor by capital are the major forces which account for both the spatial and temporal irregularities of the world economy. A city's mode of integration with the global economy cannot simply be understood by identifying its functional role in the articulation of the system. Rather, we suggest that the driving forces of competition, the need for accumulation, and the challenges posed by political struggle make the intersection of world economy/world city a point of intense conflict and dynamic change. World city integration is not a mechanical process; it involves many interconnected changes that leave few aspects of its life untouched and create the arenas for concerted action.

IV

The world city today is in transition, which is to say it is in movement. Perhaps it has always been like this. Equilibrium is not part of the experience of large cities. Structural instability manifests itself in a variety of ways: dramatic changes in the distribution of employment, the polarization of class divisions, physical expansion and decaying older areas, political conflict. We shall have a quick look at all of these to render more specific how the formation of the world cities is affecting the quality of their life experience.

World cities are the control centers of the global economy. Their status, of course, is evolving in the measure that given regions are integrated in a dominant role with the world system. And like the golden cities of ancient empires, they draw into themselves the wealth of the world that is ruled by them. They become the major points for the accumulation of capital and "all that money can buy." They are luxurious, splendid cities whose very splendor obscures the poverty on which their wealth is based. The juxtaposition is not merely spatial; it is a functional relation: rich and poor define each other.

It is not a new story, and yet its particular features are new. As we attempt to describe the changes that occur as urban regions strive for world city status several things must be borne in mind. The characteristics we are describing are merely tendencies, not final destinations. Particular cities will exhibit particular features. Still, the account we give of conditions prevailing in urban regions as they become world cities may be regarded as the best current hypothesis. In every instance, we have tried to relate it back to specific aspects of integration with global economy. Not only are world cities in themselves not uniform, there

is no definite cut-off point with other cities that belong to the same system but are not so tightly integrated with the global economy have only a national/subnational span of control, or are integrated primarily on a basis of dependency. In a way, the world economy is everywhere, and many of the features we will describe may be found in cities other than those we are discussing here.

The world city "approach" is, in the first instance, a methodology, a point of departure, an initial hypothesis. It is a way of asking questions and of bringing foot loose facts into relation. We do not have an all-embracing theory of world city formation.

Economic restructuring

A primary fact about emerging world cities is the impact which the incipient shifts in the structure of their employment will have on the economy and on the social composition of their population. The dynamism of the world city economy results chiefly from the growth of a primary cluster of high-level business services which employ a large number of professionals – the transnational elite – and ancillary staffs of clerical personnel. The activities are those which are coming to define the chief economic functions of the world city: *management; banking and finance, legal services, accounting, technical consulting, telecommunications and computing, international transportation, research, and higher education.*

A secondary cluster of employment, also in rapid ascendancy, may be defined as essentially serving the first. Its demand is largely derived, and it employs proportionately a much smaller number of professionals: *real estate, construction activities, hotel services, restaurants, luxury shopping, entertainment, private police and domestic services.* A more varied mix than the primary cluster, its fortunes are closely tied to it. Although most jobs in this cluster are permanent and reasonably well-paid, this is not true for domestic services which is the most vulnerable employment sector and the most exploited.

A tertiary cluster of service employment centers on *international tourism*. To a considerable extent, this overlaps with the secondary cluster (hotels, restaurants, luxury shopping, entertainment), and like that cluster, it is tied to the performance of the world economy.

The growth of the first three clusters is taking place at the expense *of manufacturing* employment. Although a large cluster, its numbers are gradually declining as a proportion of all employment. Some industry serves the specialized needs of local markets, while other sectors – in Los Angeles, for example, electronics and garment industries – are choosing world city locations because of the large influx of cheap labor which helps to keep the average cost of wages down. The future of manufacturing employment in the world city is not bright, however. The next two decades will see the rapid automation and robotization of many jobs. While factories will still be producing and earning large and perhaps even rising profits, they will be largely devoid of working people.

Government services constitute a fifth cluster. They are concerned with the maintenance and reproduction of the world city, as well as the provision of certain items of collective consumption: the planning and regulation of urban land use and expansion; the provision of public housing, basic utilities, and transportation services; the maintenance of public order; education; business regulation; urban parks; sanitation; and public welfare for the destitute.

Because of the uniqueness and scale of world cities, and because they are often considered national showcases, the government sector tends to be larger here than elsewhere. And because it is a political and, for the most part, technically backward sector, with uncertain criteria of adequate performance, it tends to be bloated, employing large numbers of people at relatively low levels of productivity and wages. Moreover, because the world city extends over many political jurisdictions that are contiguous with each other, there is much overlap and redundancy in employment. During periods of depression, government will often be the employer of last resort. Its internal rhythm tends to be counter-cyclical.

A sixth and, at least in some cities, numerically the largest cluster, embraces the "informal," "*floating*," or "*street*" *economy* which ranges from the casual services of day laborers and shoeshine boys to fruit vendors, glaziers, rug dealers, and modest artisans. Frequently an extension of the household economy, most informal activities require little or no overhead (though they do require start-up capital). They demand long hours, and the returns are low and uncertain. They offer no security to those who work in them. New arrivals to the city often find their first job in the informal sector, and many of them stay there. When times are bad, some makeshift income earning opportunities can always be found in the informal sector for people who are temporarily unemployed. Although informal sector work may be a choice between independence and security for some, for many more it is the only way to survive in the city. The cluster of informal activities takes up the slack in the "formal" economy, and thus despite its marginal character, it tends to be tolerated by the state.

Some informal activities are not as "unorganized" or "casual" as they might appear. Perhaps increasingly, small businesses are subcontracted by large, frequently multinational corporations who in this way are able to lower their costs of operation. Informal businesses are usually beyond the reach of government regulation. They don't pay minimum wages and their labor is often self-exploited. Much of it is done by women and children.

But essentially the informal sector exists because of the large influx of people into the world city from other cities and from the countryside, people who are attracted to the world city as to a honey pot. They don't all find legitimate employment. A significant number drift into illicit occupations which perhaps more than elsewhere appear to thrive in the large city: thieves, pickpockets, swindlers, pimps, prostitutes, drug peddlers, black marketeers . . . the list can be extended with endless refinements.

Finally, there is the undefined category of those without a steady income: the full-time unemployed, who depend on family and public charity for support. Excepting women, who manage their households but are not paid for this, and therefore do not appear in the official statistics; their numbers are surprisingly small in the order of 5–10 per cent of the labor force.

Social restructuring

The primary social fact about world city formation is the polarization of its social class divisions. Transnational elites are the dominant class in the world city, and the city is arranged to cater to their lifestyles and occupational necessities. It is a cosmopolitan world that surrounds them, corresponding to their own high energy, rootlessness, and affluence. Members of this class are predominantly males between the ages of 30 and 50. Because of their importance to the city, they are a class well served.

The contrast with the third (or so) of the population who make up the permanent underclass of the world city could scarcely be more striking. The underclass are the victims of a system that holds out little hope to them in the periphery from which they came but also fails them in the very nerve centers of the world economy where they are queuing for a job. They crowd along the edges of the primary economy – the "formal" sector – or settle in its interstices, barely tolerated, yet providing personal services to the ruling class, doing the dirty work of the city. The ruling class and its dependent middle sectors enjoy permanent employment, a steady income, and complete legality; they do not have to justify their existence. For all practical purposes, they *are the* city. The underclass lives at its sufferance.

Many, though not all, of the underclass are of different ethnic origin than the ruling strata; often, they have a different skin color as well, or speak a different dialect or language. These immigrant workers give to many world cities a distinctly "third world" aspect: Puerto Ricans and Haitians in New York, Mexicans in Los Angeles and San Francisco, barefoot Indians in Mexico City, "nordestinos" in São Paulo, Jamaicans in London, Algerians in Paris, Turks in Frankfurt, Malays in Singapore.

There is a city that serves this underclass, as suited to their own condition if not their preferences and needs, as is the city of the "upper circuit." Physically separated from and many times larger than the citadel of the ruling class, it is the ghetto of the poor. Both cities live under the constant threat of violence: the upper city is guarded by private security forces, while the lower city is the double victim of its own incipient violence and of police repression. The typical world city situation is thus for *both* the crime rate and police expenditures to rise.

Racism reinforces class contradictions, and a good deal of ethnic and racial hostility is found

within the working class itself. Under conditions of tight labor markets, "foreign" workers, whether undocumented or not, frequently occasion racist outbursts, as "national" workers (particularly among the underclass) struggle to preserve their limited terrain for livelihood. Street gangs of different ethnic origin, and the pitched battles between them, especially in the United States, are a major manifestation of this violence.

Yet racial conflict is only one facet of the general increase in violence that is brought on by class polarization. Terrorism, kidnapping, street demonstrations, and rioting are other common forms. There is an undeniable fascination with violence among residents in the world city which is picked up and amplified by the popular media. Yet for all the turmoil, world city conflicts are not a sign of an impending revolution. A good deal of the violence occurs within the working class itself and is a measure of its internal divisions. The world city is in any event immune to revolutionary action. Lacking a political center, it can only be rendered irrelevant – at the present time a rather unlikely occurrence.

Confronted with violence, the nation state responds in coin. Given the severity of its fiscal constraints, in the face of constantly rising costs, it resorts to the simplest, least imaginative alternative: the application of brute force. The response is acceptable to the new ruling class who generally prefer administrative to political solutions. But police repression can at best contain class violence; it cannot eliminate or significantly reduce it. Violence is here to stay.

Physical restructuring

Over the next generation, world cities can be expected to grow to unprecedented size. By the end of the century, the typical world city will have ten million people or more. Much of the increase will have come from migration. Obviously, a population that rivals that of a medium sized country by today's standards can no longer be considered a city in the traditional sense. It is an urbanized region, or an "urban field." In the case of Los Angeles, for example, the pertinent economic region has been defined as having a radius of about 60 miles (roughly 80 minutes commuting at normal speeds); it represents the life space for more than half the population of California!

The urban field is essentially an economic concept. Although it does not respond to the traditional political concept of the city as civitas, it imposes its own logic on the vestiges of the political city which struggles to survive in this highly charged, volatile materialization of capitalist energy. The urban field is expanding, more rapidly in most cases than even the increase in population, but it is expanding unequally, regardless of whether one applies functional, social, or economic criteria. Underlying its kaleidoscopic spatial form is the ever-shifting topography of land values which quickly and efficiently excludes all potential users who are unable to meet the price of a given parcel of land. Of course, this method does not at all correspond to any social need, least of all to the need of what Joan Nelson (1979) has called "access to power."

The concentration of activities and wealth on a world city scale imposes extraordinary strains on the natural resources on which the continued viability of the world city depends. To feed its voracious appetite for water and energy (which tends to grow at multiples of the increases in population), the city must reach further and further afield, sometimes for hundreds of miles, across mountains and deserts and even national boundaries. As it does so, it comes inevitably into conflict with competing interests and jurisdictions.

At the same time, and considering alone sheer volume, the world city faces enormous problems of waste management. Pollution at levels of concentration dangerous to human health pose a constant and growing threat. Huge areas must be set aside for low-intensity uses of the land, such as airports, water-treatment facilities, solid waste disposal, agriculture and dairying, and urban mass recreation.

This enormously varied complex of activities must be knitted together through high-speed transport devices and linked to the outside world through a system of international transport terminals and telecommunications capable of serving the entire region. And of course it must have a basic infrastructure of utilities that serves the housing and industrial needs of the city. At a scale of 10 or 20 million people each, and an area of several hundred square miles, these common basic needs of the world city can be satisfied only with the considered application of high technology. But high technology renders the city more vulnerable: the growing reliance on ever more sophisticated methods may be counterproductive in the end. Its costs will escalate even as the quality of its services deteriorates.

In its internal spatial structure, the world city may be divided, as we suggested in the preceding section, into the "citadel" and the "ghetto." Its geography is typically one of inequality and class domination. The citadel serves the specific needs of the transnational elites and their immediate retinues who rule the city's economic life, the ghetto is adapted to the circumstances of the permanent underclass.

With its towers of steel and glass and its fanciful shopping malls, the Citadel is the city's most vulnerable symbol. Its smooth surfaces suggest the sleek impersonality of money power. Its interior spaces are ample, elegant, and plush. In appropriately secluded spaces, the transnational elites have built their residences and playgrounds: country clubs and bridle paths and private beaches.

The overcrowded ghettos exist in the far shadows of the Citadel, where it is further divided into racial and ethnic enclaves; some areas are shanty towns. None are well provided with public services: garbage does not get collected, only the police in their squad cars are visibly in evidence. In many places, ghetto residents are allowed outside their zones only during working hours: their appearance in the Citadel after dark creates a small panic. With its dozens of apartheid Sowetos, South Africa is perhaps the extreme case of a country whose elites are gripped in fear of their underclass, but political manifestations of this fear are found to a degree in nearly all world cities. Not long ago, the papers in Los Angeles were filled with horror stories of "marauders" sallying forth from the black ghetto to the Citadel on the West Side to rob, rape and kill. They called it "predatory crime." The implication was clear: at night, ghetto residents belong to the ghettos. There, isolated like a virus, they can harm only themselves.

Political conflict

Every restructuring implies conflict. And when conflict occurs in the public domain, it concerns the distribution of costs and benefits, and who shall gain advantage for the next round of the struggle. Focused on the world city, political conflict is multidimensional; it is also at the heart of the matter, as if all the world's lines of force came together here, and the contradictions in the self expansion of capital are magnified on a scale commensurate with that of the world city itself. In the absence of counter-forces, complex feedback loops tend to destabilize the system, and localized conflicts may suddenly erupt into a worldwide crisis.

The emergence of world cities sets processes into motion that restructure people's life spaces and the economic space that intersects with them. Economic space obeys the logic of capital: it is profit-motivated and individualized. Life spaces are territorial: they are the areas that people occupy, in which their dreams are made, their lives unfold. They are thus the areas they really care about. For the dominant actors in economic space, life space is nothing but a hindrance, an irrational residue of a more primitive existence. Yet it can neither be denied nor circumvented. Every economic space overlaps with an existing life space, and without this, economic enterprise would perish. On the other hand, for people who are collectively the sovereign in their respective life spaces, the space of economic logic is the basis for their physical survival.

It is thus very clear that a restructuring of these two kinds of space is bound to generate deep conflicts. In practice, because so many cross currents bear down on them, the reasons for conflicts are often hard to isolate. Conflicts persist; they have a history and a future, they interconnect in place, and the system of world cities ensures their transmittal from continent to continent. Though it may not be very enlightening, it might be more accurate to speak of them as a form of social turbulence.

There are the conflicts over livelihood, as the restructuring of economic space draws jobs away from one place to resurrect them in another, or as entirely new kinds of job are suddenly ascendant for which older workers are not qualified. There are racial and ethnic conflicts, as workers battle over access to the few good jobs there are, and even over jobs that are less desirable but relatively more abundant. There are struggles between world city and the national periphery over the political auto-

nomy of peripheral regions, as the periphery sees its collective life chances systematically denied by the imperial interests of world city cores. There are conflicts over the spaces within the city, as people seek access to housing they can afford, defend their neighborhoods from the intrusion of capitalist logic, or merely struggle for turf to enjoy the freedom of following lifeways of their own. There are the political campaigns launched by concerned citizens to protect and enhance the quality of their lives as they perceive it: in struggles for the environment, against nuclear power, for child care centers, for the access of handicapped workers in public facilities ... struggles and campaigns which are incapable of being separated from the peculiar setting of world cities. Or there are the bitter and tenacious struggles of poor people, workers who belong to one class fraction or another, for greater access to the conditions for social power: the right to organize, to demonstrate, to call to account, to gain control over the conditions of their work, to keep their bodies healthy, to educate their children and themselves to higher incomes, to sources of the means for livelihood.

All these struggles are occurring simultaneously. They are centered on the restructuring process and in the contradictions that arise from the interfacing of economic and life spaces. As such, they reveal to the astute observer the true forces at work in the world city and the actual distribution of power. Also, of course, they determine the ultimate outcomes we observe ... not only the outcomes of the particular struggles being waged, but of the form and direction which world city development will take. For outcomes are not predetermined. Broad tendencies may be irreversible in the medium term, such as the integration of the world economic system under the aegis of transnational capital. But within that historical tendency, there are always opportunities for action. It is precisely at the searing points of political conflict that opportunities arise for a concerted effort to change the course of history.

Conflict between life space and economic space poses new questions for the state. More than ever, the state is faced with multiple contradictions and difficult choices. Some would say that the state itself is threatened by the dramatic appearance of transnational power. On the world periphery, the choice may be the relatively simple one between complete dependency (with all that this implies in terms of uneven development) and stone age survival. In the semi-periphery, which has already chosen a dependent development path, the problems of choice are more complex. More highly articulated than in the periphery, the state reflects within its own structure many of the conflicts within civil society and the economy, as class fraction is pitted against class fraction, territory against class, working class against capital, and citizens against the state, and as these conflicts come together in successive waves of "turbulence" in the world city, the state is merely one more actor, trying to safeguard its own specific interests.

One such interest is the political integration of its territory. Semi-peripheral states are especially concerned with regional inequalities, focal points of regional revolt, regional resources development, and rural land reform. In countries such as Mexico and Brazil, regional issues have been major concerns of state action, and civil wars have been fought over them.

Regional questions recede in the territorially more integrated countries of the world core area. Here the more important conflicts tend to be between national and transnational fractions of capital, and the state becomes a major "arena" for the conduct of this struggle. Much of it is over specific legislation and the budget allocation. In all this turmoil, a major loser is the local state. Small, isolated without financial power, and encapsulated within the world economy, it is barely able to provide for even the minimal services its population needs. And yet, instead of seeking alliances with neighboring cities and organized labor, it leaves the real decisions to the higher powers on which it is itself dependent, or to the quasi-independent authorities created by state charter that manage the infrastructure of global capital-system-wide facilities such as ports, airports, rapid transit, water supply, communications, and electric power.

REFERENCES FROM THE READING

Jacobs, J. (1984) *Cities and the Wealth of Nations: Principles of Economic Life*. New York: Random House.

Keil, R. (1998) *Los Angeles: Globalization, Urbanization, and Social Struggles*. Chichester and New York: John Wiley & Sons.

Nelson, J. (1979) *Access to Power: Politics and the Urban Poor in Developing Countries*. Princeton, NJ: Princeton University Press.

Scott, A. and Soja, E. (1996) *The City: Los Angeles and Urban Theory at the End of the Twentieth Century*. Los Angeles: University of California Press.

Soja, E., Morales, R. and Wolff, G. (1983) *Urban Restructuring: An Analysis of Social and Spatial Change in Los Angeles*. Los Angeles: Graduate School of Architecture and Urban Planning, University of California.

"The World City Hypothesis"

Development and Change (1986)

John Friedmann

Editors' introduction

In 1986, buoyed by the positive reception and creative appropriation of his earlier ideas on world city formation, John Friedmann rearticulated his initial, somewhat speculative ideas into a sharply focused set of analytical propositions on what he now termed "the world city hypothesis." While Friedmann's taxonomy has been subjected to considerable critique and reformulation, at the time of its initial publication it created a powerful hermeneutic device with which researchers could examine the consequences of globalized urbanization in a broad range of geographic contexts. The world city hypothesis was further consolidated, in the wake of Friedmann's 1986 article, as an essential tool of radical urban critique and, in some cases, social and political action. The following contribution is thus most usefully read directly alongside Reading 6 from Friedmann and Wolff's classic article. Reading 7 extends and clarifies the core contentions of global cities research, develops a number of more concrete empirical claims, and even ventures an initial mapping of the emergent world urban system.

Manuel Castells (1972) and David Harvey (1973) revolutionized the study of urbanization and initiated a period of exciting and fruitful scholarship. Only in recent years, however, has the study of cities been directly connected to the world economy. This new approach sharpened insights into processes of urban change; it also offered a needed spatial perspective of an economy which seems increasingly oblivious to national boundaries. My purpose here is to state, as succinctly as I can, the main theses that link urbanization processes to global economic forces. The world city hypothesis, as I shall call these loosely joined statements, is primarily intended as a framework for research. It is neither a theory nor a universal generalization about cities, but a starting point for political enquiry. We

would, in fact, expect to find significant differences among those cities that have become the "basing points" for global capital. We would expect cities to differ among themselves according to not only the mode of their integration with the global economy, but also their own historical past, national policies, and cultural influences. The economic variable, however, is likely to be decisive for all attempts at explanation.

The world city hypothesis is about the spatial organization of the new international division of labor. As such, it concerns the contradictory relations between production in the era of global management and the political determination of territorial interests. It helps us to understand what happens in the major global cities of the world economy and what much political conflict in these cities is about. Although it cannot predict the outcomes of these struggles, it does suggest their common origins in the global system of market relations.

There are seven interrelated theses in all. As they are stated, I shall follow with a comment in which they are explained, or examples are given, or further questions are posed.

1. *The form and extent of a city's integration with the world economy, and the functions assigned to the city in the new spatial division of labor, will be decisive for any structural changes occurring within it.* Let us examine each of the key terms in this thesis.

(a) *City.* Reference is to an economic definition. A city in these terms is a spatially integrated economic and social system at a given location or metropolitan region. For administrative or political purposes the region may be divided into smaller units which underlie, as a political or administrative space, the economic space of the region.

(b) *Integration with the world capitalist system.* Reference is to the specific forms, intensity, and duration of the relations that link the urban economy into the global system of markets for capital, labor and commodities.

(c) *Functions assigned to it in the new spatial division of labor.* The standard definition of the world capitalist system is that it corresponds to a single (spatial) division of labor (Wallerstein 1984). Within this division, different localities – national, regional, and urban subsystems – perform specialized roles. Focusing only on metropolitan economies, some carry out headquarter functions, others serve primarily as a financial center, and still others have as their main function the articulation of regional and/or national economies with the global system. The most important cities, however, such as New York, may carry out all of these functions simultaneously.

(d) *Structural changes occurring within it.* Contemporary urban change is for the most part a process of adaptation to changes that are externally induced. More specifically, changes in metropolitan function, the structure of metropolitan labor markets, and the physical form of cities can be explained with reference to a world-wide process that affects: the direction and volume of transnational capital flows; the spatial division of the functions of finance, management and production or, more generally, between production

and control and the employment structure of economic base activities.

These economic influences are, in turn, modified by certain endogenous conditions. Among these the most important are: first, the "spatial patterns of historical accumulation" (King 1984); second, national policies, whose aim is to protect the national economic subsystem from outside competition through partial closure to immigration, commodity imports, and the operation of international capital; and third, certain social conditions, such as apartheid in South Africa, which exert a major influence on urban process and structure.

2. *Key cities throughout the world are used by global capital as "basing points" in the spatial organization and articulation of production and markets.* The resulting linkages make it possible to arrange world cities into a complex spatial hierarchy. Several taxonomies of world cities have been attempted, most notably by Cohen (1981; see Reading 5). In Figure 1, a different approach to world city distribution is attempted. Because the data to verify it are still lacking, the present effort is meant chiefly as a means to visualize a possible rank ordering of major cities, based on the presumed nature of their integration with the world economy.

The complete spatial distribution suggests a distinctively linear character of the world city system which connects along an East–West axis, three distinct subsystems: an Asian subsystem centered on the Tokyo–Singapore axis, with Singapore playing a subsidiary role as regional metropolis in South-East Asia; an American subsystem based on the primary core cities of New York, Chicago and Los Angeles, linked to Toronto in the North and to Mexico City and Caracas in the South, thus bringing Canada, Central America, and the small Caribbean nations into the American orbit; and a West European subsystem focused on London, Paris, and the Rhine valley axis from Randstad and Holland to Zurich. The southern hemisphere is linked into this subsystem via Johannesburg and São Paulo.

3. *The global control functions of world cities are directly reflected in the structure and dynamics of their production sectors and employment.* The driving force of world city growth is found in a small number of rapidly expanding sectors. Major importance attaches to corporate headquarters,

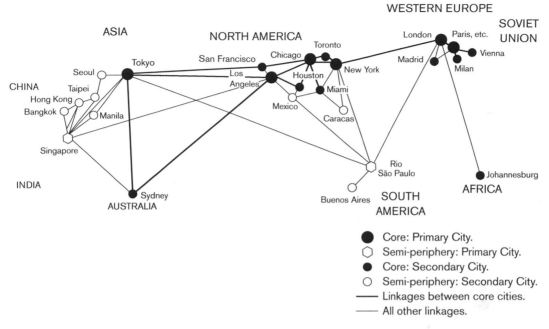

Figure 1 The hierarchy of the world cities

international finance, global transport and communications, and high level business services, such as advertising, accounting, insurance, and legal. An important ancillary function of world cities is ideological penetration and control. New York and Los Angeles, London and Paris, and to a lesser degree Tokyo, are centers for the production and dissemination of information, news, entertainment and other cultural artefacts.

In terms of occupations, world cities are characterized by a dichotomized labor force: on the one hand, a high percentage of professionals specialized in control functions and, on the other, a vast army of low-skilled workers engaged in manufacturing, personal services, and the hotel, tourist, and entertainment industries that cater to the privileged classes for whose sake the world city primarily exists (Sassen-Koob 1984).

In the semi-periphery, with its rapidly multiplying rural population, large numbers of unskilled workers migrate to world city locations in their respective countries in search of livelihood. Because the "modern" sector is incapable of absorbing more than a small fraction of this human mass, a large "informal" sector of microscopic survival activities has evolved.

4. *World cities are major sites for the concentration and accumulation of international capital.* Although this statement would seem to be axiomatic there are significant exceptions. In core countries, the major atypical case is Tokyo. Although a major control center for Japanese multinational capital, Japanese business practices and government policy have so far been successful in preventing foreign capital from making major investments in the city.

5. *World cities are points of destination for large numbers of both domestic and/or international migrants.* Two kinds of migrants can be distinguished: international and inter-regional. Both contribute to the growth of primary core cities, but in the semi-periphery world cities grow chiefly from inter-regional migration.

In one form or another, all countries of the capitalist core have attempted to curb immigration from abroad. Japan and Singapore have the most restrictive legislation and, for all practical purposes, prohibit permanent immigration. Western European countries have experimented with tightly controlled "guest-worker" programs. They, too, are jealous of their boundaries. And traditional immigrant countries, such as Canada and Australia, are attempting to limit the influx of migrants to workers possessing professional and other skills who are in high demand. Few if any countries have been as open to immigration from abroad as the USA, where both legal and illegal immigrants abound.

6. *World city formation brings into focus the major contradictions of industrial capitalism – among them spatial and class polarization*. Spatial polarization occurs at three scales. The first is global and is expressed by the widening gulf in wealth, income, and power between peripheral economies as a whole and a handful of rich countries at the heart of the capitalist world. The second scale is regional and is especially pertinent in the semi-periphery. In core countries regional income gradients are relatively smooth, and the difference between high and low income regions is rarely greater than 1:3. The corresponding ratio in the semi-periphery, however, is more likely to be 1:10.

Meanwhile, the income gradient between peripheral world cities and the rest of the national economies which they articulate remains very steep. The third scale of spatial polarization is metropolitan. It is the familiar story of spatially segregating poor inner-city ghettos, suburban squatter housing, and ethnic working-class enclaves. Spatial polarization arises from class polarization. And in world cities class polarization has three principal facets: huge income gaps between transnational elites and low-skilled workers, large-scale immigration from rural areas or from abroad, and structural trends in the evolution of jobs (see Figure 2)

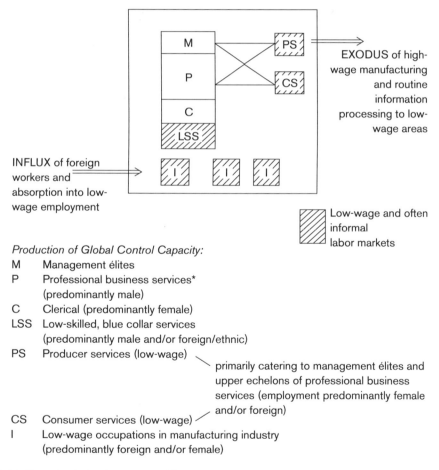

EXODUS of high-wage manufacturing and routine information processing to low-wage areas

INFLUX of foreign workers and absorption into low-wage employment

Low-wage and often informal labor markets

Production of Global Control Capacity:
M   Management élites
P   Professional business services*
      (predominantly male)
C   Clerical (predominantly female)
LSS   Low-skilled, blue collar services
      (predominantly male and/or foreign/ethnic)
PS   Producer services (low-wage)

primarily catering to management élites and upper echelons of professional business services (employment predominantly female and/or foreign)

CS   Consumer services (low-wage)
I   Low-wage occupations in manufacturing industry
      (predominantly foreign and/or female)

Figure 2 World city restructuring in core countries

* Many professional business services engage increasingly in international trade serving their clients, the transnational corporations, both at home and abroad. They include accounting, advertising, banking, communications, computer services, health services, insurance, leasing, legal services, shipping and air transport, and tourism. In 1981, US service exports equalled 50 per cent of merchandise exports and were still rising (see Sassen-Koob 1984).

The basic structural reason for social polarization in world cities must be looked for in the evolution of jobs, which is itself a result of the increasing capital intensity of production. In the semi-periphery most rural immigrants find accommodation in low-level service jobs, small industry, and the "informal" sector. In core countries the process is more complex. Given the downward pressure on wages resulting from large-scale immigration of foreign (including undocumented) workers, the number of low-paid, chiefly non-unionized jobs rises rapidly in three sectors: personal and consumer services (domestics, boutiques, restaurants, and entertainment); low-wage manufacturing (electronics, garments, and prepared foods); and the dynamic sectors of finance and business services which comprise from one-quarter to one-third of all world city jobs and also give employment in many low-wage categories.

7. *World city growth generates social costs at rates that tend to exceed the fiscal capacity of the state.* The rapid influx of poor workers into world cities – be it from abroad or from within the country – generates massive needs for social reproduction, among them housing, education, health, transportation, and welfare. These needs are increasingly arrayed against other needs that arise from transnational capital for economic infrastructure and from the dominant elites for their own social reproduction.

In this competitive struggle the poor, and especially the new immigrant populations, tend to lose out. State budgets reflect the general balance of political power. Not only are corporations exempt from taxes; they are generously subsidized in a variety of other ways as well. At the same time the social classes that feed at the trough of the transnational economy insist, and usually insist successfully, on the priority of their own substantial claims for urban amenities and services. The overall result is a steady state of fiscal and social crisis in which the burden of capitalist accumulation is systematically shifted to the politically weakest, most disorganized sectors of the population. Their capacity for pressing their rightful claims against the corporations and the state is further contained by the ubiquitous forces of police repression.

REFERENCES FROM THE READING

Castells, M. (1972) *La Question urbaine*. Paris: Maspero.

Cohen, R.B. (1981) The new international division of labor, multinational corporations and urban hierarchy. In M. Dear and A.J. Scott (eds) *Urbanization and Urban Planning in Capitalist Society*. New York: Methuen.

Harvey, D. (1973) *Social Justice and the City*. London: Edward Arnold.

King, A.D. (1984) Capital city: physical and spatial aspects of London's role in the world economy. Unpublished paper.

Sassen-Koob, S. (1984) Capital mobility and labor migration: their expression in core cities. In M. Timberlake (ed.) *Urbanization in the World System*. New York: Academic Press.

Wallerstein, I. (1984) *The Politics of the World Economy*. Cambridge: Cambridge University Press.

ONE

PART TWO

Structures, dynamics and geographies of global city formation

Plate 10 World Trade Center, New York, 1980 (Roger Keil)

INTRODUCTION TO PART TWO

The label "global city" was initially used to characterize the emergence of headquarter economies within major urban centers located primarily in western Europe (London, Paris, Frankfurt, Amsterdam, Zurich), North America (New York, Chicago, Los Angeles, Toronto) and East Asia (Tokyo, Hong Kong, Singapore). In Part Two, we survey a selection of classic and contemporary analyses of global city formation in such headquarter economies, to which Robinson has aptly referred as "new industrial districts of transnational management and control" (2002: 536; see also Reading 26). The following readings present some of the core arguments of global cities researchers on the interplay between economic globalization and, among other themes, the formation of a global urban hierarchy, urban economic restructuring, urban labor market change, immigration flows and urban sociospatial polarization.

In mainstream discourses on globalization, it has frequently been claimed that intensified capital mobility and new informational technologies are undermining the significance of local places. Increasingly, it is argued, major corporations possess the technological capacities (such as the internet) to coordinate production and investment no matter where they are located; they are thus able to avoid the high costs and negative externalities that are said to be associated with major cities by locating their management and production operations elsewhere. One of the seminal contributions of global cities research has been to dismantle this mainstream vision of contemporary globalization by demonstrating that cities provide a locationally specific, economically strategic and effectively non-substitutable infrastructure for the worldwide operations of transnational corporations. More than any other global cities researcher, Saskia Sassen (1991, 1999, 2000) has attempted to decipher the dominant socioeconomic and technological dynamics within global city economies (see also Readings 9 and 10).

For Sassen, the increasing geographical dispersal of production under contemporary capitalism has necessarily entailed an equally intense impulsion towards the centralization of command and control capacities within the agglomeration economies of major urban centers. Sassen accepts the emphasis on transnational corporate headquarters locations that is evident, for instance, in the work of Cohen (Reading 5), Friedmann (Reading 7) and Feagin and Smith (1989); however, her research is focused more closely upon a broad ensemble of advanced producer and financial services industries – for instance, law, accounting, finance, advertising and consulting – that, she argues, likewise agglomerate intensively in global city economies. While advanced producer and financial services industries have long been clustered together within major urban centers, Sassen suggests that their agglomeration tendencies have significantly intensified within the most strategic urban command and control centers of the world economy, where their outputs are in particularly high demand by transnational firms. Indeed, according to Sassen, this newly consolidated producer and financial services "complex" represents the real economic foundation for global cities, because it provides transnational corporations with a host of essential services that enable them to implement, manage and regulate their production and investment networks on a global scale. For Sassen, then, global cities provide a variety of place-bound, place-specific socioeconomic assets that are fundamental to the production of capital mobility, and more generally, to the consolidation of a globalized formation of world capitalism (see also Storper 1996). Accordingly, much of Sassen's empirical research has explored the diverse patterns of socioeconomic and sociospatial change that have unfolded in cities such as New York, London and Tokyo in light of their increasingly significant roles as command and control centers for global capitalism.

Sassen's basic insight, that industrial agglomeration, capital fixity and clusters of advanced corporate services are essential to the globalization of capitalist production and exchange, has been elaborated in some detail in the literature on global cities, through investigations of diverse aspects of urban restructuring. Although not all of the contributions to Part Two engage directly with Sassen's work, we would argue that they can all still be viewed as explorations and elaborations of this fundamental proposition. In addition to Readings 9 and 10, in which Sassen summarizes some of her own major arguments, the selections included here explore the broad constellation of institutional, socioeconomic and spatial transformations that have facilitated, and resulted from, the process of global city formation, both within and beyond global cities themselves. In so doing, these readings provide a broad snapshot of several major strands of global cities research, and the scholarly debates they have provoked. In general terms, these readings explore the interplay between geoeconomic integration and the formation of global cities in at least three central ways, as follows.

The formation of a global urban hierarchy

Global city theory postulates the formation of a worldwide urban hierarchy in and through which transnational corporations coordinate their production and investment activities. The geography, composition and evolutionary tendencies of this hierarchy have been a topic of intensive research and debate since the pioneering studies of global city formation in the 1980s by Cohen (Reading 5), Friedmann and Wolff (Reading 6) and Friedmann (Reading 7). Subsequent scholarship has explored a variety of methodological strategies and empirical data sources through which to map this hierarchy (Short et al. 1996). However, whatever their differences of interpretation, most studies of the global urban system conceptualize this grid of cities simultaneously as a fundamental spatial infrastructure for the globalization of capital, and as a medium and expression of the new patterns of global sociospatial polarization that have emerged during the post-1970s period. To represent this strand of research, we have included a selection by three members of the GaWC group at Loughborough University – Jonathan V. Beaverstock, Richard G. Smith and Peter J. Taylor – which introduces a new, "relational" approach to the study of the global urban system (see Reading 11).

The restructuring of urban space

The consolidation of global cities is expressed not only on a global scale, through the establishment of new, worldwide linkages among cities and a global urban system. As Deyan Sudjic's contribution to Part Two eloquently demonstrates (Reading 8), the process of global city formation also entails significant spatial transformations at an urban scale, within cities themselves. Most contemporary global cities served as major urban agglomerations long before being transformed into global command and control centers, and the legacies of earlier (pre-industrial and/or industrial) patterns of urban development generally leave strong, lasting imprints upon these cities' built environments and spatial configurations (Abu-Lughod 1999). However, as global cities researchers have demonstrated, the globalization of urban development still generates powerful spatial expressions in the built environment. The intensified clustering of transnational corporate headquarters and advanced corporate services firms in the city core overburdens inherited land-use infrastructures, leading to new, often speculative, real estate booms as new office towers are constructed both within and beyond established downtown areas. Meanwhile, the rising cost of office space in the global city core may generate massive spillover effects on a regional scale, as small- and medium-sized agglomerations of corporate services and back offices crystallize throughout the urban region (see Reading 34 by Keil and Ronneberger). Finally, the consolidation of such headquarter economies may also generate significant shifts within local housing markets as developers attempt to transform once-devalorized properties into residential space for corporate elites and other members of the "yuppie" milieux. Consequently, gentrification ensues in formerly working-class neighborhoods, and considerable residential displacement may be caused in the wake of rising rents and housing prices (N. Smith 1996). Global cities researchers have tracked these and many other spatial transformations at some length: the urban built environment is viewed as an arena of contestation in which competing social forces and interests, from transnational firms and developers to residents and social movements, struggle over issues of urban design, land use and public space. Of course, such issues are hotly contested in nearly all contemporary cities. Global cities researchers acknowledge this, but are concerned to explore their distinctive forms and outcomes in cities that have come to serve key command and control functions in the global system. Accordingly, several readings address the key topics of urban spatial transformation and the built environment (see, for instance, Reading 16 by Zukin, Reading 19 by Schmid, Reading 23 by King, Reading 33 by Haila, Reading 34 by Keil and Ronneberger, Reading 40 by Lehrer and Reading 44 by Marcuse). In Part Two, these issues are expertly explored in Reading 14 by Stephen Graham, who investigates the newly emergent infrastructural geographies – particularly in the fields of media and telecommunications – that have been established in global city-regions.

The transformation of the urban social fabric

One of the most provocative, if also highly controversial, aspects of global cities research involves claims regarding the effects of global city formation upon the urban social fabric. Friedmann (Reading 7) and Sassen (1991), in particular, have suggested that the emergence of a global city hierarchy has generated a "dualized" urban labor market structure dominated, on the one hand, by a high-earning corporate elite and, on the other hand, by a large mass of workers employed in menial, low-paying and/or informalized jobs. For Sassen (1991: 13), this "new class alignment in global cities" has emerged in direct conjunction with the downgrading of traditional manufacturing industries and the emergence of the advanced producer and financial services complex. Her work on London, New York and Tokyo suggests that broadly analogous, if place-specific, patterns of social polarization have emerged in these otherwise quite different cities, and suggests that these outcomes are causally linked to these cities' new roles as global command and control centers. The polarization thesis proposed by Friedmann and further elaborated by Sassen has attracted considerable discussion and debate: whereas some scholars have attempted to apply their argument to other globalizing cities, other analysts have questioned its logical and/or empirical validity (see, for instance, Hamnett 1994, 1996; White 1998). In Part Two of the Reader, we consider various positions within the polarization debate, first, through a reading of some of Sassen's own texts on global city formation (Readings 9 and 10), second, through a classic 1983 essay by Robert Ross and Kent Trachte (Reading 12), and third, through a more recent critical review and evaluation by Susan S. Fainstein (Reading 13). Whereas Ross and Trachte trace various forms of polarization and inequality in post-1970s New York City directly to the process of globalization, Fainstein reviews the polarization debate and, on this basis, examines empirical evidence on inequality levels in New York, London, Tokyo, Paris and the Dutch Randstad. Fainstein questions much of the literature on polarization in global cities, but she does suggest a number of ways in which distinctive forms of inequality have emerged within the cities' labor markets.

As the readings in Part Two indicate, global city theory has been a key analytical reference point in diverse strands of urban research. Its arguments and methods have been mobilized to examine a broad range of urban and supra-urban transformations since the mid-1970s. While certain arguments of global city theory have been challenged and revised by many scholars, it has nonetheless provided a useful methodological starting point through which to explore the interface between geoeconomic restructuring and urban transformation.

References and suggestions for further reading

Abu-Lughod, J. (1999) *New York, Chicago, Los Angeles: America's Global Cities.* Minneapolis, MN: University of Minnesota Press.

Amin, A. and Thrift, N. (1992) Neo-Marshallian nodes in global networks, *International Journal of Urban and Regional Research*, 16, 4, 571–587.

Baum, S. (1997) Sydney, Australia: a global city?, *Urban Studies*, 34, 11, 1881–1901.

Beaverstock, J.V. (1994) Re-thinking skilled international labour migration: world cities and banking organizations, *Geoforum*, 25, 323–338.

Beaverstock, J.V., Smith, R.G. and Taylor, P.J. (1999a) A roster of world cities, *Cities*, 16, 6, 445–458.

Beaverstock, J.V., Smith, R.G. and Taylor, P.J. (1999b) The long arm of the law: London's law firms in a globalizing world-economy, *Environment and Planning A*, 31, 10, 1857–1876.

Burgers, J. (1996) No polarisation in Dutch cities? Inequality in a corporatist country, *Urban Studies*, 33, 1, 99–105.

Castells, M. and Hall, P. (1996) *Technopoles of the World.* New York: Routledge.

Feagin, J.R. and Smith, M.P. (1989) Cities and the new international division of labor. In M.P. Smith and J.R. Feagin (eds) *The Capitalist City.* Cambridge, MA: Blackwell, 3–36.

Godfrey, B.J. and Zhou, Y. (1999) Ranking world cities: multinational corporations and the global urban hierarchy, *Urban Geography*, 20, 268–281.

Graham, S. (1999) Global grids of glass: on global cities, telecommunications and planetary urban networks, *Urban Studies*, 36, 5–6, 929–949.

Graham, S. (2002) Communication grids: cities and infrastructure. In S. Sassen (ed.) *Global Cities, Linked Networks.* New York: Routledge, 71–92.

Hamnett, C. (1994) Social polarisation in global cities: theory and evidence, *Urban Studies*, 31, 3, 401–424.

Hamnett, C. (1996) Social polarisation, economic restructuring and welfare state regimes, *Urban Studies*, 33, 8, 1407–1430.

Keeling, D.J. (1995) Transport and the world city paradigm. In P.L. Knox and P.J. Taylor (eds) *World Cities in a World-System.* Cambridge: Cambridge University Press, 115–131.

Leyshon, A. and Thrift, N. (1997) Sexy greedy: the new international financial system, the City of London and the south east of England. In A. Leyshon and N. Thrift, *Money/Space.* New York: Routledge, 118–164.

Li, W. (1998) Los Angeles' Chinese ethnoburb: from ethnic service center to global economy outpost, *Urban Geography*, 19, 502–517.

Logan, J. (2000) Still a global city: the racial and ethnic segmentation of New York. In P. Marcuse and R. van Kempen (eds) *Globalizing Cities: A New Spatial Order*, Oxford: Blackwell, 158–185.

Rehin, C. (1998) Globalization, social change and minorities in metropolitan Paris: the emergence of new class patterns, *Urban Studies*, 35, 429–447.

Rimmer, P.J. (1998) Transport and telecommunications among world cities. In F.C. Lo and Y.M. Yeung (eds) *Globalization and the World of Large Cities.* Tokyo: United Nations University Press, 433–470.

Robinson, J. (2002) Global and world cities: a view from off the map, *International Journal of Urban and Regional Research*, 26, 3, 531–554.

Sassen, S. (1991) *The Global City: New York, London, Tokyo.* Princeton, NJ: Princeton University Press.

Sassen, S. (1999) Global financial centers, *Foreign Affairs*, 78, 1, 75–87.

Sassen, S. (2000) *Cities in a World Economy*, 2nd edition. Thousand Oaks, CA: Pine Forge Press.

Short, J.R., Kim, Y., Kuus, M. and Wells, H. (1996) The dirty little secret of world cities research: data problems in comparative analysis, *International Journal of Urban and Regional Research*, 20, 697–717.

Smith, D.A. and Timberlake, M. (1995) Conceptualizing and mapping the structure of the world systems' city system, *Urban Studies*, 32, 2, 287–302.

Smith, N. (1996) *The New Urban Frontier.* New York and London: Routledge.

Storper, M. (1996) *The Regional World.* New York: Guilford.

Storper, M. and Scott, A.J. (1989) The geographical foundations and social regulation of flexible production complexes. In J. Wolch and M. Dear (eds) *The Power of Geography*. Boston, MA: Unwin Hyman, 19–40.

Taylor, P.J. (1997) Hierarchical tendencies amongst world cities: a global research proposal, *Cities*, 14, 6, 323–332.

Thrift, N. (1987) The fixers: the urban geography of international commercial capital. In J. Henderson and M. Castells (eds) *Global Restructuring and Territorial Development*. London: Sage, 203–233.

Waldinger, R. (2001) The immigrant niche in global city-regions: concept, patterns, controversy. In A.J. Scott (ed.) *Global City-Regions*. New York: Oxford University Press, 285–298.

White, J.W. (1998) Old wine, cracked bottle? Tokyo, Paris, and the global city hypothesis, *Urban Affairs Review*, 33–34, 451–477.

T
W
O

8

Prologue

"100-mile Cities"

from *The 100-mile City* (1992)

Deyan Sudjic

Imagine the force field around a high-tension power line, crackling with energy and ready to flash over and discharge 20,000 volts at any point along its length, and you have some idea of the nature of the modern city as it enters the last decade of the century.

The city's force field is not a linear one, however. Rather, it stretches for a hundred miles in each direction, over towns and villages and across vast tracts of what appears to be open country, far from any existing settlement that could conventionally be called a city. Without any warning, a flash of energy short-circuits the field, and precipitates a shopping center so big that it needs three or five million people within reach to make it pay. Just as the dust has settled, there is another discharge of energy, and an office park erupts out of nothing, its thirty- and forty-storey towers rising sheer out of what had previously been farmland. The two have no visible connection, yet they are part of the same city, linked only by the energy field, just like the housing compounds that crop up here and there, and the airport, and the cloverleaf on the freeway, and the corporate headquarters with its own lake in the middle of a park.

Somewhere, in a remote corner, there is no doubt a little enclave of pedestrian streets, a fringe of terraced houses that circles the crop of office towers that marks downtown. There will be a sandblasted old market hall recycled for recreational shopping. And somewhere else, there will be the social derelicts, the casualties, trapped in welfare housing or worse.

In the force field city, nothing is unself-conscious, any urban gesture is calculated. For the affluent, the home is the center of life – though given the astonishing increase in household mobility, it's not likely to be the same home for very long. From it, the city radiates outwards as a star-shaped pattern of overlapping routes to and from the workplace, the shopping center, and the school. They are all self-contained abstractions that function as free-floating elements. Each destination caters to a certain range of the needs of urban life, but they have no physical or spatial connection with each other in the way that we have been conditioned to expect of the city.

Mobility means overlapping force fields. Cities compete with each other in a grimly determined struggle to maintain the energy that keeps them working. But the lives of their citizens take in other cities to an extent that is unparalleled in history. They move from one to another constantly, to live, to visit, to do business.

In most economic areas, one or two individual cities have come to monopolise the world market. So New York, Tokyo and London are the world's financial capitals, Los Angeles is its entertainment center. Perhaps the hundred-mile city has already become the thousand-mile city. The other side to the equation is the widening gap between those cities that are successful, and those that are not. Some cities are clearly failing. Their treasuries teeter on the edge of bankruptcy, they fail to attract new employers, they have little to recommend them culturally. But beyond that there is the widening difference between metropolitan cities which draw in the ambitious and the gifted, and provincial cities which lose them.

The hundred-mile city is a model of urban life which many people find threatening even as they embrace it. By and large, those who have a choice do not care to live in cramped city center homes, nor do they see much economic future in the old downtowns. The historic pattern of the city is undoubtedly a high point in civilisation, one that is worth preserving. But even if masonry endures, its meaning has been irrevocably altered.

To accept this image of the city is to accept uncomfortable things about ourselves, and our illusions about the way we want to live. The city is as much about selfishness and fear as it is about community and civic life. And yet to accept that the city has a dark side, of menace and greed, does not diminish its vitality and strength. In the last analysis, it reflects man and all his potential.

9

"Cities and Communities in the Global Economy"

from *American Behavioral Scientist* (1996)

Saskia Sassen

Editors' introduction

Saskia Sassen is currently the Ralph Lewis Professor of Sociology at the University of Chicago and Centennial Visiting Professor in the Department of Sociology at the London School of Economics. She previously taught in the Urban Planning Department at Columbia University. Sassen is arguably the single most influential and widely cited contemporary analyst of global city formation, primarily due to the massive international impact of her path-breaking book, *The Global City: New York, London, Tokyo* (1991; republished in a revised edition in 2002). Prior to her work on *The Global City*, Sassen had written a research monograph exploring the interplay between new forms of foreign direct investment and changing patterns of labor migration (Sassen 1988). In the penultimate chapter of that work, Sassen examined the transformation of urban labor markets in major US global city-regions in conjunction with the increasing centralization of transnational capitalist management capacities (Sassen 1988: 126–170). Subsequently, Sassen decided to explore more systematically the restructuring of economic systems and sociospatial configurations in cities that had come to serve as global command and control centers for transnational corporations. Sassen's innovative inquiry into these issues culminated in her book, *The Global City*, which continues, well over a decade after its original publication, to provoke intensive discussion and debate across the social sciences.

In contrast to early contributors to global cities research (see, for instance, Reading 5 by Cohen), Sassen did not define global city economies simply with reference to the presence of transnational corporate headquarters locations. Instead, she emphasized, the specificity and dynamism of global cities stems from the broad cluster of advanced producer and financial services industries – including law, banking, accounting, advertising, insurance and consulting – that cater to the distinctive organizational and informational needs of global firms. One of the most original, enduring contributions of *The Global City* was to explore the surprisingly analogous patterns of industrial restructuring that were unfolding in three otherwise quite different urban economies – New York, London and Tokyo – due to their increasing specialization in these sectors. More controversially, Sassen postulated that new patterns of social and spatial inequality were crystallizing within global city economies due to the polarization of local labor markets among highly paid corporate elites and a large mass of poor people working in low paid, menial and insecure service jobs (for variations on this proposition, see Reading 7 by Friedmann and Reading 12 by Ross and Trachte; for a critique, see Reading 13 by Fainstein).

Reading 9 is derived from a 1996 article that summarizes some of the major arguments developed in Sassen's research during the preceding decade. Her analysis focuses, in particular, upon the following issues, each of which will be explored in greater detail in subsequent readings: the conceptualization of globalization, the central role of cities within a globalizing world system, the restructuring of local labor markets and sociospatial arrangements within global cities, the problem of social polarization within global cities, and the emergence of new forms of sociopolitical contestation within strategic urban spaces.

This is a period of massive changes for cities and communities. Economic globalization and the ascendance of information technologies are reconfiguring the spatial organization of the economy, with often devastating consequences for cities and communities. Many politicians and analysts appear convinced that these forces have rendered localities powerless to address their own situation. Substantial transitions or discontinuities such as these call for the development of new categories for analysis, new lines of theorization, and perhaps some new political and economic practices as well. I posit that the urban level and the community level need to be incorporated in the analysis of economic globalization and the study of new information technologies. This requires going beyond mainstream notions about the impact of globalization and the spread of information technologies for localities – that is, the relative powerlessness of localities confronted with hypermobile capital. This powerlessness is only one part of the story. In my reading of the evidence, a focus on cities and communities allows for a more concrete analysis of globalization, and in that regard we can think of cities and communities as strategic sites for an examination of global processes and major politico-economic processes.

Introducing cities and communities in an analysis of economic globalization allows us to reconceptualize processes of economic globalization as concrete economic complexes situated in specific places. Such a focus decomposes the national economy into a variety of subnational components, some profoundly articulated with the global economy, and others not. It allows us to see the multiplicity of economies and work cultures in which the global information economy is embedded. It also allows us to recover the concrete, localized processes through which globalization exists, and to argue, for example, that the multiculturalism evident in large cities is as much a part of globalization as is international finance.

Why does it matter to recover the local or, more generally, the specific place setting in analyses of the global economy? Mainstream accounts have the effect of evicting cities, communities, and workers from the story of today's advanced economy; and hence we need corrective action to resurrect these elements. Insofar as an economic analysis focused on cities and communities recovers the broad array of jobs and work cultures that are part of the global economy – though typically not marked as such – one can examine the possibility of a new politics of traditionally disadvantaged actors, a new politics that operates in this new transnational economic geography, whether the actors are factory workers in export processing zones or cleaners on Wall Street.

PLACE AND WORK PROCESS IN THE GLOBAL INFORMATION ECONOMY

The mainstream account about economic globalization found in media and policy circles, as well as in much economic analysis, emphasizes hypermobility, global communications, and the gradual neutralization of place and distance. There is a tendency in that account to take the existence of a global economic system as a given, to see it as a function of the power of transnational corporations and global communications. But the capabilities for global operation, coordination, and control contained in the new information technologies and in the power of transnational corporations are not the same as the actual work of coordination and control. By focusing on the actual work of running global businesses, we add a neglected dimension to the familiar issue of the power of large corporations and the broad reach of the new technologies. The emphasis shifts to the *practices* that constitute what we call economic globalization and global control, the work of producing and reproducing the organization and management of a global production system for manufacturing and a global marketplace for finance, both under conditions of economic concentration in ownership and control.

For instance, as multinational firms establish more and more operations overseas, their central management, coordination, and servicing functions multiply and become increasingly complex as they need to negotiate different national legal, accounting, and management systems. Thus, in the early 1990s, U.S. firms had about 19,000 affiliates worldwide, as did German and Japanese firms. The sheer number of dispersed factories and service outlets that are part of a firm's integrated operation creates massive new needs for central coordination and servicing. Finance and advanced corporate services are industries producing the organizational commodities necessary for the implementation and management of global economic systems. Cities are preferred sites for the production of these services, particularly the most innovative and speculative internationalized service sectors.

A focus on corporate practices draws the categories of place and work process into the analysis of economic globalization. These are two categories easily overlooked in accounts centered on the hypermobility of capital and the power of transnational. Developing categories such as "place" and "production process" does not negate the importance of hypermobility and power. Rather, it brings to the fore the fact that many of the resources necessary for global economic activities are not hypermobile and are, indeed, deeply embedded in place, notably places such as global cities and export processing zones, and that many of the communities that may appear as unrelated to global processes, such as the South Bronx in New York City or the town of Plainfield in Massachusetts, are actually part of such processes. Further, global processes are structured by local constraints, including the composition of the workforce, work cultures, and prevailing political cultures and processes.

Moreover, by emphasizing the fact that global processes are at least partly embedded in national territories, such a focus introduces new variables in current conceptions about economic globalization and the shrinking regulatory role of the state. That is to say, the space economy for major new

transnational economic processes diverges in significant ways from the global/national duality presupposed in much analysis of the global economy. The duality of national versus global foci suggests two mutually exclusive spaces – where one begins the other ends, what one gains the other loses. One of the main purposes in this chapter is to suggest that this conceptualization is fundamentally incorrect, that the global materializes by necessity in specific places and institutional arrangements – a good number of which (if not most) are located in national territories and in identifiable localities. National and local governments play an important role in the implementation of global economic systems, a role that can assume different forms depending on socio-economic development levels, political culture, and mode of articulation with global processes.

What this discussion signals is that economic globalization contains both a dynamic of dispersal and a dynamic of centralization, a condition that is only now beginning to receive recognition (Abu-Lughod 1995; Castells 1989; Friedmann 1995). The massive trends toward the spatial dispersal of economic activities at the metropolitan, national, and global level, which we associate with globalization, have contributed to a demand for new forms of territorial centralization of top-level management and control operations. The expansion of central functions results from the continuing concentration in control, ownership, and profit appropriation that characterizes the current economic system. National and global markets, as well as globally integrated organizations, require central places where the work of globalization gets done. Further, information industries require a vast physical infrastructure containing strategic nodes with hyperconcentration of facilities; we need to distinguish between the capacity for global transmission/communication and the material conditions that make this interchange of information possible. Finally, even the most advanced information industries have a production process that is at least partly place-bound because of the combination of resources it requires even when the outputs are hypermobile.

One of the central concerns of my work has been to view cities as production sites for the leading information industries of our time and to recover the infrastructure of activities, firms, and jobs that is necessary to run the advanced corporate economy. These industries are typically conceptualized in terms of the hypermobility of their outputs and the high levels of expertise of their professionals rather than in terms of the work process involved and the requisite infrastructure of facilities and nonexpert jobs that are also part of these industries. A detailed analysis of service-based urban economies shows that there is considerable articulation of firms, sectors, and workers who may appear as though they have little connection to an urban economy dominated by finance and specialized services but in fact fulfill a series of functions that are an integral part of that economy. They do so, however, under conditions of sharp social, earnings, and (often) racial/ethnic segmentation. Our cities and communities reflect these patterns. Recapturing the geography of places involved in globalization allows us to recapture knowledge of people, workers, communities, and, more specifically, the many different work cultures, besides the corporate culture involved in the work of globalization. The global city can be seen as one strategic research site for the study of these processes.

THE GLOBAL CITY: A NEXUS FOR NEW POLITICO-ECONOMIC ALIGNMENTS

In the day-to-day work of the leading services complex dominated by finance, a large share of the jobs involved are low paid and manual, many held by women and immigrants. Although these types of workers and jobs are never represented as part of the global economy, they are in fact an essential part of the multifaceted infrastructure of jobs involved in running and implementing the global economic system – including its most advanced sectors, such as international finance. The top end of the corporate economy – the corporate towers that project engineering expertise, precision, "techne" – is far easier to mark as necessary for an advanced economic system than are truckers and other industrial service workers, even though these are a necessary ingredient as well.

We see here at work a dynamic of valorization that has sharply increased the gap between the devalorized and the valorized (indeed, overvalorized) sectors of the economy. There is a growing inequality in the profit-making capacities of firms and in the earnings capacities of households. The overvalorization of specialized services and of professional workers has marked many of the "other" types of economic activities and workers as unnecessary or irrelevant to an advanced economy, even when they are actually part of the leading internationalized sector. We can see this effect in, for example, the unusually sharp increase in the beginning salaries of MBAs and lawyers hired in the corporate sector and in the precipitous fall in the wages of low-skilled manual workers and clerical workers. We can see the same effect in the retreat of many real estate developers from the low- and medium-income housing market in the wake of the rapidly expanding housing demand by the new highly paid professionals and the possibility for vast overpricing of this so-called spec housing supply.

The implantation of global processes and markets in major cities has meant that the internationalized sector of the economy has expanded sharply and has imposed a new set of criteria for valuing or pricing various economic activities and outcomes. This development has had devastating effects on large sectors of the urban economy and has been felt unevenly in urban communities. It is not simply a quantitative transformation; we see here the elements for a new economic regime (see, generally, Fainstein 1993; Knox and Taylor 1995; Sassen 1994).

The ascendance of the specialized services-led economy, particularly the new finance and services complex, engenders what may be regarded as a new economic regime because although this sector may account for only a fraction of the economy of a city, it imposes itself on that larger economy. One of these pressures is toward polarization, as is the case with the possibility for superprofits in finance, which contributes to the devalorization of manufacturing and low value-added services insofar as these sectors cannot generate the superprofits typical in much financial activity.

The super-profit-making capacity of many of the leading industries is embedded in a complex combination of new trends: (a) technologies that make possible the hypermobility of capital at a global scale and the deregulation of multiple markets that allows for implementing that hypermobility;

(b) financial inventions such as securitization that liquify hitherto unliquid capital and allow it to circulate and hence make additional profits; and (c) the growing demand for services in all industries, along with the increasing complexity and specialization of many of these inputs, which has contributed to their valorization as illustrated in the unusually high salary increases beginning in the 1980s for top-level professionals and CEOs. Globalization further adds to the complexity of these services, their strategic character, their glamour, and therewith to their overvalorization.

The presence of a critical mass of firms with extremely high profit-making capabilities leads to the bidding up the prices of commercial space, industrial services, and other business needs, and thereby makes survival for firms with moderate profit-making capabilities increasingly precarious. And although the latter are essential to the operation of the urban economy and for the daily needs of residents, their economic viability is threatened in a situation where finance and specialized services can earn superprofits. High prices and profit levels in the internationalized sector and its ancillary activities, such as top-of-the-line restaurants and hotels, make it increasingly difficult for other sectors to compete for space and investments. Many of these other sectors have experienced considerable downgrading and/or displacement; for example, the replacement of neighborhood shops tailored to local needs by upscale boutiques and restaurants catering to new high-income urban elites is now a common urban occurrence.

Inequality in the profit-making capabilities of different sectors of the economy has always existed. But what we see happening today takes place on another order of magnitude and is engendering massive distortions in the operations of various markets, from housing to labor. For instance, the polarization among firms and households and in the spatial organization of the economy contribute, in my reading, toward the informalization of a growing array of economic activities in advanced urban economies. When firms with low or modest profit-making capacities experience an ongoing if not increasing demand for their goods and services from households and other firms in a context where a significant sector of the economy makes superprofits,

they often cannot compete even though there is an effective demand for what they produce. Operating informally is often one of the few ways in which such firms can survive; illustrations are the use of spaces not zoned for commercial or manufacturing activities, such as basements in residential areas, or space that is not up to code in terms of health, fire, and other such standards. Similarly, new firms in low-profit industries entering a strong market for their goods and services may only be able to do so informally. Another option for firms with limited profit-making capabilities is to subcontract part of their work to informal operations.

The recomposition of the sources of growth and of profit making entailed in these transformations also contributes to a reorganization of some components of social reproduction or consumption. Although the middle strata still constitute the majority, the conditions that contributed to their expansion and politico-economic power in the postwar decades – the centrality of mass production and mass consumption in economic growth and profit realization – have been displaced by new sources of growth.

The rapid growth of industries with strong concentrations of high- and low-income jobs has assumed distinct forms in the consumption structure, which in turn has produced a feedback effect on the organization of work and the types of jobs being created. The expansion of the high-income workforce in conjunction with the emergence of new cultural forms has led to a process of high-income gentrification that ultimately rests on the availability of a vast supply of low-wage workers.

In good part, the consumption needs of the low-income population in large cities are met by manufacturing and retail establishments that are small, rely on family labor, and often fall below minimum safety and health standards. Cheap, locally produced sweatshop garments, for example, can compete with low-cost Asian imports. A growing range of products and services, from low-cost furniture made in basements to "gypsy cabs" and family day care, is available to meet the demand for the growing low-income population. There are numerous instances of how the increased inequality in earnings reshapes the consumption structure and how this in turn has feedback effects on the organization of work, both in the formal and in the informal economy. In the case of New York City, we see the creation of a special taxi line that services only the financial district and the increase of gypsy cabs in low-income neighborhoods not serviced by regular cabs; the increase in highly customized woodwork in gentrified areas and low-cost self-help rehabilitation in poor neighborhoods; the increase of home workers and sweatshops making either very expensive designer items for boutiques or very cheap products.

One way of conceptualizing informalization in advanced urban economies today is to posit it as the systemic equivalent of what we call deregulation when it happens at the top of the economy. Both the deregulation of a growing number of leading information industries and the informalization of a growing number of sectors with low profit-making capacities can be conceptualized as adjustments under conditions where new economic developments and old regulations enter in growing tension.

WHOSE CITY IS IT?

One way of thinking about the political implications of these developments is in terms of the formation of new claims on cities. Among the major new actors making claims are foreign firms who have been increasingly entitled to do business through progressive deregulation of national economies, as well as the large numbers of international businesspeople. These are among the new city users. They have profoundly marked the urban landscape, and their claim to the city is not contested, even though the costs and benefits to cities have barely been examined.

The new city users have made an often immense claim on the city and have reconstituted strategic spaces of the city in their image: theirs is a claim rarely made problematic. The new city of city users is a fragile one whose survival and successes are centered on an economy of high profits, advanced technologies, intensified exchanges, speculation, and innovation.

On the one hand, this raises a question of what the city is for international businesspeople and international firms; it is a city whose space consists of airports, top-level business districts, top-of-the-line hotels and restaurants, a sort of urban glamour zone. On the other hand, there is the difficult task of establishing whether a city that functions as an international business center does in fact recover the costs involved in being such a center. The costs involved in maintaining a state-of-the-art business district, and all it requires, from advanced communications facilities to top-level security (and "world-class culture") have not been adequately estimated.

Perhaps at the other extreme of conventional representations are those who use urban political violence to make their claims on the city, claims that lack the de facto legitimacy enjoyed by the new "city users." These are claims made by actors struggling for recognition and entitlement, claiming their rights to the city. These claims have, of course, a long history; every new epoch brings specific conditions to the manner in which the claims are made. The growing weight of "delinquency" (e.g., smashing cars and shopwindows; robbing and burning stores) in some of these uprisings over the last decade in major cities of the developed and third world is perhaps an indication of the sharpened inequality being witnessed in the process of economic globalization. The distance, as seen and as lived, between the urban glamour zone and the urban war zone has become enormous. The extreme transparency and high public visibility of this difference is likely to contribute to further brutalization of the conflict. The apparent indifference and visible greed of the new elites versus the hopelessness and rage of the poor are quite manifest in such things as rap music and the behavior of skinheads.

Large cities around the world are the terrain where a multiplicity of globalization processes assume concrete, localized forms. These localized forms are, in good part, what globalization is about. If we consider, further, that large cities also concentrate a growing share of disadvantaged populations – immigrants in Europe and the United States, African Americans and Latinos in the United States, masses of shanty dwellers in the megacities of the developing world – then we can see that cities have become a strategic terrain for a whole series of conflicts and contradictions (see Body-Gendrot 1993; King 1996).

CONCLUSION

A focus on place allows us to capture some of the new spatial configurations and place-specific characteristics that economic globalization produces and by which it is in turn shaped. Processes of economic globalization are thereby reconstituted as concrete production complexes situated in specific places containing a multiplicity of activities. This has the effect of adding to the common focus on the power of large corporations over governments and economies, a focus on the range of activities and organizational arrangements necessary for the implementation and maintenance of a global network of factories, service operations, and markets. These are all processes only partly encompassed by the activities of transnational corporations and global markets.

Focusing on cities and communities allows us to specify a geography of strategic places at the global scale as well as to specify the microgeographies and politics unfolding within these places. This in turn allows us to foreground the new class formations and power alignments emerging from the new forms of capital mobility and the new technologies through which production and work are organized.

REFERENCES FROM THE READING

Abu-Lughod, J.L. (1995) Comparing Chicago, New York and Los Angeles. In P.L. Knox and P.J. Taylor (eds) *World Cities in a World-system*. Cambridge: Cambridge University Press, 179–191.

Body-Gendrot, S. (1993) *Ville et violence*. Paris: Presses Universitaires de France.

Castells, M. (1989) *The Informational City*. London: Blackwell.

Fainstein, S. (1993) *The City Builders*. Oxford: Blackwell.

Friedmann, J. (1995) Where we stand: a decade of world city research. In P.L. Knox and P.J. Taylor (eds) *World Cities in a World-system*. Cambridge: Cambridge University Press, 21–47.

King, A.D. (ed.) (1996) *Representing the City*. London: Macmillan.

Knox, P.L. and Taylor, P.J. (eds) (1995) *World Cities in a World-system*. Cambridge: Cambridge University Press.

Sassen, S. (1988) *The Mobility of Capital and Labor*. New York: Cambridge University Press.

Sassen, S. (1994) *Cities in a World Economy*. Thousand Oaks, CA: Pine Forge.

Sassen, S. (2002 [1991]) *The Global City: New York, London, Tokyo*, 2nd edition. Princeton, NJ: Princeton University Press.

10

"Locating Cities on Global Circuits"

from *Environment and Urbanization* (2002)

Saskia Sassen

Editors' introduction

We now consider a more recent contribution by Saskia Sassen. Building upon her work in *The Global City*, Sassen here develops a more sustained analysis of specialized service industries and advanced digital technologies within global city economies. First, she reiterates her central contention that global city economies contain a broad complex of specialized service industries that enable transnational corporations to coordinate production, investment and finance on a world scale. For Sassen, therefore, the worldwide geographical dispersion of production is linked intrinsically to an increasing centralization of key command and control capacities within the agglomeration economies of global cities. Whereas she previously focused her research upon core cities such as New York, London and Tokyo, Sassen now argues that these trends are becoming increasingly visible in many cities located in developing countries as well (see also Sassen 2002). Second, Sassen explores one of the most perplexing paradoxes of contemporary global urbanization – namely, the consolidation of a small number of strategically dominant global financial centers even as the proliferation of new informational technologies would appear to render such places obsolete. Sassen elaborates a multifaceted explanation of this paradox that emphasizes the various organizational and logistical advantages that flow from the intensive patterns of spatial agglomeration within global cities.

A key feature of the contemporary global system is that it contains both the capability for enormous geographic dispersal and mobility as well as pronounced territorial concentrations of resources necessary for the management and servicing of that dispersal. The management and servicing of much of the global economic system takes place in a growing network of global cities and cities that might best be described as having global city functions. The expansion of global management and servicing activities has brought with it a massive upgrading and expansion of central urban areas, even as large portions of these cities fall into deeper poverty and infrastructural decay. While this role involves only certain components of urban economies, it has contributed to a repositioning of cities both nationally and globally.

WORLDWIDE NETWORKS AND CENTRAL COMMAND FUNCTIONS

The geography of globalization contains both a dynamic of dispersal and of centralization. The massive trend towards the spatial dispersal of economic activities at the metropolitan, national and global level which we associate with globalization has contributed to a demand for new forms of territorial centralization of top-level management and control functions. Insofar as these functions benefit from agglomeration economies, even in the face of telematic integration of a firm's globally-dispersed manufacturing and service operations, they tend to locate in cities. This raises a question as to why they should benefit from agglomeration economies, especially since globalized economic sectors tend to be intensive users of the new telecommunications and computer technologies. In my book, *The Global City* (Sassen 1991), I have found that the key variable contributing to the spatial concentration of central functions and associated agglomeration economies is the extent to which this dispersal occurs under conditions of concentration of control, ownership and profit appropriation.

This dynamic of simultaneous geographic dispersal and concentration is one of the key elements in the organizational architecture of the global economic system. I will first give some empirical referents and then examine some of the implications for theorizing the impacts of globalization and the new technologies on cities.

The rapid growth of affiliates illustrates the dynamic of simultaneous geographic dispersal and concentration of a firm's operations. By 1999, firms had well over half a million affiliates outside their home countries, accounting for US$ 11 trillion in sales, a very significant figure if we consider that global trade stood at US$ 8 trillion. Firms with large numbers of geographically-dispersed factories and service outlets face massive new needs for central coordination and servicing, especially when their affiliates involve foreign countries with different legal and accounting systems.

Another current instance of this negotiation between a global cross-border dynamic and territorially-specific sites is that of the global financial markets. The orders of magnitude in these transactions have risen sharply, as illustrated by the US$ 65 trillion in the value of traded derivatives, a major component of the global economy. These transactions are partly embedded in electronic systems that make possible the instantaneous transmission of money and information around the globe, and much attention has been paid to these new technologies. But the other half of the story is the extent to which the global financial markets are located in an expanding network of cities, with a disproportionate concentration in cities of the global North. Indeed, the degrees of concentration internationally and within countries are unexpectedly high for an increasingly globalized and digitized economic sector. Within countries, the leading financial centers today concentrate a greater share of national financial activity than even ten years ago, and internationally, cities in the global North concentrate well over half of the global capital market.

The specific forms assumed by globalization over the last decade have created particular organizational requirements. The emergence of global markets for finance and specialized services, the growth of investment as a major type of international transaction, all have contributed to the expansion in command functions and in the demand for specialized services for firms. By central functions, I do not only mean top-level headquarters but, rather, all the top-level financial, legal, accounting, managerial, executive and planning functions necessary to run a corporate organization operating in more than one country, and increasingly in several countries. These central functions are partly embedded in headquarters, but also in good part in what has been called the corporate services complex, that is, the network of financial, legal, accounting and advertising firms that handle the complexities of operating in more than one national legal system, national accounting system, advertising culture, etc., and do so under conditions of rapid innovation in all these fields. Such services have become so specialized and complex, that headquarters increasingly buy them from specialized firms rather than producing them in-house. These agglomerations of firms producing central functions for the management and coordination of global economic systems are disproportionately concentrated in the highly-developed countries – particularly, though not exclusively, in global cities. Such concentrations of functions represent a strategic factor in the organization of the global economy and they are situated in an expanding network of global cities.

National and global markets, as well as globally-integrated organizations, require central places where the work of globalization gets done. Finance and advanced corporate services are industries producing the organizational commodities necessary for the implementation and management of global economic systems. Cities are preferred sites for the production of these services, particularly the most innovative, speculative, internationalized service sectors. Further, leading firms in information industries require a vast physical infrastructure containing strategic nodes with hyper-concentration of facilities. Finally, even the most advanced information industries have a production process that is at least partly place-bound because of the combination of resources it requires even when the outputs are hyper-mobile.

Capital mobility cannot be reduced simply to that which moves nor can it be reduced to the technologies that facilitate movement. Rather, multiple components of what we keep thinking of as capital fixity are actually components of capital mobility. This conceptualization allows us to reposition the role of cities in an increasingly globalizing world, in that they contain the resources that enable firms and markets to have global operations. The mobility of capital, whether in the form of investments, trade or overseas affiliates, needs to be managed, serviced and coordinated. These are often rather place-bound, yet are key components of capital mobility. Finally, states – place-bound institutional orders – have played an often crucial role in producing regulatory environments that facilitate the implementation of cross-border operations for their national firms and for foreign firms, investors and markets (Sassen, forthcoming).

In brief, a focus on cities makes it possible to recognize the anchoring of multiple cross-border dynamics in a network of places, prominent among which are cities, particularly global cities or those with global city functions. This, in turn, anchors various features of globalization in the specific conditions and histories of these cities, in their variable articulations with their national economies and with various world economies across time and place. This optic on globalization contributes to identifying a complex organizational architecture which cuts across borders and is both partly deterritorialized and partly spatially concentrated in cities. Further, it creates an enormous research agenda in that every particular national or urban economy has its specific and partly inherited modes of articulating with current global circuits. Once we have more information about this variance, we may be able also to establish whether position in the global hierarchy makes a difference, and the various ways in which it might do so.

THE INTERSECTION OF SERVICE INTENSITY AND GLOBALIZATION

To understand the new or sharply expanded role of a particular kind of city in the world economy since the early 1980s, we need to focus on the intersection of two major processes. The first is the sharp growth in the globalization of economic activity, which has raised the scale and the complexity of transactions, thereby feeding the growth of top-level multinational headquarter functions and the growth of advanced corporate services. The second process we need to consider is the growing service intensity in the organization of all industries. This has contributed to a massive growth in the demand for services by firms in all industries, from mining and manufacturing to finance and consumer services industries with mixed business and consumer markets; they are insurance, banking, financial services, real estate, legal services, accounting and professional associations. Cities are key sites for the production of services for firms. Hence, the increase in service intensity in the organization of all industries has had a significant growth effect on cities, beginning in the 1980s and continuing today.

In the case of cities that are major international business centers, we are seeing the formation of a new urban economy. This is so in at least two regards. First, even though these cities have long been centers for business and finance, since the late 1970s there have been dramatic changes in the structure of the business and financial sectors, as well as sharp increases in the overall magnitude of these sectors and their weight in the urban economy. Second, the ascendance of the new finance and services complex, particularly international finance, engenders what may be regarded as a new economic regime, that is, although this sector may account for only a fraction of the economy of a city, it imposes itself on that larger economy. Most notably, the possibility for superprofits in finance has the effect of devaloriz-ing manufacturing insofar as the latter cannot generate the superprofits typical in much financial activity.

This is not to say that everything in the econ-omy of these cities has changed. On the contrary, they still show a great deal of continuity and many similarities with cities that are not global nodes. Rather, the implantation of global processes and markets has meant that the internationalized sector of the economy has expanded sharply and has imposed a new valorization dynamic – that is, a new set of criteria for valuing or pricing various economic activities and outcomes. This has had devastating effects on large sectors of the urban economy. High prices and profit levels in the internationalized sector and its ancillary activities, such as top-of-the-line restaurants and hotels, have made it increasingly difficult for other sectors to compete for space and investments. Many of these other sectors have experienced considerable downgrading and/or displacement as, for example, neighborhood shops tailored to local needs are replaced by upscale boutiques and restaurants catering to new high-income urban elites.

Although of a different order of magnitude, these trends also became evident, beginning in the late 1980s and early 1990s, in a number of major cities in low- and middle-income nations that have become integrated into various world markets: São Paulo, Buenos Aires, Bangkok, Taipei, Shanghai, Manila, Beirut and Mexico City are a few examples. Here also, the new urban core was fed by the deregulation of various economic sectors, the ascendance of finance and specialized services, and integration into the world markets. The opening of stock markets to foreign investors and the privatization of what were once public sector firms have been crucial institutional arenas for this articulation. Given the vast size of some of these cities, the impact of this new core on the broader city is not always as evident as in central London or Frankfurt, but the transformation is still very real.

It is important to recognize that manufacturing remains a crucial sector in all these economies, even when it may have ceased to be a dominant sector in major cities. Indeed, several scholars have argued that the producer services sector could not exist without manufacturing (Cohen and Zysman 1987). A key proposition for these authors is that producer services are dependent on a strong manufacturing sector in order to grow. There is considerable debate around this issue. I argue that manufacturing indeed feeds the growth of the pro-ducer services sector, but that it does so whether located in the area in question, somewhere else in the country or overseas (Sassen 1991). Even though manufacturing plants – and mining and agriculture, for that matter – feed growth in the demand for producer services, their actual location is of secondary importance in the case of global level service firms: thus, whether manufacturing plants are located offshore or within a country may be quite irrelevant as long as it is part of a multinational corporation likely to buy the services from those top-level firms. Second, the territorial dispersal of factories, especially if international, actually raises the demand for producer services. This is yet another meaning, or consequence, of globaliza-tion: the growth of producer services firms head-quartered in New York or London or Paris can be fed by manufacturing located anywhere in the world as long as it is part of a multinational corporate network. Third, a good part of the producer services sector is fed by financial and business transactions that either have nothing to do with manufacturing, as is the case in many of the global financial markets, or for which manufactur-ing is incidental, as in much merger and acquisition activity, which is centered on buying and selling firms rather than on buying manufacturing firms as such.

THE LOCATIONAL AND INSTITUTIONAL EMBEDDEDNESS OF GLOBAL FINANCE

Several of the issues discussed thus far assume particularly sharp forms in the emerging global network of financial centers. The global financial system has reached levels of complexity that require the existence of a cross-border network of financial centers to service the operations of global capital. This network of financial centers will increasingly differ from earlier versions of the international financial system. In a world of largely closed national financial systems, each country duplicated most of the necessary functions for its economy; collaborations among different national financial markets were often no more than the execution of a given set of operations in each of the countries involved, as in clearing and settlement. With few exceptions, such as the off shore markets and some of the large banks, the international system consisted of a string of closed domestic systems.

The global integration of markets pushes towards the elimination of various redundant systems and makes collaboration a far more complex matter, one which has the effect of raising the division of labor within the network. Rather than each country having its own center for global operations, the tendency is towards the formation of networks and strategic alliances with a measure of specialization and division of functions. This may well become a system with fewer strategic centers and more hierarchy, even as it adds centers. In this context, London and New York, with their enormous concentrations of resources and talent, continue to be powerhouses in the global network for the most strategic and complex operations for the system as a whole.

The financial centers of many countries around the world are increasingly fulfilling gateway functions for the circulation in an out of national and foreign capital. The incorporation of a growing number of these financial centers is one form through which the global financial system expands: each of these centers is the nexus between that country's wealth and the global market, and between foreign investors and that country's investment opportunities.

WHY THE NEED FOR CENTERS IN THE DIGITAL ERA?

What really stands out in the evidence for the global financial industry is the extent to which there is a sharp concentration of the shares of many financial markets in a few financial centers. London, New York, Tokyo (notwithstanding a national economic recession), Paris, Frankfurt and a few other cities regularly appear at the top and represent a large share of global transactions. London, followed by Tokyo, New York, Hong Kong and Frankfurt account for a major share of all international banking. London, Frankfurt and New York account for an enormous world share in the export of financial services. London, New York

and Tokyo account for over one-third of global institutional equity holdings, this as of the end of 1997 after a 32 per cent decline in Tokyo's value over 1996. London, New York and Tokyo account for 58 per cent of the foreign exchange market, one of the few truly global markets; and together with Singapore, Hong Kong, Zurich, Geneva, Frankfurt and Paris, they account for 85 per cent of this, the most global of markets.

Why is it that at a time of rapid growth in the network of financial centers, in overall volumes and in electronic networks, we have such high concentration of market shares in the leading global and national centers? Both globalization and electronic trading are about expansion and dispersal beyond what had been the confined realm of national economies and floor trading. Indeed, one might well ask why financial centers matter at all. The continuing weight of major centers is, in a way, counter sensical, as is, for that matter, the existence of an expanding network of financial centers. The rapid development of electronic exchanges, the growing digitalization of much financial activity, the fact that finance has become one of the leading sectors in a growing number of countries, and that it is a sector that produces a dematerialized, hyper-mobile product, all suggest that location should not matter. In fact, geographic dispersal would seem to be a good option given the high cost of operating in major financial centers.

There are, in my view, at least three reasons that explain the trend towards consolidation in a few centers rather than massive dispersal. I developed this analysis in *The Global City* (1991), focusing on New York, London and Tokyo, and since then events have made this even clearer and more pronounced.

Social connectivity and central functions

While the new communications technologies do indeed facilitate geographic dispersal of economic activities without losing system integration, they have also had the effect of strengthening the importance of central coordination and control functions for firms and, even, markets. Indeed, for firms in any sector, operating a widely-dispersed network of branches and affiliates and operating in multiple markets has made central functions far more complicated. Their execution requires access to top talent, not only inside headquarters but also, more generally, from innovative milieux – in technology, accounting, legal services, economic forecasting and all sorts of other, many new, specialized corporate services. Major centers have massive concentrations of state-of-the-art resources that allow them to maximize the benefits of the new communication technologies and to govern the new conditions for operating globally. Even electronic markets such as NASDAQ and E*Trade rely on traders and banks which are located somewhere, with at least some in a major financial center.

One fact that has become increasingly evident is that to maximize the benefits of the new information technologies, firms need not only the infrastructure but a complex mix of other resources. Most of the added value that these technologies can produce for advanced services firms lies in so-called externalities; and this means the material and human resources – state-of-the-art office buildings, top talent and the social networking infrastructure that maximizes connectivity. Any town can have fibre optic cables, but this is not sufficient.

Cross-border mergers and alliances

Global players in the financial industry need enormous resources, which is leading to rapid mergers and acquisitions of firms, and strategic alliances between markets in different countries. These are taking place on a scale and in combinations few would have foreseen just three or four years ago. There are growing numbers of mergers among, respectively, financial services firms, accounting firms, law firms, insurance brokers, in brief, firms that need to provide a global service. A similar evolution is also possible for the global telecommunications industry, which will have to consolidate in order to offer a state-of-the-art, globe-spanning service to its global clients, among which are the financial firms.

These developments may well ensure the consolidation of a stratum of select financial centers at the top of the worldwide network of 30 or 40 cities through which the global financial industry operates. We now also know that a major financial center needs to have a significant share of global

operations to become such. If Tokyo does not succeed in getting more of these operations, it is going to lose standing in the global hierarchy, notwithstanding its importance as a capital exporter. It is this same capacity for global operations that will keep New York at the top levels of the hierarchy even though it is largely fed by the resources and the demands of domestic (although state-of-the-art) investors.

In brief, the need for enormous resources to handle increasingly global operations, in combination with the growth of central functions described earlier, produces strong tendencies towards concentration and hence hierarchy in an expanding network.

Denationalized elites and agendas

Finally, national attachments and identities are becoming weaker for these global firms and their customers. Thus, the major US and European investment banks have set up specialized offices in London to handle various aspects of their global business. Even French banks have set up some of their global specialized operations in London, inconceivable even a few years ago. Deregulation and privatization have further weakened the need for national financial centers. The nationality question simply plays differently in these sectors than it did even a decade ago. Global financial products are accessible in national markets and national investors can operate in global markets. For instance, some of the major Brazilian firms now list on the New York Stock Exchange and bypass the São Paulo exchange.

One way of describing this process is as an incipient denationalization of certain institutional arenas (Sassen, forthcoming). It can be argued that such denationalization is a necessary condition for economic globalization as we know it today. The sophistication of this system lies in the fact that it only needs to involve strategic institutional areas – most national systems can be left basically unaltered. Major international business centers produce what we could think of as a new subculture, a move from the "national" version of international activities to the "global" version. The longstanding resistance in Europe to mergers and acquisitions, especially hostile takeovers, or to foreign ownership and control in East Asia, signal national business cultures that are somewhat incompatible with the new global economic culture. I would posit that major cities, and the variety of so-called global business meetings (such as those of the World Economic Forum in Davos), contribute to denationalizing corporate elites. Whether this is good or bad is a separate issue, but it is, I would argue, one of the conditions for setting in place the systems and sub-cultures necessary for a global economic system.

CONCLUSION

Economic globalization and telecommunications have contributed to producing a spatiality for the urban which pivots on cross-border networks and territorial locations with massive concentrations of resources. This is not a completely new feature. Over the centuries, cities have been at the crossroads of major, often worldwide, processes. What is different today is the intensity, complexity and global span of these networks, the extent to which significant portions of economies are now dematerialized and digitalized and hence the extent to which they can travel at great speeds through some of these networks, and the numbers of cities that are part of cross-border networks operating on vast geographic scales.

REFERENCES FROM THE READING

Cohen, S. and Zysman, J. (1987) *Manufacturing Matters*. New York: Basic Books.

Sassen, S. (1991) *The Global City*. Princeton, NJ: Princeton University Press.

Sassen, S. (ed.) (2002) *Global Networks, Linked Cities*. New York: Routledge.

Sassen, S. (forthcoming) *Denationalization*. Princeton, NJ: Princeton University Press.

T
W
O

11

"World-city Network: A New Metageography?"

from *Annals of the Association of American Geographers* (2000)

Jonathan V. Beaverstock, Richard G. Smith and Peter J. Taylor

Editors' introduction

As we saw in Part One of this Reader, early theorists of global city formation focused on transnational headquarters locations as the key indicator in terms of which the emergent hierarchy of global cities could be delineated (see Reading 5 by Cohen and Reading 7 by Friedmann). However, in the wake of Sassen's research on the producer and financial services complex in global cities, a number of scholars began to explore alternative approaches to the study of the global urban system. One of the most sophisticated efforts to analyze the global city hierarchy has been developed by a group of scholars based at Loughborough University in the UK, through an innovative research network known as GaWC (Globalization and World Cities: for further discussion see the group's website at http://www.lboro.ac.uk/gawc/). Led by Peter J. Taylor, a prominent radical political geographer and urbanist, the GaWC group has generated a variety of new data sources for the study of global city formation while also developing some extremely innovative methodological strategies for analyzing that data. Taylor, in particular, has also grappled extensively with the problem of *theorizing* about global cities and global urban networks (for a synthesis of his arguments, see Taylor 2003).

Reading 11 represents a relatively early, programmatic statement by three core members of the GaWC group that was published in the millennial issue of a major US geography journal. The reading opens by questioning the inherited, territorialist "metageography" of mainstream social science which conceives the world in terms of bordered, state-defined containers of political-economic life. Instead, the authors argue for greater attention to inter-city networks – which are said to be based upon flows, linkages, connections and relations. Such inter-city networks, they argue, have acquired unprecedented importance under contemporary globalization; yet our knowledge of their developmental trajectories and geographical contours remains seriously underdeveloped. While the GaWC researchers acknowledge their debt to the first wave of global cities research, they also criticize established approaches for a tendency to focus on the fixed *attributes* of global cities (such as the number of transnational headquarter locations) rather than examining the changing *relations* between globally interlinked urban centers. In this chapter, the authors introduce a relational approach to the study of the global urban system by examining the worldwide office networks of 74 major producer and financial services firms, and through a case study of the global networks of major London-based producer and financial services firms. This innovative methodology opens up a radically new theoretical and empirical perspective on the global urban system.

During the Apollo space flights, it was reported that one of the astronauts, looking back to Earth, expressed amazement that he could see no boundaries. This new view of our world as the "blue planet" contradicted the taken-for-granted, state-centric Ptolemaic model or image of world-space that most modern people carry around in their heads: a world of grids, graticule, and territorial boundaries (Cosgrove 1994). As a further jolt to the arrogance of modernity, it was soon accepted as a truism that the only "man-made" artifact visible from space was the ancient Great Wall of China. Interestingly, however, the Great Wall is not the only visible feature: at night, modern settlements are clearly visible as pin-pricks of electric light on a black canvas. The globality of modern society is clear for all to see in the photo prints, communicated back to Earth, of lights delimiting a global pattern of cities, consisting of a broad swath girdling the mid-latitudes of the northern hemisphere plus many oases of light elsewhere.

The fact that these "outside views" of Earth identified a world-space of settlements rather than the more familiar world-space of countries has contributed to the growth of contemporary "One World" rhetoric (also "Spaceship Earth" or "Whole Earth") which has culminated in "border-less world" theories of globalization. Of course, geographies do not depend solely upon visibility or metaphors. The fact that state boundaries are missing from space-flight photographs tells us nothing, therefore, about the current power of states to affect world geography. The photographs can, however, influence "metageography," or the "spatial structures through which people order their knowledge of the world" (Lewis and Wigen 1997: ix). In the modern world, this has been notably Eurocentric and state-based in character. It is this mosaic spatial structure that the night-time photographs challenge since, first and foremost, people live in settlements. In this chapter, we consider the largest pinpricks of light, "the world cities" whose transnational functions materially challenge states and their territories. These cities exist in a world of flows, linkages, connections, and relations. World cities represent an alternative metageography, one of networks rather than the mosaic of states.

Historically, cities have always existed in environments of linkages, both material flows and information transfers. They have acted as centers from where their hinterlands are serviced and connected to wider realms. This is reflected in how economic geographers have treated economic sectors: primary and secondary activities are typically mapped as formal agricultural or industrial regions, tertiary activities as functional regions, epitomized by central-place theory. Why is our concern for contemporary cities in a world of flows any different from this previous tertiary activity and its study? First, the twentieth century has witnessed a remarkable sectoral turnabout in advanced economies: originally defined by their manufacturing industry, economic growth has become increasingly dependent on service industries. Second, this trend has been massively augmented by more recent developments in information technology that has enabled service and control to operate not only more rapidly and effectively, but crucially on a global scale. Contemporary world cities are an outcome of these economic changes. The large electric pinpricks of light on space photos are actually connected by massive electronic flows of information, a new functional space that will be crucial to geographical understanding in the new millennium.

This chapter reports preliminary research on the empirical groundwork required for describing the new metageography of relations between world cities. Such a modest goal is made necessary because of a critical empirical deficit within the world-city literature on intercity relations: studies of world cities are generally full of information that facilitates evaluations of individual cities and comparative analyses of several cities; yet, the data upon which these analyses are based has been overwhelmingly derived from measures of city attributes. Such information is useful for estimating the general importance of cities and for studying intra-city processes, but it tells us nothing directly about relations between cities. Hence cities can be ranked by attributes, but a hierarchical ordering aimed at uncovering flows or networks requires a different type of data based upon measures of relations between cities (Taylor 1997, 1999). It is the dearth of relational data that is the "dirty little secret" (Short et al. 1996) of this research area. In other words, we know about the nodes but not the links in this new metageography.

Our particular solution to this data problem, for purposes of this chapter, is to focus on the global office-location strategies of major corporate-service firms. After outlining this data-collection exercise, we analyze the resultant data in two ways: the first defines a network; the second deals with the global relations of a single city, London. We claim both of these analyses to be unique, first empirical studies of their kind. In a brief conclusion, we consider the future implications of this new metageography: are we witnessing a dystopia in the making?

GLOBAL OFFICE LOCATION STRATEGIES

The only published data available for studying relations between cities at a global scale are international airline-passenger statistics. Not surprisingly, therefore, empirical studies that present *networks* of world cities have focused upon this source (Keeling 1995). There are, however, serious limitations to these statistics as descriptions of relations between world cities: first, the information includes much more than trips associated with world-city processes (e.g., tourism), and second, important intercity trips within countries are not recorded in international data (e.g., New York–Toronto does feature in the data, New York–Los Angeles does not). While the latter can be overcome by augmenting the data with domestic flight statistics, the particularities of hub-and-spoke systems operated by airlines creates another important caveat to using this data to describe the world-city network.

Studying the global location strategies of advanced producer-service firms is an alternative approach for describing world-city networks, one which overcomes these problems. Firms that provide business services on a global scale have to decide on the distribution of their practitioners and professionals across world cities. Setting up an office is an expensive undertaking, but a necessary investment if the firm believes that a particular city is a place where it must locate in order to fulfill its corporate goals. Hence the office geographies of advanced producer firms provide a strategic insight into world-city processes by interpreting intrafirm office networks as intercity relations. In this argument, world-city network formation con-

sists of the aggregate of the global location strategies of major, advanced producer-service firms.

Information on the office networks of firms can be obtained by investigating a variety of sources, such as company web sites, internal directories, handbooks for customers, and trade publications. We have collected data on the distributions of offices for 74 companies (covering accountancy, advertising, banking/finance, and commercial law) in 263 cities. An initial analysis of this data identified the 143 major office centers in these cities, and 55 of these were designated world cities on the basis of the number, size, and importance of their offices (for details of this classification exercise, see Beaverstock et al. 1999a). No other such roster of world cities exists; it is used here as the basic framework for studying the world-city network.

AN INTERCITY GLOBAL NETWORK

The roster of 55 world cities is divided into three levels of service provision comprising 10 Alpha cities, 10 Beta cities, and 35 Gamma cities. Only the Alpha cities – Chicago, Frankfurt, Hong Kong, London, Los Angeles, Milan, New York, Paris, Singapore, and Tokyo – are used in this section to illustrate how office geographies can define intercity relations. Note the geographical spread of these top 10 world cities; they are distributed relatively evenly across three regions we have previously identified as the major "globalization arenas": the U.S., western Europe, and Pacific Asia (Beaverstock et al. 1999b). World-city network patterns are constructed for these Alpha world cities, using simple presence/absence data for the largest 46 firms in the data (all of these firms have offices in 15 or more different cities).

Shared presences are shown in Table 1. Each cell in this intercity matrix indicates the number of firms with offices in both cities. Thus, London and New York "share" 45 of the 46 firms; only one firm in the data does not have offices in both of these cities. Obviously these two cities are the places to be for a corporate-service firm with serious global pretensions. This finding is not, of course, at all surprising; interest comes when lower levels of intercity relations are explored.

In Figure 1, the highest 20 shared presences are depicted at two levels of relation.

Table 1 Relations between Alpha world cities: shared firm presences, number of firms with offices in both cities

	CH	FF	HK	LN	LA	ML	NY	PA	SG	TK
Chicago										
Frankfurt	21									
Hong Kong	21	30								
London	23	32	38							
Los Angeles	21	23	29	33						
Milan	19	28	29	32	22					
New York	23	32	38	45	32	32				
Paris	21	30	32	35	27	28	34			
Singapore	20	30	34	35	26	29	35	32		
Tokyo	23	30	34	37	30	29	37	32	32	

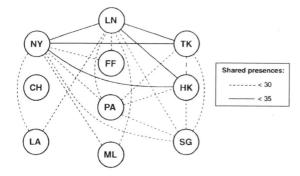

Shared presences:
----- < 30
——— < 35

Figure 1 Shared presences among Alpha world cities

The higher level picks out Sassen's (1991) trio of global cities – London, New York, and Tokyo – as a triangular relationship (but note that, in addition, Hong Kong has such a relationship with London and New York). Bringing in the lower level of relations, London and New York have shared presences with eight other cities in all, but note again the high Pacific Asia profile in this data: Singapore joins with Tokyo and Hong Kong in showing relations with five other cities, the same level as Paris. This contrasts with the U.S. world cities below New York; Los Angeles is in the next-to-bottom class of shared presences with Frankfurt and Milan, and Chicago stands alone, with no intercity relations at the minimum level for inclusion in the diagram. This pattern can be interpreted in terms of the different degrees of political fragmentation in the three major globalization arenas. In the most fragmented, Pacific Asia, there is no dominant world city, so that presences are needed in at least three cities to cover the region:

Hong Kong for China, Singapore for southeast Asia, and Tokyo for Japan. In contrast, the U.S. consists of a single state such that one city can suffice for a presence in that market. The result is that New York throws a shadow effect over other U.S. cities. In between, western Europe is becoming more unified politically, but numerous national markets remain so that London does not dominate its regional hinterland to the same degree as New York.

Shared presences define a symmetric matrix that shows sizes, but not the direction, of intercity relations. By contrast, Table 2 is an asymmetric matrix showing probabilities of connections.

Each cell contains the probability that a firm in city A will have an office in city B. Thus, Table 2 shows that if you do business with a Chicago-based firm, then there is a 0.91 probability that that firm will also have an office in Frankfurt. On the other hand, go to a Frankfurt-based firm, and the probability of it having an office in Chicago is only 0.66. Such asymmetry is represented by vectors in Figures 2 and 3.

Primary vectors are defined by probabilities above 0.95. Note that all cities connect to London and New York at this level (Figure 2). As in Figure 1, only Tokyo and Hong Kong reach this highest category of connection, but each with only one link. Again, it is also interesting to look at the lower level relations, and these are shown in Figure 3. This diagram reinforces the interpretation concerning the three globalization arenas presented above: Chicago and Los Angeles have no inward vectors from the other arenas in what is largely a Eurasian pattern of connections. Vectors to the Pacific

Table 2 Matrix of office-presence linkage indices for Alpha world cities

					Linkage to					
	CH	FF	HK	LN	LA	ML	NY	PA	SG	TK
Chicago		89	89	100	91	79	100	89	83	100
Frankfurt	67		93	100	72	87	100	95	94	95
Hong Kong	60	82		100	80	80	100	85	92	90
London	59	77	87		78	78	98	83	83	86
Los Angeles	67	73	89	100		70	97	84	81	89
Milan	59	88	93	100	67		100	88	91	93
New York	59	77	87	98	77	77		79	83	85
Paris	64	85	90	100	80	81	97		90	90
Singapore	60	87	98	100	78	83	100	92		95
Tokyo	64	84	93	100	83	81	100	87	88	

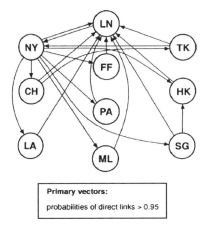

Figure 2 Primary vectors (probabilities of links) among Alpha world cities

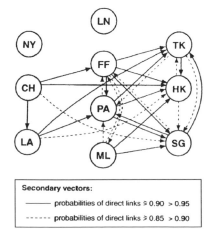

Figure 3 Secondary vectors (probabilities of links) among Alpha world cities

Asian cities dominate, but Frankfurt and Paris also have a reasonable number of inward vectors.

This is the first time intercity relations on a global scale have been studied in this way. As expected of such initial research, several opportunities for further investigations are suggested, not least using more cities and more sophisticated network analysis to tease out further features of the contemporary world-city network. But the most urgent task is to go beyond this cross-sectional analysis and study changes over time in order to delineate the evolution of world-city network formation. Only in this way will we be able to make informed assessments of how the network will develop in the new millennium and how this will affect different cities. For instance, is the New York shadow effect growing or declining? We simply do not know.

CASE STUDY: LONDON'S GLOBAL REACH

There is no published study assessing the global capacity of a world city in terms of its relations with other world cities. The producer-service-office geography dataset is particularly suited for such an exercise; here we illustrate this with a brief case study of London (Figure 4).

The data we employ for London differs from that used in the last section in three ways. First, it is obvious that since we will consider only London-

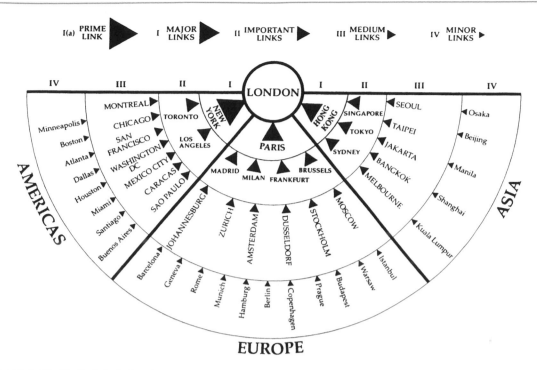

Figure 4 World city links to London

based firms, one of the firms used previously is dropped. In addition, we add data for smaller London-based firms, creating a total of 69. Second, we consider all 55 world cities in our roster. Third, for many firms, there is richer information than simply whether they are present or absent in a city. Further information provides interval-level measurements, on the numbers of practitioners or professionals employed by a firm across all its offices, as well as ordinal-level measurements in which the importance of offices was allocated to ranked classes on the basis of given functions. In order to combine this data into a single, comparable set of measures, all three levels – interval, ordinal, and nominal (presence/absence) – were combined as a single ordinal scale. For every world city, each firm is scored as one of the following: (0) indicating absence; (1) indicating presence, or where additional information is available, indicating only minor presence; (2) when additional information indicates a medium presence in a city; and (3) when additional information indicates a major presence. In delineating these data when additional information is available, we were careful to be

sensitive to the range of data; for large accountancy firms, for example, "minor presence" was defined as less than 20 practitioners in a city, while "major presences" required more than 50 professionals; yet, for law firms, the equivalent figures were 10 and 20. Through this approach, we can move beyond simple geographies of presence to geographies of the *level* of producer services available in a city.

For each of the producer-service sectors represented in our data, levels of service are summed for London-based firms in each of the other 54 world cities. This provides an estimate of the level of external service that can be expected when doing business in another world city from London. In Table 3, the top 10 world cities are ranked in terms of service available for each of the producer services represented in our data. As would be expected, the Alpha cities, identified in the last section, figure prominently in these rankings, with New York first or first equal in all four sectors. Yet, other world cities now make an appearance: the notable examples are the key political cities of Washington, DC and Brussels featuring prominently in law,

Table 3 Top ten office linkages to London by advanced producer services

Accountancy	Advertising	Finance	Law services	Average
1. Düsseldorf, New York, Paris, Tokyo, Toronto	1. New York 2. Brussels, Madrid, Sydney, Toronto	1. New York 2. Singapore 3. Hong Kong, Tokyo	1. New York 2. Washington 3. Brussels, Hong Kong	1. New York 2. Paris 4. Hong Kong
6. Chicago, Milan, Sydney, Washington	6. Milan, Paris	5. Frankfurt 6. Paris, Zurich	5. Paris 6. Los Angeles	5. Tokyo, Brussels
10. Atlanta, Brussels, Frankfurt, San Francisco	8. Los Angeles, Singapore, Stockholm	8. Sydney 9. Madrid 10. Milan, Taipei	7. Tokyo 8. Singapore 9. Moscow 10. Frankfurt	6. Singapore 7. Sydney 8. Milan 9. Frankfurt, Los Angeles, Toronto

Dusseldorf easily out-ranking Germany's Alpha world city Frankfurt in accountancy, and Britain's imperial links being represented by Sydney and Toronto in several lists. Average levels of linkage have been computed from standardized sector scores (city totals as percentage of maximum possible; i.e., 3 [maximum score] × number of firms in a given sector), showing all the Alpha cities in London's top 10 except Chicago. Given also Los Angeles's bottom ranking in the list, this can be interpreted as the New York shadow effect operating even from London.

Average levels of linkage with London, when computed for all other world cities, provide an illustration of London's global reach within the world-city network. These average percentages range from a top score of 87 for New York, followed by Paris (68) and Hong Kong (64), to the lowest score for Minneapolis of only 15, with Osaka (21) and Munich (22) just above the bottom. Using these averages, world cities can be divided into five groups in terms of the intensity of their relations with London. Out on its own is New York, the "prime link," followed by Paris and Hong Kong as the other two "major links." Below these three and all with scores over 50, come nine "important links." The remainder of the cities are divided between 18 "medium links" (36–50) and the remaining 24 "minor links." These links are arrayed in Figure 3, showing a relatively even distribution across the three globalization arenas and their adjacent regions, the New York shadow effect notwithstanding.

The big question for London, as we enter the next millennium, is whether it can retain its position as Europe's leading world city. With the European Central Bank located in Frankfurt, the scene is set for some intense intercity competition. Currently, as we showed in the previous section, London is clearly preeminent in relation to all its European rivals but, as before, in order to see which way this competition is moving, we need to supplement our cross-sectional analysis with evolutionary data and analysis, and repeat the exercise across several cities.

CONCLUSION: METAGEOGRAPHIC DYSTOPIA?

Riccardo Petrella, sometimes referred to as the "official futurist of the European Union" (1995: 21), has warned of the rise of a "wealthy archipelago of city regions . . . surrounded by an impoverished *lumpenplanet*." He envisages a scenario in which the 30 most powerful city regions (the CR-30) will replace the G-7 (the seven most powerful states), presiding over a new global governance by 2025.

Such a scenario is given credence by the fact that contemporary world cities are implicated in the current polarization of wealth and wages accompanying economic globalization. World-city practitioners and professionals operating in a global labor market have demanded and received "global wages" (largely in the form of bonuses) to create a new income category of the "waged rich."

Petrella sets out his global apartheid dystopia as a warning about current trends so as to alert us to the dangers ahead. But cities do not have to play the *bête noire* role of the future. It is within cosmopolitan cities that cultural tensions can be best managed and creatively developed. Certainly modern states, in their ambition to be nation-states, have an appalling record in dealing with matters of cultural difference. But the key point is that this is not a simple matter of cities versus states (Taylor 2000). World cities are not eliminating the power of states, they are part of a global restructuring which is "rescaling" power relations, in which states will change and adapt as they have done many times in previous restucturings (Brenner 1998). The "renegotiations" going on between London's world role and the nation's economy, between New York's world role and the U.S. economy, and with all other world cities and their encompassing territorial "home" economies, are part of a broader change affecting the balance between networks and territories in the global space-economy. In this chapter, we have illustrated how empirical analysis of city economic networks might be undertaken to complement traditional economic geography's concern for comparative advantage between states. Our one firm conclusion is that in the new millennium, we cannot afford to ignore this new metageography, the world-city network.

REFERENCES FROM THE READING

Beaverstock, J.V., Smith, R.G. and Taylor, P.J. (1999a) A roster of world cities, *Cities*, 16, 6, 445–458.

Beaverstock, J.V., Smith, R.G. and Taylor, P.J. (1999b) The long arm of the law: London's law firms in a globalizing world-economy, *Environment and Planning A*, 31, 10, 1857–1876.

Brenner, N. (1998) Global cities, "glocal" states: global city formation and state territorial restructuring in contemporary Europe, *Review of International Political Economy*, 5, 1–37.

Cosgrove, D. (1994) Contested global visions: one world, whole earth, and the Apollo space photographs, *Annals of the Association of American Geographers*, 84, 270–294.

Keeling, D.J. (1995) Transportation and the world city paradigm. In P.L. Knox and P.J. Taylor (eds) *World Cities in a World-system*. Cambridge: Cambridge University Press, 115–131.

Lewis, M.W. and Wigen, K.E. (1997) *The Myth of Continents: A Critique of Metageography*. Berkeley, CA: University of California Press.

Petrella, R. (1995) A global agora vs. gated city-regions, *New Perspectives Quarterly*, Winter, 21–22.

Sassen, S. (1991) *The Global City*. Princeton, NJ: Princeton University Press.

Short, J.R., Kim, Y., Kuss, M. and Wells, H. (1996) The dirty little secret of world cities research, *International Journal of Urban and Regional Research*, 20, 697–717.

Taylor, P.J. (1997) Hierarchical tendencies amongst world cities: a global research proposal, *Cities*, 14, 323–332.

Taylor, P.J. (1999) The so-called world cities: the evidential structure of a literature, *Environment and Planning A*, 31, 1901–1904.

Taylor, P.J. (2000) World cities and territorial states under conditions of contemporary globalization, *Political Geography*, 19, 1, 5–32.

Taylor, P.J. (2003) *Global City Network*. New York: Routledge.

TWO

12

"Global Cities and Global Classes: The Peripheralization of Labor in New York City"

from *Review* (1983)

Robert Ross and Kent Trachte

Editors' introduction

Robert Ross and Kent Trachte are among the pioneers of global cities research. During the late 1970s and early 1980s at Clark University, Massachusetts, Ross and Trachte began to investigate the degradation of labor conditions and the impoverishment of everyday life for many working people within some of the most globally integrated US metropolitan regions. Their work culminated in an important monograph, entitled *Global Capitalism: The New Leviathan* (1990), which provided a detailed critical examination of various theoretical approaches to the study of global capitalism and a sustained empirical analysis of various US industrial regions – including Detroit, New York City and Massachusetts – that were undergoing globally induced economic crises during the 1970s and early 1980s. In Reading 12, which is excerpted from an article derived from that book, Ross and Trachte focus on the case of New York City, examining, in particular, the shrinking of the manufacturing sector, the expansion of unemployment, the spread of sweatshops and informalized labor arrangements, and the decay of housing conditions for significant segments of the urban working class. According to Ross and Trachte, US workers by the 1980s had become increasingly vulnerable to capitalist exploitation due to the combined impact of internationalized production and declining union power. In this context, Ross and Trachte speak of the "peripheralization of the core" in order to underscore the ways in which the highly degraded conditions of work and social reproduction that were once found primarily in peripheral, "Third World" zones are now being imposed in many core industrial regions of the North. While the causal assumptions underpinning their arguments remain controversial (see Reading 13 by Fainstein), Ross and Trachte's analysis usefully illuminates the ways in which global city formation entails significant, often highly detrimental, transformations of everyday social, living and employment conditions for working people (see also Reading 28 by Buechler).

In the analysis of the global system of capitalism, a special place is reserved for those cities and regions in which are located the command centers of financial and corporate decision-making.

Referred to as global cities, such places are the location of the institutional heights of worldwide resource allocation [. . .] From these headquarters radiate a web of electronic communications and air travel corridors along which capital is deployed and redeployed, and through which the fundamental decisions about the structure of the world-economy are sent [. . .]

The concentration of the institutional forms of capital in banks, financial institutions, and global enterprises, and the centralization of their head-quarters in global cities, are widely understood and noted in the political economy of global capitalism. What is not so widely acknowledged or understood is the paradox, that is, the contradictions, of the existence of such physical concentration of capital and control over it and the conditions of the working-class resident in such places [. . .] Developing an analysis of the class structure, and in particular of the working class, in a global city constitutes the project of this chapter. This must be done from a perspective that attends to the role of global cities within an economic system characterized by the global accumulation of capital. Determination of the conditions of the labor force in a global city cannot limit its conceptual horizons to the urban, regional, or national economies. Capital in these cities operates on a world scale, and therefore capital–labor relations in global cities must be analyzed from a perspective that conceives of capitalism as a global system.

A THEORY OF CAPITALISM AS A SYSTEM OF GLOBAL PRODUCTION

[. . .] We explicitly conceive of a capitalist social formation at a given point in time as the articulation of various forms of capitalism: competitive, monopoly, global. This articulation involves relations of dominance understood as flows of surplus value from subordinate forms of capitalist organization to more dominant ones. In particular, although the economy of a core state or a global city may be dominated by monopoly or global capital, this does not mean that its economy is composed of a homogenous monopoly sector of capital–labor relations. Rather, a competitive sector, in which the conditions of labor are typically and dramatically different than those depicted in images of core regions as a whole, has not disappeared despite its subordination. Labor in this competitive sector is neither well paid nor power-

ful. But, more than that, in the competitive sector, the existence of large pools of unemployed or underemployed workers acts as a depressant on wages and autonomous organization.

Also central to the contemporary era of capitalism are flows of capital that are creating a changed international structure of production. In the core countries a dominant firm in a given market can no longer casually accept wage demands that its monopoly power allows it to pass on to competitive sector capital and labor. The aggressive redeployment of productive capital on a world scale is both a cause and a symptom of structural change. Driven by foreign competition in price or quality terms, even the older nationally based monopolies find their markets invaded and their profitability threatened. They are forced to find ways to cut their costs, to deal with more aggressive foreign governments concerned with development or employment, and to maintain markets conceived in world terms. This has frequently meant the shifting of productive facilities to low-wage sites in the periphery or semiperiphery. In turn, this has had sharp impact on the employment and security of formerly more or less secure core, monopoly sector workers. Thus, [. . .] workers in some of the core countries, among them North America, find themselves in competition for access to capital and to buyers of their labor power across the world span. This competition provides a new tool for capital in its relation to labor, the threat to move.

Specifically, we consider that two related movements of capital are decisively altering the conditions of class struggle in the core. One of these circuits has been the redeployment of productive facilities from core areas to Third World and backward European sites. This first circuit may then reestablish the conditions for profitable accumulation in the older areas and thus a second moment in the circuit of investment. This second circuit favors economic sectors where the working class organization is underdeveloped or especially vulnerable. The entries of at least two kinds of capital, in the wake of the exit of some types of manufacturing capital, are of particular importance for global cities. The first type is attracted by the administrative specialization of the global city; office and janitorial workers, clerks and computer operators are employed in increasing numbers. A

second kind of capital finds niches in the manufacturing structure where low barriers to entry and
a pool of cheap labor allow for high (peripheral) rates
of exploitation. Both forms of capital employ
women, minorities, and other vulnerable or unorganized workers in proportions larger than many
of the jobs that have been lost. Together, these
circuits represent aspects of capital flow and the
emergence of global capitalism as a phase in
the continuing evolution of the capitalist mode of
production.

In the global phase of capitalism, capital has new
instruments in its struggle with labor over the
conditions of production. The employer now
possesses the ability to disaggregate further the production process over space. Research and development can be located in one geographic region,
skilled machining and fabrication in another
region, administration and services still elsewhere,
and semi-skilled assembly in yet another area. The
shifting of production out of a geographic region
leads to unemployment and the growth of a
reserve army in that region. The threat to move to
another production site where wages are lower
becomes a viable mechanism for extracting con-
cessions from the labor force.

The spatial mobility of capital also contributes
to a worsening of the situation of workers in the
sphere of reproduction. Regions experiencing capital outflow suffer an erosion of their tax bases and
are forced to adopt austerity budgets that reduce
the scope of government services and allow for the
deterioration of the physical infrastructure of public transportation, roads, energy, etc. This deterioration then becomes exacerbated as these regions
compete to retain capital investment and attract new
investment through the extension of subsidies to
investors. As a result, the public budget is further
strained and revenues available to government
for other functions are further reduced. Capital
mobility also negatively affects the general living
conditions of the resident labor force.

Once a region has been the victim of capital
outflows, the conditions for some kinds of
profitable accumulation may be reestablished.
First, the bargaining power of the resident labor force
has been weakened. Secondly, the site often
becomes the recipient of labor migration from the
periphery. Both these segments of the labor force
become vulnerable to offers of employment from

new capital investing in sectors where the conditions of production are favorable to capital.

The internationalization of production under
the current phase of capitalism, we contend,
fosters a peripheralization of segments of the
working class in global cities. A reserve army of labor
not only continues to exist but is apparently
expanding. The material conditions of laborers are
not improving, but rather are stagnating or declining. Standards of reproduction of the working
class may be high in comparison to that of the
periphery, but the changing conditions of class
conflict under global capitalism may erode that
comparison. We expect that the acceleration of
international capital flows away from the core
regions and toward the peripheral and semiperipheral formations will erode the bargaining
power and ultimately the level of living of formerly
privileged labor in global cities.

This approach can be applied to an analysis of
New York City. In particular, we contend that
there is a local reserve army of labor in New York
and that it is joined in a global reserve, any
national or local fraction of which capital can
choose to favor with investment. Further, the
restructuring of New York as more specialized in
global administrative functions has a negative
impact on its numerous vulnerable workers [. . .]
We anticipate a complex working class, sizeable
fractions of which are vulnerable to recent structural changes in the deployment of global capital.
These fractions suffer from low levels of wages and
of reproduction and are also threatened by the insecurity implied by growing reserves of unemployed
or marginalized labor.

NEW YORK: THE EMPIRE CITY

Throughout this century New York has been the
financial capital of the nation and, certainly by the
end of the First World War, the world. Indeed, one
of the intentions and triumphs of New York's
financial leadership was the use of the mechanism
of the Federal Reserve Bank System to maintain that
position [. . .] New York specializes in corporations
and banks at the centers of the international command and control network. Yet for large segments
of New York's residents, New York as Empire City
means New York as a place having within it all the

variations in conditions of labor that can be observed on a world scale.

Along with its historic role as the centers of finance and corporate administration, New York was a major industrial location for manufacturing, particularly of non-durable goods [. . .] But the high prices of land, of some labor, and of services and housing, among other factors, first made other U.S. locations preferred, and now make off-shore locations even more desirable.

Operating a vast network of business services, retail and eating places, and the labor-intensive industries, New York has always been a stream of immigrants. Extremes of poverty and wealth are part of the city's history and culture. What is new in the current situation is that what might once have been an accepted standard of minimum reproduction, what Katznelson (1981) has called the "social democratic minimum," now appears to have collapsed.

CIRCUIT ONE CAPITAL MOBILITY AND THE VULNERABILITY OF LABOR IN NEW YORK CITY

We have argued that the first circuit of the global capitalist economy, the movement of capital from core regions and core cities to alternative sites, has altered the structure of power relations between core workers and core capital. Workers have become more vulnerable to the strategies, plans, and interests of globally mobile firms.

Workers in New York City have suffered an extraordinary loss of job opportunities in the manufacturing sector in the last several decades. More than one-half of all jobs in manufacturing, a total of approximately 545,000 employment opportunities, disappeared during 1950–81 [. . .] Overall, the capacity of capital to shift production to alternative sites has eliminated 52 per cent of all manufacturing jobs, and 59 per cent of legal apparel and textile jobs in the Empire City (U.S. Department of Labor 1979).

While jobs in the unionized, manufacturing sector have been decreasing, growth has been occurring in the less unionized, lower-wage competitive sector industries such as retail trade and services. Employment by general merchandise stores, financial insurance and real estate firms,

banks, hospitals, and governmental agencies has been rising. More and more of New York City's visible workers are bank tellers, file clerks, secretaries, janitors, sales clerks, and accountants, while fewer and fewer are skilled machinists, garment workers, and construction workers. A larger percentage of workers in the global city now labor for competitive sector employers. Therefore, they frequently lack protection afforded by unions, and/or they must accept lower wages and face competition from those who are unemployed and underemployed (U.S. Department of Labor 1979).

THE RESERVE ARMY IN NEW YORK CITY

The official unemployment rate in New York City rose from 4.8 per cent to 9.2 per cent over the period 1970–81. As a consequence, by 1980, more than 250,000 people in New York were on the unemployment rolls. Moreover, real unemployment and underemployment, which has been estimated at 50 per cent higher than recorded figures, would yield a reserve army of approximately 400,000 people [. . .]

Two factors account in large measure for the size and growth of the reserve army in New York. First, capital migration has led to the disemployment of many of the resident labor force. Second, immigration, both legal and illegal, has played an important role.

Immigration has also increased the size of the reserve army in New York [. . .] As with the disemployed, not all of the immigrants enter the ranks of the unemployed. Many of the legal migrants obtain employment in the service sector. Businesses such as messenger services, hand laundries, restaurants, gourmet food stores, and repair and domestic services are examples of those that frequently employ legal migrants. Similarly, thousands of illegal migrants are absorbed by the illegal underground economy of sweatshops, gambling, and the like. Nevertheless, migrants, legal and illegal, have swelled the ranks of the reserve army in New York.

New York as a global city has then become a site of a reserve army of labor that is global in recruitment. The expansion of this reserve army, combined with the continuing threat of capital migration, a weakened municipal government, and declining conditions in the sphere of reproduction, alter the investment conditions in New York. Precisely because the working class has been rendered less powerful in relation to capital, and the municipal authorities more pliable, New York and its labor force become an attractive alternative for reemployment by certain fractions of capital. A working class that has been peripheralized now emerges as more acceptable to both globally mobile firms and local capital in competitive sector industries.

CIRCUIT TWO CAPITAL MOBILITY: THE PERIPHERY IN THE GLOBAL CITY

Global capitalism may thus lead to a deterioration of the conditions of the working class in the global cities of the core regions. In addition, we contend that the situation of New York supports the proposition that the periphery is reproduced within the global city. By the late 1970s in New York City, in the garment industry, more than 50,000 workers toiled for wages under conditions closely resembling those associated with the labor force in the periphery. Moreover, the existence, expansion, and fate of these workers can be linked directly to the circuits of capital mobility that have been outlined above.

The birthplace of the garment industry in North America was New York City. In the nineteenth and early twentieth centuries, the sweatshop was the typical site of production; and, as the name suggests, the relations of production were characterized by labor's extreme vulnerability to capital. By the 1930s, however, the situation had been altered as militancy, union organization and collective bargaining, and state regulation had eliminated many of the sweatshops, raised wages, and improved working conditions. Then, after the Second World War, the globalization of production in the garment industry began undermining the position of garment workers. Jobs in the legal garment industry were lost, and by the late seventies the number of workers employed in illegal sweatshops rose dramatically.

The appearance of sweatshop labor in New York responds to pressures originating in the structure of capital–labor relations on a world scale. Indeed, the existence of a vast surplus population in dependent economies serves as a magnet for labor-intensive stages of the production process [. . .] There exists within New York, the global city, a substantial growing segment of the labor force whose conditions of production resemble those of the labor force in the Third World. This segment cannot be considered a mere aberration [. . .] Sweatshop labor is a necessary condition of global competition. Sweatshops in New York are the logical consequence of the globalization of production in the garment industry and the consequent competition for jobs between segments of the global reserve of labor.

THE INCOME AND WAGE IMPLICATIONS OF NEW YORK'S STRUCTURAL SHIFT

The outflow of manufacturing capital from New York City has not been totally matched by job-creating investment in the various service sectors. From 1969 (its post-1950 employment peak) to 1980, New York lost almost 500,000 jobs (U.S. Department of Labor 1979). Even if the total number of jobs in the city had been maintained, however, the structural shift from manufacturing to service employment would be likely to result in lower wages and incomes because service sector jobs are generally paid less than manufacturing jobs. Furthermore, the city's cost of living is among the highest in the country and the cost of living in New York appears to be linked to factors other than the purchasing power of its workers. Thus, through the three mechanisms of job loss, structural shift, and a cost of living that, indeed, reflects New York's global status, relative purchasing power in the city has shrunk, especially for its working-class population.

From 1964 to 1977, the median annual income of New York renters fell by over 11 per cent. In 1967 constant dollars the city median income in 1964 was $5,900, while in 1977 it was $4,800. The range across the boroughs of the city was wide: Queens, experiencing racial and ethnic transition in formerly middle-income areas, experienced the largest loss, over 17 per cent, while Manhattan renters registered a 1.9 per cent gain in purchasing power over the 13-year period. It should be noted that about two-thirds of New York City's populace are renters; thus, these data capture both large-scale realities of the income situation, and they also focus on those most apt to be working class.

THE SPHERE OF REPRODUCTION: HOUSING CONDITIONS

Living at the nerve centers of world capitalism, workers in New York City consume goods and services in a market structure that, at its top, serves the discretionary spending of the most affluent and discriminating consumers in the corporate world. New York's cost of living, perennially among the highest in the country – and in the world – reflects the pool of buying power concentrated in a high-density market. And this density also contributes to high land prices and high demand for housing.

Despite public subsidies and various degrees of rent control, and despite population decline, pressure on rental housing in the city is extremely high. This is reflected in relatively low vacancy rates and high rents, which claim increasing proportions of the income of the city's renters [. . .]

Rents in the era following the Second World War rose faster than the cost of living, and rents now consume higher proportions of renter income. By 1978, over 28 per cent of renters' aggregate income was going toward rents compared to 19 per cent in 1950. The distribution of this aggregate burden, however, is even more revealing; among the city's renters, 57 per cent now spend more than 25 per cent of their income on rent, the level conventionally used to measure "excessive" rent burdens. Though this is an arbitrary figure, its relative increase is consistent with other measures of a declining real level of living among New York's workers. Additionally, there has been an even more rapid increase in the number of renter households paying more than 35 per cent of their incomes in rent, doubling from 19 per cent in 1960 to 38 per cent in 1978. And in the Bronx and Brooklyn, the areas of the city with the lowest incomes, over a third of renter households pay over 40 per cent of their incomes for rent.

In summary, renters in the city are paying large and increasing proportions of income on housing, and for a significant fraction of renters this purchases substandard dwellings. For whatever it is worth as a subjective report, 44 per cent of renter households consider their neighborhoods fair or poor as distinct from good or excellent (Marcuse 1979).

SUMMARY AND REPRISE

In the core countries, certain cities emerge as the location of headquarters' functions for the worldwide system. Multinational corporations and financial institutions with far-flung activities induce growth in the advanced corporate service sector and other related service industries in such global cities.

We have addressed the following question: do the conditions of working-class life in the global cities of the core regions correspond to expectations that might be derived from structural analyses of core regions in general, or some other pattern? Predictions based on the perspective we call global capitalism focus on:

- capitalism as a global system of production;
- the existence in capitalist development of variants in the form of dominant capitalist organization and variation in the distribution of dominant forms across industrial sectors;
- the working class in global cities as part of a worldwide reserve of labor; and
- mobile capital (and thus investors) joined in a global allocation system permitting investors to induce competitive bargaining between workers located throughout the world.

Implied by our view of global capitalism are some expectations not tested here. For example, we would expect the general processes at work in New York to have negative impact on the formerly high wage, unionized workers in the monopoly sectors, and on their influence in community affairs as well. Were this shown to be true, then the structural analyses of world inequality would require, not just specification and disaggregation within core regions, but perhaps revision in their views of the relevance of new institutional forms of capital and capital mobility. In any case, in the global city, one finds jobs, wages, and levels of living reflecting the range of working-class life and work throughout the world, including the world's poor regions.

REFERENCES FROM THE READING

Katznelson, I. (1981) A radical departure? Social welfare and the election. In T. Ferguson and J. Rogers (eds) *The Hidden Election: Politics and Economics in the 1980 Presidential Campaign*. New York: Pantheon, 313–340.

Marcuse, P. (1979) *Rental Housing in the City of New York: Supply and Condition 1975–78*. New York: Housing and Development Administration.

Ross, R. and Trachte, K. (1990) *Global Capitalism: The New Leviathan*. Albany, NY: State University of New York Press.

U.S. Department of Labor (1979) *Employment and Earnings, U.S., 1909–1971*. Washington, DC: Bureau of Labor Statistics, Bulletin, 1312–1318.

13

"Inequality in Global City-regions"

from A.J. Scott (ed.), *Global City-regions: Trends, Theory, Policy* (2001)

Susan S. Fainstein

Editors' introduction

Since the early 1970s, Susan S. Fainstein has made significant contributions to the study of local political-economic restructuring in relation to broader, national and global transformations. For much of her career, Fainstein taught in the Urban Planning Department at Rutgers University. She is currently Professor of Urban Planning at Columbia University in New York City. Fainstein's most recent writings have focused on urban mega-projects and the politics of urban real estate development in New York City and London (Fainstein 2001). Additionally, she has explored the role of tourism in urban economic and spatial development (Judd and Fainstein 1999). While Fainstein has never explicitly identified herself as a global cities researcher, many of her major writings deal with themes that are quite central to the concerns of this literature, including local political restructuring, urban spatial transformation and urban socioeconomic inequality. It is to the latter topic that Reading 13 is devoted.

Fainstein provides an assessment of recent debates on social inequality within global city-regions. She begins by summarizing Sassen's widely discussed argument (see Readings 9 and 10; see also Reading 7 by Friedmann) that global cities are characterized by highly polarized labor markets. In light of this proposition, Fainstein then provides an empirical assessment of the empirical evidence in five major global cities – New York, London, Tokyo, the Dutch Randstad and Paris. While Fainstein does find that levels of inequality are intensifying in these cities, she contends that this is happening largely because the upper strata of the class structure are receiving a greater proportion of total earnings, and not due to the purported shrinking of the "middle class" or an expansion of the working class. Moreover, as her treatment of empirical research on each city shows, it is deeply problematic to generalize about global cities "as such," for individual cities are situated within different national political systems and are thus subject to divergent, contextually specific types of labor, social and economic policy regimes. Indeed, Fainstein insists that poverty levels in global city-regions are mediated powerfully through public policy arrangements; thus national governmental institutions continue to play a central role in producing urban sociospatial outcomes (see also Reading 20 by Hill and Kim, Reading 30 by Brenner and Reading 48 by Olds and Yeung). Finally, Fainstein shows that immigration patterns also influence patterns of inequality within each city – a theme addressed by other contributors to this Reader (see Reading 47 by Samers). While Fainstein's work does not invalidate the claim that global cities are characterized by distinctive, and relatively high, levels of inequality, her contribution suggests that more contextually grounded case studies and explanations will be required in order to decipher the precise nature of this connection.

Within the developed countries, business and governmental leaders of large cities typically aspire to reach global-city status. Yet no convincing evidence shows that the inhabitants of global cities and their surrounding regions fare better than the residents of lesser places. Indeed "the global-city hypothesis" argues that these metropolises are especially prone to extremes of inequality (Friedmann 1986). Despite being, in aggregate, the wealthiest areas of their respective nations, global city-regions tend to have large, dense groups of very poor people, often living in close juxtaposition with concentrations of the extraordinarily wealthy. According to Sassen (1991), the particular industrial and occupational structure of global cities produces a bifurcated earnings structure that in turn creates the outcome of the "disappearing middle." This chapter shows that global city-regions in wealthy countries do display high levels of income inequality (although not necessarily of class polarization) but that the explanation given by global-city theorists in terms of earnings is not wholly satisfactory [. . .]

THE SKEWED EARNINGS CURVE ARGUMENT

Crudely put, global-city theory makes the following argument: Global flows of capital produce similar economic structures in those few cities where the industries that control these flows have their home. Such control rests preeminently with financial institutions, business-services firms, and corporate headquarters; thus, the mark of a global city is its disproportionate share of finance and business services and corporate headquarters. In Sassen's (1991) terms, global cities constitute "strategic sites" in which leading-edge global functions are performed. Moreover, the most significant linkages between these cities' leading industries and other economic enterprises are international rather than national. Because the sectors of the economy performing global roles dominate the economic base of the affected cities, the cities display similar labor markets. In turn, these produce similar occupational and earnings hierarchies resulting in similar social outcomes. In Sassen's view, the economic structure of the global city leads to social polarization, as the leading sectors, on the one hand, employ a group of extraordinarily high-earning individuals and, on the other, create a demand for low-paid, low-skilled service workers. Global-city theory implies that such cities will have similar social characteristics, despite differing culture, history, governmental institutions, or public policy. This outcome is not just the product of a globalized world in which all cities increasingly look alike, but rather of ineluctable economic forces that impose a particular economic and social structure on these nodal sites.

THE EMPIRICAL EVIDENCE

Analysis of the social structures of the global city-regions of New York, London, Tokyo, Paris, and the Randstad supports an argument of increasing inequality, but it is much more ambiguous on the issues of the declining middle and of growth at the bottom. Global-city theory, as noted above, predicts that the income distribution in global cities will become increasingly polarized – i.e. not only will the relative shares of total income attained by different strata of the population shift, but the numbers of people at the top and bottom will increase relative to the modal point in the income distribution curve [. . .]

What is clear is that incomes within the five city-regions are becoming more unequal as the upper strata receive an increasingly large share of earnings. In all five regions most of the cause of growing inequality is very large increases in both individual earnings and household incomes at the top (if unrealized capital gains were included, the skew would be even greater). To put this another way, only a small proportion of households is profiting very much from the growth in aggregate income that has occurred in these regions over the last twenty years. The argument of the disappearing middle does not seem to hold up except in the sense that the shares of all income quintiles are declining relative to the top. In New York, Tokyo, and the Randstad, where there has been a steady outmigration of middle-class families to the suburbs, this statement applies more to the metropolitan region as a whole than to the central cities [. . .] A brief individual look at each city-region permits a better picture of their social structure.

THE CITY-REGIONS

New York

During the 1980s the trend of declining median income that had characterized the previous decade in the city was reversed. Real median household income grew more than 28 per cent compared to a national rate of 6.5 per cent (Mollenkopf 1997). The total real income of the bottom decile fell, but all other deciles gained, with the gains increasing as one went up the income scale. The principal reason for income gains was the upward movement of earnings in different economic sectors: health, education, and social service employment produced improved incomes in the middle ranks, while finance and business service earnings pushed up the top. Only the bottom decile had no increase in labor force participation. Most importantly the bottom suffered from a decline in the real value of retirement benefits and welfare payments [. . .] A more recent study of the New York region (New York City Council 1997) shows general improvement during the 1980s but reversal in the 1990s. In the period 1977 through 1989, poor households increased in number due to overall population growth, but they decreased as a proportion of total households. At the same time upper-income households increased numerically as well as in the relative share of total income they received. The size of the middle class, defined in relation to a middle-class standard of living, also increased. Between 1989 and 1996, however, a period of sharp recession and slow recovery, almost all gains went to the upper-income group, resulting in an aggravation of regional inequality and a hollowing out of the middle, especially in Manhattan [. . .]

London

The massive increase in income inequality that occurred in London during the period 1979–1993 resulted almost wholly from increases in earnings at the top rather than loss at the bottom of the income distribution. Within London the share of the top income decile increased from 26 per cent to 33 per cent. Hamnett (1997) argues that even while the very top was increasing its share, there

was an upward socioeconomic shift whereby the number and proportion of professional, managerial, and technical workers have been increasing, while the number and proportion of manual workers have been falling dramatically. Even within the service sector there has been a decline in clerical and blue-collar occupations. Hamnett finds that between 1979 and 1993 the number of people with a high standard of living (i.e. earning an amount that would have put them in the top 25 per cent of earners in 1979, adjusted for inflation) had increased by fifteen times. Moreover, this growth was substantially greater for London than for Britain as a whole.

Tokyo

Central Tokyo displays strong evidence of polarization. Within central Tokyo there has been growth in both the top and bottom occupations within fast-growing industries; at the same time middle-level clerical and manufacturing jobs are disappearing. Nevertheless, overall income distribution remains less skewed in central Tokyo than in the core areas of the New York, London, or Paris regions. Sonobe and Machimura (1997) relate trends in earnings and income distributions to the effect of the 1986–1991 bubble economy. During that period income inequality grew both inside Tokyo and between Tokyo and the rest of the nation; after the bubble burst, the reverse occurred as a result of losses at the top. Occupational polarization, however, has continued in the aftermath of the period of excessive speculation. Foreign workers who immigrated during the expansionary time remain in menial jobs within the service sector, and homelessness has emerged.

Paris

Paris has also seen occupational restructuring but in ways somewhat different from that experienced in New York, Tokyo, and London. As elsewhere there has been a decline in the stable working class as a result of economic restructuring but not directly as a consequence of globalization. Earnings have increased for all but the lowest decile of households, with the strongest growth at the top; again, as in the other cities, the household income distribution, which includes the unemployed, is more unequal than the earnings distribution. Although some industries associated with globalization have increased their size, as in Tokyo upper-level occupations have not expanded. Thus, Preteceille (1997) does not find in Paris the professionalization of the workforce that Hamnett discovered in London. Preteceille finds that the most important reason for income growth at the top results from the return on assets. Nevertheless, the range of salary inequality has also increased; the lowest level has received a decreasing share but stable wages, while the top has made disproportionate gains. Unexpectedly, within the context of both globalization and regulation theory, the size of the public sector has increased. There has been a limited but noticeable enlargement of intermediate groups and increased stability of their share. Opposite to the other four regions, salary inequality has increased more quickly outside the core area of the Ile-de-France than within it.

The Randstad

The Randstad, a conurbation of nearly seven million people that includes the cities of Amsterdam, Rotterdam, The Hague, and Utrecht and their surrounding suburbs, is smaller than the other global city-regions discussed here, lacks a primate city, and is not a major site for headquarters of multinational firms. It is less dependent on finance than the others, although like them it has a large and rapidly growing business services sector. Its significance as a global center depends on its key role as a transit point; the port of Rotterdam and Amsterdam's Schiphol Airport are the biggest transshipping points in continental Europe, and they have spawned a huge, trade-dependent commercial sector [. . .] Like the other global city-regions, the Randstad endured severe loss of manufacturing jobs and above-average growth in producers services during the 1970s and 1980s. Changes in the workforce paralleled the London pattern, with a trend toward increasing professionalization during the 1980s. Within both service and manufacturing industries, the categories of technical and execu-

tive personnel grew rapidly, while unskilled work diminished. The central Dutch state provides a high level of support for the poor of the Netherlands, who are primarily clustered in the Randstad's four cities. High direct subsidy to individuals combines with large governmental supply side subventions to support the provision of housing, transport, medical services, and education. Consequently the cost of living for those on low incomes is substantially less than is the case in other countries, a fact that is not revealed by income distribution data [. . .]

COMMON PATTERNS, DISSIMILARITIES AND CAUSAL FACTORS

The five regions are most alike in containing a very wealthy segment of their nations' population and supporting the economic and cultural institutions that sustain this group. Their economic bases are similar but by no means identical, and the similarities are only partly a consequence of globalization. New York and London have the strongest resemblance to each other of the five in economic structure, but they differ in social composition, with New York having a population with much larger racial and ethnic minorities and London having seen a greater professionalization of its labor force. At the end of the decade, Tokyo and Paris seemed to be diverging from the trajectory of the other three regions, largely as a result of changes in the economic position of their respective nation-states, which are themselves the consequence of both global and internal factors [. . .] All five regions had seen increasing income inequality, but polarization was not the most accurate description of this phenomenon in that, as a term, it failed to capture the improving position of the largest proportion of the middle mass.

Global-city theory associates the widening gap between the (shrinking) middle and the (growing) bottom of the income distribution with the demand by upper-income people for the services of casual, very poorly paid workers and the resurgence of sweatshop manufacturing using immigrant labor. The studies referred to here, however, do not support this argument but rather show that exclusion from the labor force is the principal cause of poverty. Moreover, the severity of poverty is largely a consequence of public policy [. . .] The growth in income of high-end households, as well as arising from individual earnings and equity market gains, reflects the presence of women in the upper reaches of the workforce, making possible the existence of families or housemates with two high-earning individuals [. . .]

If labor force exclusion is the principal impetus causing downward pressure on incomes, and returns from financial markets a main cause of growth in wealth, then we can expect income distribution to respond to cyclical economic changes. This makes comparisons of the regions quite difficult if their business cycles are not synchronous. During the 1980s boom it appeared as if there was a single business cycle that operated in all these regions simultaneously, but in the 1990s their courses have diverged. The studies that focused on global cities during the 1980s boom found increasing inequality in all of them and concluded that trends in income distribution were common to all global cities. More refined analysis, however, shows a number of cross-cutting forces that make generalizations about social outcomes extremely complicated.

First, the particular sectoral composition of the region affects its response to economic forces. During the 1970s and 1980s, workers in American and European cities suffered severely from loss of jobs in manufacturing while Japanese workers did not. In the 1990s, the New York, London, and Randstad regions, which had already lost a substantial proportion of their manufacturing jobs, were fairly immune to downturns in the production of manufactured goods. In contrast, Tokyo and, to a lesser extent, Paris, which had until recently remained major manufacturing centers, continued to be particularly susceptible. Second, geographic location matters. The New York region is currently benefiting from its location within the expanding American market, while the Tokyo region is suffering from being within the Asian Pacific zone, still emerging from the sharp recession of the last several years. Third, governmental policy considerably affects the situation of people in the bottom deciles of the income curve.

GLOBALIZATION AND GOVERNMENTAL POLICY

Global-city theory considers governmental policy as a response to global economic pressures. In this light, the increasing deregulation of financial markets and privatization of industry that occurred in the five regions can be viewed as the consequence of global forces. And, in turn, that deregulation and marketization have stimulated the growth of the financial services and international-trade-related sectors that are the leading edge of the economies of the global cities. Deregulation of labor markets can similarly be viewed as a response to international pressure. The effects of these moves toward freer markets on income distribution and job security can therefore be attributed to globalization. But we also see that the extent of national state withdrawal from intervention in the economy varies among the five regions with differing consequences for industrial structure, labor markets, and income and earnings distributions. Furthermore, there is a question of the direction of causality. Does (national) governmental policy create the conditions for the development of global cities? Or do global forces cause governments to act as they do? And, is the growth initially stimulated by deregulation primarily speculative and part of a longer term destabilization that will ultimately result in increased economic volatility and insecurity? The great dependence of global city-regions on international financial flows makes them particularly susceptible to fluctuations in the increasingly volatile global financial system.

The policies that matter the most for the economic situation of global-city residents are made by their national governments and have to do with securities regulation, interest rate levels, and labor-market restrictions. The greater income equality of Tokyo and the Randstad and the better situation of groups at the bottom in Paris than in New York or London indicate that ideology and governmental policy mediate tendencies toward the worsening of poverty. Nevertheless, policy-makers within the five regions have deliberately embarked on courses that have been given the name "global-cities strategy." By this is meant subsidies and loosened land-use regulation to encourage office development and intensive place marketing to international business, especially financial and producers services firms.

Is there a usefulness in adopting such a program? Tokyo has lately retreated from its global-cities strategy and its government is now emphasizing quality of life. Policymakers in the other regions, however, continue to emphasize global-city functions. London and Paris have mounted a number of projects aimed at increasing their cultural prominence and improving infrastructure for the conduct of business. New York during the 1980s was strongly committed to development projects that enhanced the role of finance and business services in the city's economy, while in the 1990s the main emphasis has switched to tourism and entertainment. In the Randstad considerable rhetoric has emphasized the global-city ideal, and massive infrastructure investment has aimed at improving the area's importance in international shipping and communication.

Emphasis on global-city status means a reinforcement of the tendencies toward dependence on international finance rather than an attempt to counterbalance them. To the extent that being a global city results in increased instability and income inequality, a global-cities strategy may worsen conditions for those worst off [. . .] The four-region comparison of New York, London, Tokyo, and Paris found that the first two of these regions saw a considerably greater increase in inequality and worsening of the situation of those at the bottom than in the latter two. Adding the Randstad to the analysis reveals even greater variation. Despite the small size of the Dutch state, it demonstrates the possibility of overcoming global ideological pressures in favor of the dismantling of the welfare state. During the nineties the Dutch have managed simultaneously to stimulate economic growth, reduce unemployment, protect sectors of the economy from full-scale marketization, and maintain the standard of living of those at the bottom through welfare state policies.

IMMIGRATION, RACE AND ETHNICITY

In all five regions there is a correlation between low income and membership in marginal ethnic or racial groups. One aspect of the global character of these regions is that, to a considerably greater extent than other parts of their nations, they have been a destination for immigrants. Nevertheless, there is not a simple relationship between the existence of large foreign-born populations and inequality. The five regions differ considerably in their level of racial/ethnic difference, the amount of immigration they have sustained, the length of time that foreign-born residents have resided within them, and the ability of immigrant groups to improve their economic situation. New York has received by far the largest number of immigrants both absolutely and proportionally of the five regions, and it continues to be a major recipient. On the whole, foreign-born residents who have lived in the region have done quite well. In New York race is highly correlated with income, yet within the black population West Indian immigrants fare better than African Americans. In London and the Randstad ethnicity is strongly associated with low income, with East Indians in London and Turks and Moroccans in the Randstad being relatively poorly educated and having a high unemployment rate. In Paris changes in income distribution are not attributable to immigration. The rate of immigration has remained more or less stable, with the total foreign-born population increasing by 11 per cent from 1982 to 1990, a rate that has changed little in the past several decades (Preteceille 1997). Although the Tokyo metropolitan area had a rise in immigration during the 1980s, its level of immigration has always been quite low in comparison to the other four regions. Its current economic troubles have discouraged increases in immigration, and foreigners living in Japan have been the group most severely affected by rising unemployment.

LINKAGES BETWEEN GLOBAL CITY-REGION STATUS AND INEQUALITY

There do appear to be some significant links between global-city status and inequality. First, global city-regions encompass particularly high-earning individuals resulting in an upward skew in the income distribution curve. The second correlation is a spatial one: the high cost of living in the core areas of these global-city-regions either forces low-income people into unaffordable housing at the center or pushes them, along with industries not associated with the global economy, to the periphery [...] To the extent that they contribute to a spatial mismatch that reinforces labor-market exclusion, global-city characteristics may then be an indirect cause of income inequality. Third, those global cities whose fortunes are particularly tied to financial markets are supersensitive to swings in those markets, with the consequence of serious instability in the livelihoods of their residents [...] The well-to-do, of course, are also vulnerable to loss of jobs and income, but their superior asset position and educational credentials insulate them from the extreme insecurity that affects the bottom strata.

The nation-state continues to matter [...] Although international forces clearly do shape the economic possibilities open to any region, national policy mediates the impact of those forces in ways that strongly affect the life chances of urban residents. And, from this perspective, the nation is far more potent than the city-region. National governments may not be able to affect the global economy, but they can shield their citizens from the most pernicious effects of that economy. Without such an effort occurring at the national level, local governments are largely helpless. As the Randstad illustrates, a combination of national initiative and local commitment can provide the basis for a region that both enjoys economic growth and sustains the well-being of the poorer section of the population.

REFERENCES FROM THE READING

Fainstein, S. (2001) *The City Builders: Property, Politics and Planning in New York and London*, 2nd edition. Kansas City, KS: University of Kansas Press.

Friedmann, J. (1986) The world city hypothesis, *Development and Change*, 17, 69–83.

Hamnett, C. (1997) Social change and polarization in London. Paper presented at the Workshop on Global Cities, CUNY Graduate Center, New York.

Judd, D. and Fainstein, S. (eds) (1999) *The Tourist City*. New Haven, CT: Yale University Press.

Mollenkopf, J. (1997) Changing patterns of inequality in New York City. Paper presented at the Workshop on Global Cities, CUNY Graduate Center, New York.

New York City Council (1997) *Hollow in the Middle: The Rise and Fall of New York City's Middle Class*. New York: City Council.

Preteceille, E. (1997) Social restructuring of the global city: the Paris case. Paper presented at the Workshop on Global Cities, CUNY Graduate Center, New York.

Sassen, S. (1991) *The Global City*. Princeton, NJ: Princeton University Press.

Sonobe, M. and Machimura, T. (1997) Globalization effect or bubble effect? Social polarization in Tokyo. Paper presented at the Workshop on Global Cities, CUNY Graduate Center, New York.

14

"Global Grids of Glass: On Global Cities, Telecommunications and Planetary Urban Networks"

from *Urban Studies* (1999)

Stephen Graham

Editors' introduction

Stephen Graham is a pioneering scholar of late twentieth and early twenty-first century urbanism. Throughout much of the 1990s, Graham worked in the Urban Planning Department at the University of Newcastle, where he co-founded the Centre for Urban Technology. Currently, Graham is Professor of Human Geography at the University of Durham. A major strand of Graham's work has focused on the historical, political, economic and cultural geographies of urban infrastructural systems. Rather than treating these urban infrastructural arrangements – from water, heating, sewage and electrical systems to transportation and communications grids – in technocratic terms, as simple instruments for the provision of public and private goods, this new scholarship has attempted to decipher their complex social, political and cultural dimensions and their role as expressions of historically specific forms of urban power. With his long-time collaborator, Simon Marvin, Graham has been at the forefront of this new research on urban infrastructural networks. Their two major books, *Telecommunications and the City* (1996) and *Splintering Urbanism* (2001), have explored at length the worldwide reorganization of urban infrastructural arrangements under contemporary globalizing capitalism, and the role of such transformations in a broader rearticulation (or "splintering") of urban space.

Reading 14 introduces Graham's approach to the study of urban infrastructure and applies it to the restructuring of telecommunications systems in global cities (see also Graham 2004). The global urban system is linked together not only through the activities of transnational corporations, but also through dedicated, high-performance informational networks based upon transplanetary optic-fiber connections. Global cities contain the highest concentrations of advanced telecommunications infrastructures within their respective national urban systems. This trend is being further accelerated and intensified through the deregulation and privatization of telecommunications systems, a process that has triggered an intensive reorganization of telecom, media and cable firms across Europe and North America, as ever-larger, global conglomerates are formed in order to compete on a global scale. Finally, Graham considers the intra- and inter-urban spatial transformations that flow from the construction of such transplanetary technological networks. On the one hand, the effective operation of planetary telecommunications grids depends upon concrete physical installations within the dense built environments of global cities: optic-fibers must be "looped" under roads and sidewalks, into and out of buildings. On the other hand, "global grids of glass" must be constructed that link major metropolitan centers across oceans and continents: this entails creating transoceanic optic-fiber networks and transterrestrial satellite linkages that wire distant cities together to maintain instantaneous connectivity. Graham concludes by underscoring the persistent problem of the "digital divide." While the new informational and telecommunications technologies may link global cities more closely together, they have to date failed to include significant zones of the world economy within their web. At present, therefore, such technologies appear to serve primarily the needs of hegemonic economic actors and organizations.

The growing centrality of key large urban regions, or global cities, to the economic, social, political and cultural dynamics of the world presents a particularly potent example of the reconfiguration of space through telecommunications. In such cities, the most sophisticated electronic infrastructures ever seen are being mobilised to reconfigure space and time barriers in a veritable frenzy of network construction. Such processes seem likely to maintain the electronic competitive advantages of the largest global cities for some time to come. But the wiring of cities with the latest optic-fibre networks is also extremely uneven. It is characterised by a dynamic of dualisation. On the one hand, seamless and powerful global–local connections are being constructed within and between highly valued spaces, based on the physical construction of tailored networks to the doorsteps of institutions. On the other hand, intervening spaces – even those which may geographically be cheek-by-jowel with the favoured zones within the same city – seem, at the same time, to be largely ignored by investment plans for the most sophisticated telecommunications networks. Such spaces threaten to emerge as "network ghettos", places of low telecommunications access and concentrated social disadvantage. As with many contemporary urban trends, then, uneven global interconnection via advanced telecommunications becomes subtly combined with local disconnection in the production of urban space.

Global cities research, in particular, has detailed at length how an interconnected network of such cities has recently grown to attain extraordinary status. Such cities bring together the greatest concentrations of control, finance, service, cultural, institutional, social, informational and infrastructural industries in the world. All aspects of the functioning of global cities are increasingly reliant on advanced telecommunications networks and services; such cities concentrate the most communications-intensive elements of all economic sectors and transnational activities within small portions of geographical space. It is no surprise, therefore, that there is growing evidence that such city-regions heavily dominate investment in, and use of, these technologies.

The diversifying electronic infrastructures that girdle the planet have very specific geographies and spatialities. These counter the prevailing "information age" rhetoric suggesting that advances in telecommunications somehow prefigure some simple "end of geography" [. . .] Indeed, the current international shift towards liberalised, privatised and internationalised telecommunications regimes seems to be accentuating the centrality of global cities within telecommunications investment patterns.

Many research challenges remain to be faced before we can satisfactorily understand the complex interlinkages between telecommunications grids, global cities and planetary urban networks. The focus of this chapter is to explore the linkages between the growth of a planetary network of global financial, corporate and media capitals, and the emerging global and urban information infrastructures that interlink, and underpin, such centres. Very little is known about how the global wiring of the planet with a new generation of optic-fibre grids interconnects with the development of intense concentrations of new communications infrastructures within global cities. This chapter aims to develop such an understanding by attempting to address intra-urban, inter-urban and transplanetary optic-fibre connections (and disconnections) in parallel. Such an approach is necessary given the logics inherent within the "network society", which force us to collapse conventional hierarchical

T
W
O

notions of scale – building, district, city, nation, continent or planet. As a result, it is difficult to be a specialist on urban landscapes, intra-urban shifts or urban systems in separation. Discussions of restructuring *within* cities increasingly must address the changing relations *between* them, whilst also being cognisant of the importance of these changing relations within broader dynamics of geopolitics and geoeconomics.

GLOBAL CITIES: THE SOCIAL PRODUCTION OF LANDSCAPES OF MULTIPLE RELATIONAL CENTRALITY

It is now clear that global cities grow by cumulatively concentrating the key assets which corporate headquarters, high-level service industries, global financial service industries, national and supranational governance institutions and international cultural industries rely on, within a volatile, globalising, operating environment. The growing extent of globally stretched corporate, financial and media webs, mediated by telecommunications and transport networks, seems to support a parallel need for the social production and management of places of intense centrality. This is especially so when one adds the volatilities thrown up by global shifts towards financial globalisation, economic liberalisation and the opening up of regional blocs to "free" trade. The complex mediation of economic activity in extremely volatile contexts necessitates high levels of face-to-face interaction within the high-level managerial and control functions that concentrate in global cities. Place-based social relations become central to the economic survival of high-level corporate, media and financial organisations [. . .]

The economic, social and cultural dynamics of global cities rely in essence on the control, co-ordination, processing and movement of information, knowledge and symbolic goods and services (advertising, marketing, design, consultancy, finance, media, music, etc.). Within a post-Fordist context of vertical disintegration, niche marketing, precise logistical co-ordination, internationalisation, pervasive computerisation and the powerful growth of symbolic exchange, all such activities are generating booming demands for voice, computer and image communications of all types. Above all,

such conditions are supporting an enormous growth in economic *reflexivity* placing a premium on both the specific socioeconomies offered by particular global cities and intensifying, electronically mediated connections between them. The reflexive nature of global city functions thus demands on-going social relations based on trust and reciprocity. These are supported by both intense mobility and sophisticated telemediated exchange. In a nutshell, such dynamics help explain why perhaps the two most dominant social trends in the world today are an unprecedented urbanisation and a growing reliance on telecommunications-based relations.

As mediators of all aspects of the reflexive functioning and development of global cities, convergent media, telecommunications and computing grids (known collectively as telematics) are basic integrating infrastructures underpinning the shift towards intensely interconnected planetary urban networks. Inter-urban telecommunications networks (both transoceanic optic fibres and satellites) comprise a vital set of hubs, spokes and "tunnel effects" linking urban economies together into real or near real time systems of interaction which substantially reconfigure the production of both space and time barriers within and between them. Such technologies help to integrate distant financial markets, service industries, corporate locations and media industries with virtual instantaneity and rapidly increasing sophistication. But they underpin the enormously complex communications demands *within* global cities, generated by the intense clustering of reflexive practices in space [. . .] Such dynamics mean that the very small geographical areas of the main global finance, corporate and media capitals dominate the emerging global political economy of telecommunications.

Two sources of data can help to give an indication of this dominance. First, we can see how the economic sectors that are overwhelmingly located in global cities tend to dominate international telecommunications flows as a whole. For example, over 80 per cent of international data flows are taken up by the communications, information flows and transactions in the financial services sector. Over 50 per cent of all long-distance telephone calls in the US are taken up by only 5 per cent of phone customers, largely transnational corporations whose control functions still cluster in the global metropolitan areas of the nation. Secondly, there is a small amount of available data on the dominance of national telecommunications patterns by particular global cities. A recent survey (Finnie 1998) found that around 55 per cent of all international private telecommunications circuits that terminate in the UK do so within London. And about three-quarters of all advanced data traffic generated in France come from within the Paris region.

TELECOMMUNICATIONS LIBERALISATION, GLOBAL CITIES AND URBAN COMPETITIVENESS

Central business districts (CBDs) within global cities play a predominant role within fast-moving communications landscapes. They provide leading foci of rapid technological change and concentrated patterns of investment in new telecommunications infrastructures, from multiple competing providers. Global financial service industries, in particular, are especially important in driving telecommunications liberalisation. With telecom costs taking up around 8 per cent of expenditure on goods and services in global financial firms, a world-class telecommunications infrastructure and a business friendly, fully liberalised regulatory environment are thus becoming key assets in the competitive race between global cities to lure in financial and corporate operations and their telecommunications hubs.

For transnational companies (TNCs) of all kinds, liberalised telecommunications markets allow the benefits of competition to be maximised. The spread of liberalisation also minimises the transaction, negotiation and interconnection costs that stem from constructing global networks within the diverse regulatory and cultural contexts of multiple national post, telegraph and telephone systems (PTTs) across the globe. Global telecoms liberalisation is thus critical given the strategic centrality of private telecommunications networks to the functioning of all TNCs, financial and media firms. The emerging global, private regime allows lucrative corporate and financial market segments to benefit from intense, customised and often very localised competition in high-level telecommunications infrastructure and services, when this was impossible through old-style PTTs.

In short, for leading finance, transnational and media firms, sophisticated telecommunications are becoming central to business success or failure. Such firms demand a seamless package of broadband connections and services, within and between the global cities where they operate. Such demands include leased lines, private optic-fibre connections, dedicated satellite circuits, video conferencing and, increasingly, highly capable mobile and wireless services.

In addition, such powerful corporate users are pressing hard to enjoy the fruits of competition between multiple network providers in global cities. Such competition tends to increase discounts, improve efficiency and innovation and, above all, reduce risks of network failure by increasing network resilience. Meanwhile, peripheral regions and marginalised social groups often actually lose out and fail to reap the benefits of competition, as they are left with the rump of old PTT infrastructures, relatively high prices and poor levels of reliability and innovation. Highly uneven micro-geographies of "splintered" telecommunications development thus replace the relatively integrated and homogeneous networks developed by national telephone monopolies.

GLOBAL CITIES AND INTERNATIONAL TELECOMMUNICATIONS LIBERALISATION

As a result of such liberalising pressures, a global wave of liberalisation and/or privatisation is transforming national telecommunications regimes. The key lobbying pressures driving this shift derive from TNCs, the World Trade Organisation (WTO), the G7 countries, telecommunication industries and, especially, the corporate, financial and media service industries that concentrate in global cities. In the context of the recent WTO agreement to move towards a global liberalised telecommunications market and with most regions of the globe now instigating regional trading bloc agreements (EU, NAFTA, ASEAN, Mercosur, etc.) based on removing national barriers to telecoms competition, a profound reorganisation of the global telecommunications industry is taking place.

The initial example of metropolitan network competition was set by New York in the 1980s.

Plate 11 DGTAL HOUSE, London (Roger Keil)

Then, a whole new competitive urban telecommunications infrastructure was developed in the city by the teleport company and port authority. As the pressures of the global neo-liberal orthodoxy in telecommunications have grown, even the most resistant nation-states (such as France, Germany, Singapore and Malaysia) are now succumbing to calls for national PTT monopolies to be withdrawn, to be replaced by uneven, multiple, telecoms infrastructures which inevitably articulate centrally around large metropolitan markets.

In the shift to a seamlessly interconnected global telecoms industry, national postal, telegraph and telephone monopolies are thus being rapidly privatised and/or liberalised right across the globe. Private telecom firms are aggressively "uprooting" to make acquisitions, strategic alliances and mergers with other firms, in a global struggle to build truly planetary telecommunications service firms, geared towards meeting the precise needs of TNCs and financial firms within the whole planetary network of global cities, on a "one stop, one contract" basis.

For example, in an effort to position themselves for this "one-stop" market, BT and AT&T, both trying to reposition themselves from national to global players, have recently announced a global alliance. AT&T have also merged with the huge US cable firm Tele-Communications to improve their position within home US markets. The Swedish, Swiss, Spanish and Dutch PTTs, meanwhile, have their Global One umbrella. France Telecom and Deutsch Telekom have an alliance with the US international operator Sprint. A myriad of other firms are offering services based on reselling network capacity leased in bulk from other firms' telecommunications networks. Finally, there is a process of alliance formation between telecoms, media and cable firms. All are jostling to position themselves favourably for corporate and domestic markets, especially in the information-rich global city-regions, for the future where many envisage a globally liberalised telecommunications environment with perhaps four or five giant, truly global, multimedia conglomerates.

The position of global cities is therefore being substantially reinforced by the shift from national telecommunications monopolies to a globalised, liberalised communications marketplace. As cross-subsidies between rich and poor and core and periphery, spaces within nations are removed, infrastructure developments now reflect unevenness in communications demand in a much more potent way than has been witnessed for the past half century. Prices and tariffs are being unbundled at the national level, revealing stark new geographies which compound the advantages of valued spaces within global cities as attractive and highly profitable telecommunications markets, dominating telecommunications investment patterns within nation-states.

THE MICROGEOGRAPHICAL IMPERATIVES OF "LOCAL LOOP" CONNECTION

It is paradoxical, however, that an industry which endlessly proclaims the "death of distance" actually remains driven by the old-fashioned geographical imperative of using networks to drive physical market access. The greatest challenge of these multiplying telecommunications firms in global cities is what is termed the problem of the "last mile": getting satellite installations, optic-fibre drops and whole networks through the expensive "local loop", under the roads and pavements of the urban fabric, to the buildings and sites of target users. Without the expensive laying of hardware, it is not possible to enter the market and gain lucrative contracts. Fully 80 per cent of the costs of a network are associated with this traditional, messy business of getting it into the ground in highly congested, and contested, urban areas.

This is why precise infrastructure planning (through the use of sophisticated geographical information systems) is increasingly being used to ensure that the minimum investment brings the highest market potential. It also explains why any opportunity is explored to string optic fibres through the older networks of ducts and leeways that are literally sunk deep within the archaeological "root systems" of old urban cores. Mercury's fibre grid in London's financial district, for example, uses the ducts of a long-forgotten, 19th-century, hydraulic power network. Other operators thread optic fibres along rail, road, canal and other leeways and conduits. Whilst satellite infrastructures are obviously more flexible, they are nowhere near as capable or secure as physical optic-fibre "drops" to a building. And they, too, ultimately rely on having the hardware in place to deliver services to the right geo-economic "foot print" [...]

GLOBAL GRIDS OF GLASS: INTERCONNECTING URBAN FIBRE NETWORKS

We now come to our final scale of analysis: that of planetary interconnection and the geopolitics of infrastructure construction across oceans and continents. Of course, dedicated optic-fibre grids *within* the business cores of global cities are of little use without interconnections that allow seamless corporate and financial networks to piece together to match directly the hub and spoke geographies of international urban systems themselves. To this end, WorldCom and the many other internationalising telecommunications companies are currently expending huge resources within massive consortia, laying the satellite and optic-fibre infrastructures necessary to string the planet's urban regions into a single, highly interconnected communications landscape. Such efforts are especially focusing on the high-demand corridors linking the geostrategic, metropolitan zones of the three dominant global economic blocs: North America, Europe and East and South-eastern Asia. Wiring North America and the North Atlantic is proceeding at a particularly rapid pace, with AT&T, Sprint, WorldCom and US Regional "Baby Bells" fighting it out with newcomers like Quest to provide the "pipes" that will keep up with demand, especially from the Internet. Other lesser infrastructures are also being laid, designed to link southern metropolitan regions in Australia, South Africa and Brazil into the global constellation of inter-urban information infrastructures.

Currently, transoceanic and transterrestrial optic-fibre and satellite capacities concentrate overwhelmingly on linking North America across the North Atlantic to Europe and across the North Pacific to Japan, reflecting the geopolitical hegemony of these three regions in the late 20th century. The first transoceanic fibre networks, developed since 1988 across the Atlantic, Pacific and Indian oceans, with AT&T playing a leading role, tended to stop at the shorelines, leaving terrestrial networks to connect to each nation's markets. Increasingly, however, transoceanic fibre networks are being built explicitly to link metropolitan cores, in response to their centrality as generators of traffic and centres of investment.

Once again, WorldCom, in its efforts to connect together its metropolitan infrastructures, is leading this process. As well as constructing a transatlantic fibre network known as Gemini between the centres of New York and London, WorldCom is building its own pan-European Ulysses network linking its city grids in Paris, London, Amsterdam, Brussels and major UK business cities beyond London. Elsewhere in the world, too, it is exploring the construction of transnational and transoceanic fibre networks to connect its globalising archipelago of dedicated city networks.

At the strategic global scale, globalisation and the rapid growth of newly industrial countries means that much effort is being spent filling in gaps in the global patterns of optic-fibre interconnection, particularly between Europe and Asia. One project, for example, known as FLAG (fibre optic link around the globe) will provide a new ultra high-capacity (120,000 simultaneous phone calls) telecoms grid over 28,000 km from London to Japan, via many previously poorly connected metropolitan regions and nation-states. The route first goes by sea from England via the Bay of Biscay and the Mediterranean to Alexandria and Cairo. Then it crosses Suez and Saudi Arabia overland to Dubai; crosses the Indian Ocean to Bombay and Penang and Kuala Lumpur; traverses the Malay peninsula to Bangkok; and finally goes by sea again to Hong Kong, Shanghai and Japan. Nynex, the New York phone company, is organising the project with support from a huge range of private investors. FLAG will also connect with the 12,000 km Pacific Cable Network linking Japan, South Korea, Taiwan, Hong Kong, the Philippines, Thailand, Vietnam and Indonesia. Other similar projects are proliferating across the globe. A 30,000-km China–US cable network, the first direct linkage between the two nations, was announced by a 14-firm consortium in December 1997. Other transoceanic cables are also being constructed: in the Caribbean (the Eastern Caribbean Fibre System); from Florida, through the Strait of Gibraltar, to the Mediterranean urban system; by AT&T around the African continent (offering a high level of security whilst providing fibre drops to all the main cities on the African coast); and by Telefonica of Spain from the US and Europe to the main metropolitan regions of Latin America (Fortaleza, Rio, São Paulo, Buenos Aires), again with the involvement of WorldCom. But the largest project of them all, Project Oxygen, was announced in 1998: a $15 billion global network, built by 73 multinational telecommunications firms, linking the largest urban regions in 171 countries by 2003, via over 300,000 km of optic-fibre cable.

CONCLUSIONS

This chapter has sought to extend our empirical and conceptual understanding of the ways in which global cities and global city networks are related to rapid advances in both intra-urban and inter-urban information infrastructures. It has emerged that current advances in telecommunications are a set of phenomena which tend overwhelmingly to be driven by large, internationally oriented, global metropolitan regions. The activities, functions and urban dynamics which become concentrated in global city-regions rely intensely on the facilitating attributes of advanced telecommunications for supporting relational complexity, distance links and snowballing interactions, both within and between cities. Such telereliance is particularly high in internationally oriented industries whose products and services are little more than telemediated flows of exchange, information, communication and transaction, backed up also by intense face-to-face contact and supporting electronic co-ordination and transportation.

It should be no surprise, then, that, in an increasingly demand-driven media and communications landscape, global cities should dominate investment in, and use of, advanced telecommunications infrastructures and services. Global cities are the main focus of the highly uneven emerging geographies of network competition. They dominate all aspects of telecommunications innovation. As we have seen, such cities are now developing their own superimposed and customised fibre networks, seamlessly linked together across the planet into a global cities communicational fabric, whilst often (at least in the initial stages of development) separating them off from traditional notions of hinterland, urban and regional interdependence and national infrastructural sovereignty. The WorldCom city networks explored above, for example, demonstrate that the emerging urban communicational landscape is rapidly becoming dominated by tailored, customised infrastructures. Through these, dominant corporate, media and financial players can maintain and extend their powers over space, time and people. Such infrastructures are carefully localised physically to include only the users and territory necessary to drive profits and connect together global corporate, financial and media clusters.

But, in so doing, such networks seem likely to further urban socio-spatial polarisation by carefully by-passing non-lucrative spaces within the city. They are, in other words, physical, infrastructural embodiments of the splintering and fracturing of urban space. The combination of intra- and inter-urban fibre networks thus materially supports the dynamic and highly uneven production of space-times of intense global connectivity, made up of linked assemblies of high-level corporate, financial and media clusters and their associated socio-economic elites, within the spaces of global cities.

Finally, it seems inevitable that the customised combinations of intra- and inter-urban fibre networks analysed in this chapter will also drive uneven development at the national level. They will be excluding non-dominant parts of national urban systems from the competitive advantages that stem from tailored, customised urban information infrastructures. Such infrastructures lay to rest any prevailing assumptions that there necessarily remains any meaningful connection between the sovereignty of nation-states as territorial "containers," and patterns of infrastructural development, especially for telecommunications infrastructures. The modernist notion that nationhood is partly defined by the ability to roll-out universally accessible infrastructural grids to bind the national space is completely destroyed by the new infrastructural logic. Instead, the logic of the emerging electronic infrastructure is to follow directly the global city networks, through direct global–local interconnection. Consequently, relatively cohesive, homogeneous and equalising infrastructure grids at the national level are now being "splintered" into tailored, customised and global–local grids, designed to meet the needs of hegemonic economic and social actors (Graham and Marvin 2001).

REFERENCES FROM THE READING

Finnie, F.G. (1998) Wired cities, *Communications Week International*, May 18, 19–22.

Graham, S. (ed.) (2004) *The Cybercities Reader*. London and New York: Routledge.

Graham, S. and Marvin, S. (1996) *Telecommunications and the City*. London: Routledge.

Graham, S. and Marvin, S. (2001) *Splintering Urbanism*. London: Routledge.

PART THREE

Local pathways of global city formation: classic and contemporary case studies

Plate 12 Tokyo (Takashi Machimura)

INTRODUCTION TO PART THREE

The 1970s were widely considered a decade of urban crisis in western Europe and North America. Cities seemed to have largely lost the battle against the suburban "flight" of populations and industries; many older industrial cities, such as New York, were on the brink of fiscal collapse; urban renewal and redevelopment schemes had generated negligible, even disastrous, results; and urban social problems and social unrest were proliferating. Under these circumstances, policy makers and academics tended to assume that positive change for urban centers was not likely to occur any time soon, particularly in light of ongoing developments in transportation and communication technology that appeared to be promoting an increasing decentralization of work, housing and recreation. The end of the city seemed unavoidable.

This situation changed dramatically during the 1980s. An urban "renaissance" was widely observed as "gentrification" became a household term and major speculative investments by international real estate firms reinvigorated office space markets in downtown centers (for instance, by the notorious Reichmann brothers in New York City). Meanwhile, even as older industrial centers continued to bleed jobs and people (for example, in cities such as Detroit, Buffalo, Hamburg, Dortmund and Liverpool), a variety of so-called "new industrial spaces" (Scott 1988) emerged in urban regions located in the "sunbelts" of North America and western Europe. City-regions such as Los Angeles, Toulouse, Munich and Barcelona were among the core sites of this dynamic process of reindustrialization. This rejuvenation of urban centers assumed a variety of forms. In San Francisco, a resurgent gay middle class community promoted a qualitatively new form of gentrification. In Zurich, a radical youth movement revived the staid conservative financial center and forced a new "territorial compromise" (see Reading 19 by Schmid). In Los Angeles, the 1984 Olympics put the city on the map of an emerging Pacific Rim economy. And in Frankfurt, a visionary but troubled conservative local government developed an ambitious world city development project (Keil and Lieser 1992). Taken together, these diverse processes of urban transformation helped rejuvenate public interest in, and debate about, city life. Crucially, however, this latest wave of urban revitalization was now understood above all in relation to the newly consolidated global urban system, rather than with reference to nationally self-enclosed systems of cities.

These developments were analyzed in detail by critical urban scholars, who now began to move beyond their earlier preoccupations with declining collective consumption, failing urban welfare states, unequal real estate markets and other trends associated with the crisis of the Fordist form of urban development. Increasingly, the urban crises of the 1970s and early 1980s were viewed as expressions of multifaceted processes of restructuring, rather than as symptoms of linear, catastrophic decline. Global city theory was among the key approaches that were introduced by urbanists in order to decipher these developments.

Accordingly, the readings in Part Three examine the applications of global city theory to the investigation of processes of urban restructuring that were unfolding during the 1970s and 1980s in some of the most globally integrated metropolitan centers of western Europe (London, Amsterdam, Zurich), North America (New York, Detroit, Houston, Los Angeles) and East Asia (Tokyo, Seoul). While the founders of global cities research, such as Friedmann and Wolff (Reading 6), Friedmann (Reading 7) and Sassen

(Readings 9 and 10), ventured a number of broad generalizations regarding the dynamics of industrial change, labor market dualization, sociospatial polarization and local politics within global cities, others have been concerned to study the distinctive, place-specific ways in which cities are articulated to the world economy. For, beyond their underlying commonalities as basing points for global capital accumulation, global cities are extremely diverse:

- their built environments and political-economic structures have been shaped by distinctive historical legacies;
- they have been transformed into command and control centers through place-specific political strategies and along place-specific institutional pathways;
- they generally contain a wide range of locationally specific economic functions;
- their evolutionary trajectories are in turn shaped through nationally and locally specific institutional arrangements, regulatory frameworks and sociopolitical struggles.

Early contributors to world city theory attempted to explain certain key developments and processes *internal* to a given city with reference to its *external* positioning in the global urban system. Subsequent research, including the contributions to Part Three, has effectively reversed this line of causality, demonstrating that the history, spatiality, institutional configuration and sociopolitical environment of a global city exercise a powerful influence upon its mode of insertion into the world economy. From this point of view, a city's position in the world economy cannot be taken for granted, but must itself be analyzed as an expression and outcome of complex, contested socioeconomic processes and strategies within that city and the broader (generally regional and national) space-economies in which it is embedded. One of the basic methodological challenges of global cities research, therefore, is to uncover, simultaneously, not only the general roles that cities have come to play in the contemporary world economy but also their distinctive histories, geographies, political economies and developmental trajectories.

For this reason, we would argue, global city theory should not be viewed as postulating a "convergence" of urban political-economic and spatial structures towards a single, generic model. Instead, we believe, this conceptual framework is better understood as a means to decipher the *interplay* between general, cross-national trends of urban restructuring and place- and territory-specific outcomes within major cities. While confronting this intellectual task presents many complex methodological and empirical challenges, the readings in Part Three demonstrate that doing so can generate highly illuminating analyses of processes of urban restructuring in cities throughout the world economy.

Following the pioneering research of the 1980s, case study-based research on the process of global city formation was pursued with enhanced methodological sophistication and with reference to an increasingly broad range of cases. Global city theory thus came to serve as a key analytical reference point for the investigation of contemporary urban restructuring. At the same time, the empirical extension of global cities research opened up a variety of new debates regarding its appropriate theoretical foundations and methodological applications. Was the concept of the global city to be viewed as a description of a particular type of city (e.g., a transnational headquarters location or a global command and control center), or was it better understood as a basis for examining the more general trend towards globalized urban restructuring? Were analogous patterns of political-economic and sociospatial restructuring unfolding in cities that fulfilled similar roles in the world economy, or was the globalization of urbanization contributing to a further differentiation of urban institutional configurations and sociospatial forms? Were local political institutions and social movements helpless victims of transnational capital, or could they find ways to defend local interests and priorities? As we shall see in subsequent parts of this Reader, debates on these and other key issues facilitated the further development and differentiation of global cities research, as well as any number of sustained critiques of this approach.

In Part Three, we explore a representative selection of case studies and comparative analyses of global city formation in the three major zones of the world economy – North America, western Europe and East Asia. These empirical analyses are intended, above all, to illuminate how the process of global city formation has unfolded in a variety of otherwise very different cities and national territories. At the same time, whether implicitly or explicitly, each of the following readings necessarily engages with the theoretical foundations of global city theory in order to decipher the puzzles and paradoxes of the particular place(s) under investigation. The selections included in Part Three cover a broad range of themes and draw attention to an extremely diverse constellation of transformations, strategies and struggles that have been associated with global city formation in each of the places under investigation. Yet, despite this diversity, several common threads of investigation, analysis and interpretation can be gleaned from these readings, as follows:

Remaking the built environment: producing global connectivity

The case studies track various ways in which the urban built environment is transformed in and through the process of global city formation (see also Reading 14 by Graham). New transportation and telecommunications infrastructures are established to ensure global connectivity; large-scale property developments, mega-projects and office towers are constructed to meet the perceived requirements of transnational capital; central business districts and luxury housing development projects are extended into older working class neighborhoods; and a new "landscape of power" (see Reading 16 by Zukin) is constructed to market the city, both to its inhabitants and to potential investors, as a global command and control center.

Growth coalitions and the politics of global city formation

Several of the case studies underscore the role of growth-oriented political alliances – generally composed of national, regional and/or local state officials, transnational corporate elites and local property owners or "rentiers" (Logan and Molotch 1987) – in establishing the local preconditions for global city formation. Such alliances serve not only to mobilize the tremendous infrastructural, technological and financial resources that are required to ensure a city's seamless global connectivity, but also to blunt local opposition to the large-scale mega-projects and real estate developments that are generally associated with global city formation. Reading 17 by Machimura, for instance, shows how global city formation in Tokyo hinged upon the active support of diverse public agencies (at both national and local scales) as well as transnational, national and local corporations. As Machimura indicates, however, this growth coalition excluded many local interests from key decisions about the form and geographical distribution of urban economic development.

State strategies and spatial outcomes

The case studies illustrate the central role of state institutions, at once on national, regional and municipal scales, in promoting global city formation, whether through regulatory realignments, financial subsidies or some combination thereof. From London, Tokyo and Seoul to Houston, Detroit and Zurich, it is evident that states actively facilitate the reorganization of urban space to accommodate the demands of globally oriented firms, real estate developers and political elites. This often entails the construction of new, quasi-public institutions or public–private partnerships, lacking in democratic accountability, that oversee and finance the development of globalized urban spaces while protecting private investors from "excessive" risk (Zukin, Reading 16). In an interesting variation on this theme, Hill and Kim's study of Tokyo and Seoul (Reading 20) suggests that urban development in East Asia represents a key spatial outcome of "developmentalist" state policies oriented towards both national *and* local economic growth. However, local social forces may also attempt, with varying degrees of success, to harness state institutions in order to block particular pathways of global city formation, leading in turn to the negotiation of "territorial compromises" that may protect certain local interests while nonetheless facilitating (a modified form of) globalized urban development (Schmid, Reading 19). State institutions thus become a key political terrain in which the form, pace and geography of globalized urbanization are fought out by diverse local and supralocal social forces (see the contributions to Part Five).

Uneven development, territorial inequality and regionalization

In addition to their attention to urban built environments, the case studies also map out some of the wide-ranging sociospatial transformations that have ensued in conjunction with global city formation. Whereas Zukin (Reading 16) and Machimura (Reading 17) focus on the reconfiguration of downtown urban cores through large-scale office developments and luxury residential developments, Hill and Feagin (Reading 18), Schmid (Reading 19) and Soja (Reading 21) examine some of the broader, region-wide expressions of global city formation. Existing patterns of uneven development and sociospatial inequality are exacerbated (Hill and Feagin, Reading 18) and new clusters of globally oriented firms may be established beyond the traditional city core (Schmid, Reading 19; Soja, Reading 21; see also Reading 34 by Keil and Ronneberger). Consequently, as Soja (Reading 21) forcefully suggests, polycentric patterns of urban development appear to be superseding traditional, monocentric models of urban spatial form. While these trends cannot be explained entirely with reference to the forces of globalization, the readings in Part Three do suggest that they are being articulated within global city-regions around the world.

Comparative perspectives

The readings in Part Three can be viewed as engaging in a common effort to develop comparative perspectives on the process of global city formation. Some readings, such as Machimura's study of Tokyo (Reading 17) and Schmid's study of Zurich (Reading 19) make use of global city theory through individualizing case studies that are intended to illuminate the particularities of the places under investigation, albeit with reference to a broader geoeconomic context. Other readings, by contrast, mobilize global city theory in order to explore the similarities and/or differences among two distinct cases of globalized urban development; they thus adopt a more explicitly comparative methodology. This comparative mode of analysis is elaborated, for instance, in Zukin's study of New York and London (Reading 16), Hill and Feagin's investigation of Detroit and Houston (Reading 18), Hill and Kim's account of Tokyo and Seoul (Reading 20) and Soja's classic analysis of Amsterdam and Los Angeles (Reading 21; see also Reading 27 by Grant and Nijman). Hill and Feagin (Reading 18) and Soja (Reading 21) present a particularly reflexive approach to comparative analysis. For, precisely by highlighting the persistent differences among these cities, these authors are also able to develop theoretically informed interpretations of their quite striking commonalities. Most of the readings engage in comparative analysis in order to concretize and extend some of the general propositions of global city theory. It should be noted, however, that Hill and Kim's analysis of Tokyo and Seoul (Reading 20) mobilizes a comparative approach in order to call into question some of this theory's core tenets.

Everyday life and social struggle

Globalization-induced urban restructuring has occurred in vastly diverse patterns. This diversity has also entailed markedly different everyday social conditions for the residents of rapidly globalizing centers. Globalized urbanization has threatened established working class lifeworlds and everyday routines as new "lifestyles" based upon loft living, conspicuous consumption and "yuppie" tastes have proliferated.

In addition, processes of gentrification and urban restructuring have threatened many "alternative" urban communities that had been consolidated during the 1970s, such as squatted buildings, "liberated" cultural quarters and artistic enclaves. The signature movements of this new period of confrontation over the everyday lifespaces of global cities were presciently described in a path-breaking article by Gilda Haas and Allan Heskin (1981), who underscored the new dividing line between the global economy and the local community in a series of struggles in Los Angeles in the early 1980s. Schmid's analysis of the Zurich political scene in the wake of the youth revolt (Reading 19) and Soja's comparative analysis of urban struggles and spatial outcomes in Amsterdam and Los Angeles (Reading 21) likewise usefully illustrate the new lines of conflict that have been emerging within globalizing cities (see also Reading 34 by Keil and Ronneberger and Reading 35 by Mayer).

In sum, Part Three provides readers with a more concrete perspective on how the general arguments developed by the contributors to Parts One and Two of this book have been mobilized in concrete research on globalized urbanization. To be sure, the contributors to Part Three leave open many significant questions – for instance, regarding the concept of the global city itself, regarding the preconditions and consequences of global city formation, and on the appropriate research methodology for the study of globalizing cities. Nonetheless, the readings included here provide a fruitful starting point for further inquiry into these and many other issues associated with globalized urbanization. In subsequent parts of this volume, we shall continue this exploration with reference to a broader range of empirical cases and a number of substantive themes – such as governance, state restructuring, social mobilization, representation, identity and culture – that are addressed only in passing in the readings presented in Part Three.

References and suggestions for further reading

Abu-Lughod, J. (1999) *New York, Chicago, Los Angeles: America's Global Cities*. Minneapolis, MN: University of Minnesota Press.

Beauregard, R. (1991) Capital restructuring and the new built environment of global cities: New York and Los Angeles, *International Journal of Urban and Regional Research*, 15, 90–105.

Castells, M. (1994) European cities, the informational society, and the global economy, *New Left Review*, March/April, 19–32.

Cybriwsky, R. (1991) *Tokyo: The Changing Profile of an Urban Giant*. London: Belhaven.

Davis, M. (1990) *City of Quartz: Excavating the Future in Los Angeles*. London: Verso.

Douglass, M. (1989) The transnationalization of urbanization in Japan, *International Journal of Urban and Regional Research*, 12, 425–454.

Douglass, M. (1993) The "new" Tokyo story: restructuring space and the struggle for place in a world city. In K. Fujita and R.C. Hill (eds) *Japanese Cities in the World Economy*. Philadelphia, PA: Temple University Press, 83–119.

Fainstein, S. (1994) *The City Builders: Property, Politics and Planning in London and New York*. Cambridge, MA: Blackwell.

Fainstein, S., Gordon, I. and Harloe, M. (eds) (1992) *Divided Cities: New York and London in the Contemporary World*. Cambridge, MA: Blackwell.

Gladstone, D. and Fainstein, S. (2001) Tourism in US global cities: a comparison of New York and Los Angeles, *Journal of Urban Affairs*, 23, 1, 23–40.

Grosfoguel, R. (1995) Global logics in the Caribbean: the case of Miami. In P.L. Knox and P.J. Taylor (eds) *World Cities in a World-system*. New York: Cambridge University Press, 156–170.

Haas, G. and Heskin, A.D. (1981) Community struggles in L.A., *International Journal of Urban and Regional Research*, 5, 4, 546–564.

Hamnett, C. (2003) *Unequal City: London in the Global Arena*. London: Routledge.

Hitz, H., Schmid, C. and Wolff, R. (1994) Urbanization in Zürich: headquarter economy and city-belt, *Environment and Planning D: Society and Space*, 12, 2, 167–185.

Jacobs, J.M. (1994) The battle of bank junction: the contested iconography of capital. In S. Corbridge, N. Thrift and R. Martin (eds) *Money, Power and Space*. Cambridge, MA: Blackwell, 356–382.

Keil, R. (1998) *Los Angeles: Globalization, Urbanization and Social Struggles*. New York: John Wiley & Son.

Keil, R. and Lieser, P. (1992) Frankfurt: global city–local politics. In M.P. Smith (ed.) *After Modernism: Global Restructuring and the Changing Boundaries of City Life*. New Brunswick, NJ: Transaction, 39–69.

King, A.D. (1991) *The Global City: Post-Imperialism and the Internationalization of London*. New York: Routledge.

Krätke, S. (2001) Berlin: towards a global city?, *Urban Studies*, 38, 10, 1777–1799.

Logan, J. and Molotch, H. (1987) *Urban Fortunes: The Political Economy of Place*. Berkeley and Los Angeles: University of California Press.

Machimura, T. (1998) Symbolic use of globalization in urban politics in Tokyo, *International Journal of Urban and Regional Research*, 12, 183–194.

Markusen, A. and Gwiasda, V. (1994) Multipolarity and the layering of functions in world cities: New York City's struggle to stay on top, *International Journal of Urban and Regional Research*, 18, 2, 181–194.

Mele, C. (1996) Globalization, culture and neighborhood change: reinventing the Lower East Side of New York, *Urban Affairs Review*, 32, 3–33.

Mollenkopf, J. and Castells, M. (1992) *Dual City: Restructuring New York*. New York: Russell Sage.

Olds, K. (1998) Globalization and urban change: tales from Vancouver via Hong Kong, *Urban Geography*, 19, 360–385.

Pryke, M. (1991) An international city going global: spatial change in the City of London, *Environment and Planning D: Society and Space*, 9, 197–222.

Rimmer, P. (1986) Japan's world cities: Tokyo, Osaka, Nagoya or Tokaido Megalopolis?, *Development and Change*, 17, 1, 121–158.

Sassen, S. (2001 [1991]) *The Global City: New York, London, Tokyo*, 2nd edition. Princeton, NJ: Princeton University Press.

Scott, A.J. (1988) *New Industrial Spaces*. Berkeley and Los Angeles: University of California Press.

Soja, E., Morales, R. and Wolff, G. (1983) Urban restructuring: an analysis of social and spatial change in Los Angeles, *Economic Geography*, 59, 195–230.

Thrift, N. (1994) On the social and cultural determinants of international financial centres: the case of the City of London. In S. Corbridge, R. Martin and N. Thrift (eds) *Money, Power and Space*. Cambridge, MA: Blackwell, 327–355.

Todd, G. (1995) "Going global" in the semi-periphery: world cities as political projects. The case of Toronto. In P.L. Knox and P.J. Taylor (eds) *World Cities in a World-system*. New York: Cambridge University Press, 192–214.

15

Prologue

"Cities, the Informational Society and the Global Economy"

from Centrum voor Grootstedelijk Onderzoek, University of Amsterdam, *Working Papers Series* (1993)

Manuel Castells

An old axiom in urban sociology considers space as a reflection of society. Yet life, and cities, are always too complex to be captured in axioms. Thus the close relationship between space and society, between cities and history, is more a matter of expression than of reflection. The social matrix expresses itself into the spatial pattern through a dialectical interaction that opposes social contradictions and conflicts as trends fighting each other in an endless supersession. The result is not the coherent spatial form of an overwhelming social logic – be it the capitalist city, the pre-industrial city or the ahistorical utopia – but the tortured and disorderly, yet beautiful patchwork of human creation and suffering.

Cities are socially determined in their forms and in their processes. Some of their determinants are structural, linked to deep trends of social evolution that transcend geographic or social singularity. Others are historically and culturally specific. And all are played out, and twisted, by social actors that impose their interests and their values, to project the city of their dreams and to fight the space of their nightmares [. . .] Sociological analysis of urban evolution must start from the theoretical standpoint of considering the complexity of these interacting trends in a given time-space context [. . .]

In recent years, a new trademark has become popular in urban theory: capitalist restructuring. Indeed it is most relevant to pinpoint the fundamental shift in policies that both governments and corporations have introduced in the 1980s to steer capitalist economies out of their 1970s crises. Yet more often than not, the theory of capitalist re-structuring has missed the specificity of the process of transformation in each area of the world, as well as the variation of the cultural and political factors that shape the process of economic restructuring, and ultimately determine its outcome.

Thus the deindustrialization processes of New York and London take place at the same time that a wave of industrialization of historic proportions occurs in China and in the Asian Pacific. The rise of the informal economy and of urban dualism takes place in Los Angeles, as well as in Madrid, Miami, Moscow, Bogota and Kuala Lumpur, but the social paths and social consequences of such similarly structural processes are so different as to induce a fundamental variegation of each resulting urban structure.

The new spatial logic, characteristic of the Informational City, is determined by the pre-eminence of the space of flows over the space of places. By space of flows I refer to the system of exchanges of information, capital, and power that structures the basic processes of societies, economies and states between different localities, regardless of localization. I call it "space" because it does have a spatial materiality: the directional centres located in a few selected areas of a few selected localities; the telecommunication system, dependent upon telecommunication facilities and services that are unevenly distributed in the space, thus marking a telecommunicated space; the advanced transportation system, that makes such nodal points dependent on major airports and airlines services, on freeway systems, on highspeed trains; the security systems necessary to the protection of such directional spaces, surrounded by a potentially hostile world; and the symbolic marking of such spaces by the new monumentality of abstraction, making the locales of the space of flows meaningfully meaningless, both in their internal arrangement and in their architectural form. The space of flows, superseding the space of places, epitomizes the increasing differentiation between power and experience, the separation between meaning and function.

Urban life muddles through the pace of history. When this pace accelerates, cities – and their people – become confused, spaces turn threatening, and meaning escapes from experience. In such disconcerting yet magnificent times, knowledge becomes the only source to restore meaning, and thus meaningful action.

16

"The City as a Landscape of Power: London and New York as Global Financial Capitals"

from L. Budd and S. Whimster (eds),
Global Finance and Urban Living (1992)

Sharon Zukin

Editors' introduction

Sharon Zukin is Broeklundian Professor of Sociology, Brooklyn College, at the CUNY Graduate Center in Manhattan. She has written extensively on the changing built environments and cultural landscapes of US cities undergoing major economic transformations. Her most influential publications include the award-winning books, *The Cultures of Cities* (1995) and *Landscapes of Power: From Detroit to Disney World* (1991). Zukin's contribution here was written while she was completing the latter book and was published in an edited volume focused on the local sociospatial changes that were unfolding in global financial centers. Zukin's comparative case study analyzes the built environments of London and New York City, where a variety of state-subsidized mega-projects were constructed during the course of the 1980s in order to promote downtown redevelopment and to reinforce each city's position as a global financial center. Whereas entirely new quasi-governmental bodies were established in London to build King's Cross and Docklands, the construction of Battery Park City and the redevelopment of 42nd Street/Times Square in New York City continued a longer trend of governmental support for large real estate capital. Zukin frames her analysis of these mega-projects around the crucial distinction between *vernacular* – the spaces of everyday life – and *landscape* – the spaces of power dominated by capital and state institutions. For Zukin, the built environments of global financial centers are zones of intense contestation in which the forces of vernacular and landscape continually clash. The basic question, for Zukin, is this: should the space of the global city be organized to facilitate the everyday social reproduction of working people, or should it be reshaped to serve the demands of global corporations and financial elites? As Zukin shows, the local governmental institutions of New York and London appear to favor the latter goal, but in so doing, they also expose the city to the turbulence of global markets. Thus the "landscape of power" forged in each city is never entirely secure or stable; it may be threatened both from within, through social unrest and resistance, and from without, through externally imposed geoeconomic dislocations. Reading 16 is a useful example of a case study that draws upon empirical material from two major global cities in order to develop a broader set of arguments regarding the politics of urban sociospatial restructuring.

The global cities that have captured attention in recent years owe much to intense competition in international financial services. Partly a matter of providing world-class commercial facilities, and partly a matter of image-creation, the effort to attract geographically mobile investment activity changes a city's perspective. The old, diversified urban center is cleaned up for new offices and cultural consumption; in the process, it becomes more expensive. Not surprisingly, governmental priorities shift from public goods to private development.

Local officials in New York and London have pursued this sort of growth with ambivalence. Forced to shed traditional allies from labor unions and the left, they have been disciplined by fiscal crisis, co-opted by property developers, taken in hand by financial institutions and quasi-public authorities, and scolded by national political leaders. For their part, national governments have eagerly anticipated the smaller world and larger cities that follow global financial markets. In the United States, the tradition of local autonomy precludes central government's taking a key role in directing New York's growth. In the UK, however, the national government has intervened to spur the City's eastward expansion. From 1980, activist conservative governments in both Britain and the United States broadened official support for the expansion of financial markets – and for an inevitable struggle between New York and London over priority of place. Competitive efforts to capture global markets in a single place would lead cities to the same general strategies [. . .]

It's all very well to describe such changes in terms of a traditional agglomeration economy. Yet the degree of conscious competition between cities that is involved, in a political context of seeking private resources for public problems, creates a new version of "cities in a race with time." Municipal officials of London, New York and even Tokyo must worry not only about the competitive ability of firms that are based within their realms – and the employment opportunities and tax revenues that might slip from within their grasp; they must also build an infrastructure that attracts and retains world-class financial actors. This infrastructure includes advanced telecommunications facilities,

Plate 13 New York City (Roger Keil)

a computer-literate workforce, and new sky-scraper office buildings that lift urban identity from the modern to the spectacular [. . .] The inter-related effects of economic structure, institutional intervention and cultural reorganization are most directly perceived in change in the landscape: creating the city as a landscape of power.

LANDSCAPE AND VERNACULAR

Cities always struggle between images that express a landscape of power and those that form the local vernacular. While power in modern times is best abstracted in the skyscraper outline of a city's financial wealth, the vernacular is most intimately experienced in low-lying residential neighbor-hoods or *quartiers* outside the commercial center. Much of the social quality of urbanity on which a world city depends reflects both the polished land-scape and the gritty vernacular as well as the ten-sion between them. New Yorkers point with pride, for example, to their "city of neighborhoods," whose architectural and economic diversity suggests

the cultural and social heterogeneity of its popu-lation. Londoners praise their city's informal domesticity that rests on relatively small houses and a mixture of social classes in each area. By contrast, each global city has at least one densely built, centrally located, high-rise district that drives both property values for the metropolitan region as a whole and office employment in finan-cial and other business services.

The coherent vertical landscape of this center – the Wall Street area in downtown Manhattan and the City of London – is circumscribed by a segmented horizontal vernacular of working-class districts and low-income, immigrant ghettos. Yet the opposition between landscape and vernacular refers to more than mere architectural or art-historical categories. Just as *landscape* shows the imprint of powerful business and political institutions on both the built environment and its symbolic representation, so does *vernacular* express the resistance, autonomy and originality of the powerless. Their opposition, moreover, suggests an important *asymmetry* of power. Since elites are capable of imposing multiple perspectives on the

Plate 14 London (Roger Keil)

surrounding vernacular, landscape implies their special contribution as well as the entire material and symbolic construction (Zukin 1991).

The juxtaposition between landscape and vernacular in New York is easily visualized by taking the subway line between Manhattan and Brooklyn across the Manhattan Bridge. On the Manhattan side of the East River, at the southern tip of the island, several generations of twentieth-century skyscrapers raise a profusion of gothic, flat-top and mansard roofs against the sky. The Woolworth Building was the tallest skyscraper in the world when it was built in 1913. Now [1992] it is dwarfed by the elongated twin cigar-box towers of the World Trade Center, completed in the early 1970s, and the lower yet no less massive commercial and residential development of Battery Park City. From the train windows one also sees in miniature along the waterfront the few remaining early nineteenth-century merchants' warehouses and the wholesale fish market that make up South Street Seaport, a commercial redevelopment of trendy bars, retail stores and tourist shops in a municipally designated historic landmark zone. Directly beside the elevated tracks are the red-brick, late nineteenth-century tenements and loft buildings of Chinatown. The train rushes past open windows where ceiling fans revolve and fluorescent lights shine on Chinese garment workers, who are engaged in the only expanding manufacturing sector left in Manhattan. North of the bridge, where the subway re-enters the earth, a dense array of public housing projects is aligned along the shore.

In London's East End, a more horizontal landscape is also chronologically segmented by financial capital and social vernacular, juxtaposing rich and poor in close proximity. The Docklands Light Railway connects the City's stern office buildings with the brighter, more reflective steel-and-glass or restored commercial centers at Canary Wharf, Tobacco Dock and the Royal Albert Dock, two miles away. The railway rushes past council flats whose building walls are striped with graffiti. It also passes clusters of small single-family council houses that stand as solitary out-croppings on ground cleared for new construction. Docks that have not yet been demolished or redeveloped are stark monuments to Victorian industry: huge, empty, lacking a function in the eight-square-mile,

purpose-built financial quarter around them, where Canary Wharf alone occupies 71 acres.

Yet none of this is permanent. Just as landscape has often been transformed into low-rent quarters for the poor or artisanal workshops, so has vernacular been changed into a new landscape of power.

This spatial metaphor unifies some of the disparate changes we see around us in the city, and couches our unease with an unfamiliar material reality. But change has not been so sudden as developments like Docklands and Battery Park City imply. Much of the landscape has been re-made incrementally. After all, London and New York have always had high property values, especially downtown and in the City. Further, gentrification proceeds by individual houses and streets; although it changes the character of an area, it rarely occurs on so dense a scale as greatly to raise aggregate income or educational levels. And despite the perennial tension between the developers' urge to tear down and reconstruct and the residents' desire that things remain as they are, most large redevelopment projects, although often compromised in size and style, have in fact been built. Seeing the landscape of the City is believing in its mission, for visual consumption holds the key to the old geographical center's creative destruction. The very syntax, the rules by which we perceive it, have been changed and, as the rules have changed, so has the way we use the city.

A structural precondition of these shifts has been the abandonment of most urban manufacturing. Since the 1960s, New York and London have lost half their blue-collar jobs to automation, regional decentralization and the internationalization of industrial production. Certainly jobs in processing industries (e.g. chemicals, foods) and product assembly had located years earlier on the periphery or in the suburbs. Yet low rents and easy access to markets kept the garment and printing industries in the center, along with a decreasing number of traditional crafts and, of course, the docks. The displacement of these jobs, however, freed the areas in which they were located for other uses. Property values that had been restricted by manufacturing plants and working-class housing could now reflect the inherent utility of a central location. Higher land values exerted pressure, in turn, on remaining industrial uses. The social meaning of this space was, moreover, freed for reinterpretation and new appropriation. The previously "closed" vernacular of working-class districts, immigrant ghettos and industrial areas that were always regarded as dangerous and suspicious, especially after dark, were now "open" as landscape to those who could consume it. These factors fed a wave of private-market property investment from new office construction to gentrification.

Historic building and district classifications have greatly contributed to opening up the vernacular to a broadly defined upper middle class. So have restaurants, gourmet groceries and stores that sell artisanal or artist-designed products. These amenities lure consumers with cultural capital, i.e. the experience, education and time to seek them out. And artists' districts – notably, Manhattan's loft district of SoHo – have also opened up the dingy charms of the vernacular by incorporating them into a landscape of cultural consumption (Zukin 1988).

While individuals change their use of the center, geography reflects the center's new functions. Clerkenwell and parts of Hackney lose their somewhat heterogeneous social character, emerging as enclaves of history that can best be apprehended by means of a guide. Spitalfields sheds the uneasy shadow of artisans' workshops, the wholesale market and Jack the Ripper to be presented as a prime site for commercial redevelopment. Docklands is re-created as a direct expansion of the financial center on the one hand, and an urban frontier, on the other. And if the railroad yards and canals at King's Cross are unified by Norman Foster's new urban center, they will lose their "inaccessible," somewhat desolate and inchoate quality: King's Cross will take its place between Camden and Islington in a coherent North London landscape.

A TALE OF TWO CITIES

With big projects like King's Cross and Docklands, London seems to escape traditional institutional bounds, i.e. the instrumental roles that are constituted by various territorial levels of government, social communities and public–private distinctions. In New York, on the other hand, big projects like Battery Park City or the 42nd Street Redevelopment at Times Square seem only to extend institutional precedents by which developers get to build what they want. For over 150 years New York's high land values have driven low-rent, low-class users out of the center, including working-class neighbourhoods, manufacturers and small crafts shops. But isn't each city really more complex? London's growth has been restricted by the Green Belt since the 1930s, as well as by political pressure by central government and an extraordinarily high percentage of housing stock in council housing; yet one observer of recent urban changes says that London is geographically and politically more mutable than New York. For its part, although New York has always been in thrall to business elites, a change of mayors (albeit usually within the Democratic Party) and extraordinary mobilization by citizens' groups, often in the law courts, have sometimes shifted the purpose and reduced the size of new construction, and occasionally slowed the pace of change.

Despite these differences, the landscapes of power that are emerging in New York and London are strikingly similar. This is hardly surprising, for redevelopment of the center in both cities is commissioned and designed by the same worldly superstars, including developers, architects and private-sector financial institutions. Just as skyscrapers have become the *sine qua non* of place in the global hierarchy of cities, so do US, Japanese and Canadian builders and bankers represent the basis of global market rank. A city that aims to be a world financial center makes deals with Olympia & York and Kumagai Gumi, welcomes Citibank and Dai-Ichi Kangyo, and transplants Cesar Pelli as well as Skidmore Owings and Merrill and Kohn Pederson Fox.

Developers tend to use the same strategic tools. They seek to diversify their portfolios by spreading investment projects around the world, and find these investment sites according to computerized projections made by the same accounting, property and management firms, which are also organized on a worldwide basis [. . .] Wrongly or not, the progressive lowering of barriers to multinational trade in services, as in London's Big Bang of 1986 and the consolidation of Western European markets in 1992, has encouraged this foreign participation [in local property markets]. Yet unified markets do not require a large number of first-tier financial capitals. And the close links between trading practices in global market centers probably lead to a coordinated overexpansion of office space, like the coordinated retrenchment of personnel in New York and London following the stock exchange crash of 1987 and the increased volatility in the market in 1989. In other words – despite the common wisdom – it may no longer be a good idea to use property development in London as countercyclical hedging against loss of investment value in New York.

New York's office market should at any rate provide a cautionary tale. While foreign banks cut investments and employment less in New York than in London following the 1987 crash, and domestic financial firms either delayed massive layoffs or limited them to specific low-yield activities, office vacancy rates rose as US firms trimmed their payrolls selectively and physically consolidated operations [...] Even the large money center banks have added slack to New York's office market. While generally profitable and still expanding overseas, these financial institutions decided to sell some of the remarkable corporate headquarters buildings with which they were identified and move back-office operations outside Manhattan. Chase Manhattan Bank, long the linchpin of and major property investor in the Wall Street area, moved back-office operations to a new building in downtown Brooklyn. Citibank sold most of its diagonal-roof headquarters to a Japanese insurance company, building a distinctive new skyscraper across the East River in Long Island City. Storing assets in face of Third World debt and Japanese competition may have been only one among many reasons for Citibank's action, for the bank is playing an important part in the projected commercial redevelopment of the waterfront in that area of Queens.

But the pattern of playing musical chairs with office relocations has long been typical of New York's commercial property markets. City government subsidies and zoning changes have recently lured major developers from the overbuilt East Side to the West Side of midtown Manhattan, and office building owners attract large corporate tenants by offering special facilities, rent reductions, build-out allowances for tenants' improvements, and even a share of equity. These conditions allowed developer William Zeckendorf Jr to sign up the Madison Avenue advertising firms Ogilvy & Mather and N.W. Ayer and the Wall Street law firm Cravath Swain & Moore as tenants in Worldwide Plaza, his new mixed-use complex on Eighth Avenue north of Times Square.

Governmental connections also aid developers' efforts to create a locational advantage. During the decades that David Rockefeller directed Chase Manhattan Bank, he formed a business group, the Downtown-Lower Manhattan Association, in 1957. The association's desiderata for protecting Chase's property investment in the area began with relocating the wholesale food markets to the Outer Boroughs, and included an ambitious program to rebuild the municipal administrative center, develop middle-income housing, attract university facilities and establish new bellwether skyscrapers that could compete with the more modern offices in midtown. The city government implemented most of this wish-list during the 1960s. Moreover, when David's brother Nelson was elected governor of New York State – a post he held from the 1960s to the mid-1970s – powerful state institutions were created and utilized for large-scale urban redevelopment. A new public authority, the New York State Urban Development Corporation, superseded local powers to condemn property, design projects and finance their construction by issuing bonds. The Battery Park City Authority controlled new waterfront construction. Meanwhile, the older Port Authority of New York and New Jersey, which manages most of the mega-scale transportation systems in the metropolitan region, used its jurisdiction over the port to commission its redevelopment. Mainly due to these institutions, the World Trade Center was conceived and built on waterfront landfill. To aid the process of marketing this mammoth space, some say, a local law requiring new fire protection devices was passed, hastening tenant relocation from older office buildings throughout lower Manhattan into the World Trade Center. The Rockefellers plied a winning strategy in the downtown office market with their influence over government, ability to move both public and private investment capital into new construction, and fine-grained understanding of the value of architecture as symbolic capital. This was in any event a period of manufacturing decline and business service boom, accentuated in property markets by the levers of public and private power.

Until recently, in fact, commercial property development in New York was seldom constrained by popular protest or political opposition. In contrast to housing and highway construction, where community groups made their pressure felt, the business district in the center was both materially and symbolically regarded as the province of business elites. Industrial decentralization toward the Outer Boroughs, New Jersey and Long Island had reduced manufacturers' and workers' influence over land use in housing over commercial development. By the sixties, large-scale removal of Manhattan residents occurred on only two, non-commercial project sites: the cultural complex at Lincoln Center and the primarily residential Upper West Side urban renewal area, both of which were intended to complement rather than expand the midtown business center. Thus by 1970 no residential community remained to voice complaints about a "change of character" in the center.

During the 1980s, however, the enormous expansion of New York's landscape of power was seen as having taken a toll on both "quality of life" and "affordable housing," two shibboleths of liberal opposition to business strategies. Yet without a tenacious working-class presence and pressure groups, as in Tower Hamlets and nearby boroughs of London, institutional bargaining focused mainly on trade-offs. On the one hand, the city government fully acknowledged the right to create private value by permitting taller buildings with more rentable space in congested parts of the center. On the other hand, city officials tried to impose an obligation on developers to restore or renew public value, by adding plaza-like spaces open to public use, incorporating theatres or historic landmark structures instead of tearing them down, or refurbishing parks and subway stations near their building sites [. . .]

IN THE LONG RUN

The material and symbolic reconstruction of the center is a long-term historical process. The sudden appearance of a Docklands or Battery Park City masks the gradual effects of structural change, notably, a shift from organizing the city as an assortment of concrete production spaces toward visualizing it as a coherent space of abstract financial processes and consumption. By the same token, the seeming abruptness with which urban redevelopment schemes are adopted and abandoned reflects temporary booms and busts in property markets and changing political alliances, as well as financial market advances and declines. Yet, in addition to redeveloping capitalist space, the landscape of power reflects both the incremental institutional interventions of the national and local state and the conscious cultural reorganization of business, political and artistic elites – those who wield *all* the levers of social power.

Battery Park City and Docklands offer concise examples of the morphological, geographical and aesthetic processes that are involved. They are primarily sites that government has had to market many times: first to private investors and developers, then to banks and public and private bodies with the authority to approve large-scale financing (e.g. bonds), also to a restive citizenry that presses for both quality-of-life improvements (e.g. low-rise building, low-density zoning) and affordable housing, and finally to a broad segment of the public that has both financial and cultural capital to consume the space the city has produced. These processes lead, on the one hand, to an aestheticization of the project at the expense of a focus on critical social needs such as housing and jobs. On the other hand, they lead to image-making that obscures the removal or incorporation of the segmented vernacular by the landscape of power of a world financial center [. . .]

Despite great differences between the political systems of New York and London, and different relations between their financial communities and the national economy, both major projects of contemporary new construction – Battery Park City and Docklands – show the same trajectory of a landscape of power. Significant to globalization, the redevelopment of the center in New York and London has proceeded almost in tandem. While Docklands required, and still requires, the building of more transportation infrastructure, Battery Park City has merely spawned new streets and parks, essentially consumption amenities on the water (plazas, marinas, fountains). Each project has been delayed by the continued belief among investors and developers that the project might be unmarketable, although the offices of the World Financial Center are over 90 per cent filled. Each crisis of investors' confidence was overcome by expanding the project's commercial facilities and reducing or removing the proportion of low-income housing. But the major obstacle to construction in both cases was eased by maximizing the *private* role in development. Battery Park City Authority relinquished most of its instrumental role to private developers in 1979, and London Docklands Development Corporation proceeded the same way when it replaced the earlier, borough-dominated London Docklands Joint Commission in 1981. A belated interest in community development on the part of LDDC in 1989–1990 led the corporation to lobby central government for more social services without, however, denying the priority of private-sector demands.

A multi-centered city that would somehow dilute the arbitrary expansion of the center negates the cheek-by-jowl juxtaposition of neighborliness and imperial power that characterized New York and London at their commercial height in the not-too-distant past. The old contrast between a singular landscape of power at the center and a segmented vernacular has deliberately been destroyed.

REFERENCES FROM THE READING

Zukin, S. (1988) *Loft Living: Culture and Capital in Urban Change*, 2nd edition. London: Radius/Century Hutchinson.

Zukin, S. (1991) *Landscapes of Power: From Detroit to Disney World*. Berkeley and Los Angeles: University of California Press.

Zukin, S. (1995) *The Cultures of Cities*. Cambridge, MA: Blackwell.

17

"The Urban Restructuring Process in Tokyo in the 1980s: Transforming Tokyo into a World City"

from *International Journal of Urban and Regional Research* (1992)

Takashi Machimura

Editors' introduction

Takashi Machimura is Professor in the Graduate School of Social Sciences at Hitotsubashi University in Tokyo. Machimura was a key figure in introducing neo-Marxian approaches into Japanese urban sociology during the 1980s. Subsequently, in a series of influential papers, he applied this perspective to the study of global city formation in Tokyo (Hashimoto 2002). While most of Machimura's work has been published in Japanese academic journals, he has also written several prominent English-language articles, including the selection here, for the prestigious *International Journal of Urban and Regional Research*, where much of the international debate on global city formation has taken place.

In this widely read contribution, Machimura provides a classic case study of urban restructuring in one of the paradigmatic global cities of East Asian and global capitalism: he situates the case of Tokyo in relation to other major examples of global city formation, while also forcefully emphasizing its contextually specific preconditions, dynamics and consequences. Machimura begins by outlining the massive expansion of transnational investment by major Japanese corporations during the course of the 1980s, a trend that simultaneously reinforced Tokyo's dominance within the Japanese urban system and engendered significant spatial and socioeconomic transformations within it. Machimura also shows how the process of global city formation in Tokyo was actively promoted by a rather heterogeneous growth coalition composed of governmental bodies (at both national and local scales), various factions of capital (transnational corporations, real estate firms, landowners) and other sectorally specific interests (see also Reading 19 by Schmid). Finally, Machimura explores various profound sociospatial transformations that have resulted from Tokyo's globalization – including land-use reorganization, residential recomposition and displacement, and the proliferation of struggles over housing conditions. Like Friedmann and Wolff (see Reading 6) and Zukin (see Reading 16), Machimura prominently emphasizes the deep tension within globalizing cities between capitalistic and everyday uses of urban space, and he carefully traces the articulation of this tension in the form of diverse sociopolitical conflicts. Machimura's case study thus provides an illuminating snapshot into some of the disruptive political-economic and sociospatial changes that were triggered through the process of global city formation in Tokyo during the course of the 1980s.

Since the 1970s, the economic growth of East and Southeast Asian countries has transformed regional structures within the capitalist world economy. The fact that Japan has gained a powerful influence not only as a regional center but also as a world economic center is one of these transformations. Japanese status in the global system has changed so drastically that its domestic structure has also been shaken in various ways. Above all, rapid transnationalization of Japanese capital has produced a new domestic spatial division of labor, and in effect a domestic urban and regional system. These changes have promoted the monocentric expansion of Tokyo as a global economic center.

The main aim of this chapter is to describe and interpret the process and characteristics of Tokyo's urban restructuring in the 1980s, and by doing so to contribute to the comparative study of world cities. Tokyo, like New York and London, is often regarded as one of the primary world cities in core countries. Certainly Tokyo has gained global control capability as a major world financial center, but there are nevertheless essential differences

Plate 15 Tokyo (Takashi Machimura)

between Tokyo and the other primary world cities in core countries. First, Tokyo is a purely economic world center, unsupported by a political or military hegemony of the state. In comparison, New York is a center not only of globalizing economic activities but also of global political power, and so initially was London. Second, Tokyo, in two important respects, is more dependent on a global network system than either New York or London. Because Tokyo is physically far away from the western countries that are traditional centers of the world economy, advances in telecommunications and transportation technology are indispensable for it to overcome this geographical disadvantage. In addition to this, recent Japanese prosperity depends, to a greater degree, on the wealth produced by the electronics and computer industries, which serve as the backbone to the establishment of a global network system for telecommunications. Third, Tokyo has been closed to the influx of foreign immigrants for a long time. In comparison, both New York and London have long histories as destinations for large numbers of international immigrants, which gave them an ethnically and culturally diverse nature. Their histories as centers of a hegemonic state explain this fact to a large extent. Recently, however, an increase in the number of immigrant laborers entering Japan from other Asian countries has began to gradually change this situation.

My intention here is not an exhaustive investigation of the urban restructuring of Tokyo. Since this is still an ongoing process, we cannot help but arrive at only a temporary conclusion. The focus in this chapter will be to explain what kinds of factors have transformed Tokyo into a world city, and are continuing to transform it, by describing the economic, political and social changes that Tokyo has experienced. Both internal and external factors are indispensable for understanding the formation of a world city.

TRANSNATIONALIZATION OF JAPANESE CAPITAL AND TOKYO

It is impossible to understand the urban restructuring of Tokyo in the 1980s without referring to its increasing influence on both global and national levels. Central functions on a national level have traditionally been overconcentrated in Tokyo, while Tokyo has also recently assumed a more important role on a global level. This means that the Japanese urban and regional system has been directly linked to a global system. Many factors gave rise to this link, but the most decisive one is the rapid transnationalization of Japanese capital.

The purpose of overseas investment has changed gradually as a result of Japanese economic growth and its rising position in the world economy. In the early stage of postwar economic development, a large part of Japanese investment was aimed at securing raw materials such as oil, wood and pulp. In the period of high economic growth from the late 1950s to 1973, corporations increased direct foreign investment both to support export to advanced countries and to manufacture products such as textile goods and simple electric goods in developing countries. In the early 1970s, capital export was liberalized almost completely. Since then, Japan has tried to develop its economic activities on a global scale. In the 1980s, when Japan experienced trade disputes with the United States and the EC, the rapid valuation of the yen and financial globalization, Japanese capital began to transnationalize on a greater scale.

The number of Japanese transnational manufacturing corporations (TNCs) increased from 35 in 1975 to 90 in 1987. Countries in which their subsidiaries are located fall into two groups: the less developed countries in East and Southeast Asia; and western industrialized countries, especially the United States. The decision to locate in the former areas reflects the strategy of shifting production to new low-cost industrial sites. Major electric or semiconductor companies established factories in newly industrializing economies [NIEs], ASEAN countries and China (especially in the export processing zones). But a distinctive character of recent globalization is increasing direct investment in advanced countries. For example, the opening of factories in the United States by major Japanese automobile companies was particularly important, not only because cars had been among the most important products of the United States, but also because such moves were usually accompanied by the relocation of other diverse Japanese companies producing various automobile components such as tyres, glass and bearings.

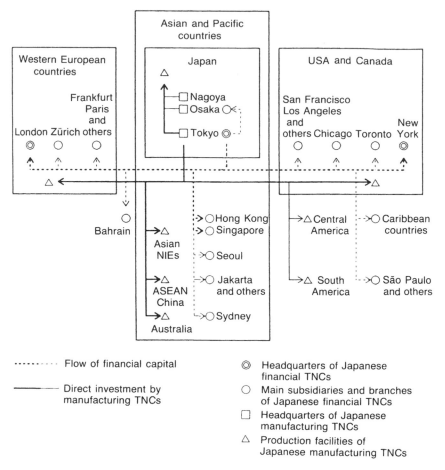

Figure 1 Global flow of capital controlled by Japanese TNCs

The recent globalization of Japanese financial and service corporations has been even more drastic than that of manufacturing corporations. As the international realtime network among major financial centers has been developed, Japanese financial corporations, including commercial banks, security companies and insurance companies, have transnationalized rapidly. By 1987, most of the large corporations had subsidiaries and branch offices in New York and London, primary international financial centers, and many also have them in secondary centers such as Zurich, Frankfurt, Toronto, San Francisco and São Paulo. It is also characteristic of Japanese corporations that western Pacific financial centers such as Hong Kong, Singapore, Sydney and Jakarta are also important sites of their activities.

As a result of this transnationalization of Japanese capital, Tokyo gained a role as an economic center both on a global level and on a western Pacific level, because most of the Japanese TNCs had headquarters there (Figure 1). Tokyo is in a commanding position as regards the two different capital networks, that is both among core capitalistic countries and among western Pacific ones. These changes have reinforced the monocentric structure of the Japanese urban system in recent years.

As a result of this new role, the necessity of urban development in Tokyo was often emphasized, especially by business leaders and the Japanese government. For example, the National Land Agency, in its reform plan for the national capital area made in 1985, estimated the total new demand of office space in central Tokyo to be more than 50 million square metres during the next 18 years; the total of newly built office space in the same area in 1985 was only 1 million square metres, so we can see the magnitude of the proposed growth. However, urban restructuring policy was not automatically begun to meet these economic demands. The choice of this policy depended upon more diverse conditions at both national and local levels.

THE POLITICAL BACKGROUND OF URBAN RESTRUCTURING

National policy for regional development in postwar Japan was oriented mainly to non-metropolitan areas, because its basic purpose was to resolve problems of regional economic imbalance. Thus, adopting the urban development of Tokyo as a major national policy meant a drastic alteration of a traditional course, and therefore, needed some rationale that would justify the new policy.

The change in the Japanese position within the world economy set fundamental conditions for domestic urban development in two ways. One is concerned with a process of world city formation. Because of the sharp trade disputes with the United States and the EC, the Japanese government had been forced to adopt a policy for increasing domestic demand. And in the early 1980s, the major field of policy that the Nakasone administration chose to increase domestic demand was urban development. The fact that the first and most direct impetus to urban restructuring had come not from inside the city but by the choice of national government put a specific limit on the development of the policy. Second, the general trend toward an "information society" provided an important background to the recent urban transformation. In particular, the establishment of infrastructures for information networks, such as the Tokyo Teleport, was thought to be essential for sustaining the economic activities of Tokyo as a network-dependent world city.

The fiscal crisis of the Japanese state in the late 1970s led to the formation of private–public partnerships in many fields of policy. In order to introduce private capital into the sphere of public development, many incentives to investment – such as low tax, large subsidies, deregulation etc. – were regarded as indispensable. Among the reasons why urban development was selected as one of the main areas for the formation of public–private partnerships, its high profitability was undoubtedly a major one. In order to realize this partnership, many restrictions on construction and urban development were established, and some nationally owned land within central Tokyo was sold to private real estate companies. As a result, the urban space of Tokyo has become an arena for capital accumulation on a huge scale (Figure 2).

Although a city usually has various kinds of actors, who all aim at different goals, a coalition for urban growth is often formed in capitalist cities (Logan and Molotch 1987). According to Logan and Molotch, a basic goal shared by the members of such a coalition is usually the pursuit of exchange value in the form of rent. We find this type of coalition in Tokyo, but here it was formed for a more

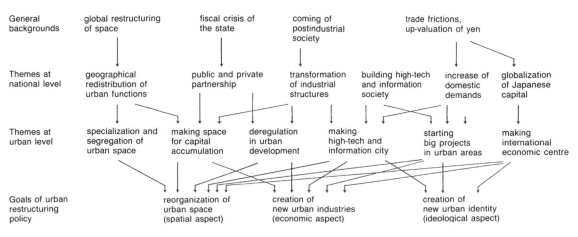

Figure 2 The dominant logic of urban restructuring: Tokyo in the 1980s

general goal – urban restructuring for world city formation.

The members of the coalition divide broadly into two major groups. The public sector of the coalition consists of the national government, local governments (Tokyo metropolitan government and the city or ward governments) and other special public corporations. However, since each ministry of the national government (e.g. MITI, the Ministry of Construction, the Ministry of Posts and Telecommunication, the Ministry of Transport and National Land Agency) tends to defend its own interests in the urban field, the national government cannot always be regarded as a unified body. The actors in the private sector are divided into groups according to their positions in the urban restructuring process.

- The first are direct beneficiaries, making profits from urban development actions themselves. They include: (1) manufacturing companies whose products are used for construction and development (e.g. steel and cement companies); (2) construction companies and architectural firms; (3) developers and real estate companies which buy and develop land for sale, lease and rent; (4) financial companies which make loans for land purchase and construction; and (5) landowners who can make profits by selling land or by becoming developers.
- The second group consists of those who need more space in central Tokyo to extend their own businesses. They include foreign firms, financial companies, electronics and computer companies, business services, hotels and leisure services.
- The third group covers those who try to influence the symbolic and semiotic structure of urban space and to diffuse urban ideology: big commercial capital, advertising agencies and various media institutions play important roles as "space directors."

[...] While the coalition linked the specific interests of selected groups and individuals, its latent function was the exclusion of many important urban agents from the decision-making process of urban restructuring. The criteria of member selection were not obvious, but there were tacit understandings about common interests among members. For example, enlarging an opportunity of capital accumulation in order to introduce big companies into projects is one of the basic presuppositions. Moreover, the decision-making processes in such organizations were usually not open to the urban public. Therefore tenants, small-size landowners and small enterprises in central Tokyo, although they were often influenced profoundly by the projects, were usually excluded from the coalition.

However, the fact that the coalition based on those organizations was not perfectly well-organized or monolithic should be strongly emphasized. In each stage of planning, decision-making, or execution, we can find many types of conflicts or controversies. In addition to this, there are many kinds of social tensions outside of the coalition. This point will be explained in the following section.

MAKING TOKYO A WORLD CITY

To understand how urban restructuring has been carried out in Tokyo, attention has to focus on its spatial, economical, social and cultural sides. The first and the most obvious change in the 1980s was a sharp increase in space production. In general there are three basic types of space production: horizontal expansion, vertical expansion and renewal of old business space. Needless to say, space production itself is not a unique phenomenon, but what was important in the 1980s was the size and effect of it.

Secondly, as Tokyo's global control functions have expanded, the relationships between urban functions and land use patterns have been forcibly transformed. One of the main characteristics of recent world cities is the expansion of global control functions, which are supported by various commercial activities such as banks, legal and accountants' offices, advertisement agencies and information services [. . .] On the whole, the basic pattern of change was, first, to select the urban functions which are worth locating in central districts; second, to remove old and unnecessary functions from them; and third, to make new space for new functions. It is easy to find behind this pattern the basic intention of turning Tokyo into a world city, oriented primarily to various economic functions. As a result of these processes, many districts

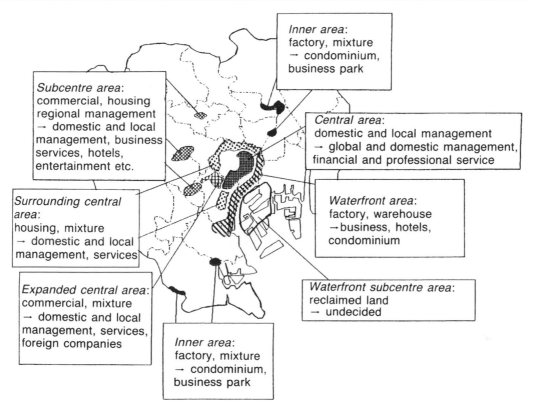

Figure 3 The geographical pattern of functional change in central Tokyo in the 1980s

within Tokyo have been forced to change their economic functions and spatial forms (Figure 3).

The third and unexpected outcome of restructuring is that the number of foreign residents has increased considerably during the last decade in Tokyo, and this has changed the ethnic composition of the population. This seems to be an entirely different process from those mentioned above, but it is an important part of world city formation. Japan is often said to be a single-nation state, but actually it has some ethnic minority groups, such as Koreans and Chinese. Most of them are people who came spontaneously or were forced to come before the end of the Second World War and their descendants, who now desire to be permanently settled in Japan. By contrast with these old minorities, the characteristics of newcomers in 1980s become clearer. Broadly speaking, most of them are migrant laborers from peripheral countries in East, Southeast and South Asia. Male laborers are employed as manual workers in sweatshops or construction industries,

as cleaners, waiters etc., while females work as waitresses, hostesses, dancers, maids and in some cases as prostitutes [. . .]

The process of Tokyo's urban restructuring started from an economic dimension and has extended to spatial and social dimensions, then further to a cultural or ideological dimension. I will cite two examples here. First, since the mid-1980s the Tokyo metropolitan government has formulated the policy of internationalization (*kokusaika*), which in this context means providing both physically and culturally appropriate conditions for the activities of foreign capital and foreign residents. Above all, an attempt to increase the cultural tolerance of Japanese society is thought to be the most important part of this policy. In addition to this, an attempt to create a new urban identity for Tokyoites has become an important cultural policy during the period of urban restructuring. The social change that Tokyo has experienced during the past decades has been so drastic that the Tokyo metropolitan government has eagerly sought an official new image. Now two types of image-making process can be found there: one is through a re-emphasis of the historical continuity from Edo (the old name of Tokyo before the Meiji Restoration) to Tokyo; another is through a comparison with other world cities such as New York, Paris and London. A series of such attempts is often summarized by the Tokyo metropolitan government as the "Tokyo Renaissance" or "Tokyo Frontier" policy. In both cases, holding attractive events, building symbolic memorials and publishing books and Journals are common means for achieving their goals. The effects of such ideological mobilization have not been obvious, but it has become an important plan of urban restructuring, just as in many other cities – for instance, the "I Love New York" campaign of New York City.

SOCIAL TENSION AND THE FUTURE IMAGE OF THE CITY

There are many types of tensions brought about by the urban restructuring of Tokyo. Yet, among these, the most important is the social conflict between the coalition which pursues exchange value of space, and ordinary urban residents, who pursue its use value as living space. In the 1980s the following three forms of tension, both manifest and latent, were important in Tokyo.

First, the expansion of business space into traditional urban neighborhoods caused the most direct type of conflict in the central districts and inner areas. Enormous and disorderly land purchases by real estate companies adversely affected living conditions and broke close and stable relationships and community consciousness which had supported the integrity of neighborhood societies for a long time. Second, the rapid increase in land prices deepened the split between urban residents all over Tokyo. For those who could pursue the exchange value of space, it meant a considerable increase in the value of fixed assets. This enabled landowners to finance their enterprises on an enormous scale and to invest capital to buy land and building, not only in Japan but also in foreign countries. On the other hand, the rise in land prices increased the cost of living, such as rent or property tax, for ordinary residents who make use of space only for housing, and forced a lot of

people to move out from central Tokyo. Third, the fact that short-range economic interests took precedence over the grand design for a future Tokyo produced new sources of urban problems. Consequently, sites of urban redevelopment were restricted mostly to places where there were less obstacles to its progress or where making profits was easier: publicly owned land, reclaimed land, sites of unused factories etc. Most of the physically or economically deteriorated residential areas, which actually needed improvement or renewal, were neglected because of the difficulty or lower profitability of developing them. Moreover, the remarkable rise in land prices made it more difficult for local governments to control land use patterns and to start a new urban reform project. Overall, then, the urban restructuring process has tended to increase rather than decrease social and spatial imbalance, or the unevenness of development, in the total area of Tokyo.

In such a situation, the number of organized social movements related to urban development slightly increased in Tokyo, according to the statistical data (Machimura 1987). Certainly this may reflect the social tension caused by recent urban restructuring. Yet, generally speaking, social movements which directly confronted urban developments were not influential enough to change the main trends of the urban restructuring. Compared with the political power of urban social movements in the 1960s and early 1970s, when they were at their peak, they presently seem to be rather weak. This is due to the fact that internal conflicts about interests among residents made it difficult for them to form influential social bases of activities. In addition to this, the consciousness of political conservatism spreading throughout the nation in the 1980s had some deleterious effects on these movements.

CONCLUSION

This chapter has described the process of the urban restructuring of Tokyo in the 1980s. From a long-term point of view, it might be only the first stage of a continuing process, but the fact that Tokyo in the 1980s was at a turning point in its urban development cannot be denied. Will Tokyo continue to take a course to becoming a giant growth machine, solely for processing global capital and information? At present, much is uncertain, but it is necessary to consider the limit of growth which Tokyo as a network-dependent world city may face in the near future.

As we have mentioned earlier, Tokyo's global functions will be basically limited to the economic sphere. This specialization will give Tokyo many weak points. First, continuing globalization of Japanese TNCs may possibly lead to the relocation of their headquarters functions to other world cities, thus damaging Tokyo's present position. Second, Tokyo's position can relatively easily change in response to international and domestic factors. Therefore, a severe competition for the primary or secondary position of urban hierarchies in the western Pacific Rim has already been fought and will continue to be fought among major Asian cities such as Tokyo, Osaka, Singapore, Hong Kong, Seoul etc. In addition to these external factors, Tokyo itself is beginning to face many internal problems specific to world cities.

Tokyo and other Japanese cities are now entering a new stage of urban social change. But the ongoing urban restructuring policy in Japan has inclined too much to economic development. It is necessary to reconsider a future course for Japanese cities in a wider context, and this will lead us to examine the future form of the urban network in East and Southeast Asia. The present urban transformation can be seen as a kind of urban crisis, but at the same time we should not forget that it is also a good starting point for reconstructing the city from a new point of view.

REFERENCES FROM THE READING

Hashimoto, K. (2002) New urban sociology in Japan: the changing debates, *International Journal of Urban and Regional Research*, 26, 4, 726–736.

Logan, J. and Molotch, H.L. (1987) *Urban Fortunes*. Berkeley, CA: University of California Press.

Machimura, T. (1987) Teiseichouki niokeru toshi shakai undo no tenkai (Development of urban social movements in the period of stable economic growth). In A. Kurihara and K. Shoji (eds) *Bunka Keisei to Shakai Undo*. Tokyo: University of Tokyo Press.

18

"Detroit and Houston: Two Cities in Global Perspective"

from M.P. Smith and J.R. Feagin (eds), *The Capitalist City* (1989)

Richard Child Hill and Joe R. Feagin

Editors' introduction

We have already encountered Joe R. Feagin's work earlier in this book (see Reading 3 by Rodriguez and Feagin). In this reading, originally published in an influential edited volume on cities and economic restructuring under global capitalism, Feagin collaborates with Richard Child Hill, Professor Emeritus of Sociology at Michigan State University, in order to compare the process of urban restructuring in two major US cities, Detroit and Houston. Their goal is to decipher the developmental dynamics of each city under investigation by examining its specialized economic niche within world-scale divisions of labor; on this basis, they attempt to decipher the long-term evolution of each city with reference to a combination of geoeconomic pressures and internal political-economic struggles. In this sense, Hill and Feagin's analysis provides an elegant illustration of Friedmann's hypothesis (see Reading 7) that a city's position in the world economy has decisive implications for its internal sociospatial development.

Hill and Feagin frame their analysis around the opening observation that, as of the 1970s, each city had adopted very different approaches to political-economic regulation. Whereas Detroit had embraced a welfarist model in which state institutions attempted to correct market failures and organized labor exercised considerable political influence, Houston had embraced a neoliberal model of "free market" capitalism that privileged big business and imposed only minimal regulatory constraints upon capital accumulation. However, despite these significant regulatory differences, the authors argue, both cities experienced similar structural crises, albeit at different moments, in the face of economic fluctuations within their niche industry (automobiles in Detroit; oil and petrochemicals in Houston). In each case, industrial specialization during the post-World War II period fueled massive urban and regional growth. But, following the global economic recession of the 1970s and the oil shocks of the 1980s, this sectoral specialization rendered both local economies particularly vulnerable to shocks and fluctuations within the international division of labor. In addition, the authors emphasize a number of specific features of North American urban development that emerged during the postwar period in both cities – including extensive residential suburbanization, industrial decentralization and institutionalized racism. Thus, while Hill and Feagin's study is highly sensitive to the contextually specific pathways of industrial restructuring that each city has undergone, their analysis ultimately uncovers a variety of commonalities among them. For Hill and Feagin, it is the embeddedness of cities such as Detroit and Houston within an integrated, worldwide division of labor that renders such similarities sociologically significant.

INTRODUCTION

Detroit and Houston became urban archetypes in the United States in the 1970s. Detroit was the snowbelt city in decline; Houston, the booming sunbelt metropolis. Each city was held up as an object lesson for the other. It was commonly argued, for example, that if Detroit wanted to revitalize, the city should embrace something like Houston's freewheeling, boomtown philosophy with all that implied: aggressive business promotion, weak labor unions, social inequality, low taxes, few social services and unplanned urban sprawl. Others thought that as Houston matured, the Motor City's brand of welfare state capitalism would take hold and chart the Oil City's future. If so, Houston would experience growth in worker organization, an institutional partnership between Big Business and Big Labor, more political power among minority groups, tax increases, expanded social services and eventually, business disinvestment and fiscal crisis. But the 1980s held something different in store from what is implied in these laissez-faire versus welfare state contrasts.

Houston, in fact, was to experience an economic crisis in the 1980s not unlike the one which began to confront Detroit a decade earlier. Ironically, by the mid-1980s, it was the two cities' similarities, not their differences, that were most obvious. This suggests that answers to questions about a city's political-economic future are to be found as much beyond as within its local boundaries. Cities are spatial locations in a globally interdependent system of production and exchange. That global system is in crisis and transition. So the path a city follows in the future will depend upon the niche it comes to occupy in a changing international division of labor. It seems fruitful, therefore, to conceptualize how the city as a "localization of social forces" (Zukin 1980) is articulated with the city as a nodal point in the world capitalist system (Friedmann and Wolff 1982).

Here we take a holistic methodological approach. We assume that cities are not discrete and independent entities, but rather are interconnected parts of a world system of cities. Explanation, in this scheme of things, comes from locating parts within a larger whole in such a way as to render their complementary and contradictory relationships meaningful.

DETROIT: CRISIS IN THE MOTOR CITY

Specialization and growth

Detroit grew with the automobile industry. By importing raw materials and semi-finished products and converting them into durable finished goods to be exported throughout the world, Detroit's factories formed the heart of a vibrant international production system. Apart from auto-parts suppliers, three complementary industries have played a particularly salient role in Detroit's "metal-bending" economy: non-electrical machinery, fabricated metals and primary metals [. . .]

The Motor City's "Big Three" – General Motors, Ford and Chrysler – came to number among the world's largest corporations [. . .] The United Auto Workers (UAW) became the nation's biggest industrial union. Confrontations between the "Big Three" and the UAW ushered in a postwar era of collective bargaining. Productivity bargaining, cost-of-living adjustments and group insurance plans brought Detroit's industrial workers the highest standard of living to be found in any major North American metropolis. But it was also in Detroit that ethnic minority unemployment, inner city poverty, decaying neighborhoods and violent street unrest came to symbolize the urban crisis of the 1960s.

Crisis and reorganization

Suburbanization, institutionalized racism and uneven urban growth set the terms of political discourse in Detroit during the 1960s. By the early 1970s, however, it was capital flight to other regions and abroad that most focused public attention. The Great Lakes manufacturing empire was crumbling in the face of regional and international shifts in business investment and employment growth. In 1972, Detroit's civic elite commissioned a study of the exodus of capital from the Motor City. The study's conclusion was alarmingly simple: with outdated production facilities, a public infrastructure in poor condition, high taxes, and a strongly unionized and aggressive labor force, Detroit would be hard-pressed to retain business activity, let alone attract new investment.

Plate 16 Renaissance Center, Detroit (Roger Keil)

In 1980, the US auto industry experienced an economic slide unparalleled since the Great Depression [...] Economic recession, rising energy prices, a saturated market for energy-inefficient cars, and increased foreign competition sent the Big Three's profits plummeting. The way the auto production system is laid out in the United States meant hardship piled upon hardship for people in Detroit and the industrial cities of the Great Lakes.

With their survival at stake, the auto giants introduced changes in product design; more global concentration and centralization of capital; redesign of the labor process in relation to new technologies; and transformation in the industry's international division of labor [...] The auto transnationals are also working out a new inter-national division of labor designed to maximize global profits by minimizing production costs through global resourcing: that is, by locating different segments of the production process in different regional and national locations according to the most favorable wage rates and government subsidies.

Decentralization and uneven development

Viewed along the dimensions of time and space, the trajectory of economic development in metropolitan Detroit during this century can use-fully be divided into three periods: (1) the era of city building, 1910–1949; (2) the era of sub-urbanization, 1950–1978; and (3) the era of regional competition, 1979 to the present. One era does not give way to the next so much as each new period forms a layer upon the ones that came before [...]

The era of city building in Detroit coincided with the creation of the assembly-line and mass production. When Henry Ford built his "Crystal Palace" in Highland Park in 1913, the modern factory system was born. From then on the auto industry expanded according to a well-defined spa-tial logic: a factory, then complementary plants and residential development clustered along industrial corridors following railroad lines.

The era of suburbanization can be dated from 1951, the year the central city's population peaked at 1.85 million residents. The United States experienced unparalleled economic growth during the early postwar years and the logic of industrial expansion stayed much the same, but now it extended beyond the city's limits. The Big Three built 20 new auto plants in the Detroit area during the decade following World War II, all beyond the boundaries of the central city. Complementary industries, commercial development and residential enclaves followed, like metal shavings drawn to a magnet, but this time it was the suburbs that boomed, not the central city [. . .] As industrial growth extended to the suburbs, and as commercial capital concentrated on the urban periphery, the principal axis of uneven development shifted from cities within the city to the line that divided the city from its suburbs. In Detroit that line became a racial barrier as the division between central city and suburb came to coincide all too closely with that between black and white.

The Detroit metropolitan area was a thriving economy, stimulated by high levels of capital investment. It contained nearly half of Michigan's population. Residents of this Detroit were mostly white; they lived in single-family houses located in the suburbs; and they earned an income above the state average. But the other Detroit, the central city, had become more and more like a segregated urban enclave during the decades following World War II. Home to hundreds of thousands of poor and unemployed people, the city was now pitted against the suburbs in a dual pattern of uneven urban development (Taylor and Peppard 1976). Deindustrialization now spread out from the central city, and down river, into white suburban Detroit. Now residents in Detroit's industrial, working-class white suburbs came to share many problems with their central city neighbors. But even as capital flight and automation were dealing a hard blow to Detroit's industrial suburbs, a new type of regional growth pole was emerging. Epitomized by Silicon Valley outside of San Francisco, and Route 128 outside of Boston, it is the science city, or the technopolis, as the Japanese like to call it. At the core of the technopolis are universities with strong science and engineering faculties, government-subsidized research parks, and closely linked high-technology companies specializing in high-value production. Oakland County's Technology Park in suburban

Detroit is billed as the "workplace of the 21st Century," and fits this blueprint precisely. The era of regional competition is dominated neither by industrial nor by commercial capital. Rather, the driving force seems to be the creation of new information technologies and their application to all sectors of the economy.

For their part, Detroit officials have tried to revitalize the Motor City by following a corporate center redevelopment strategy. The linchpin in Detroit's redevelopment effort is the Renaissance Center (RenCen), a towering riverfront office, hotel and commercial complex meant to compete with outlying office centers and symbolize the city's corporate future [. . .] But this attempt to revitalize the central city got into serious trouble from the beginning. The heart of the matter was Detroit's depressed downtown real estate market – the most salient indicator of the central city's weak position in the regional economy. A region of independent yet relatively autonomous cities is emerging. Knit together by a regional division of labor, these cities remain deeply divided along lines of race, class and municipal boundary. The principal fault line of uneven development no longer runs among areas within the city, nor between the city and its suburbs; rather it travels among competing cities within a region of cities. In the era of regional competition there is no longer one Detroit, nor two Detroits; there are many Detroits.

HOUSTON: THE CAPITAL OF THE SUNBELT

Specialization and growth

[Since the 1920s] Houston has become the center of a world oil and petrochemical production system: 34 of the nation's 35 largest oil companies have located major administrative, research and production facilities in the metropolitan area. In addition to these corporate giants, there are 400 other major oil and gas companies there. Thousands of smaller oil-related companies have attached themselves to these major petroleum companies [. . .]

The expanded flow of profits to the oil–petrochemical sector has provided the direct capital and borrowing capacity for other capital which lies behind much of Houston's industrial and real estate (spatial) growth. At the heart of investment decision-making by these companies is business leader concern for a "good business climate," a codeword for an area with lower wages, weaker unions, lower taxes and conservative politics. Companies that function as locators, such as the Fantus Company, have advertised Houston as having one of the best business climates in the US. Houston has grown because of cheaper production costs (e.g. weaker unions, lower wages), weaker physical and structural barriers to new develop-ment (e.g. no ageing industrial foundation) and tremendous federal expenditures on infrastructure facilities (e.g. highways) and high-technology defense industries.

In 1973 the OPEC countries gained control over their oil, and once-dominant US companies became primarily suppliers of technology and marketing agents for OPEC oil. US company profits on Middle Eastern oil fell, but the sharp rise in world prices brought great increases in profits on oil controlled by US companies elsewhere [. . .] In the 1973–1975 recession employment in goods-producing industries dropped 6 per cent in cities such as Dallas, but grew by 18 per cent in Houston, because its manufacturing firms produce for the oil world's industry. The rise in OPEC oil prices in 1973–1974 gave a boost to oil exploration and drilling, thus stimulating the Houston economy in a time of national recession. Between 1968 and 1980 the percentage of Houston employment in oil exploration, drilling and machinery expanded.

Prior to the 1973–1974 price-rise an economic diversification trend was underway, with growing investment in non-oil projects. With the sharp rise in the oil price, oil companies and allied bankers moved away from diversification to a heavier emphasis on investments in oil projects. In the late 1970s there was yet another rise in the OPEC oil price, which further stimulated companies to invest in oil.

Crisis and reorganization

Yet, in 1982–1987, Houston was looking a lot like Detroit. Job announcements brought long lines; tax revenues had plummeted; public sector workers were laid off; bond ratings had slipped; firms were going bankrupt in increasing numbers; and cor-porations were closing plants and shifting work overseas. World-wide recession had led to an oil glut [. . .] The downturn rippled its way through drilling pipe and oil rig production; through con-struction and trucking; and eventually through retail stores and real estate. Industrial production declined in Houston more rapidly than the national average. The unemployment rate grew more rapidly in Houston than in the nation. In 1984, the number of bankruptcies continued to escalate [. . .]

Decentralization and uneven development

The oil industry has brought periods of rapid growth to Houston. Coupled with the commitment of the local elite to auto-centered transit and private enterprise in housing – and a fierce opposi-tion to mass transit and public housing – this rapid growth created a decentralized city. Commercial and industrial corporations have commissioned or leased a vast array of megastructures (industrial parks, shopping malls, multiple-use projects and office towers built in business centers). Scattered between and beyond these business centers are residential areas, including condominium apart-ment buildings and sprawling suburban subdivisions [. . .]

Houston's developers have been pioneers in multiple-use developments (called MXDs in developers' publications). These megastructure projects illustrate the central role of oil and gas companies in Houston's physical development. The large office and MXD complexes scattered in the seven "downtowns" in Houston are populated primarily by oil and gas companies and by the legal, accountancy and other business service firms serving the oil industry.

Yet, not all Houston residents profit from growth. Houston's low-income and minority homeowners and tenants have suffered greatly from market-oriented growth. The central city houses large numbers of black and Mexican-American, low- and middle-income families. Many areas of the central city have suffered from gen-trification, the replacement of poorer families with better-off professional, technical and managerial families. Gentrification has displaced residents of ethnic areas as well as elderly whites. The Fourth Ward is one of Houston's oldest black communities, with the misfortune of being in the path of expansion of the central business district. The area is populated by tenants living in single-family dwellings and in a major public housing project. Because of its proximity to downtown, developers have their eye on the area. A number of prominent consultant reports have suggested that the area should be redeveloped. The major public housing estate there is scheduled for demolition, significantly reducing the amount of housing for middle-income families.

Houston is also facing a major infrastructure crisis. The hidden side of its "good business climate," its low taxes and scaled-down government, has been a neglect of sewerage, water, flood prevention and other infrastructure facilities. This neglect has simply postponed the cost of paying for decaying or seriously inadequate facilities. Hundreds of billions of dollars will be required to meet Houston's escalating infrastructure costs. And that does not include the human costs of this neglect. Paying for infrastructure repair will require massive tax increases, which are even less likely in an era of slow economic decline. The "free enterprise" city has cost, and will continue to cost, its citizenry heavily in monetary and social expenses not normally enumerated in promotional and news media accounts of Houston's growth.

CONCLUSION

Because Detroit and Houston are spatial locations in a global system of production and exchange, the forces shaping their convergent and divergent paths have not been bounded by municipal, regional or even national lines. Three themes have ordered our tale of these two cities: (1) specialization and growth; (2) crisis and reorganization; and (3) decentralization and uneven development. It is to these reference points that we turn to draw our concluding comparisons.

Specialization and growth

Detroit and Houston evolved as specialized nodes in internationally organized production systems,

one centered on the auto industry, the other on oil and petrochemicals. The Motor City grew and prospered, expanded horizontally and vertically, all in time with the beat of plant, warehouse and office investment for the production of cars. Detroit developed as a one-industry town located near supplies of labor and raw materials. The mass production of automobiles spurred a vast and growing demand for fuel, and at the other end of an auto–oil "pipeline," more than 1,000 miles away, Houston began to prosper not long after Detroit's emergence as the Motor City. Experiencing a series of growth spurts, particularly in the 1920s, the 1940s and the 1960s, Houston became the oil capital of the world. It too expanded horizontally and vertically, with its seven business centers and hundreds of major plants, office towers and shopping centers. The Oil City too developed as a one-industry town located near crucial raw materials and labor pools.

Detroit's auto specialization generated a leapfrog logic of spatial development. Industry, commerce and residences decentralized into suburban rings, assisted by federal highway and housing finance programs. But even in the midst of postwar prosperity, large numbers of blacks were confined to blighted areas in the inner city. Houston's own distinctive specialization created a complex geographical network of oil-gas centers encircled by suburban belts. Houston's spatial sprawl was assisted by the same federal highway and housing programs that conditioned suburbanization in the Motor City. The auto and housing industries played the most aggressive role in the postwar "Highway Lobby" that pressured the federal government to support new highway and housing programs, while the oil companies remained more in the background. Yet Houston's oil interests were just as firmly committed to a low-density, decentralized urban environment. Black and Mexican-American workers and their families were a large island of poverty in a central city bordered by huge megastructures, like Houston Center.

Crisis and reorganization

The era of prosperity for the city of Detroit was temporary. Detroit's economic hegemony in world auto markets and its expanding employment in auto firms proved transitory. Global crisis and reorganization in the auto industry brought massive economic decline to the central city [...] As its industry left town, its human problems grew.

Yet even the "shining buckle of the Sunbelt," once thought to be immune to urban crisis, experienced the fundamental contradictions of a capitalist world system. The 1982–1987 recession was Houston's worst ever; the Oil City had even boomed through most of the Great Depression. The 10 per cent unemployment, over 1,000 bankruptcies, oil refinery and other oil company layoffs, and cutbacks in the petrochemical industry indicate that Houston too is showing the effects of crisis and reorganization in the world oil-gas industry [...]

Capital moves on a world market stage, and with modern modes of transportation and communication, big businesses can accelerate investment at a high velocity (Bluestone and Harrison 1983). Working people and their families, on the other hand, move in locally-bounded communities. They cannot chart a new course with capital's velocity. Corporation's investment space outdistances people's living space and that is the fundamental urban contradiction in the world capitalist system.

Decentralization and uneven development

Capitalists themselves are caught up in the investment-living space contradiction, as indicated by Henry Ford II's failed attempt to revitalize Detroit's central city with a multi-million dollar investment in the Renaissance Center. The same contradiction is revealed in Houston's oversupply of office towers and residential blocks which now have record-setting vacancies. During the next two decades Houston may come to bask in the same light that is now refracted through the "Rust Belt." And given Houston's starved social services sectors, the human and monetary costs of Houston's decline could well surpass Detroit's own experience.

So, ironically, the declining center of a crumbling Great Lakes manufacturing empire and the booming buckle of the Sunbelt both turn out to be one-industry towns whose export industries are going through a global reorganization that bodes well

for neither's economic future. Officials in Detroit and Houston recognize the need to restructure and diversify their local economies. And both are emphasizing hi-tech complexes and suburban office–commercial parks; but that development policy presages further inequality and uneven development along lines of race, class and territory.

The extent to which each city can muster new comparative advantages and revitalize its economic base will continue to be affected by the kind of relationship each establishes with the federal government. And here another irony comes into play. For one big advantage Houston, the self-proclaimed bastion of free enterprise, retains over Detroit, a fading outpost of the welfare state, is Houston's greater ability to garner largesse from the federal government; not the kind distributed by the Department of Health, Education and Welfare; but the sort passed out by the Pentagon. Today's massive defense spending on the Pentagon's production system, the military-industrial complex, is spawning new industries, setting the direction for future economic development, and enriching or impoverishing regions. Even so, the critical issues always seem to remain the same. New urban development continues to be targeted to the privileged few. Power over urban development continues to be concentrated among a handful of individuals and corporations whose reach spans far beyond the metropolis. And urban development continues to be uneven, unpredictable and precarious, since the most important development decisions are made in private offices not in publicly accountable places.

REFERENCES FROM THE READING

Bluestone, B. and Harrison, B. (1983) *The Deindustrialization of America*. New York: Basic Books.

Friedmann, J. and Wolff, G. (1982) World city formation: an agenda for research and action, *International Journal of Urban and Regional Research*, 6, 3, 309–344.

Taylor, M. and Peppard, D. (1976) *Jobs for the Jobless: Detroit's Unresolved Dilemma*. East Lansing, MI: Institute for Community Development.

Zukin, S. (1980) A decade of the new urban sociology, *Theory and Society*, 9, 575–601.

19

"Global City Zurich: Paradigms of Urban Development"

Christian Schmid

Editors' introduction

Christian Schmid is a Swiss geographer who has worked at the Institute of Geography at the University of Berne and the Department of Architecture at the Swiss Federal Institute of Technology (ETH) in Zurich. He is a founding member of the International Network for Urban Research and Action (INURA) and the Ssenter for Applied Urbanism (SAU) in Zurich. Schmid has conducted much of his research on Zurich and, along with several German and Swiss collaborators, has played a key role in introducing global city theory to a German-language readership (see Hitz et al. 1995). Schmid has written a number of incisive analyses of urban development in Zurich that draw upon various interpretive influences, including global city theory, neo-Marxian political economy and Henri Lefebvre's approach to urban theory (Hitz et al. 1994; Schmid 1998).

In this contribution, Schmid situates urban development in Zurich in relation to its role as a "head-quarter economy" within European and global circuits of capital. Schmid traces the evolution of Zurich into a global city during the 1970s, emphasizing the clash between a modernizing growth coalition and various locally rooted oppositional forces that opposed the expansion of the central business district (CBD) into surrounding residential neighborhoods (for an analogous analysis of Tokyo, see Reading 17 by Machimura). As Schmid indicates, a "territorial compromise" was established in the wake of these struggles that slowed down inner city restructuring while pushing many global city functions outwards into the city's peripheries, such as Zurich North. Schmid's case study provides a nuanced perspective on a number of aspects of global city formation in Zurich – including conflicts between parochial and cosmopolitan forms of urban culture, the transformation of urban form through the decentralization of global city functions, the degradation of the built environment in the suburban periphery, and the formation of new cultural milieux in revitalized inner city neighborhoods such as Zurich West. Schmid also devotes considerable attention to the contested politics of global city formation, emphasizing both the periodic reworking of municipal and regional growth coalitions and the everyday contestation of urban restructuring through diverse, neighborhood-based social movements. Schmid concludes by situating global city formation in Zurich in global perspective. For Schmid, the process of urban restructuring in Zurich is necessarily contextually specific, but it is also indicative of a number of more general trends and conflicts that can be witnessed in global cities throughout the world.

Zurich today is a global city, one of a group of global control centers of the world economy. In international comparisons, Zurich has been routinely placed at the second or third rank in global city hierarchy, together with cities like San Francisco, Sydney or Toronto (cf. e.g. Friedmann 1995).

In the mid-1950s, Zurich was an industrial town with a strong position in the machine-building and armament industries. The transformation of Zurich into a global city began in the 1970s, with the increasing deregulation and globalization of financial markets. Zurich became the undisputed center of Switzerland as a location for finance, and a headquarter economy established itself, which specialized in the organization and control of global financial flows. In 2001 only around 7 per cent of all jobs were still in the manufacturing sector (not counting construction), while 36 per cent were in the core sectors of the global city economy (financial industries, insurance and business services). Yet, even though it directly depended on global lines of development, the transformation of Zurich into a global city was still a contradictory process, which was also strongly marked by local conflicts.

This radical economic transformation has caused fundamental changes in the urban development of Zurich. Two differing historic models of urbanization can be distinguished. The first model developed in the 1970s and was characterized by the process of global city formation. It was growth oriented, but it was also grounded upon a strong regulation of urban development and by the conservation of inner-city areas. In the 1990s, with the process of metropolitanization and the expansion of the global city into the region, a second model of urbanization established itself that has been characterized by a neoliberal policy of urban development, the emergence of new urban spatial configurations, and a new definition of the urban.

GLOBAL CITY FORMATION: TERRITORIAL COMPROMISE AND URBAN REVOLT

In the decades after World War II, urban development in Zurich was defined by an encompassing growth coalition consisting of right-wing and left-wing forces, following a relatively moderate strategy of modernization. The conditions of this policy changed in the early 1970s: the protest movement of 1968 led to a radical questioning of functionalistic approaches to urban development; meanwhile, the global economic crisis ended the "golden age" of Fordism. Subsequently, in the mid-1970s, the transformation of Zurich into a global city began.

Through the process of global city formation, globally defined strategies to establish a "headquarter economy" collided with the locally defined everyday concerns of many residents. The growth coalition fell apart, and the city was subsequently divided into two camps quarreling about the appropriate model of urban development. On the one hand, a new "modernizing coalition" took shape, consisting of right-wing parties and the growth-oriented sections of the trade unions, which promoted the development of Zurich as a financial center, the extension of the central business district (CBD) and the construction of new traffic infrastructure. On the other hand, in the wake of the social movements of 1968, an alternative position emerged which was critical of urban growth. Left-wing parties and various action groups and neighborhood organizations united in a heterogeneous and fragile "stabilization alliance"; their goal was to fight for a livable city, low rents and the preservation of residential neighborhoods in the inner-city. Occasionally, this alliance was also supported by conservative forces. Through the Swiss system of direct democracy, in which many questions and projects must be decided by referenda, these opposing positions were transferred directly to the level of institutional politics. Both parties had their victories and defeats in this conflict, but neither side was ultimately able to win decisively. Thus, for two decades, from the mid-1970s to the mid-1990s, urban development in Zurich was in fact determined by a precarious political stalemate that generated a specific type of "territorial compromise." This territorial compromise entailed a rejection of large-scale modernization strategies and it considerably slowed down the transformation of inner-city residential neigborhoods. Yet, global city formation and the dynamics of urbanization were not fundamentally challenged.

However, this territorial compromise also encompassed a second line of conflict, which did not immediately reveal itself. At the level of everyday life, the demands of cosmopolitan open-mindedness and urban culture created by global city formation clashed with the highly localized forms of social regulation which had been inherited from the Fordist period, and which were oriented towards social control and conformism. As of the 1970s, public life in Zurich was still characterized by a crushing parochialism that left very little space for the development of new lifestyles or alternative forms of cultural expression. This situation eventually caused a social explosion: on May 30, 1980, an urban revolt began. With riots, happenings and actions of all kinds, a new cosmopolitan urban generation demanded what Henri Lefebvre (1968) once called "the right to the city." Although the urban revolt collapsed after two years, its consequences subsequently became evident; the urban movement had changed the city's everyday life, its cultural sphere and its public spaces. A cosmopolitan ambience emerged. The city government began to promote diverse types of cultural projects, and a cultural and artistic milieu established itself, radiating far beyond Zurich. This created the basis for a successful economic sector of "cultural production," including design, image production, events, etc. This economic sector today plays a key role in international competition between global cities. The urban revolt thus became an important catalyst for the process of global city formation in Zurich.

As a result of these two lines of conflict, a new model of urbanization was established which combined the goals of modernization, stabilization and both economic and cultural globalization. This model contained a concept of the city that was simultaneously metropolitan and exclusive, derived from the classical European image of the city as a coherent, dense and innovative whole. This concept ultimately reduced the focus of urbanity to a narrow fraction of urban reality, to downtown Zurich. Seen from the urban center, all areas outside this restricted district were considered to be elements of a drab and uninteresting urban periphery.

While the inner-city evolved into a culturally and socially pulsating urban center, the opportunities for the construction of new offices and the expansion of the central business district were massively restricted. Service and financial enterprises were compelled to establish their additional offices at other locations. In various places outside Zurich City,

the new strategic centers of the global city headquarter economy were developed. This process can be seen as an "explosion of the center" – global city functions were increasingly spread over an extended region, which was now structured as a center (Sassen 1994). Thus, a new urban configuration evolved, characterized by the regionalization of economy and society.

EXOPOLIS: THE CASE OF ZURICH NORTH

As in many other cases, the demarcation of this new global city region is quite difficult, since it is not formed as a coherent unit. A growing number of towns and villages in the densely populated lowlands of Switzerland have come under the influence of Zurich's headquarter economy and have become metropolitan in character. Therefore, depending on the criteria selected, a great variety of regions can be delineated around Zurich. Whereas the municipality of Zurich (Zurich City) has a population of 360,000, the global city region can be estimated at approximately 2 million inhabitants (Figure 1).

Analysis and deconstruction of this urban universe has only just begun. It is an amoebae-like urban space, which is characterized by floating centralities and by the constant emergence of ever new and surprising urban configurations. A showcase example for these new urban configurations is the Glatt Valley, north of Zurich, where the airport is also located. A series of "edge cities" (Garreau 1991) have developed here, forming a kind of fragmented twin city of Zurich. This "new" city is called simply "Zürich Nord" (Zurich North). It is one of these amorphous implosions of archaic suburbia that Soja (1996: 238) named "'Exopolis' – 'the city without' – to stress their oxymoronic ambiguity, their city-full non-cityness. These are not only exo-cities, orbiting outside; they are ex-cities as well, no longer what the city used to be" (Figure 2).

In a narrow sense, Zurich North includes eight municipalities and two districts of Zurich City. With 147,000 inhabitants and 117,000 jobs, this area is today the fourth largest city in Switzerland – it is even bigger than Berne, the capital. The emphasis is on activities of the headquarter

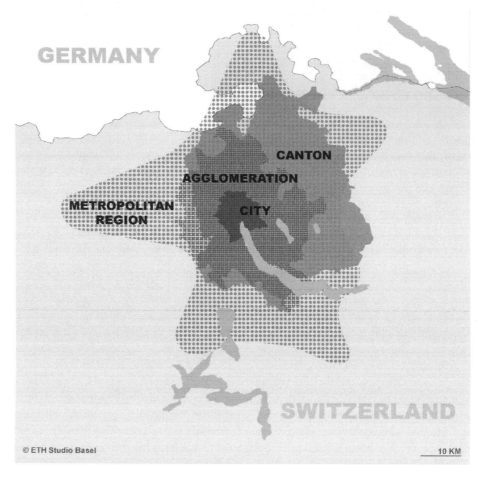

GERMANY

CANTON

AGGLOMERATION

METROPOLITAN
REGION

CITY

SWITZERLAND

© ETH Studio Basel   10 KM

Figure 1 The Zurich region

economy, predominantly producer services, bank-ing, and information and telecommunications (IT) industries. The concentration of these activities in this area is based less on the effect of the immediate (physical) vicinity than on the oppor-tunities for flexible interconnections in a broader logistical space that stretches from the airport to the national highway system and electronic networks.

In the mid-1980s, Zurich North was composed almost entirely of classical middle-class suburbs. Planning was in the hands of the individual munic-ipalities, which as a rule followed a simple planning concept: they tried to preserve the historic core of the settlement, expanding the housing zone con-centrically around the core and placing industrial zones at the outskirts of the municipal territory. But, in the wake of deindustrialization and global city

formation, it was not industrial operations, but the headquarter functions of global corporations, that were located within these formerly industrial zones. Consequently, satellites of the headquarter economy developed, consolidating in an odd kind of belt located at the periphery of the old cores of the settlement. Here, high-quality business inter-mingles closely with highways or even waste incineration plants. The geographic center of this belt is a forest, and so a kind of circular town emerged with an "empty" center. This shape corres-ponds exactly to the "doughnut model" that Soja (1996) developed to describe Orange County.

Fascinating as this urban patchwork may appear, it nevertheless produces severe problems. Since the new centralities are dispersed over a wide area, this fragmented non-city is largely dependent on private cars, a situation that produces traffic

SCHAFFHAUSEN

FRAUENFELD

BADEN   WINTERTHUR

AARAU

ZÜRICH

ZUG

LUZERN

© ETH Studio Basel   10 KM

Figure 2 Commuters to Zurich

jams and air pollution. In addition, the environment is often not very attractive. In most of the new centers, there is a lack of urban infrastructure, restaurants, meeting places and cultural establishments, and there are few places that create a sense of identification or an urban atmosphere. Accordingly, many residents and employees are not at all happy with the quality of everyday life in this urban patchwork, which still has yet to overcome its peripheral status.

In contrast to many other urban peripheries, in Zurich North this lack of urban character has increasingly been perceived as a deficiency. Attitudes have gradually been changing, and there is now an explicit agenda of creating a "real city" from this patchwork by achieving a certain architectural and social coherence. The first initiatives for coordinated planning among municipalities already appeared in the beginning of the 1990s, and in 2001 an asso-

ciation of eight municipalities was created. The new label for Zurich North was "Glattstadt" (City of Glatt Valley). This name is meant to stand for a new region with its own identity, while also underscoring its separation from Zurich City. Accordingly, Zurich City is not included in this new organization, even though its northern neighborhoods formally belong to Zurich North.

The most important project in this new cooperation is the construction of a tram line, which was approved in a referendum in spring 2003. This line not only is meant to open up and connect the various new centers of Zurich North, but also is the symbol of the newly discovered self-confidence of this new "city of the future." This is why it is officially called "Stadtbahn" (city train) instead of tram, the traditional term for streetcars in Zurich.

Thus the "model of exopolis" remains delicately balanced: on the one hand there are attempts

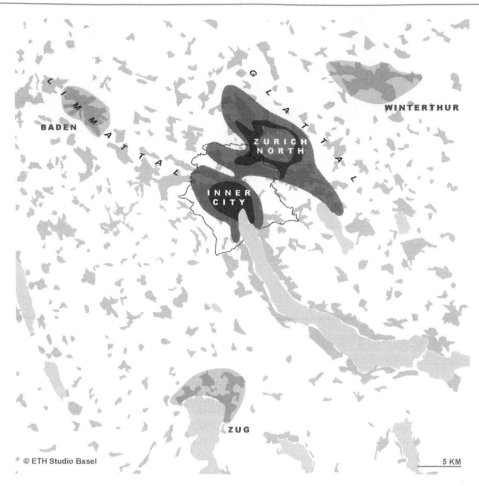

Figure 3 Zurich's spatial configurations

to make Zurich North into a "true city." On the other hand, however, the underlying historically developed patchwork structure of the area is continually creating new difficulties and surprises (Figure 3).

CONTRADICTIONS IN THE GLOBAL CITY: THE METROPOLIS ALLIANCE

While the global city has been expanding into the region, the situation in Zurich City has also undergone fundamental changes. During the course of the 1990s, a major paradigm shift in urban development has become apparent.

This paradigm shift originated in a double crisis. On the one hand, in the beginning of the 1990s Switzerland – like most west European countries – was in the grip of a long-term economic crisis, and growing deficits forced public authorities to impose new budgetary restrictions. On the other hand, the social consequences of globalization became visible: like many other global cities, extensive economic and social polarization and fragmentation became apparent. These developments were accompanied by fundamental shifts in the political landscape. In reaction to globalization and urbanization, an aggressive right-wing populism emerged for the first time in Switzerland. It grew first at the margins of the unravelling metropolis, in the suburban and periurban areas, but soon it began to take hold in Zurich City as well. Right-wing populist forces mobilized against the social and cultural open-mindedness of the 1980s. Through an aggressive campaign, these forces succeeded in transforming drug policy and the asylum question into central political issues. In subsequent years, the political and social climate clearly deteriorated.

Against these right-wing populist political activities, the positions of moderate right-wing and social-democratic forces drew closer. The antagonisms between the modernizing coalition and the stabilization alliance softened, and the moderate forces began to seek out pragmatic solutions across ideological and partisan boundaries. Thus the territorial compromise, which had existed for two decades, was broken and a new hegemonic political alliance emerged – the metropolis alliance. For a number of the main issues, trail-blazing compromises were reached, first over the drug question and then over traffic policy. This political shift also resulted in a fundamental change of urban development policy. While urban planning in Zurich City up to the 1990s had attempted to protect the quality of everyday life, to preserve the historical built environment and to defend the "city of residents" against the headquarter economy, now the focus was on competition: international investors, global capital and affluent residents were to be attracted to Zurich. Viewed in historical context, this development entailed a disavowal of the basic principles of urban planning that had dominated the development of Zurich City for about a century, which had aimed to establish a clearly defined, coherent urban structure for the entire city. At the level of urban planning, the city was increasingly assimilated into the region.

THE RECONSTITUTION OF THE CITY: THE CASE OF ZURICH WEST

While in the course of the 1990s the processes of urbanization in the city and the region were interlinked and the disintegration of the city into the region advanced, a remarkable reversal occurred – the reconstitution of the city and the reproduction of the old center/periphery dichotomy.

Based on the analysis of national referenda since the early 1980s, this tendency can be illustrated in detail at the political level (Hermann and Leuthold 2003). While in the region there was a strong tendency towards right-wing populist positions, Zurich City, and in particular the inner-city neighborhoods, showed an increasingly left-liberal

orientation. This tendency was evident in the entire German-speaking part of Switzerland, but most distinctly in Zurich. The pattern of political polarization does correspond to socioeconomic disparities; rather, it reflects different preferences in everyday life. Within the Zurich region, different lifestyles have evolved. While suburban life still has a great attraction for many, others seek a distinctly urban lifestyle. The proximity of cultural facilities, a cosmopolitan milieu and a trendy image have become important location factors, not just for the lifestyle-conscious urban professionals, but increasingly for companies as well. These factors are still to a large extent concentrated in the center (Figure 4).

The new urban feeling has been manifested most visibly in the trendy new neighborhood "Zurich West." As late as in the 1980s, this inner-city area was still one of the main centers of Swiss engineering industry. Due to the process of de-industrialization, a growing number of industrial activities relocated. At that time, the area was earmarked for the expansion of the financial sector. Because of the stalemate in urban planning, and the consequences of the economic and real estate crisis, development projects remained frozen for years. Eventually, the impressive industrial landscape, with its imposing warehouses and its austere charm, became a utopian place, a projection zone for fantasies, a promise of opportunity. Small businesses, illegal or semi-legal bars and discos, theaters, hang-outs, artists' studios and projects of all sorts began to appear throughout the area.

From the beginning, this transformation was also linked to market-based developments. One of the first projects was a condominium with luxury lofts and a multiplex movie theater. The result was a highly urban blend of both the commercial and the ephemeral, something extraordinary, and not only for Switzerland. The new combination of working, living and entertainment as well as the unconventional atmosphere of the new neighborhood attracted a wide range of additional business activities, from hotels to international consulting firms. A veritable cultural zone developed, housing several renowned institutions of arts and culture. The brownfield was thus transmuted into an elegant urban neighborhood, which was presented to astonished visitors as a "Swiss Greenwich Village." Many pioneer projects from the early

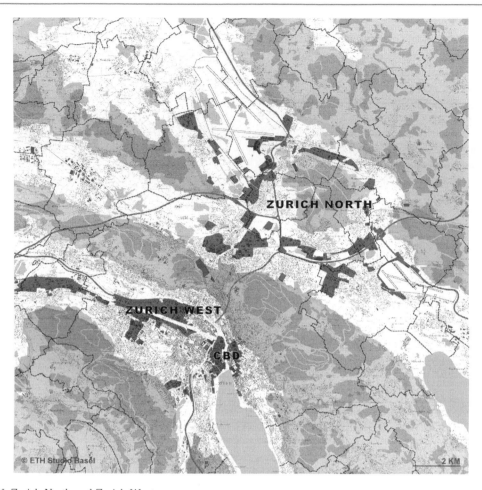

Figure 4 Zurich North and Zurich West

days have more recently been displaced, but a number of remarkable alternative projects succeeded in securing land while real estate prices were still low.

So Zurich West today represents a new inner-city development model. Nevertheless, the new neighborhood differs radically from the existing downtown area. It presents an amazingly high density and diversity of varying utilizations and social groups. These are, however, hardly interrelated, but rather live side by side in an overlay of social and economic networks extending over the entire metropolitan region. The area basically consists of individual islands belonging together less on the basis of interactive processes than by sharing an urban milieu and a metropolitan image. This is not only an effect of the large-scale structure of the built

environment, originating in the area's industrial history, but also the result of the changed every-day routines of the metropolitan population.

PARADIGMS OF URBAN DEVELOPMENT

In the early 1990s, urban researchers discovered fundamental transformations and postulated a new urban era in urban development. They stated that contemporary cities did not resemble the traditional cities of the past (cf. e.g. Garreau 1991; Soja 1996). Los Angeles, with its polycentric and excentric development, was declared the "paradigmatic industrial metropolis of the modern world" (Soja and Scott 1986). In the meantime, things have settled down at Exopolis. What is it that's new? What does the paradigm shift consist of? The example of Zurich reveals some reference points for confronting these questions.

First, the polycentric development of cities has become a general phenomenon. Even smaller cities follow this developmental trend. At the same time, the example of Zurich also illustrates that specific local traditions, contradictions and fields of conflict may have a decisive influence on urban development. The specific form of urbanization not only is determined by economic development, but also is a result of debates on the concept of the "city," of both struggles and compromises.

Second, the process of metropolitanization breaks up the unity of the city. It is no longer possible to define the urban clearly. It is composed of overlaying configurations and unexpected constellations. In Zurich, two ideal-typical configurations can be distinguished, giving an impressive illustration of the change – on the one hand the "exopolis model," as exemplified in Zurich North; on the other hand the "inner-city model," as manifested in Zurich West. Both areas stand for differing urban forms that are developing simultaneously. Yet the two models do not differ as much as might initially seem to be the case. In Zurich North, there is an attempt to reintroduce a classical conception of urbanity into the excentric urban chaos and to create new, coherent urban structures. On the other hand, the new "inner-city model" does not correspond to the traditional image of a downtown neighborhood, with its dense network of social interaction. Indeed, it represents a conjunction of regional networks that are barely interlinked on an everyday level.

Third, in spite of the trend towards polycentricity, the relationship between center and periphery remains highly contradictory. In the case of Zurich, the dichotomy between center and periphery did not weaken in this process, but was actually strengthened. Politically and culturally, center and periphery have drifted further apart. While the center exploded and disintegrated into the region, the city was meanwhile reproducing itself at the level of everyday life.

These are only some aspects of the paradigm shift in urban development that are currently transforming living conditions in Zurich. The new model of urbanization has proved to be contradictory and indeterminate.

REFERENCES FROM THE READING

Friedmann, J. (1995) Where we stand: a decade of world city research. In P.L. Knox and P.J. Taylor (eds) *World Cities in a World-system*. Cambridge: Cambridge University Press, 21–47.

Garreau, J. (1991) *Edge City: Life on the New Frontier*. New York: Doubleday.

Hermann, M. and Leuthold, H. (2003) *Atlas der politischen Landschaften: Ein weltanschauliches Porträt der Schweiz*. Zurich: vdf-Verlag.

Hitz, H., Schmid, C. and Wolff, R. (1994) Urbanization in Zurich, *Environment and Planning D: Society and Space*, 12, 167–185.

Hitz, H., Keil, R., Lehrer, U., Ronneberger, K., Schmid, C. and Wolff, R. (eds) (1995) *Capitales Fatales: Urbanisierung und Politik in den Finanzmetropolen Frankfurt und Zürich*. Zurich: Rotpunktverlag.

Lefebvre, H. (1968) *Le Droit à la ville*. Paris: Anthropos.

Sassen, S. (1994) *Cities in a World Economy*. Thousand Oaks, CA: Pine Forge Press.

Schmid, C. (1998) The dialectics of urbanisation in Zurich: global city formation and social movements. In INURA (ed.) *Possible Urban Worlds: Urban Strategies at the End of the 20th Century*. Basel: Birkhauser, 216–225.

Soja, E.W. (1996) *Thirdspace*. Cambridge, MA and Oxford: Blackwell.

Soja, E.W. and Scott, A.J. (1986) Los Angeles: capital of the late twentieth century, *Environment and Planning D: Society and Space*, 4, 249–254.

20

"Global Cities and Developmental States: New York, Tokyo and Seoul"

from *Urban Studies* (2000)

Richard Child Hill and June Woo Kim

Editors' introduction

In this reading, Richard Child Hill and June Woo Kim elaborate a comparative analysis of economic restructuring and urban development in Tokyo and Seoul. Whereas Hill's work on Detroit and Houston (see Reading 18 by Hill and Feagin) applied certain key arguments of global city theory to the comparative investigation of those two cities, his more recent analysis of Tokyo and Seoul adopts a more critical perspective on this approach. Hill has written extensively on East Asian urbanization, and in this reading he collaborates with South Korean sociologist June Woo Kim, who completed his PhD in Sociology at Hill's home institution, Michigan State University, and is now employed at the National University of Singapore. In a section of their article that could not be reproduced here, Hill and Kim develop a detailed critique of major global city theorists, such as Saskia Sassen and John Friedmann, for a purported tendency to overgeneralize the effects of globalization upon urban structures. Against this alleged assumption of "convergence," Hill and Kim insist upon the highly variegated national and local pathways of urban restructuring that are crystallizing under contemporary capitalism. In this context, Hill and Kim also insist that the developmental trajectories of East Asian cities such as Tokyo and Seoul are profoundly shaped by national state institutions, which continue to channel significant resources into urban industrial growth (see also Reading 30 by Brenner and Reading 48 by Olds and Yeung). For Hill and Kim, the existence of these activist, "developmental states" in East Asian cities differentiates them qualitatively from the supposedly paradigmatic cases of New York and London, and thus undermines the applicability of global city theory beyond the "market centered" urban systems of the Anglo-American world.

Tokyo and Seoul, according to Hill and Kim, diverge from the standard, New York- and London-centric model of global city formation due to the lack of extensive urban sociospatial polarization, the embeddedness of local financial institutions within the national industrial fabric, the persistence of manufacturing industries in the city, the continued role of activist national governmental policies and national political elites in guiding urban development, and the continued contribution of metropolitan economic dynamics to the national economy as a whole. Hill and Kim conclude by arguing that contemporary capitalism is composed of multiple, competing national and regional institutional systems, and not an overarching global regime.

In a subsequent issue of the journal in which Hill and Kim's article was originally published, John Friedmann and Saskia Sassen contributed brief responses that sharply challenged, among other issues, the authors' presentation of their respective theoretical positions and their broader interpretive claims regarding current trends in the cities under discussion. Readers interested in this rather pointed, but informative, exchange are encouraged to consult Sassen (2001) and Friedmann (2001), as well as Hill and Kim's (2001) response.

The "world city paradigm" is the most important contribution by urbanists to the contemporary globalization literature. Oddly, however, the world city hypothesis has not generated the vigorous debate among globalists, statists and those attempting to bridge the two camps that so enlivens most current work on globalization. John Friedmann and Saskia Sassen, the best-known architects of world city theory, take a globalist view – that is, they believe that a single global system is becoming superimposed on nation-states which are losing importance as a result. Globalization produces a world city system that transcends national institutions, politics and culture, they argue. Such a view assumes a convergence in "economic base, spatial organization and social structure" among the world's major cities, especially New York, London and Tokyo (Sassen 1991: 4).

In fact, however, fundamental differences in "economic base, spatial organization and social structure" persist between major cities in the North Atlantic and East Asian regions. Most telling for the paradigm, Tokyo, center of the world's second-largest national economy and the world's largest urban agglomeration, departs from the world city paradigm on most salient dimensions. Seoul, center of East Asia's second OECD member and the region's second-largest metropolis, exhibits the same anomaly. An awareness of these differences has been creeping into the world city literature. John Friedmann (1995) has acknowledged that Tokyo does not fit the world city paradigm in some respects, but he does not address the implications for the world city hypothesis. Saskia Sassen (1999: 86), on the other hand, explains away Tokyo's differences as a temporary function of "Japan's uniqueness" and continues to assume that convergence among the world's major financial centers is the overall trend.

We disagree. Understanding Tokyo and Seoul necessitates a different conception of the world system from that underlying the globalist version of the world city argument. World cities differ from one another in many salient respects because they are lodged within a non-hegemonic and inter-dependent world political economy divided among differently organized national systems and regional alliances (Stallings and Streeck 1996).

GLOBAL CITIES AND DEVELOPMENTAL STATES

Western neo-classical economists enquire mostly into markets, and occasionally into organizational hierarchies. They recognize the state as a third means for economic governance but confine it to defining property rights, enforcing contracts, overseeing the general rules of competition and (sometimes) providing collective goods; that is, to setting the minimal conditions without which markets and hierarchies could not function. More state involvement in the economy than this would interfere with the market mechanism, and by deviating from efficiency and productivity, would ultimately give way to competitive pressures. In this view, capitalist societies are bound to converge upon a market-driven corporate model.

In contrast to their neo-classical brethren, Western development economists have long been interested in comparing national paths to development. In their view, latecomers to the industrialization process must forge their own development institutions and ideologies because they invariably face a different set of problems and possibilities from those of their technically more advanced predecessors (Gershenkron 1962). A poor country's fledgling firms, for example, confront formidable competition from transnational corporations possessing far greater economies of scale, advanced technologies and global networks. But less developed countries also have hidden reserves of labor, savings and entrepreneurship. Nations wishing to overcome the penalties and realize the possibilities of late development require a strong state. The real issue, development economists often conclude, is not whether the state should or should not intervene, but rather "the art of getting something done with intervention" (Amsden 1989: 140).

Japanese thinkers developed their own art of late development between the Great Depression and the end of World War II. In contrast to liberal capitalism, Japanese developmentalism addressed industrialization at the level of the nation-state. Strengthening national production was the top priority of industrial policy. The economy was viewed strategically with the aim of building an industrial structure that would maximize Japan's gains from international trade. State regulations and non-market governance mechanisms were designed to restrain competition in order to concentrate resources in strategic industries and maintain orderly economic growth. And the quest for short-term profits was rejected to secure workers' cooperation in promoting productivity. The Japanese first institutionalized these principles between 1931 and 1945, but a number of studies (Gao 1997) suggest that these tenets continued to underlie Japan's post-war industrial policy, despite changes in Japan's political institutions and national purpose.

The transformation of Japanese developmentalism from militarism to trade was largely accomplished by the end of the 1960s. The state emerged with a strong capacity to sustain economic growth in contrast to the more free-wheeling role played by the market in Anglo-American capitalist economies. Japanese managers emphasized cooperative industrial relations in contrast to their conflict-prone Western counterparts. Family-based *zaibatsu* business groups were reorganized into management-controlled *keiretsu* networks as a powerful weapon in market competition.

Given the differences between Anglo-American liberalism and East Asian developmentalism, it seems reasonable to expect related contrasts in the role each sphere's major cities play in the world economy. Table 1 offers such a contrast by hypothesizing two world city types: a market-centered, bourgeois type, modeled on New York City, and a state-centered, political-bureaucratic type, modeled on Tokyo. We provide a brief sketch of Tokyo to establish the contrasting type, and then focus our empirical investigation on another state-centered world city, Seoul, Korea.

The global economy is spatially imbedded in Tokyo, to be sure, but Tokyo is not primarily a global basing-point for the operations of stateless TNCs. Rather, Tokyo is mainly a national basing-point for the global operations of Japanese TNCs. Tokyo's relationship to the world economy is not driven in the first instance by market efficiency, but by a strategic concern to preserve national autonomy through global economic power (Johnson 1995). In Japan, economic power is indexed by the world market shares held by the nation's industries, not by quarterly dividends and privately accumulated wealth.

Tokyo offers corporations global control capability, but the primary vehicle is not private financial and producer services clustered into complexes by market forces. Tokyo's global control apparatus resides in financial and industrial policy networks among public policy companies, banks and industrial enterprises, under the guidance of government ministries like the Ministry of Finance (MOF) and the Ministry of International Trade and Industry (MITI). Indeed, by emphasizing reinvestment and employment rather than high profits and individual consumption, Japanese policies have actively discouraged growth in the kinds of services distinguishing New York City.

The practice of global control in Tokyo has not resulted in a social regime characterized by massive loss in manufacturing jobs, high levels of foreign immigration, extreme wealth concentration and social and spatial polarization. One-quarter of Tokyo's labor force continues to work in manufacturing (as against less than 10 per cent in New York City), primarily in high-tech, research-intensive pilot plants, and the headquarters of Japan's major manufacturing companies continue to concentrate in Tokyo to be near government ministries. The state tightly controls foreign immigration with an eye to available employment, and the foreign-born represent a minuscule 1.8 per cent of Tokyo's population, compared to 28 per cent in New York City. In contrast to New York City's dualism, Tokyo's occupational structure is compressed around the median, the middle strata encompass most city residents, and the extremes in wealth concentration and impoverishment found in New York are missing.

Tokyo's commanding place in the world urban hierarchy is not determined by the city's ability to

Table 1 Two world city types

	Market-centered bourgeois	State-centered political-bureaucratic
Prototypical city	New York	Tokyo
Regional base	West Atlantic	East Pacific
Leading actors	TN capitalist class	State bureaucratic elite
Group organization	Finance TNCs	State ministries tied to business
Economic ideology	Vertically integrated firms	networks via main banks
	Liberalism	Developmentalism
	Self-regulating market	Strategic national interest
Trade, investment and production	Market-rational	Plan-rational
Relation to world economy	Private wealth	National power
Prime objectives	Profit-maximizing	Market-share, employment-maximizing
Global control capability	Private-producer service complexes	Government ministries Public corporations Policy networks
Industrial structure	Manufacturing HQs and production dispersed Services emphasized	Manufacturing HQs and high-tech production concentrated Services de-emphasized
Occupational structure (social and spatial)	Polarized Missing middle High inequality High segregation	Compressed Missing extremes Low inequality Low segregation
Foreign immigration	Weak controls High	Strong controls Low
Culture	Consumerist Yuppie, ethnic	Productionist Salaryman, officelady
City–central state relationship	Separation	Integration
Source of urban contradictions	Short-term profit Market volatility Polarization	State capital controls Overregulation Centralization
Competitive advantages	Fluidity Mobility	Stability Planning

attract global investments, but by the ability of Tokyo companies to generate earnings from abroad. In 1990, for example, Japanese TNCs controlled 12 per cent of world FDI, while foreign investment in Japan represented only 1 per cent of the total world stock. The comparable figures in 1980 were 4 per cent and 1 per cent (Ostry 1996: 334). While Japan's *keiretsu* networks and the main bank system enabled the MOF to influence big-firm strategies via the supply and cost of capital to network banks, the system was also explicitly designed to protect Japanese companies against foreign penetration and short-term profit pressures.

Control over investment equals power in Tokyo as in New York City, but Tokyo is under the sway of a political-bureaucratic elite, not a transnational capitalist class. Control in New York City is in the hands of a private investor class. The stock market is the barometer of New York's economy. Control in Tokyo is exercised through management-run corporate networks centering upon main banks which in turn are guided by government ministries. Employment is the barometer of Tokyo's economy. Tokyo's elite possesses a "productionist" not a consumerist ideology. The clash between classes in the state-centered world city is not between transnational and local capitalist classes but between bourgeois and political-bureaucratic elites.

Finally, Tokyo is not parting company with the Japanese nation and central state. Japan is a unitary state, and the relationship between the city of Tokyo and the central government is bureaucratically integrated in a myriad of ways, and especially through the Ministry of Home Affairs. Tokyo is, in effect, a national champion (Hill and Fujita 1995).

SEOUL

Seoul is Korea's command post for government planning and business management. While 24 per cent of the nation's population resides in Seoul City, virtually all of Korea's central government agencies (96 per cent) and top corporate headquarters (48 out of the top 50) are located there; 61 per cent of Korea's business managers and 64 per cent of the nation's research scientists work in the city. The city of Seoul combines with surrounding satellites in Kyonggi-do Province to form the Seoul Metropolitan Region (SMR). Koreans often refer to the SMR as the "Seoul Republic" because it is so dominant over other regions of the country. With 17 million residents, the capital region contains 39 per cent of South Korea's population (Kim and Choe 1997: 2, 43).

Seoul is also Korea's window on the world. All but one of the nation's foreign embassies, and 15 out of Korea's 22 foreign consulates, are in Seoul. All of the nation's stock brokerages (76), foreign bank offices (66), offices of foreign media (25) and broadcasting networks (8) are in the city. Seoul hosts 71 per cent of Korea's overseas-based service

industries, half the nation's international hotels and trading companies, and nearly all of its communication services (Hong 1996). Seoul now ranks 7th among the world's cities in the number of industrial Fortune 500 transnational corporations (TNCs) headquartered there, 13th in the number of TNC bank headquarters, 17th in number of international organizations and 23rd in frequency of international conferences (Jo 1992). The outward indicators certainly point to Seoul's world city status, but how well in fact does the Seoul Republic fit the world city paradigm?

Seoul's global base

With scarce natural resources and a small domestic market, the Korean state subsidized export-oriented industries and South Korea industrialized by exporting to overseas markets. Seoul, as home to the central government, was the place to be for all who desired contact with government ministries and exposure to international markets. Little wonder then that Korea's major firms and business associations headquartered in Seoul. Following the *chaebols*, related industries and supporting services also clustered in Seoul in pursuit of close contacts with the major business groups, central government agencies, trade and industrial associations.

Seoul is certainly a basing-point for TNCs, as emphasized by world city theory. Seoul hosted four companies on the list of the world's 100 largest in 1997 – only six cities in the world had more. Ten of the global Fortune 500 companies are located in Seoul. But, contrary to world city theory, Seoul's TNCs are industrial not finance or producer service companies, and the contrast with New York City in Table 2 is revealing. Ranked by sales, 16 out of the top 20 New York City-based firms are in the finance and producer services sector, but only three out of the top 20 Seoul corporations are in that category; the rest are manufacturing and construction firms. Lest one think this difference is explained by New York City's more advanced economy, Table 2 also reveals that Tokyo, the capital of global capital according to the Global Fortune 500 list, resembles Seoul not New York City – just four out of Tokyo's top 20 firms are in the finance and producer service sector.

Table 2 The distribution and ranking (by net sales) of producer services (inns in New York, Seoul and Tokyo), 1997

New York	Seoul	Tokyo
1. Philip Morris, Inc.	1. Samsung Corporation	1. Mitsui & Co. Ltd
2. **AT&T Corporation**	2. Daewoo Corporation	2. Mitsubishi Corporation
3. **Citicorp**	3. Hyundai Corporation	3. **Nippon Tel. & Tel. Corporation**
4. **Chase Manhattan Corporation**	4. LG International Corporation	4. Hitachi Ltd
5. **American International Inc.**	5. Hyundai Motor Co. Ltd	5. Sony Corporation
6. **Merrill Lynch & Co., Inc.**	6. Korea Electric Power	6. **Dai-Ichi Mutual Life Insurance**
7. **ITT Corporation**	7. Yukong Ltd	7. Toshiba Corporation
8. **Travelers Group, Inc.**	8. LG Electronics Co. Ltd	8. Honda Motor Corporation
9. **Loews Corporation**	9. Hyundai Engineering & Construction	9. **Bank of Tokyo-Mitsubishi**
10. **American Express Co.**	10. Kia Motors Corporation	10. Tokyo Electric Power Co.
11. RJR Nabisco Holdings Corporation	11. Ssangyong Corporation	11. NEC Corporation
12. **Morgan J.P. & Co.**	12. Hyundai Motor Service	12. Fujitsu Ltd
13. Bristol-Myers Squibb Co.	13. Ssangyong Oil Refining	13. Japan Tobacco, Inc.
14. **Lehman Brothers Holdings**	14. **Korea Exchange Bank**	14. Mitsubishi Motors Corporation
15. **Nynex Corporation**	15. Daewoo Electronics Co.	15. **Meiji Mutual Life Insurance**
16. **Morgan Stanley Group Inc.**	16. **Hanil Bank Ltd**	16. Mitsubishi Electric Corporation
17. **Viacom, Inc.**	17. **Cho-Hung Hank, Ltd**	17. Kanematsu Corporation
18. Pfizer Incorporated	18. Korean Air Lines Co. Ltd	18. Mitsubishi Heavy Industries
19. **Chase Manhattan**	19. Ssangyong Cement	19. Nippon Steel Corporation
20. **Time Warner, Inc.**	20. LG Chemical Co. Ltd	20. Ito-Yokado Co.

Source: Disclosure (1998).

Note: Producer services firms are shown in bold.

Contrary to the world city model, Seoul is primarily a *national* basing-point for the global operations of *Korean* TNCs. The Korean state has controlled the inward and outward flow of foreign investment until recently; the largest foreign holdings in Seoul as of 1990 were minority equity shares in major Korean corporations, mostly held by Japanese TNCs, rather than subsidiaries that were wholly or majority owned by companies headquartered abroad. Seoul hosts many fewer branches of foreign headquartered companies (161) than comparable world cities in the Americas, like Mexico City (266) or São Paulo (380) (Hoopes 1994).

Seoul's global control capability

Seoul certainly hosts an infrastructure for global control. But, contrary to world city theory, Seoul's global control apparatus is anchored in central government ministries and the state's continuous channels of communication with business leaders and organizations that monitor industrial performance, not in private finance and producer service firms. Because export marketing requires substantial fixed costs and externalities in the initial stages of any industry, the expanded credit made available by the state crucially enabled Korean exporters to fill foreign orders and explore foreign markets. Without government intervention in the allocation of credit, it is unlikely that Korea's rapid transformations in industrial composition and level of industrial development would have been possible. Firms which finance their investments primarily through bank credit and foreign loans, instead of through stock issues, accumulate heavy debt. Through its control over finance, the Korean government became a risk partner for industrialists, enabling new export ventures and entrepreneurship. And by controlling the banks, the government created incentives for firms to maximize their assets and growth, rather than to strive for immediate profitability. As long as they satisfied the government by expanding exports and constructing new plants, firms ensured their access to credit.

Economic organization

According to the world city paradigm, the decentralization of manufacturing and the associated shift to a service-based economy occasion a massive loss in a world city's manufacturing employment, the exodus of manufacturing headquarters and a downgrading in the manufacturing jobs that remain in the city (Sassen 1991). But Seoul does not fit this profile, either. As we have seen (Table 2), the headquarters of Korea's major manufacturing companies continue to concentrate in Seoul, one-quarter of Seoul's labor force continues to work in manufacturing and the city's manufacturing base is the most advanced in the nation.

Social and spatial polarization

You can certainly find class-segregated residential areas in Seoul. Squatters, most of whom work in construction, as street vendors, housemaids and taxi drivers, or in small factories, have settled on the hillsides surrounding the city. But, contrary to world city theory, most are migrants from Korea's countryside not from abroad. Seoul's foreign residents numbered 39,246 in 1994, only 0.4 per cent of the city's population, much lower even than Tokyo's 1.8 per cent, let alone New York City's 28 per cent (Crahan and Vourvoulias-Bush 1997). Several wards south of the Han River, on the other

hand, are well planned, mixed residential and business districts for the middle classes.

Still, Seoul has few of the plywood shanty towns visible in many Asian cities. And, contrary to world city theory, income disparities among Seoul's wards are hard to discern. In 1993, the average monthly household income of the poorest of Seoul's 26 wards was 97 per cent that of the wealthiest. By contrast, per capita income in the Bronx, New York City's poorest borough, was just 38 per cent that of Manhattan, the wealthiest. Indeed, even with the extreme concentration of capital in the hands of a few business groups, the distribution of income among Korean families is among the most egalitarian in the world.

Seoul's bureaucratic elite

Seoul is not under the sway of a transnational class, as posited by world city theory. Rather, the large political power of Korea's small bureaucracy continues to the present day. As in Japan, Korea's state officialdom, not the bourgeoisie, led the industrialization effort. And political bureaucrats controlled the state. Most parliamentary statutes originated with the bureaucracy, not with legislators, and administrative policies were also orchestrated within the bureaucracy.

Seoul's integration with the nation-state

Korea, like Japan, is a unitary state. Until recently, the city of Seoul was simply an appendage of the central government. There is no material basis for arguing that Seoul is severing economic ties with the rest of Korea, as world city theory would predict. Just the opposite is true. Seoul's gross product per capita was almost twice that of the nation in 1960, but the gap has steadily diminished, and by 1991 the city was about on a par with the nation as a whole.

CONCLUSION

Like Tokyo, Seoul does not conform to the world city model. Seoul, like Tokyo, is a national basing-point for the global operations of Korean transnational corporations, not a global basing-point for the global operations of borderless firms. Like Tokyo, Seoul's industrial policy and social structure are geared less to attracting investments from abroad than to facilitating the foreign trade and investments of Korean corporations. Command and control functions are concentrated in Seoul, but so too is industrial production, particularly knowledge-intensive manufacturing, as in Tokyo. Seoul, like Tokyo, has not experienced severe manufacturing decline, rapid expansion in producer service employment, extensive foreign immigration or much social and spatial polarization. Like Tokyo, Seoul is under the sway of a political bureaucratic elite, not a transnational capitalist class. And, as with Tokyo, it would be senseless to claim that Seoul is severing ties of mutual interest with the nation-state; if anything, the capital city is becoming even more integrated with the rest of South Korea.

Tokyo and Seoul challenge world city theory's assumptions about the nature of globalization and the role of world cities in the globalization process. How damaging are the East Asian anomalies to the validity of the world city paradigm? It depends upon how one interprets the discrepancies.

One could argue, for example, that Tokyo and Seoul do not fit the world city definition and therefore their anomalous characteristics have no bearing on the model. This resolution hardly seems satisfactory, however, for it would drastically reduce the geographical scope and empirical testability of the theory. In any case, Tokyo and Seoul do fit the world city definition since both cities provide an infrastructure that enables TNCs to control their global operations. However, both cities emphasize the global operations of indigenous not foreign companies, and their international infrastructure is primarily rooted in state ministries and bureaus, not in private finance and producer service firms.

One could also argue that while Tokyo and Seoul may not have conformed to the world city model in the past, they are being forced by global pressures to move in that direction today and will continue to do so in the future [. . .] However, it is premature to equate the very real crisis in the East Asian developmental state with the end of East Asian developmentalism let alone with the transition to liberal market capitalism. There is considerable popular support in Japan and Korea for more state decentralization, deregulation and policy transparency, but there is no similar groundswell support for market-driven capitalism. Neither Japan nor Korea has the historical, ideological or political underpinnings for Western neo-liberalism. Indeed, there is entrenched opposition to market liberalism in the state bureaucracies, business groups and trade unions. The current restructuring is more likely to result in a new phase of East Asian developmentalism than anything approximating Anglo-American liberalism.

A third way to resolve the East Asian anomaly is to conclude that Tokyo and Seoul are a different type of world city from that conceptualized by Friedmann and Sassen. The world city paradigm makes sense in market-centered New York and London, but not in state-centered Tokyo and Seoul. But this resolution would put the state square in the center of world city analysis, and that clashes with two of the paradigm's central assumptions: that globalization diminishes the power and integrity of the nation-state, and that cities are replacing states as central nodes in the world economy. This approach is compatible, however, with comparative findings that national institutions, politics and culture mediate the impact of global processes to produce diverse urban outcomes.

We believe Tokyo and Seoul's divergence from the world city model reflects more than national variation within a common global context, however. Japan and Korea have developed a different kind of political economy from Western market capitalism, one nurtured, ironically, under the US geo-political umbrella during the Cold War. IMF pressure for financial reforms indicates "system friction" between Anglo-American and East Asian political economies, a kind of economic Cold War [. . .] Understanding Tokyo and Seoul necessitates a different conception of the world system from that underlying the globalist world city argument. Tokyo and Seoul differ from New York in so many salient respects because these cities are lodged within a non-hegemonic and interdependent world political economy divided among differently organized national systems and regional alliances.

Countries are attempting to open their markets to foreign competition *and* to pursue national and regional industrial policies *simultaneously*.

Concepts like non-hegemonic interdependence (Stallings and Streeck 1996) better capture the trajectory of cities in today's world political economy than claims about cities "abandoning national ties" in order to embrace supranational alliances and "denationalized expertise." In short, the economic base, spatial organization and social structure of the world's major cities are strongly influenced by the national development model and regional context in which each city is embedded.

REFERENCES FROM THE READING

Amsden, A. (1989) *Asia's Next Giant*. New York: Oxford University Press.

Crahan, M. and Vourvoulias-Bush, A. (eds) (1997) *The City and the World: New York's Global Future*. New York: Council on Foreign Relations.

Disclosure (1998) *World Scope CD-ROM*. Available online: http://www.disclosure.com.

Friedmann, J. (1995) Where we stand: a decade of world city research. In P.L. Knox and P.J. Taylor (eds) *World Cities in a World-System*. New York: Cambridge University Press, 21–47.

Friedmann, J. (2001) World cities revisited: a comment, *Urban Studies*, 38, 13, 2535–2536.

Gao, B. (1997) *Economic Ideology and Japanese Industrial Policy: Developmentalism from 1931 to 1965*. New York: Cambridge University Press.

Gershenkron, A. (1962) *Economic Backwardness in Historical Perspective*. Cambridge, MA: Harvard University Press.

Hill, R.C. and Fujita, K. (1995) Osaka's Tokyo problem, *International Journal of Urban and Regional Research*, 19, 181–193.

Hill, R.C. and Kim, J.W. (2001) Response to Friedmann and Sassen, *Urban Studies*, 38, 13, 2541–2542.

Hong, S.W. (1996) Seoul: a global city in a nation of rapid growth. In F. Lo and Y. Yeoung (eds) *Emerging World Cities in Pacific Asia*. New York: United Nations University Press, 144–178.

Hoopes, D. (1994) *Worldwide Branch Locations of Multinational Companies*. Detroit, MI: Gale Research, Inc.

Jo, S-J. (1992) The world city hierarchy and the city of Seoul. Unpublished PhD thesis, University of Delaware.

Johnson, C. (1995) *Japan: Who Governs?* New York: W.W. Norton.

Kim, J. and Choe, S. (1997) *Seoul: The Making of a Metropolis*. New York: John Wiley & Sons.

Ostry, S. (1996) Policy approaches to system friction: convergence plus. In S. Berger and R. Dore (eds) *National Diversity and Global Capitalism*. Ithaca, NY: Cornell University Press, 333–352.

Sassen, S. (1991) *The Global City*. Princeton, NJ: Princeton University Press.

Sassen, S. (1999) Global financial centers, *Foreign Affairs*, 78, 1, 75–87.

Sassen, S. (2001) Global cities and developmental states: how to derail what could be an interesting debate, *Urban Studies*, 38, 13, 2537–2540.

Stallings, B. and Streeck, W. (1996) Capitalisms in conflict? The United States and Japan in the post-cold war world. In B. Stallings (ed.) *Global Change, Regional Response*. Cambridge: Cambridge University Press, 67–99.

21

"The Stimulus of a Little Confusion: A Contemporary Comparison of Amsterdam and Los Angeles"

from L. Deben (ed.), *Understanding Amsterdam: Essays on Economic Vitality, City Life and Urban Form* (2000)

Edward W. Soja

Editors' introduction

Edward W. Soja is one of the most creative and influential contemporary urbanists. Since the early 1970s, Soja has worked in the Department of Urban Planning at the University of California, Los Angeles (UCLA), where he developed his own perspective on the globalization of urbanization amidst ongoing discussions of such matters within the so-called "LA School" of urban studies (see Scott and Soja 1996; Dear 2002). Soja's major books have been focused extensively on Los Angeles and Southern California (Soja 1989, 1996, 2000), which he has used as a geographical focal point for the development of broader theoretical arguments regarding the contemporary global urban condition. Much of Soja's work has advanced what he has termed a "postmodern" viewpoint, but in so doing, he has also drawn upon the tools of urban political economy, including neo-Marxian urban theory and global city theory. Soja's brilliantly energetic writing style, his creative theoretical eclecticism and his astute powers of observation have given his work a broad, interdisciplinary appeal both within and beyond the field of urban studies.

In his contribution here, Soja embarks upon an astute comparative analysis of Los Angeles and Amsterdam, two cities that would appear to be incomparably different. Indeed, Soja begins the chapter by briefly reviewing these differences at once on structural and experiential levels with reference to the downtown centers of each city. However, as Soja indicates, once one shifts to the regional or metropolitan scale, and examines the sprawling mega-urban galaxies (or "100-mile cities" – see Reading 8 by Sudjic) in which both of these cities are embedded, any number of similarities between them are suddenly brought into focus. Soja examines, in particular, the restructuring of urban form, the internationalization of the regional economy, the consolidation of post-Fordist forms of industrial organization, and the intensification of socioeconomic polarization within each of these urban regions. Soja's claim is not that Amsterdam and Los Angeles have become identical, or that they are converging towards a single model of global city formation. Rather, Soja is suggesting that these cities are both undergoing broadly analogous forms of restructuring due to their embeddedness within an emergent global urban system, and for this reason, their analytical juxtaposition can illuminate important aspects of urban life under contemporary capitalism. In this sense, Soja's elegant comparison of two otherwise radically different metropolises provides an interesting counterpoint to the critical arguments developed by Hill and Kim in Reading 20.

Plate 17 Spui Straat, Amsterdam (Roger Keil)

At first glance, a comparison of Los Angeles and Amsterdam seems as impossible as comparing oranges and potatoes. These two extraordinary cities virtually beg to be described as unique, incomparable, and of course to a great extent they are. But they are also linkable as opposite and apposite extremes of late twentieth-century urbanization, informatively positioned antipodes that are almost inversions of one another yet are united in a common and immediate urban experience. First I will annotate the more obvious oppositions.

Los Angeles epitomizes the sprawling, decentered, polymorphic, and centrifugal metropolis, a nebulous galaxy of suburbs in search of a city, a place where history is repeatedly spun off and ephemeralized in aggressively contemporary forms. In contrast, Amsterdam may be the most self-consciously centered and historically centripetal city in Europe, carefully preserving every one of its golden ages in a repeatedly modernized Centrum that makes other remnant mercantile capitalist "Old Towns" pale by comparison. Both have downtowns of roughly comparable area, but only one of 100 Angelenos live in the City's center, whereas more than 10 per cent of Amsterdammers are Centrum dwellers.

Many residents of the City of Los Angeles have never been downtown and experience it only vicariously, on television and film. Very few now visit it to shop; and surprisingly few tourists take in its local attractions, at least in comparison to more peripheral sites. Amsterdam's Centrum receives nearly 8 million tourists a year and is packed daily with many thousands of shoppers. Amsterdammers may not be aware of the rest of the city, but they certainly know where the center can be found.

It has been claimed that nearly three-quarters of the surface space of downtown Los Angeles is devoted to the automobile and to the average Angeleno freedom and freeway are symbolically and often politically intertwined. Here the opposition to Amsterdam's Centrum, second only to floating Venice in auto-prohibition, is almost unparalleled.

It is not the car but the bicycle that assumes, for the Amsterdammer, a similarly obsessive symbolic and political role, but it is an obsession filled not with individualistic expression and automaniacal freedom as much as with a collective urban and environmental consciousness and commitment. This makes all the contrasts even more stark.

Amsterdam's center feels like an open public forum, a daily festival of spontaneous political and cultural ideas played at a low key, but all the more effective for its lack of pretense and frenzy. Its often erogenously-zoned geography is attuned to many different age groups and civically dedicated to the playful conquest of boredom and despair in ways that most other cities have forgotten or never thought possible. Downtown Los Angeles, on the other hand, is almost pure spectacle, of business and commerce, of extreme wealth and poverty, of clashing cultures and rigidly contained ethnicities. Boredom is assuaged by overindulgence and the bombardment of artificial stimulation, while despair is controlled and contained by the omnipresence of authority and spatial surveillance. Young house-holders are virtually non-existent. In their place are the homeless, who are coming close to being half the central city's resident population despite vigorous attempts at gentrification and dispersal.

In compact Amsterdam, the whole urban fabric is clearly readable and explicit. From its prime axis of the Damrak and Rokin, the city unfolds in layers like a halved cross-section of an onion, first in the "old side" and "new side," then in the neat crescents of the ringing canals from the inner to the outer Singel girdles, and finally in segments and wedges of inner and outer suburbs. This morpho-logical regularity binds Amsterdammers to traditional concepts of urban form and function.

In comparison, Los Angeles seems to break every rule of urban readability and regularity, challenging all traditional models of what is urban and what is not. One of America's classic suburbias, the San Fernando Valley, is almost wholly within the jig-sawed boundaries of the monstro-City of Los Angeles, while many inner city barrios and ghettoes float outside on unincorporated county land. There is a City of Industry, a City of Commerce, and even a Universal City, but these are not cities at all. Moreover, in an era of what many have called post-industrial urbanization, with cities being emptied of their manufacturing

Plate 18 Homeless camp, Los Angeles, 1987 (Roger Keil)

employment, the Los Angeles region has continued its century-long boom in industrial growth in both its core and periphery. It is no surprise, then, that Southern California has become a center for inno-vative and non-traditional urban theory for there seems little from conventional, established schools of urban analysis that any longer makes sense.

And then there is that most basic of urban functions, housing. One of the most interesting features of the success of the squatter movement in Amsterdam was the absence of a significant housing shortage. Although much of the Centrum is privately owned, the rest of the city is a vast checkerboard of public, or social housing. Even what the Dutch planners consider the worst of these projects, such as the huge Bijlmermeer high-rise garden suburb, served effectively to accommodate the thousands of migrants from Surinam and other former colonies during the 1970s.

The squatter movement was more than just an occupation of abandoned offices, factories, warehouses, and some residences. It was a fight for the rights to the city itself, especially for the young and for the poor. Nowhere has this struggle been more successful than in Amsterdam. Nowhere has it been less successful than in Los Angeles. In the immediate post-war period, Los Angeles was poised to become the largest recipient of public housing investment in the country, with much of this scheduled to be constructed in or around downtown. In no other American city did plans for public housing experience such a resounding defeat by so ferociously anti-socialist campaigners. The explosion of ethnic insurrections in the 1960s and early 1970s cruelly accelerated the commercial renewal of downtown at the expense of its poor residential inhabitants. On the central city's "new side" grew a commercial, financial, governmental, and high cultural fortress, while the "old side," beyond the skyscraper walls, was left to be filled with more residual land uses, from the tiny remnants of El Pueblo de Nuestra Senora de Los Angeles to the fulsome Skid Row of cardboard tenements and streetscapes of despair, to the Dickensian sweatshops and discount marts of the expansive Garment District.

The core of my oppositional comparison is thus amply clear. But what of the periphery? Are there comparative dimensions that are missed when we focus on the antipodal centralities of Amsterdam and Los Angeles? For the remainder of this chapter, I will set the two cities in a larger, more generalizable context that focuses on contemporary processes of urban restructuring. Here, the cities follow more similar paths than might initially seem possible. These similarities are not meant to contradict or erase the profound differences that have already been described, but to supplement and expand upon their emphatic and extreme particularity.

My own research and writing on the urban restructuring of Los Angeles has identified a series of intertwined trends that have become increasingly apparent not only in Los Angeles but in most of the world's major urban regions. Each trend takes on different intensities and forms in different cities, reflecting both the normality of geographically uneven development and the social and ecological particularity of place. More important than their individual trajectories, however, is their correlative interconnectedness and the tendency for their collective impact to define an emerging new mode of urbanization, significantly different from the urbanization processes that shaped the industrial capitalist city during the long post-war boom period.

It is appropriate to begin with the *geographical recomposition of urban form*. As with the other trends, there is a certain continuity with the past, lending credence to the argument that restructuring is more of an acceleration of existing urban

Plate 19 Los Angeles (Roger Keil)

trajectories than a complete break and redirection. The current geographical recomposition, for example, is in large part a continuation on a larger scale of the decentralization and polynucleation of the industrial capitalist city that was begun in the last half of the nineteenth century.

There are, however, several features of the recent round of polynucleated decentralization that suggest a more profound qualitative shift. First, the size and scale of cities, or more appropriately of urban regions, has been reaching unprecedented levels. The older notion of "megalopolis" seems increasingly inadequate to describe a Mexico City of 30 million inhabitants or a "Mega-York" of nearly 25, stretching from Connecticut to Pennsylvania. Never before has the focus on the politically-defined "central" city become so insubstantial and misleading. Complicating the older form still further has been the emergence of "Outer Cities," amorphous agglomerations of industrial parks, financial service centers and office buildings, massive new residential developments, giant shopping malls, and spectacular entertainment facilities in what was formerly open farmland or a sprinkling of small dormitory suburbs. Neither city nor suburb, at least in the older senses of the terms of these reconcentrated poles of peripheral urban and (typically "high-tech") industrial growth, have stimulated a new descriptive vocabulary.

The growth of Outer Cities is part of the recentralization of the still decentralizing urban region, a paradoxical twist that reflects the ability of certain areas within the *regional metropolis* to compete within an increasingly globalized economy. Over the past twenty-five years, the decentralization of manufacturing and related activities from the core of the older industrial capitalist cities broke out from its national containment. Jobs and factories continued to move to suburban sites or non-metropolitan areas within the national economy, but also, much more than ever before, to hitherto non-industrialized regions of the old Third World, creating a new geographical dynamic of growth and decline that not only has been changing the long-established international division of labor but also the spatial division of labor within urban regions.

The geographical recomposition is paradigmatically clear in Greater Los Angeles. Within a radius of 60 miles (100 kilometers) from the booming Central Business District of the misshapen City of Los Angeles there is a radically restructured regional metropolis of nearly 15 million people with an economic output roughly equivalent to that of the Netherlands. At this scale, a comparison with Amsterdam seems totally inappropriate. But if we shift scales, a different picture emerges. A 100-kilometer circle from Amsterdam's Centrum cuts through Zeeland, touches the Belgian border near Tilburg, curves past Eindhoven to touch the German border not far from Nijmegen, and then arcs through the heart of Friesland to the North Sea. Most of the nearly 15 million Dutch live within this densely urbanized region and its scale and productivity come remarkably close to matching its Southern California counterpart.

The southwest quadrant of this "Greater Amsterdam" coincides rather neatly with the Randstad, which can be seen as a kind of Outer City in itself, but with the defining central core being not an old urban zone but the determinedly preserved rural and agricultural "Green Heart." Around the Green Heart are the largest cities of the Netherlands: Amsterdam, Rotterdam, The Hague, and Utrecht, each experiencing a selective redistribution of economic activities between central city, suburban fringe, and more freestanding peripheral centers. As a whole, the Randstad contains the world's largest port (Los Angeles-Long Beach is now probably second), Europe's fourth largest international financial center (after London, Zurich, and Frankfurt) and fourth largest international airport (Schiphol, surpassed in traffic only by Frankfurt, Paris, and London).

Like the Greater Los Angeles regional metropolis, Greater Amsterdam has been experiencing a complex decentralization and recentralization over the past 25 years. How useful this larger-scale regional comparison of geographical recomposition might be to a further understanding of urban restructuring I will leave to others to determine. For present purposes, however, it at least forms a useful antidote to the tendency of urban observers to persist in seeing the contemporary period of restructuring too narrowly due to an excessive focussing on the long-established central city, thereby ignoring a dimension of an entirely different order from the one they traditionally know.

The recomposition of urban form is intricately connected to other sets of restructuring processes.

Already alluded to, for example, has been the *increasing internationalization of the regional metropolis*, leading to the formation of a new kind of world city. Amsterdam in its Golden Age was the prototypical model of the world city of mercantile capitalism and it has survived various phases of formation and reformation to remain among the higher ranks of contemporary world cities, whether combined in the Randstad or not. What distinguishes the global cities of today from those of the past is the *scope* of internationalization, in terms of both capital and labor. To the control of world trade (the primary basis of mercantile world cities) and international financial investment by the national state (the foundation of imperial world cities) has been added the financial management of industrial production and producer services, allowing the contemporary world city to function at a global scale across all circuits of capital. First, Second, and Third World economies have become increasingly integrated into a global system of production, exchange, and consumption that is sustained by an information-intensive hierarchy of world cities, topped today by the triumvirate of Tokyo, New York, and London.

Los Angeles and Amsterdam are in the second tier of the restructured world city hierarchy, but the former is growing much more rapidly and some predict it will join the top three by the end of the century. Amsterdam is more stable, maintaining its specialized position in Europe on the basis of its concentration of Japanese and American banks, the large number of foreign listings on its Stock Exchange, the strong and long-established export-orientation of Dutch companies, and its control over Dutch pension funds. The banking and financial services sector remains a key actor in Amsterdam's Centrum, feeding its upscale gentrification and drawing strength from the information-rich clustering of government offices, university departments, cultural facilities, and specialized activities in advertising and publishing.

A characteristic feature of increasing internationalization everywhere has been an erosion of local control over the planning process, as the powerful exogenous demands of world city formation penetrate deeply into local decision-making. Without a significant tradition of progressive urban planning, Los Angeles has welcomed foreign investment with few constraints. Its downtown "renaissance"

was built on foreign capital to such an extent that today almost three-quarters of the prime properties in the Central Business District are foreign-owned or at least partially controlled by overseas firms. The internationalization of Amsterdam has been more controlled, but the continued expansion of the city as a global financial management center is likely to pose a major threat to many of the special qualities of the Centrum.

The other side of internationalization has been the attraction of large numbers of foreign workers into almost every segment of the local labor market, but especially at lower wage and skill levels. Los Angeles today has perhaps the largest and most culturally diverse immigrant labor force of any major world city, an enriching resource not only for its corporate entrepreneurs but also for the cultural life of the urban region. Amsterdam too is fast approaching becoming a "majority minority" city, a true cosmopolis of all the world's populations. With its long tradition of effectively absorbing diverse immigrant groups, Amsterdam appears to have been more successful than Los Angeles in integrating its immigrant populations into the urban fabric. One achievement is certain: they are better housed in Amsterdam, for Los Angeles is currently experiencing one of the worst housing crises in the developed world. As many as 600,000 people, predominantly the Latino working poor, now live in seriously overcrowded conditions in dilapidated apartments, backyard shacks, tiny hotel rooms, and on the streets.

Intertwined with the geographical recomposition and internationalization of Los Angeles and Amsterdam has been a pervasive *industrial restructuring* that has come to be described as a trend toward a "Post-Fordist" regime of "flexible accumulation" in cities and regions throughout the world. A complex mix of both deindustrialization (especially the decline of large-scale, vertically integrated, often assembly-line, mass-production industries) and reindustrialization (particularly the rise of small and middle-size firms flexibly specializing in craft-based and/or high technologically-facilitated production of diverse goods and services), this restructuring of the organization of production and the labor process has been associated with a repatterned urbanization, a new dynamic of geographically uneven development.

A quick picture of the changing Post-Fordist industrial geography would consist of several characteristic spaces: older industrial areas either in severe decline or partially revived through adaptation of more flexible production and management techniques; new science-based industrial districts or technopoles typically located in metropolitan peripheries; craft-based manufacturing clusters or networks drawing upon both the formal and informal economies; concentrated and communications-rich producer services districts, especially relating to finance and banking but also extending into the entertainment, fashion, and culture industries; and some residual areas, where little has changed. It would be easy to transpose this typology to Greater Los Angeles, for much of the research behind it has been conducted there. Although the Post-Fordist restructuring has not gone nearly as far in Amsterdam, the transposition is also quite revealing.

The Centrum has been almost entirely leached of its older, heavier industries and 25 per cent of its former office stock has been lost, primarily to an impressive array of new subcenters to the southeast, south, and west and to the growing airport node at Schiphol. One might argue that this dispersal represents a sign of major decline in the inner city, due in part to a shift from a concentric to a more grid-like pattern of office and industrial development. Just as convincing, however, is a restructuring hypothesis that identifies the Centrum as a flexibly specialized services district organized around international finance and banking, university education, and diverse aspects of the culture and entertainment industries (fashion, especially for the twenty-somethings, television and film, advertising and publishing, soft drugs and sex, and, of course, tourism).

A fourth trend needs to be added, however, before one goes too far in tracing the impact of Post-Fordist industrial restructuring. This is the tendency toward *increasing social and economic polarization* that seems to accompany the new urbanization processes. Recent studies have shown that the economic expansion and restructuring of Los Angeles has dramatically increased poverty levels and hollowed out the middle ranks of the labor market, squeezing job growth upward, to a growing executive-professional-managerial "technocracy" (stocked by the largest urban concentrations in the world of scientists, engineers, and mathematicians), and downward, in much larger numbers, to an explosive mix of the "working poor" (primarily Latino and other immigrants, and women, giving rise to an increasing "feminization of poverty") and a domestic (white, African-American, and Mexican-American, or Chicano) "urban underclass" surviving on public welfare, part-time employment, and the often illegal opportunities provided by the growing informal, or underground economy. This vertical and sectorial polarization of the division of labor is reflected in an increasing horizontal and spatial polarization in the residential geography of Los Angeles. Old and new wealth is increasingly concentrated in protected communities with armed guards, walled boundaries, "neighborhood watches," and explicit signs that announce bluntly: "Trespassers will be shot"; while the old and new poor either crowd into the expanding immigrant enclaves of the Third World City or remain trapped in murderous landscapes of despair. In this bifurcating urban geography, all the edges and turf boundaries become potentially violent battle-fronts in the continuing struggle for the rights to the city.

Here again, the Amsterdam comparison is both informative and ambiguously encouraging, for it too has been experiencing a process of social and economic polarization over the past two decades, and yet, it has managed to keep the multiplying sources of friction under relatively successful social control. The Dutch "Job Machine," for example, shows a similar hollowing out of the labor market, with the greatest growth occurring in the low-paid services sector. Official unemployment rates have been much higher than in the U.S., but this difference is made meaningless by the contrasts in welfare systems and methods of calculating the rate itself. Overall job growth has been much lower than in the U.S. and, except for the producer services sector, there has been a decline in high-wage employment thus limiting the size of the executive-professional-managerial "bulge." Increasing flexibility in the labor market, however, is clearly evident in the growth of "temporary" and "part-time" employment, with the Netherlands having the largest proportion of part-time workers in the EEC and perhaps the highest rate (more than 50 per cent) in the Western World for women.

With its exceptional concentration of young, educated, often student households, high official levels of unemployment, still solid social security system, and distinctive patterns of gentrification, an unusual synergy has developed around the personal services sector and between various age and income groups. Income polarization has been producing a growing complementarity between the higher and lower income groups with respect to the flexible use of time and place, especially in the specialized provision of such personal services as domestic help and babysitting, late-night shopping, entertainment and catering, household maintenance and repair, educational courses and therapies, fitness centers, body-care activities, etc. Such activities in Amsterdam take place primarily in the underground economy and are not captured very well in official statistics. But they nonetheless provide a legitimate and socially valuable "survival strategy" for the poor and unemployed that has worked effectively to constrain the extreme effects of social polarization that one finds in Los Angeles or New York City. Moreover, it is a strategy that draws from the peculiar urban genius of Amsterdam, its long tradition of grass roots communalism, its sensitive adaptation to locality, its continuing commitment to libertarian and participatory social and spatial democracy, and its unusual contemporary attention to the needs of the twenty-something generation.

I had originally intended to conclude by addressing *postmodernism and post-modernization* as a fifth restructuring theme and to explore the extent to which this restructuring of the "cultural logic" of contemporary capitalism can be traced into the comparison of Amsterdam and Los Angeles. In my own recent research and writings, I have argued that a neoconservative form of postmodernism, in which "image" replaces reality and the simulated and "spin-doctored" representations assume increasing political and economic power, is significantly reshaping popular ideologies and everyday life all over the world and is fastly becoming the keystone for a new mode of social regulation designed to sustain the development of (and control the resistance to) the new Post-Fordist regimes of "flexible" and "global" capitalist accumulation and the accompanying "new urbanization processes" discussed on the preceding pages. After experiencing Amsterdam, where resistance to the imposition of this neoconservative restructuring seems exceptionally strong, it is tempting just to add another polar opposition to the comparison with Los Angeles where this process is probably more advanced than almost anywhere else on earth. But I will leave the issue open for future research and reflection.

REFERENCES FROM THE READING

Dear, M. (ed.) (2002) *From Chicago to LA: Making Sense of Urban Theory.* Thousand Oaks, CA: Sage.

Scott, A.J. and Soja, E. (eds) (1996) *The City: Los Angeles and Urban Theory and the End of the Twentieth Century.* Berkeley and Los Angeles: University of California Press.

Soja, E. (1989) *Postmodern Geographies.* Cambridge, MA: Blackwell.

Soja, E. (1996) *Thirdspace.* Cambridge, MA: Blackwell.

Soja, E. (2000) *Postmetropolis.* Cambridge, MA: Blackwell.

PART FOUR

Globalization, urbanization and uneven development: perspectives on global city formation in/from the global South

Plate 20 Cape Town (Roger Keil)

INTRODUCTION TO PART FOUR

All mappings of the global urban hierarchy are also representations of the global economy. However, since the development of global city theory in the 1980s, mappings of the global urban system have been articulated primarily from the point of view of the older industrialized world (the "global North"): most global cities research has been conducted in the global North, most global cities researchers are based at universities located in the global North, and most global cities are themselves located in the global North. Thus, all of the major maps of the global urban system must be understood as representations of power, centrality and exclusion in the contemporary global economy. It is hardly surprising, for example, that with Johannesburg as the only African city to be listed on many of these maps, it appears as if an entire continent has been sidestepped by contemporary forms of globalization. The contributions to Part Four address this gross bias within the global cities literature by exploring the impacts of globalized urban development in what Josef Gugler (2004) has termed "world cities beyond the West."

Particularly during the early phase of global cities research, we would argue, it was logically consistent and intellectually legitimate to view global cities as the dominant command and control centers for global capitalism. This conceptualization was quite appropriate in light of global cities researchers' interest in the role of transnational corporate headquaters, subsidiary and regional management offices, the financial and producer services complex, and associated spin-off industries, in the new international division of labor. Accordingly, early mappings of the global city hierarchy ranked cities primarily in terms of their role as headquarters locations for transnational firms (see, for instance, Reading 5 by Cohen and Reading 7 by Friedmann) and as agglomerations for advanced producer and financial services industries (Sassen 1991). In his classic 1986 article, John Friedmann presented a general map of the global urban system that usefully encapsulated this initial conceptualization (see Figure 1 in Reading 7, p. 70). In Friedmann's mapping, the global urban system is dominated by a number of major cities in each of the world-economy's major super-regions – Tokyo in East Asia, New York, Chicago and Los Angeles in North America, and London and Paris in western Europe. Each of these global cities is in turn surrounded by a network of "secondary" global cities, several of which are positioned outside the core zones of each region, in locations such as Sydney, Buenos Aires and Johannesburg. For Friedmann, therefore, the consolidation of a global urban hierarchy starkly embodied the polarizing tendencies within post-1970s, globalizing capitalism: an archipelago of networked urban cores was viewed as the spatial infrastructure of a worldwide economic system that systematically excluded large segments, both urban and rural, of the developing world and the global periphery.

Friedmann's initial mapping of the global urban system inspired a wave of energetic research initiatives, including detailed case studies of urban sociospatial restructuring in the "headquarter cities" that had been demarcated by global city theorists, as well as more general explorations of the global urban system as a whole. Following from Friedmann's work and Sassen's (1991) pioneering comparative analysis of New York, London and Tokyo in *The Global City*, the case-study-based literature illuminated the contextually specific political-economic and sociospatial pathways through which cities were transformed into centers of global command and control (see the readings in Parts Two and Three).

Meanwhile, the rapidly expanding literature on the global urban system put forward a number of new empirical indicators through which cities' roles as articulators of global economic relations could be explored more systematically. Thus, as Short et al. (1996) have demonstrated in a comprehensive review article, several additional criteria for world city formation were subsequently proposed – including the number of banks and the stock market capitalization levels of firms within a given city, the role of cities in global telecommunications networks, the role of cities in global transportation networks, and the role of cities as sites for global cultural events or "spectacles." However, as Short et al. (1996) argue, much of this literature has remained impressionistic due to chronic data insufficiencies and a pervasive failure to synthesize the results of studies conducted on the basis of divergent empirical indicators.

More recently, the GaWC research team at Loughborough University in the UK has conducted a series of innovative, systematic and empirically detailed investigations of the world city network (see Taylor 2003; see also Reading 11 by Beaverstock, Smith and Taylor, as well as the research bulletins and data sets included at http:www.lboro.ac.uk/gawc/). In an early presentation of this research, Beaverstock et al. (1999) suggested a new "roster" of 55 world cities based upon their levels of advanced producer services. Their data focused on the locations of global firms in accountancy, advertising, banking and law, and on this basis classified global cities among "alpha," "beta" and "gamma" levels of significance. This procedure generated a more empirically precise mapping of the global urban hierarchy derived from Sassen's (1991) conception of global cities as production sites for advanced corporate services. In this manner, the GaWC researchers were also able to recognize the global roles of a broader range of cities than had previously been considered, both within North America, western Europe and Pacific Asia, as well as in Latin America and eastern Europe. More recent studies by Taylor (2003), Beaverstock, Smith and Taylor (Reading 11) and other GaWC researchers have attempted to transcend the conventional emphasis on the fixed "attributes" of global cities in favor of a relational approach that focuses on the levels of connectivity among global cities. This methodology has enabled the GaWC researchers to produce a number of innovative mappings of inter-city relations within the global urban system, which is now viewed as a dynamically evolving network of linkages and interdependencies among global cities, rather than as a fixed hierarchy of central places. The work of the GaWC research team is thus grounded upon a far more robust set of empirical indicators and a more sophisticated conceptualization of the global urban system than were available during the 1980s, when scholars such as Cohen, Friedmann, Sassen and others initiated their studies of global city formation.

Crucially, however, the work of the GaWC research group has tended to preserve the assumption, which is implicit within all previous analyses of the global urban system, that global cities represent a distinctive class of cities that serve uniquely important roles in the management and coordination of geoeconomic processes. Against this viewpoint, a number of scholars have recently criticized the literature on global cities for perpetuating what Brenda Yeoh (1999: 608) has described as "an image of the world that is empty beyond global cities, a borderless space which can be reordered, neglected or put to use according to the demands of globally articulated capital flows." While such an image would surely be rejected by all of the authors mentioned above, Yeoh's formulation highlights a recurrent problem that has underpinned global cities research since its inception – namely, a pervasive tendency to bracket the distinctive ways in which cities located *outside* the core zones of the world economy have been restructured in and through contemporary processes of globalization. This issue is the central focus of Part Four of the Reader, which is explored by our contributors at once on a methodological and empirical level. Methodologically, the authors critically examine some of the underlying assumptions about the geographies of global capitalism that have led researchers to exclude consideration of globalized urbanization beyond the global North (western Europe, North America and East Asia) (see, in particular, Reading 26 by Robinson). Empirically, the contributors examine a number of cases of globalized urbanization – including, among others, Phnom Penh, Cambodia; Accra, Ghana; Mumbai, India; and São Paulo, Brazil – that are not situated in the upper tiers of the global urban hierarchy.

Gavin Shatkin (1998; see also Reading 25) has proposed to explore the globalization of urban development outside the core zones of the world system as follows:

Rather than treating the experience of cities in LDCs [less developed countries] as an unfortunate footnote to the phenomenon of globalization and economic restructuring, the role of such cities in the process of capital accumulation, and the impact of globalization on their development, should be a topic for research and debate.

(Shatkin 1998: 378)

In recent years, a number of scholars have extended global city theory to explore a much broader range of cases of urban restructuring throughout the world economy. This has necessarily entailed a more sustained consideration of other ways of "becoming global," particularly in the South, that do not entail the attraction of command and control functions and advanced corporate services. Thus, as Robinson (2002: 532; see also Reading 26) argues, "There is a need to construct (or promote) an alternative urban theory which reflects the experiences of a much wider range of cities." In a closely analogous formulation, Shatkin (1998; see also Reading 25) suggests that

it is . . . necessary to broaden the view of the process of globalization beyond the process of relocation of industry from the developed to the developing countries to include the various social, political and economic forces at work in integrating the world.

(Shatkin 1998: 381)

The concept of "globalizing cities," introduced by Marcuse and van Kempen (2000), among other authors, has been used to describe these multifaceted *processes* of urban globalization, which are unfolding in place-specific ways throughout the world economy, including in cities that have heretofore been positioned "off the map" of global city theory (see Reading 26 by Robinson; see also Reading 48 by Olds and Yeung; Yeoh 1999).

The readings in Part Four focus on processes of globalized urban development in a range of cities located in sub-Saharan Africa, Southeast Asia, South Asia and Latin America. Like the contributions to Part Three, the readings included in Part Four underscore the diverse ways in which major cities have been restructured under conditions of intensified globalization. At the same time, the readings also reveal several common trends among globalizing cities located in the global South, as follows.

The legacy of colonialism

Many cities of the global South are situated in formerly colonized territories. As King, in particular, argues (Reading 23), the legacy of colonialism continues to have a major impact upon the built environments, political economies and developmental trajectories of such cities (see also Reading 24 by Simon, Reading 27 by Grant and Nijman and Reading 38 by King).

Neoliberalization

Since the mid-1970s, many developing countries have adopted, or have been forced to adopt, "free-market" policy regimes that promote deregulation, liberalization and a rolling back of public services. Often, such policies are imposed through World Bank structural adjustment programs (SAPs), but they may also be promoted through domestic and local political alliances. As Simon (Reading 24), Shatkin (Reading 25), Grant and Nijman (Reading 27) and Buechler (Reading 28) demonstrate, such policies have "opened up" many cities in the developing world to new forms of foreign direct investment (FDI), but, at the same time, they have also generated massively polarizing sociospatial outcomes.

International organizations

Various international organizations – including the World Bank, the United Nations and a range of IGOs (international governmental organizations) and NGOs (nongovernmental organizations) – frequently exercise a powerful influence on patterns of urban development in the global South (see Reading 24 by Simon and Reading 25 by Shatkin).

Rapid industrialization, territorial polarization and mega-city formation

The globalization of urbanization in the developing world has frequently occurred in a context of extremely rapid industrial development and rural-to-urban migration, processes that have generally intensified intra-national patterns of spatial polarization (e.g., the "urban–rural divide") while also contributing to explosive population growth within major urban centers. Consequently, the most globalized urban spaces of the developing world are frequently "mega-cities" that chronically lack basic social and spatial infrastructures, such as housing, public transportation and social services, for their rapidly expanding populations (see Reading 28 by Buechler).

Governance challenges

As we shall see in Part Five of this Reader, all globalizing cities are confronted with a variety of regulatory problems and governance challenges that are linked to their role as sites for transnational capital investment. However, as a number of contributors to Part Four demonstrate, specific types of governance problems – including extreme poverty, chronic housing shortages, mass unemployment and severe environmental devastation – have emerged in the globalizing cities in the developing world and pose unwieldy challenges for national and local political actors, social movements and NGOs.

In sum, as the readings in Part Four indicate, global city theory no longer serves simply as a basis for exploring a particular type of urban agglomeration under contemporary capitalism; it is instead increasingly mobilized in order to decipher the globalization of urban development throughout the contemporary world system. At the present time, the theory's analysis of developmental trends and sociospatial outcomes in the command and control centers of the global North is still more advanced than studies of other types of globalizing cities or of other pathways of globalized urbanization. Nonetheless, as the readings here demonstrate, a large number of researchers are now actively engaged in fruitful investigations of globalization processes in other types of urban centers as well, including those located in developing countries and in the global periphery (see also Lo and Yeung 1998; Gugler 2004). This burgeoning literature is likely to generate a more nuanced mapping of the global urban system, and a more differentiated conceptualization of globalization itself, than has previously been accomplished in the literature on global city formation.

References and suggestions for further reading

Beall, J. (2002) Globalization and social exclusion in cities: framing the debate with lessons from Africa and Asia, *Environment and Urbanization*, 14, 1, 41–51.

Beaverstock, J.V., Smith, R.G. and Taylor, P.J. (1999) A roster of world cities, *Cities*, 16, 6, 445–458.

Beavon, K. (1998) "Johannesburg": coming to grips with globalization from an abnormal base. In Fu-chen Lo and Yue-Man Yeung (eds) *Globalization and the World of Large Cities*. Tokyo: United Nations Press, 352–390.

Brade, I. and Rudolph, R. (2004) Moscow, the global city?, *Area*, 36, 1, 69–80.

Brown, E., Catalano, G. and Taylor, P.J. (2002) Beyond world cities: Central America in a global space of flows, *Area*, 34, 2, 139–148.

Gilbert, A. (1998) World cities and the urban future: the view from Latin America. In Fu-chen Lo and Yue-Man Yeung (eds) *Globalization and the World of Large Cities*. Tokyo: United Nations Press, 174–202.

Gugler, J. (1996) *The Urban Transformation of the Developing World*. Oxford: Oxford University Press.

Gugler, J. (ed.) (2004) *World Cities beyond the West: Globalization, Development and Inequality*. New York: Cambridge University Press.

Harris, N. (1995) Bombay in a global economy: structural adjustment and the role of cities, *Cities*, 12, 3, 175–184.

Jenkins, P., Robson, P. and Cain, A. (2002) Local responses to globalization and peripheralization in Luanda, Angola, *Environment and Urbanization*, 14, 1, 115–127.

Kowarick, L. and Campanario, M. (1986) Sao Paulo: the price of world city status, *Development and Change*, 17, 1, 159–174.

Lo, F-C. and Yeung, Y-M. (eds) (1998) *Globalization and the World of Large Cities*. Tokyo: United Nations Press.

McGee, T.G. (1995) Eurocentrism and geography: reflections on Asian urbanization. In J. Crush (ed.) *Power of Development*. London: Routledge, 192–207.

McGee, T.G. (1998) Globalization and urban–rural relations in the developing world. In Fu-chen Lo and Yue-Man Yeung (eds) *Globalization and the World of Large Cities*. Tokyo: United Nations Press, 471–496.

Marcuse, P. and van Kempen, R. (eds) (2000) *Globalizing Cities: A New Spatial Order?* Oxford: Blackwell.

Olds, K. (1997) Globalizing Shanghai: the "Global Intelligence Corps" and the building of Pudong, *Cities*, 14, 2, 109–123.

Olds, K. (1998) Globalization and urban change: tales from Vancouver via Hong Kong, *Urban Geography*, 19, 360–385.

Robinson, J. (2002) Global and world cities: a view from off the map, *International Journal of Urban and Regional Research*, 26, 3, 531–554.

Sassen, S. (1991) *The Global City*. Princeton, NJ: Princeton University Press.

Shatkin, G. (1998) 'Fourth world' cities in the global economy: the case of Phnom Penh, Cambodia," *International Journal of Urban and Regional Research*, 22, 3, 378–393.

Short, J.R., Kim, Y., Kuus, M. and Wells, H. (1996) The dirty little secret of world cities research: data problems in comparative analysis, *International Journal of Urban and Regional Research*, 20, 697–717.

Stren, R. (2001) Local governance and social diversity in the development world: new challenges for globalizing city-regions. In A.J. Scott (ed.) *Global City-Regions*. New York: Oxford University Press, 193–213.

Taylor, P.J. (2003) *World City Network*. New York: Routledge.

Yeoh, B. (1999) Global/globalizing cities, *Progress in Human Geography*, 23, 4, 607–616.

Yeoh, B. and Cheng, T.C. (2001) Globalising Singapore: transnational flows in the city, *Urban Studies*, 38, 1025–1044.

Yulong, S. and Hamnett, C. (2002) The potential and prospect for global cities in China in the context of the world system, *Geoforum*, 33, 121–135.

Yusuf, S. and Weiping, W. (2002) Pathways to a world city: Shanghai rising in an era of globalisation, *Urban Studies*, 39, 7, 1213–1240.

22

Prologue

"A Global Agora vs. Gated City-regions"

from *New Perspectives Quarterly* (1995)

Riccardo Petrella

It is possible to imagine two mental maps of the world system in the times ahead. One map is that of a world dominated by a hierarchy of 30 city-regions linked more to each other than to the territorial hinterlands to which the nation state once bound them. This wealthy archipelago of city-regions – with manageable populations of 8–12 million – will be run by an alliance between the global merchant class and metropolitan governments whose chief function is supporting the competitiveness of the global firms to which they are host. These disassociated islands will be surrounded by an impoverished *Lumpenplanet* where peasants have been uprooted from the land by global free trade and try to eke out an existence in violence-ridden mega-urban settlements with populations of 20 million or more. Of the eight billion people expected to populate the earth by 2020, five billion will live in Asia, and of this one billion will reside in 50 cities with more than 20 million inhabitants each.

Excluded from the contest for wealth, these megasettlements will be the source of criminalization of the world economy. Its residents, not unlike the marginalized outerclass that resides in the core city-regions, will make a living from smuggling drugs, children, human organs for transplant, illegal immigrants and weapons.

The other map is one in which the global civil society that has emerged with the information age

in all the major city-regions links together across fading national boundaries to balance the myopic commercialism of the merchant class with a global social contract. Much like the New Deal in America or the social welfare policies in Europe devised earlier in the 20th century, that contract would stabilize the world system by incorporating the *Lumpenplanet* through a redistribution of wealth. Instead of a world where the purely competitive or merely fortunate are forced to hole up in gated city-regions for fear of the crime and pandemonium all about them, the order based on a global contract would give rise to a vital, multicultural civilization on a planetary scale. A kind of plural, global *agora* rather than the medieval moated castle would symbolize this new civilization. This agora, or global civil society, would be the interlinkage between local territoriality and the consciousness of the first planetary generation, the first generation with a global dimension.

The first scenario is most likely in the immediate future; the second must inevitably arise on the agenda over time. The present trends of widespread privatization, deregulation of national safety nets and the globalization of production, trade, financial services and capital markets are clearly leading to the model suggested by the first map. However, in 20 years time, some semblance of the model suggested by the second map must emerge because, for all its undisputed dynamism, the global feudalism of the first model is not sustainable over the long term any more than the feudalism of the Middle Ages.

For now, the global merchants seeking consumer markets and production sites to garner the highest return on their capital – together with accommodating metropolitan governments seeking local jobs and tax revenues – are linking up the world's major city-regions in one giant web of low-orbit satellites, electronic networks such as the Internet, seaports, airports and tax-free industrial parks. As a result, the hierarchy of city-regions that will dominate the world system by 2025 – let's call them the CR-30 that will replace the G-7 leading industrial democracies as the core entities of the world system – is already being established.

The CR-30 includes the following: Rotterdam/ Amsterdam; the Ruhr zone around Düsseldorf; Frankfurt; Stuttgart–Baden–Württemberg; Munich–Bavaria; Oresund–Copenhagen–Malmo; London–South East England; Greater Paris; Lyon–Grenoble; the Zurich and Geneva–Lausanne regions; Barcelona–Catalonia; Montreal–Toronto–Chicago; the New York region; Los Angeles–Orange County; Miami; Vancouver; Istanbul; Johannesburg–Cape Town; the Tokyo area; Osaka; Shanghai; Hong Kong; Singapore; Kuala Lumpur; Jakarta, Sydney; and the São Paulo area. Although some industries will dominate in certain regions, each of the 30 top city regions will end up being generally well positioned in all competitive industries. In fact, "competitiveness" will become more and more a battle between city-regions than nation-states.

At the same time, new media developments such as CNN or Sky-TV, worldwide newspaper syndicates or the Internet as well as transnational non-governmental organizations like the Red Cross, Médecins sans Frontières or Greenpeace are building a global awareness that understands we cannot live in prosperous city-regions apart from the rest of the planet; that it is not possible to live in Milan or Barcelona and look only north and not to the hundreds of millions of poor and resentful people living along the Mediterranean and in North Africa. Excluded, they will turn to Muslim fundamentalism or try to migrate north. Either choice will challenge the illusion of the comfortable that the marginalized will disappear from their concerns.

In some ways, the struggle is already on between the global civil society and the worldwide merchant class to draw a new map of the world for the next century. The more successful the merchant class is in drawing the boundaries of the new order over the coming decade, the more difficult it will be for the global civil society to alter that map. That is the danger, now, in 1995.

23

"Building, Architecture, and the New International Division of Labor"

from *Urbanism, Colonialism and the World-economy* (1991)

Anthony D. King

Editors' introduction

Anthony D. King is Bartle Professor of Art History at the State University of New York at Binghamton. Since the mid-1980s, King has been one of the most influential and prolific authors in the fields of global cities research, urban cultural studies and urban design. In addition to his monograph on London as a global city (King 1990) and an innovative study of the bungalow as a globally produced building form (1984), King's work includes a number of major books and articles on colonial and postcolonial urbanism, globalization and culture, critical sociospatial theory, and urban space and architecture (including, most recently, King 2004). Reading 23 is drawn from a chapter in one of King's early contributions to global city theory, a short book of essays in which he forcefully underscored the centrality of imperialism and colonialism to the long-term history of global urbanization. For King, the contemporary round of global city formation represents only the latest in a long history of transnational influences on processes of urban development (see also Reading 2 by Braudel and Reading 4 by Abu-Lughod).

In our selection, King discusses the development of the international division of labor during the eighteenth and nineteenth centuries under Britain's imperialist system. King focuses, in particular, upon the transformation of building form both in the core cities of the UK and in various urban settlements scattered throughout the colonial periphery. King sketches, in broad strokes, the vast, worldwide inter-urban network that formed, quite literally, the spatial infrastructure of the British Empire. As of the early twentieth century, King shows the British colonial urban system encompassing major cities in India, Ceylon, Singapore, Hong Kong, South Africa, East Africa, West Africa, Canada, various Caribbean islands and Australia. The social and spatial evolution of these cities, King argues, can be understood only in terms of their changing roles within the British colonial economic system. King thus provides an analysis of urban development within the British Empire and a more programmatic research agenda on the interplay between urban development, colonialism and global capitalism.

A THEORETICAL FRAMEWORK

Whilst it is now fully recognized that contemporary urban change in Britain results from changes in world-market conditions, the historical phenomenon of urbanization is still generally treated as a nationally autonomous process. It is as though cities somehow developed independently of the world outside Britain, of the sources of raw materials that were the prerequisites for early and subsequent industrialization, the overseas (often urban) markets for which urban manufactured goods were exported or the distant destinations, both urban and rural, to which Britain exported its surplus labor. Although the contribution of colonial expansion to industrialization and capital formation is acknowledged, the urban and environmental implications of this are not followed through.

Yet, just as the emerging industrial system of Britain assumed its place in a developing international division of labor that is both social and spatial, so also the urban and environmental forms that result from this single, international system of production become component parts of a single and global system of settlement and built environment. There is, in short, a spatial hierarchy of production processes that is expressed in the built environment. The theoretical assumptions behind the analysis that follows, therefore, are these:

1. Any system of socioeconomic organization (or mode of production) has a social division of labor.
2. This social division of labor is spatially and, generally, physically expressed in terms of building and, ultimately, urban form (King 1984).
3. Historically, a spatial division of labor, which in the early stages is expressed locally, later comes to be expressed regionally or nationally and subsequently, is expressed at an international scale.

For illustrative purposes, the following example is deliberately oversimplified. Spinning and weaving in the peasant, pre-industrial economy of England is undertaken domestically by a household of, say, a woman and her husband and children in one room of a two-roomed dwelling that becomes adopted to this use: a necessary spatial division of labor results from the spatial requirements and location of the means of production (spinning and weaving equipment). Dietary requirements are supplied partly from food cultivated on the plot of land adjacent to the dwelling and cultivated by their own labor, and partly from the market. The development of factory production under capitalism leads, in the earliest stage, not only to the classical spatial division between work and residence with the development of functionally specialized, industrialized towns, but also to the development of specialized building forms – factories for the production process and particular forms of accommodation for labor. At first, these may be the densely packed courtyard dwellings of early industrialism, or row houses, and subsequently, the back-to-back dwellings or by-law terrace housing of the later industrial city. Subsequently, diversification of the production process and markets leads to a spatial division of labor, which is expressed at a regional level. This implies functionally different urban regions and towns, each with functionally different building forms developing: mills, factories, foundries, and extensive working-class housing in production-oriented industrial towns; market halls, banks, offices, and more socially differentiated housing in more commercially oriented settlements; and theatres, assembly rooms, promenades, piers, and a variety of dwellings in the more spacious, consumption-oriented spas and resorts. Likewise, the emerging social structure of the regional and national society is manifest in a particular expression of residential building forms.

With the maturing of the international division of labor in the eighteenth and nineteenth centuries, the development of capital-intensive industrial production at the core (England) and labor-intensive agricultural production at the periphery (say India, South America, Africa), there is a resultant expression, not simply in the proportion of population urbanized in various parts of this division of labor but also, in the type of built environment that results. This means the dense concentration of living space (in back-to-backs, row houses, terraces) situated close to the factory at the point of production, in a heavily urbanized area, with no provision for self-provisioning in terms of food. The land or garden on which was grown the means of sustenance of the early-twentieth-century Lancashire or Yorkshire textile operative were not, as with their pre-industrial ancestors, located by their cottage or obtained from a regionally based market: as their staple diet now consisted of tea, sugar, cocoa, and wheat ("voluntarily" imported from the colonies), at least part of their "garden" was located in India, Ceylon, the West Indies, West Africa, and Canada. Likewise, the cotton cloth worn by tea-plantation workers in Assam or Ceylon ("involuntarily" imported from Britain) was manufactured in the mills of Lancashire, and the machine tools and railway carriages used in Egypt or Natal, constructed in the workshops of Birmingham.

In short, an adequate understanding and explanation of the built forms and urban structure of any mode of production requires the simultaneous consideration of all elements of the social division of labor, irrespective of their geographical location. And this task is made easier when data on building form, function, and style are incorporated with information on the economic, social, and urban structure of particular places: social divisions of labor are architecturally expressed. The mansion of the West Indian plantation owner is a modified reproduction of his country house in Britain. Banking houses round the Empire initially express the style of the time they were built. British urban development must be studied as part of, complementary to, and simultaneously with the total system of production and consumption in those parts of the world with which its economy, society, and polity were principally connected. One part is only comprehensible by reference to the other [. . .]

BUILDINGS AND THE INTERNATIONAL DIVISION OF LABOR

This framework can now be applied to an understanding of some of the changes taking place in contemporary Britain. In the old international division of labor, Britain became the most urbanized country in the world, with 80 per cent of its population in towns at the height of the imperial connection (1914). This development presupposed the existence of particular raw materials (especially cotton and wool, but also rubber, tin, timber, minerals, sugar, and others), which were processed in urban (especially port) centers, as well as food imports and assured markets for manufactured products. Specific sectors of the economy and specific regions and towns exhibited the phenomenon of

dependent urbanism more than others. In one sense, the most dependent city was London, which largely monopolized many of the financial and banking functions of the world.

While the proportion of British trade, with its overseas Dominions and Empire, varied over the period, in the 1930s, two-thirds in value of British exports were destined for dominion and colonial markets and significant proportions of the exports of individual colonies came to Britain. In addition, something under half of overseas direct investment was placed in the Empire. British urbanization, therefore, was a specialized and symbiotic part of a colonial space economy as well as the world-economy in general.

This meant not only a particular economic and occupational structure, but also, the existence of a vast building stock, supported and maintained by the profits of industrial production derived from Britain's pre-eminent position in the world-economy. Whilst much of this stock was in the form of urban working- and middle-class housing, a significant proportion of this globally (as well as nationally) derived surplus had been creamed off over the previous century and invested in substantial aristocratic and bourgeois dwellings, as well as other building forms – in the metropolis, on the edge of cities, and especially in "the country." Without Britain's privileged place in the world-economy, and without its imperial advantage, it is unlikely that such a large stock of such building would have developed. The archetypical example is the large country house of an industrial baron deriving his profits from imported Peruvian guano [. . .]

COLONIAL URBANIZATION AND BRITAIN

[. . .] Between 1650 and 1750, four out of five of Britain's largest cities were port cities (London, Edinburgh, Bristol, Newcastle). In the later eighteenth and early nineteenth centuries, the simultaneous and complementary growth of Britain's ten largest industrial and port cities (the majority in the north) not only suggests a powerful interdependence between production and import/export but also, their collective interdependence with another set of port cities and inland towns overseas. The critical role of these ten cities in the emergence of the old international division of labor, based on the exchange of raw materials from underdeveloped countries for British manufactured goods, is sufficiently manifest by their economic decline from the 1930s and the collapse of much of their industry from the 1970s. If it was not previously acknowledged that the rapid growth of these cities in the eighteenth and nineteenth centuries was dependent on Britain's privileged role in the international economy, their subsequent decline has provided the evidence. By contrast, some of the largest towns of the mid-sixteenth century (Norwich, York, Exeter, Worcester, and Coventry), the products of an earlier mode of production and part of a local, regional, and mainly national (and European) spatial division of labor (including relative self-sufficiency in food supplies), are now flourishing. Having been bypassed by the ravages of colonially related industrialization, these historic centers are now bases for the new service-oriented economy.

It is, however, London and the large port cities of 1800 (the "colonial port cities" of Britain) that demand our attention. Given the massive growth of London, the port cities and industrial cities that both supplied and were supplied through them (Manchester, Birmingham, Leeds, and Sheffield) in the late eighteenth and nineteenth centuries, it is clear that, by 1800, a whole new set of other port and inland cities were equally, if not more part of an emerging British colonial urban system than were particular cities in Europe: they would include Calcutta, Bombay, Madras, Dacca, Nassau, Kingston, Sydney, Halifax, Montreal, Toronto, Port of Spain, Bridgetown, Gibraltar, and others. A century later (1900) this colonial urban system had expanded to include Aden, Hong Kong, Cape Town, East London, Durban, Pretoria, Johannesburg, Salisbury, Blantyre, Mombasa, Salisbury, Kampala, Zanzibar, Lagos, Accra, Nikosia, Suez, Port Louis (Mauritius), Mahé, Kuching, Georgetown (Guiana), Melbourne, Brisbane, Adelaide, Perth, Hobart, Christchurch, Wellington, Port Moresby, and Port Stanley. What is significant about these cities is that their built and spatial environments (as well as other phenomena) begin to have more in common with each other than each has with the economically, politically, and culturally very different environments of the interior of the countries and continents where such ports were located.

[. . .] As much of the present structure of Britain's (and London's) urban development was already in place by 1900, it is worth examining Britain's place in a larger imperial division of labor at this time, looking at the colonial urban system of which London was the apex and in which other major British cities (for example, Liverpool, Glasgow, and Manchester) played a major, but essentially dependent role. Of course, this colonial urban system was neither autonomous nor exclusive but was embedded in a larger world-economy: the colonial economy accounted for only a portion of Liverpool's or Glasgow's economic base (though a large one) and London had existed prior to the rise of the Empire just as had other cities in the Far East. It is, however, the extent to which any city was created by, or became dependent upon, the colonial political economy that needs to be examined.

THE BRITISH COLONIAL URBAN SYSTEM: A TOPOGRAPHICAL DESCRIPTION

By the beginning of the twentieth century, this colonial urban system was extensive. It linked the interior of countries to their ports and the ports both to each other and the metropolis. It was the system by which many countries were brought into the capitalist world-economy. It provided the nodal transportation links for the import and export of goods and services and was the network for distinctively colonial forms of international labor migration (including indentured labor). It established labor markets, but also, with the transformation of old and the creation of new environments, it provided the physical and spatial infrastructure for the restructuring of the social, cultural, and political order, creating centers for new modes of consumption, and the transformation, through "modernization" and commodification, of social, cultural, and political consciousness.

In the early twentieth century, the population of India (including independent and feudatory states) was 297 million, of whom 121,000 were European, principally British, and largely resident in the main cities or other urban areas. As India was incorporated into the world-economy, it became locked into a different international division of labor, importing manufacturing goods (principally from Britain) and exporting raw materials, to the benefit of British interests. The expansion of commercial crops, largely produced for the world market, coincided with a long-term stagnation in foodgrain production. Between 1850 and 1900, one-third of all India's exports were taken by the UK but three-quarters of India's imports came from there, mainly cotton goods, bullion, machinery, metals, railway carriages, ships, boats, and woolens. In return, India sent raw cotton, cotton goods, jute, rice, hides and skins, tea, wheat, and coffee. As the great jute-producing country of the world, India sent most of its exports to be processed in Dundee.

Loans raised on the London market were either to the government of India, for the Indian railways, or to the British agency houses that owned or managed a large number of plantations, factories, transport services, or banks. In 1911 Europeans (mainly British) owned some 88 per cent of tea plantations, 93 per cent of indigo plantations, and 35 per cent of collieries. European agencies managed 58 per cent of Indian-owned industrial enterprises. These agencies, as well as the whole apparatus of colonial management and economic and political control were located in the cities, principally, Calcutta.

Bombay, the third-largest city of the Empire, was the busiest port in Asia, responsible for shipping three-quarters of India's cotton; by 1901, there were some 190 cotton mills in India, two-thirds of them in Bombay. By the mid-nineteenth century, Bombay had also become the financial center for British India and in the next decades, its urban structure and port facilities were remodelled to orient its economic function more closely to the needs of British colonial interests. The change was also to be seen in its architecture, where new colonial institutions (a railway station, a town hall, banks, and buildings of commerce) replicated the forms and styles of those in the metropolis as well as expressing imperial power.

The third largest city, Madras, exported largely coffee, sugar, indigo, dye, and cotton. Here, too, urban remodelling had taken place with a pier, a new harbour, and railway facilities constructed at the end of the century by the British administration. This economic and trading interdependence was manifest, therefore, in the urban and especially port structure of both countries, which were linked by the main steamship routes, in the metropole, running from London and Liverpool.

In Ceylon (Sri Lanka), over half of the trade was with the UK, which took nine-tenths of its tea, and two-thirds of its coffee, trade that largely accounted for the rapid growth of Colombo. In 1900, Ceylon had most of its agriculture (coconut, paddy, tea, and coffee) in plantations largely serviced by "Tamil coolies" brought in by the colonial administration from South India.

Within the imperial system, the Straits Settlements (Singapore, Penang, and Malacca) performed a key economic and political role. Singapore's main function was strategic, commanding commercial channels to the East Indies, China, and Japan and therefore hosting the headquarters of British military and native forces. It was also, however, a vast commercial city, the entrepôt for the produce of the surrounding countries (the Malay Peninsula, the Dutch East Indies, Japan, Borneo, Siam, and the Philippines).

Like Singapore, Hong Kong's two major imperial functions were as military and naval station and as a huge commercial entrepôt. The port of Victoria (Hong Kong) did more trade than any other in Asia, controlling commerce with China and Japan. In 1899, British trade with Hong Kong was worth £3$\frac{1}{2}$ million, exporting cotton, woolen, and iron goods, copper and lead, and importing silk, tea, hemp, and copper [. . .]

What is now South Africa included, in 1901, the separate colonies of the Cape, Natal, and Transvaal [. . .] Natal exported gold, wool, and hides, with three-fifths of its entire trade done with the UK and most of the foreign commerce undertaken through Durban. In less than fourteen years, Johannesburg had risen "from a few scattered huts made from paraffin tins to its present position as one of the largest gold mining centers of the world. It has some remarkable fine streets and the buildings, shops and stores compare favourably with many old towns" (Gill c.1901). In Rhodesia (Zimbabwe), with the capital at Salisbury (Harare), the British South African Company had invested in gold production, as well as silver, copper, and land for cereal production.

The principal exports from Mombasa, British East Africa (Kenya) were tropical products – bananas, arrowroot, casa, coffee, india rubber, etc. – and these were exchanged for Lancashire and Bombay cotton cloths. From Freetown, Sierra Leone, kola nuts, ground nuts, india rubber, and hides were sent to Liverpool in exchange for "Manchester goods." Likewise, from Accra on the Gold Coast (Ghana) were sent india rubber, palm oil, gold dust, timber, three-quarters of the trade being with the UK. Again, it is worth pointing out that "the commerce of the West African coast is chiefly in the hands of Liverpool merchants" (Gill c.1901).

In 1901, Northern and Southern Nigeria were separate provinces from Lagos. The Royal Niger Company had established over 100 factories on the main river as far as Egga, with other trading stations established at Yola, Lokoja and Loko. The chief exports were rubber, ivory, palm oil, kernels, gum arabic, etc.; the main imports, textile fabrics, hardware, earthenware, tobacco, guns and powder.

Though the majority of Canada's trade was with the USA, in 1901, 40 per cent of it was still with the UK. Britain mainly imported timber, grain, cheese, horses, and other live animals, meat, etc., and exported to Canada woolens, metals, cottons, apparel, silks, spirits, books, stationery, railroad bars and engines, etc. The major ports of Montreal, Toronto, Halifax, and Quebec were supplemented by Vancouver, linked to Hong Kong by the Canadian Pacific steamboat.

Britain's oldest colonies were in the Caribbean and, although economically underdeveloped in the early twentieth century, and with the largest part of their trade with the USA, their importance was in providing British shipping with a lucrative business and again, establishing the link with Liverpool. Jamaica, first acquired in 1629, had trade worth some £4 million with the UK, exporting fruit, coffee, sugar, rum, and importing cotton goods; half of Dominica's trade was with the UK; one-third of that of the Windward, Trinidad and Tobago Islands, exporting, through the Port of Spain, raw sugar, cocoa, and asphalt; a quarter of Barbados trade was with the UK, one-half of that from British Honduras (exporting logwood and mahogany).

The final major unit in this imperial division of labor was Australia. Half of its wool exports went to the UK as well as one-tenth of its gold. The commerce of the UK with New South Wales in 1900 was worth £19.5 million. Likewise, one-third of Victoria's trade was with Britain. Queensland, nearer to China, India, and California, exported only half of its annual exchange of £18.5 million to Britain but imported telegraphic wire, nails, metal goods, etc. Three-quarters of New Zealand's trade (worth about £21 million) was with the UK.

The justification for this account of Britain's trading relations with its imperial possessions is to demonstrate how far British industrial urbanization – its degree, location, and distinctive built environment – was, within the larger world-economy, strongly influenced by and, in places, dependent upon a colonial system of production: it was largely produced by it and cannot be understood except as part of it. Admittedly, much of the argument made here is by inference rather than proof. But if concrete evidence is needed to demonstrate the complementary and interdependent development of this urban system it can be found, literally, on the ground, in the urban infrastructure of the building, architecture, and urban forms of the imperial and colonial system, irrespective of more conventional economic and social histories. It is this that gives these colonial cities more in common with each other than any of them has with the "traditional," pre-capitalist forms of built environment that exist in the country's indigenous interior. And the infrastructure of these cities formed the bases for a later transformation to the present world-economy.

Moreover, the colonial system was not just an economic but also a political, ideological, social, and cultural system, dimensions that need to be recognized if we wish to understand the way in which strands of yesterday's colonialism are woven into, and influence the fabric of today's world-economy and politico-cultural system. And whilst the colonial urban system was not "sealed," nor did it operate in territories that had previously been a tabula rasa, it none the less continues to exist as a set of powerful influences, not least in terms of language, institutions, and practices that influence the contemporary world and its international system of cities.

In historical studies of the planning and architecture of Canadian, American, South African, or Australian cities, it is conventional wisdom for scholars to look at contemporary developments in Britain as a major (though not the only) source of understanding early mercantile and industrial capitalist forms of development. Such analyses, however, go only half-way; colonial urban planning or architectural forms (or, for that matter, legal, constitutional, literary, or social and cultural forms) are not simply derivations of core forms but rather, functionally interdependent parts of a single system.

CONCLUSION

The assumptions behind this chapter are twofold. First, that urban and other phenomena can only be adequately understood by treating them as part of a larger world-system, of economy, society, and culture, of which they are an integral part. And second, that the built environment, in all its various conceptualizations, is both a product of and a major resource for understanding these global processes. However, no theory develops in a vacuum;

research needs to be grounded in data collection, informed, of course, by hypotheses and theory, the construction and reconstruction of which must be utilized to suggest new frameworks, theories, problems – and solutions.

REFERENCES FROM THE READING

Gill (c.1901) *The British Colonies, Dependencies and Protectorates.* London: George Gill & Sons.

King, A.D. (1984) *The Bungalow: The Production of a Global Culture.* London: Routledge & Kegan Paul.

King, A.D. (1990) *Global Cities: Post-Imperialism and the Internationalization of London.* New York: Routledge.

King, A.D. (2004) *Spaces of Global Cultures.* New York: Routledge.

24

"The World City Hypothesis: Reflections from the Periphery"

from P.L. Knox and P.J. Taylor (eds),
World Cities in a World-system (1995)

David Simon

Editors' introduction

David Simon is Professor in Development Geography at Royal Holloway, University of London and has published widely on urbanization in the developing world. His research has dealt with a variety of theoretical, empirical and policy-related issues in development geography, with particular reference to issues of environmental change and sustainability in the urban centers of East and West Africa, Sri Lanka, Thailand and the Philippines. During the 1990s, Simon engaged extensively with global city theory in a variety of publications, including a major book (Simon 1992) and a number of articles. Reading 24 is derived from this phase of Simon's work and was originally published in an influential edited volume on the state of the art in global cities research.

Simon's goal in this reading is to explore whether, and to what degree, the arguments of world city theory illuminate political-economic conditions and developmental trajectories in cities located in peripheralized zones of the world economy. Whereas King in Reading 23 focused on colonial cities, Simon is concerned primarily with postcolonial cities, that is, with cities located in formerly colonized territories that have now become sovereign states. Simon begins by surveying a number of features of sub-Saharan African cities: they formerly served as relay points for raw materials export; most efforts to promote industrialization have to date generated only limited results; population continues to expand due to intensive rural in-migration; and, in some cases, the service sector is growing due to the prominent role of governmental agencies in the local economy. With the exception of Johannesburg and Nairobi, Simon argues, few sub-Saharan African cities contain any of the major features of global cities, as defined by writers such as Sassen or Friedmann. On this basis, Simon examines more closely the criteria in terms of which global city status is to be determined, and argues for the addition of a new variable − number of headquarters locations for international (governmental and nongovernmental) organizations. From this point of view, Simon argues, cities such as Nairobi do indeed fulfill significant transnational networking and hub functions. In the remainder of the reading, Simon surveys a number of emergent governance problems (in the spheres of employment and housing, in particular) and political and ethnic cleavages that are emerging in sub-Saharan African cities. Simon concludes by arguing for a further analytical differentiation of the world city hypothesis to include a greater number of types of cities, from all zones of the world system, within its purview.

THE POLITICAL ECONOMY OF WORLD CITY FORMATION

Underpinning the world city hypothesis is the political economy of the global urban system. In developing this theme, it is necessary to examine the interplay between national political economies and the world system, i.e. between national and international capital, states and international organizations. Since world cities are generally primate cities (and, where they exist, also other national metropolises), these represent one important arena in which the issues can fruitfully be examined. This reflects the roles of such cities as political capitals and transactional crossroads between internal and external relations, as well as the related factors of their size, economic complexity and centrality, and their attractiveness to new migrants from within and beyond the country. Irrespective of how internationally marginal a country is, its capital and/or principal industrial and port centers are far more closely integrated into the world economy than its intermediate and small centers. Primate, and especially world, cities might also be revealing elements of international convergence in the way that production, circulation/distribution, and consumption are organized and regulated as these processes become increasingly globalized. This is not to argue, however, that convergence is all-embracing, uniform, or ubiquitous; aspects of divergence will continue to be evident. Primate cities are therefore reliable barometers of national and international politico-economic processes, and the (dis)-articulation between them.

In my recent book (Simon 1992), I sought to advance our understanding and conceptualization of postcolonial urban development by examining capital and other major cities in terms of their mode of incorporation into the world economy. This represents arguably the most fruitful approach to elucidating how they operate and the processes through which their built environments have been socially constructed and reconstructed over time. On account of its unwanted status as the poorest, most marginal continental region within the current world economy, sub-Saharan Africa (SSA) provided the case study.

In much of sub-Saharan Africa the dominant mode of production since the colonial period has been capitalism. Colonization of the continent in the late nineteenth century was predicated on the needs of European industrial capitalism and the associated imperial ambitions of the major powers. Within the evolving world economy, sub-Saharan Africa's principal roles became those of a raw material supplier and a captive market for manufactured goods; functions that, apart from some import substituting industrialization, the continent has generally been unable to transcend in the few short decades since decolonization. This is central to Africa's progressive peripheralization and current crisis.

It was in the European settlement colonies (South Africa, Namibia, Kenya, Zimbabwe, and to a far lesser extent Senegal) that indigenous land dispossession was most extensive and labor exploitation most intense. Ironically, these countries are among the most diversified and sophisticated economies today, but even they still rely mainly on primary and semi-processed, rather than on manufactured, goods for the bulk of their export revenue. In other words, Africa remains the only large and populous continent without a true newly industrializing country (NIC) which, seemingly, is another reason for its marginality to global circuits of commercial, industrial, and financial capital. South Africa has many characteristics of a NIC and, but for the distortions of the apartheid system and the effects of its increasing international isolation on that account until 1989–90, would already have attained true NIC status. In most African countries independence was followed by progressively greater direct and indirect state involvement in the economy, in an effort to catalyze development and to promote greater localization (indigenization), a politically popular strategy intended to reduce external economic control, create employment, and (usually also) to enlarge the indigenous capitalist classes. Even these measures failed to promote sophisticated industrialization, and they are now generally viewed as important contributors to Africa's economic plight, and are being reversed in favor of market forces. This backdrop of regional and national modes and relations of production within the world economy informs my analysis of the continent's cities.

Most African cities have experienced rapid growth but only limited economic transformation since independence. There are few industrial cities on the continent, and those which do exist remain

Plate 21 Walvis Bay, Namibia (Roger Keil)

geared primarily to the import substitution of consumer non-durables and durables rather than to large-scale export or the production of capital goods. Given the absence of NICs this is not surprising. In most cases commerce is still more important than manufacturing.

At the same time, however, the service sector has been growing in significance and sophistication in most capital cities. While much of this growth has been in producer services, the insurance and personal service element has also been dynamic. Perhaps more than any other, it is the communications subsector which has assumed progressively greater importance in the age of high technology and telecommunications. This provides the instant linkages to other capital cities and centers of the world economy, and has entrenched the sharpening distinction between the system of core cities which controls the international circuits of commercial, industrial, and financial capital, and the cities of the semi-periphery and periphery. So, paradoxically, just as the world is becoming ever more tightly integrated and interconnected by virtue of advances in aviation and telecommunica-

tions technology, it is also becoming increasingly differentiated in relative terms and, in the case of most of sub-Saharan Africa, also in absolute terms.

Inextricably bound up with the communications subsector is the range of activities we could best classify as producing, consuming, and reproducing culture in the broadest sense. These include the print and broadcasting media, film, video and associated industries, music, theatre, art, and other forms of formal culture production. As elsewhere, these are heavily concentrated in Africa's primate cities, although their reach or scope remains overwhelmingly national or subcontinental.

The few exceptions in SSA to the general trends just outlined are the cities which, like the metropolitan complex centered on Johannesburg (the Pretoria–Witwatersrand–Vereeniging (PWV) region) and Nairobi, have developed significant service sectors and levels of international connectivity. Given assertions about the growing importance of non-state actors in organizing and regulating human activities, one interesting variable is the extent to which individual cities serve as headquarters for such bodies. African cities compare reasonably well with those in other regions of the Third World in terms of the total number of secretariats based there. However, when the data are disaggregated, Africa loses out heavily. Unlike Latin American and Asian cities, Nairobi (Africa's most popular location) has no headquarters of global membership organizations. At the same time, however, we must remember the overwhelming extent to which such organizations are still headquartered in world cities of the North, symbolizing their continued grip on global power.

The progressive expansion of civil aviation reflects continued growth in business and international tourism. Cairo is Africa's busiest airport, while Johannesburg is SSA's primary gateway, reflecting the airport's hub functions for the southern African region and its importance for both business and tourist travelers, reflected in the large number of domestic passengers and volume of freight and post handled. Nairobi has gained its position by virtue of being East Africa's gateway and a stopover *en route* to South Africa from Europe. This latter function has fallen away significantly since the recent introduction of non-stop flights to South Africa with the new Boeing 747–400 series, while the phasing out of sanctions against South Africa since 1991 has greatly increased the number of foreign airlines and flights serving Johannesburg. Lagos is surprisingly unimportant, given the country's vast population and considerable potential in view of its economic situation. It is important to note the scores on the different variables.

The PWV's pre-eminence reflects South Africa's economic sophistication and the wealth of the white minority and new black middle classes, while Nairobi's position is due to its relative sophistication, the consequent location there of the headquarters of two UN agencies, and the attractive effect this has had on other international organizations, both international governmental organizations (IGOs) and non-governmental organizations (NGOs). In other words Nairobi has assumed a supranational role in the sphere of information flows and associated diplomatic and financial transactions. However, on account of a wide range of other problems and limitations, the city is still a considerable way short of becoming even a continental city, let alone a true world city,

despite being dubbed the "world capital of the environment" by some IGO officials.

THE DEFINITION AND NATURE OF WORLD CITIES

At this point it is appropriate to consider the question of what the prerequisites for world city status actually are. In the contemporary world it is becoming increasingly clear that the three most important criteria relate to:

(a) the existence of a sophisticated financial and service complex serving a global clientele of international agencies, transnational corporations (TNCs), governments and national corporations, and NGOs;
(b) the development of a hub of international networks of capital and information and communications flows embracing TNCs, IGOs, and NGOs; and
(c) a quality of life conducive to attracting and retaining skilled international migrants i.e. professionals, managers, bureaucrats, and diplomats. In this sense, quality of life embraces not only physical and aesthetic aspects of the environment but also broader considerations such as perceived economic and political stability, cosmopolitanism, and cultural life.

The first and second criteria just mentioned are very closely related and correspond to what Gottmann (1989: 62) calls concentrations of "brainwork-intensive" industries which he perceives as the hallmark of transactional cities. His list of current world cities is revealing and comprises seven on which there has long been agreement, i.e. London, Paris, Moscow, New York, Tokyo, Randstad Holland, and the Rhine-Ruhr (presumably with Frankfurt as core), together with three relative newcomers, Washington DC, Beijing, and Geneva. He sees another set of cities (Chicago, Los Angeles, San Francisco, Montreal, Toronto, Osaka, Sydney, and debatably also Zurich), as approaching world city status, but specifically excludes Mexico City, São Paulo, and Seoul, three of the most dynamic metropolises of the South. Surprisingly, Singapore and Hong Kong do not even rate a mention in this schema. With the possible exception of Beijing, therefore, all Gottmann's world cities are in the North, a fact clearly related to his rather loaded and contestable assertion that Third World cities are only now beginning their industrial revolution (Gottmann 1989: 62). The contrast with Friedmann's schema and list (1986) is notable, but reflects their very different conceptual underpinnings.

[...] The issue of classification inevitably raises two vital questions. Firstly, as discussed above, is a heavy industrial base a prerequisite for world city status? Put differently, in terms of national modes of production, can a world city arise in a state which has not attained at least NIC status? Secondly, in the context of key features of the contemporary world economy (international mobility of capital and new technologies, changing global divisions of labor, etc.) is it possible for major cities to develop tertiary and quaternary functions without a heavy industrial base? The discussion above would seem to indicate that world cities are found only in core and semi-peripheral countries. However, the second question requires a positive answer, exemplified best by Singapore and Hong Kong. These admittedly atypical city states have become continental, and in some respects also global, centers of accumulation and communication, with the emphasis increasingly on selected hi-tech sectors, transport, financial markets, and the business and personal services sectors.

Nairobi is one (by no means outstanding) example of how a significant supranational hub of transactional functions can evolve without a major industrial base. Moreover, at least in terms of the total number of international organizational headquarters, an important indicator of transactional networking and hub functions, Montreal and Ottawa do not rank more highly than a number of Third World metropolises. However, when the data are disaggregated, a crucial difference becomes evident, namely, that Montreal has more headquarters of global membership organizations. In terms of telecommunications and air passenger links, together with quality of life, the Canadian cities almost certainly also have an advantage. Such ambiguities simply underline the inadequacy of using any single criterion as the basis for definition. At least two of the factors discussed above, together with a favourable image and quality

of life, are likely to be necessary. Finally, one can concur with Gottmann's (1989: 64) assertion that globalization in urban terms is increasing, in that every substantial city nowadays aspires to a world role, at least in some specialty. This makes them expand linkages abroad, participating in more networks. All these trends contribute, little by little, to building up and intensifying the global weave of urban networks.

There is also every likelihood that the number of world cities will increase over time, as he suggests, but the mere development of a particular, specialized global role is clearly *not* a sufficient condition for becoming a world city. For various reasons many major cities, and even mega-cities, will not attain such status.

This leads neatly back to the world city hypothesis and the work of Friedmann (1986). He is careful to describe the seven theses comprising the hypothesis as constituting primarily "a framework for research. It is neither a theory nor a universal generalization about cities, but a starting-point for political enquiry." While extremely useful in this context, the theses do not actually *define* world cities. Rather, they provide an admirable summary of the features of all major cities with significant supranational roles, of which true world cities are only a subset. Given Friedmann and Wolff's (1982: 310–311) emphasis on the form and strength of integration into the world economy and the extent of such spatial dominance, this is surprising, but perhaps reflects their concern with the evolutionary process of "world-cities-in-the-making" rather than with a finite set of existing world cities at a given point in time.

However, even Friedmann's (1986) refinement of distinguishing primary and secondary world cities within both the global core and semi-periphery does not adequately avoid this drawback. For example, all the theses apply to Nairobi, which remains some way short of either continental or world city status. Moreover, although Kenya is the dominant economic power in East Africa, it does not yet even come close to forming part of the semi-periphery. Its status remains unambiguously peripheral in politico-economic and other terms. This is more generally true of SSA's primate cities. For example, as stated earlier, the cultural services produced or located there seldom reach continental, let alone global, markets [...]

In some respects, urban diversity in sub-Saharan Africa has actually increased since independence, rather than decreased. However, similar conflicts over access to the means of production and especially of social reproduction are occurring throughout the continent. Class, ethnicity, religion, and other potential cleavages combine or are operationalized in different contexts as means of engendering both unity and division, harmony and conflict. Urban form reflects the intertwined social relations of production and reproduction. Thus rapid urban growth, inadequate shelter provision and the consequent proliferation of irregular housing, unemployment and underemployment, and the mushrooming of informal or petty commodity activities, are the norm in sub-Saharan Africa and beyond. At the same time, capitalists continually seek out the most profitable avenues for rapid accumulation which, given the prevailing conditions in many African countries, may well be through the commercialization of irregular housing markets. This process often exploits the poorest and weakest urban dwellers.

A DIALECTIC OF CITIES: WHOSE WORLD CITIES AND FOR WHOM?

This forces us to consider the issue of conflicting and divergent interests in the city. Whose cities are they and whom do they serve? How is control or regulation organized and by whom? How are different identities and representations scripted? Globalization is obviously of fundamental import to analysis of world cities and of the way in which the various parts of the Third World are embedded and incorporated on the basis of differing – and changing – degrees of exploitation and unequal exchange. However, much of the earlier writing on the subject was vague and used "globalization" in simplistic, all-encompassing terms. The importance of specifying the context and precise meanings of analysis has now become abundantly clear.
[. . .] As with the concept of "globalization," we need to avoid slipshod terminology and the imputing of all-embracing characteristics to world cities. The nature of space and place is contested, both within and between individual cities. We must interrogate our assumptions, disaggregate our categories, and address the questions of whose

city and for whom. For many, the disjuncture between aspirations and opportunities is – and is likely to remain – great. While it is difficult to postulate a firm figure for the percentage of urban poor in world cities, it is likely to be significantly higher – perhaps even the majority – in Third World metropolises. This situation again reflects the strength of their respective national economies and the nature of their insertion into the world system. The aspirations of the poor to a more adequate standard and quality of life cannot be ignored. Their perceptions of what is appropriate and desirable are likely to be very different to those of even progressive and activist planners and populist politicians.

CONCLUDING REMARKS

Postcolonial cities in sub-Saharan Africa and other continental regions of the South represent a rich arena for research geared to practical issues of policy and planning, as well as at the level of theory. Insights derived from the complexity and dynamism of many Third World urban societies deserve to inform general urban and social theory far more than hitherto. The value of research into development paths diverging out of a common colonial experience is enhanced by the differential changes being wrought to the inherited structures and forms by increasingly powerful forces of globalization, instant communications, and the growing importance of international financial capital.
    The Third World is becoming increasingly differentiated. On the one hand, some countries (the NICs and proto-NICs) are expanding the semi-periphery of the world economy, while some key cities in the NICs are becoming important continental and global financial centers in their own right (e.g. Singapore, Hong Kong and, to a lesser extent, São Paulo). On the other hand, with very few exceptions, the countries of sub-Saharan Africa seem destined to remain very largely in the new global periphery, suffering growing urbanization and poverty together with economic stagnation, marginalization, and environmental degeneration. Such a fate is contested, with whatever means available, by Africa's nation-states. However, there is an inherent contradiction between the interests of individual nation-states, the territories

of which are spatially bounded, and the international capitalist system, led by TNCs and global economic institutions.

However, this is too simple a dichotomy; the inequality and poverty at all scales *within* the core, semi-periphery, and periphery are just as important as those *between* these entities, however construed and constructed. Deprivation and disempowerment within primate and world cities go hand in hand with wealth and power. The fortunes of individual cities rise and fall, as do the divides between rich and poor within and between them.

In the context of current, more nuanced, writings on globalization, we should ponder the "construction" of the world city hypothesis, both to sharpen its definitional clarity and to examine how its political economy articulates with the social and cultural realms of plurality, difference, meaning, and otherness. Hence we should ponder whether the notion of a world city has any meaning – and if so, what – for different groups of world city inhabitants, especially the impoverished shanty dwellers and pavement people of Mexico City, Bombay, or Johannesburg. Yet, even if the answer should be entirely negative (which it may well be), this would not negate the validity of our enterprise in this regard or, indeed, preclude these cities from functioning as world cities. It would only underscore the need for us to be conscious of the boundedness of our constructs in every sense. Colleagues working on world cities in the North so easily make unjustified implicit and explicit universalizing assumptions about the global relevance or uniqueness of our constructs. This does not, however, negate the value of such work; rather it requires that we become more self-conscious and less ready to impute or ascribe universal "truths" based on Euro/Americo – and ethnocentric – experience, paradigms, and research.

REFERENCES FROM THE READING

Friedmann, J. (1986) The world city hypothesis, *Development and Change*, 17, 1, 69–84.

Friedmann, J. and Wolff, G. (1982) World city formation: an agenda for research and action, *International Journal of Urban and Regional Research*, 6, 3, 309–314.

Gottmann, J. (1989) What are cities becoming the centers of? Sorting out the possibilities. In R.V. Knight and G. Gappert (eds) *Cities in a Global Society*. London and Newbury Park, CA: Sage, 58–67.

Simon, D. (1992) *Cities, Capital and Development: African Cities in the World Economy*. London: Belhaven.

25

" 'Fourth World' Cities in the Global Economy: The Case of Phnom Penh, Cambodia"

from *International Journal of Urban and Regional Research* (1998)

Gavin Shatkin

Editors' introduction

Gavin Shatkin teaches Urban Planning in the Taubman College of Architecture and Urban Planning at the University of Michigan. Shatkin's work focuses on urban development issues in globalizing Southeast Asian city-regions such as Bangkok and Metro Manila. In particular, Shatkin has explored the consequences of globalization for issues such as urban infrastructure provision, the role of non-governmental and community-based organizations, the restructuring of urban governance, and patterns of urban inequality. In this reading, Shatkin develops a powerful critique of the assertion that sub-Saharan African, Latin American and Asian city-regions have become "structurally irrelevant" to the world economy under contemporary conditions. This claim, which has been articulated frequently by global cities researchers, implies that contemporary globalization has produced a "Fourth World" characterized by economic stagnation, marginalization and social upheaval, and which contributes only minimally to world-scale processes of capital accumulation. Against such arguments, Shatkin demonstrates various ways in which cities in the less developed zones of the world economy contribute to, and are in turn being reshaped by, global economic processes. Even in the absence of large-scale flows of foreign direct investment (FDI), Shatkin argues, there are significant technological, economic, organizational and socio-cultural linkages between cities in the less developed world and the cities of the global North. Shatkin illustrates this argument with reference to the case of Phnom Penh, Cambodia, which had been insulated from global economic forces during the era of socialist rule from the mid-1970s to the late 1980s. As Shaktin demonstrates, the 1990s witnessed the city's rapid reintegration into the world economy, leading in turn to a variety of local sociospatial transformations. State spending on public infrastructure was reduced, a private real estate market was established, foreign capital flowed into the built environment, the presence of large numbers of United Nations peacekeeping workers created new markets for various consumer amenities, and new technological and trade linkages to other Southeast Asian cities were established. In addition, new sociospatial inequalities emerged as both squatter settlements and various types of luxury facilities (hotels, casinos, restaurants) proliferated throughout the city. Shatkin uses this case study as the basis for a more general claim about cities and globalization in the less developed world: global economic forces have shaped and reshaped such cities, but in contextually specific ways that cannot be adequately understood by positing their "exclusion" from the world economy.

The development of innovations in transportation, telecommunications and information technologies, and the consequent emergence of a global economy, has presented opportunities for nations of the developing world. However, many countries, particularly much of sub-Saharan Africa, and parts of Latin America and Asia, have largely not participated in the globalization of the economy. As a consequence, Castells (1996) contends that globalization and consequent economic restructuring has resulted in the disappearance of the third world and the emergence of a "fourth world" of regions that are increasingly excluded and "structurally irrelevant" to the current process of global capital accumulation. Cities in such excluded countries are often characterized as being distinguished by their economic stagnation, the increasing marginality of their populations, and their potential for social upheaval.

While it is true that participation in the global economy is unequal among nations, this chapter argues that the rather simplistic depiction of the process of urbanization in the so-called "fourth world" cities in much of the global cities literature is inaccurate. First, although the relative share of many least developed countries (LDCs) in global trade and investment is decreasing, these countries nevertheless remain integrated into the global system in important ways. The diffusion of new technologies, global and regional economic change, the increased regulatory power of international aid and lending institutions, and changes in the flows of information, goods and people affect LDCs in ways that have major social and spatial consequences for cities. Second, the idea of a "fourth world" of excluded nations is prone to many of the same criticisms as the concept of a "third world" which it is meant to replace. LDCs experience quite different patterns of urbanization based on place-specific factors such as local histories (particularly colonial histories), social, political and cultural systems, and modes of integration into the global economy. Thus the depiction of LDCs as places of uniform marginalization and despair is misleading, and the relevance of globalization for cities in LDCs extends beyond the dynamic of exclusion. Rather than treating the experience of cities in LDCs as an unfortunate footnote to the phenomenon of globalization and economic restructuring, the role of such cities in the process

of capital accumulation, and the impact of globalization on their development, should be a topic for research and debate.

This chapter examines the case of Phnom Penh, Cambodia, a city that has undergone dramatic spatial and social restructuring in recent years despite the relatively low levels of foreign direct investment and industrial growth in the country's economy. These changes will be reviewed in the context of recent historical developments in Cambodia as well as the country's interaction with the global economy.

GLOBALIZATION AND CITIES IN LEAST DEVELOPED COUNTRIES

Discussions of the impacts of globalization on developing countries generally focus on the relocation of industry from the developed to the developing nations and the subsequent industrialization of parts of the developing world based on a "new international division of labor." One consequence of this phenomenon has been the rapid increase in the amount of foreign direct investment (FDI) flowing to developing countries, especially since the late 1980s. While the amount of FDI flowing to developing countries continues to constitute a small percentage of total flows, this nonetheless has led to a significant increase in industrial growth in many parts of the world. In particular, the emergence of the Asian newly industrialized countries (NICs) and the rapid industrial growth of China and the countries of the Association of Southeast Asian Nations (ASEAN) have heralded an era in which Asia has become a major industrial region.

Analysis in studies on the impact of globalization on cities in developing countries has focused overwhelmingly on such rapidly industrializing areas despite the fact that FDI has flowed disproportionately to a handful of geographic regions. Between 1989 and 1992, 72 per cent of FDI flowing to developing countries went to only 10 countries, while the 48 "least developed" countries received only 2 per cent of global FDI (Broad and Landi 1996). In particular, Asia accounted for almost half of FDI to developing countries from 1983–91, and nearly two-thirds of such investment in the late 1980s (Halfani 1996). Meanwhile, parts of Asia and Latin America, as well as much of sub-Saharan Africa, have not experienced rapid export-led industrialization, and have seen their share of world trade decline.

It is this relative exclusion of certain parts of the world from participation in global trade and industry-led economic development that has prompted some commentators to write of the phenomenon of "marginalization" resulting from globalization, and to speculate about the emergence of a "fourth world" (Friedmann 1995; Castells 1996). Cities in such contexts are often discussed primarily in relation to the extremes of poverty they contain. However, the idea of exclusion is misleading for two reasons. First, a focus on declining shares in global trade and FDI in least developed countries disregards the many important ways in which LDCs interact with and remain integrated in the global economy, for example through the export of raw materials, the tourism industry, and the underground economy. Second, to ascribe observed social, political and cultural changes to the phenomenon of economic marginalization is too economistic. In order to arrive at a proper understanding of the process of urbanization in LDCs, it is necessary to examine the ways in which countries interface with the global economy, as well as the social, cultural and historical legacies that each country carries into the era of globalization, including their colonial heritage and geopolitical situation. It is also necessary to broaden the view of globalization beyond the process of relocation of industry from the developed to the developing countries to include the various social, political and economic forces at work in integrating the world.

Thus, while LDCs are not major targets for FDI, and have not become important industrial producers, they nevertheless play a role in globalization and are heavily impacted by it. In order to understand the implications globalization has had for cities in LDCs, this chapter will now turn to a case study of urbanization in a city in an LDC – Phnom Penh, Cambodia.

THE CASE OF PHNOM PENH

Since the beginning of the French colonial era, and even before, Phnom Penh's development has reflected Cambodia's role in the world economy. This role has changed over time – from provider of raw materials during the colonial and immediate post-colonial era, to "sideshow" in one of the major hotspots in the cold war, to pariah state during the era of socialist rule from 1975 to 1989. Since 1989, however, the country has reintegrated into the global economy, and this has had dramatic implications for the city's social and spatial development.

In the modern era, Phnom Penh has been the predominant urban center in the country. Yet, in the late 1980s, Cambodia was still in the process of reconstruction and rehabilitation following the utter devastation of the country during the years of the Khmer Rouge regime, which had lasted from 1975 to early 1979. The government of the People's Republic of Cambodia (PRC), which replaced the Khmer Rouge regime, was nominally socialist and depended heavily on Vietnam and the Soviet Union. In the interests of facilitating the reconstruction of the country in the short term, the government initially limited attempts at communalization of agriculture, allowed markets to function largely unregulated, and condoned a lively cross-border trade with neighboring Thailand and Vietnam. By 1989 the agricultural sector had reached production levels approximately equal to the prewar levels of the late 1960s, and basic goods and services were available on the market (FitzGerald 1993). Until 1989, however, the country remained largely cut off from the global economy, in part due to a United Nations embargo imposed on the country from 1982 to 1987.

During this period of isolation, Phnom Penh continued to play its traditional role as the political and economic center of the country. Yet the singular form of administration chosen by the government, dictated by the imperative of rehabilitation and reconstruction, gave the city a unique dynamic of development. The government owned rights to all property in the city, although residents retained relatively unrestricted usership rights. There was no real estate market, and land development was largely confined to state construction or renovation of state-owned properties. Markets functioned relatively unrestricted, yet the lack of large sources of capital or a substantial market for consumer goods inhibited large-scale commercial development. It was in this context that a series of sweeping economic and political changes transformed the urban landscape.

Today, Cambodia's low gross national product, its tiny share of world trade, and its near irrelevance as a site of manufacturing production, place the country firmly within the "fourth world." Yet the tremendous social and spatial transformation Phnom Penh is experiencing is very much a consequence of the particular attributes of the global economy – the advent of new technology and resulting hypermobility of capital, the increasing difficulty governments face in regulating capital flows, the increasing pressure on states to conform to international norms of economic practice, and the rapid economic development of Cambodia's neighboring countries. Globalization has changed Phnom Penh's meaning for its inhabitants, creating opportunities for some while immiserating others.

ECONOMIC REFORM AND REINTEGRATION INTO THE GLOBAL ECONOMY

Cambodia's reintegration into the global capitalist economy began with a number of dramatic changes that took place in the late 1980s (FitzGerald 1993). In 1989, Soviet bloc aid to Cambodia declined significantly in the wake of the collapse of the Soviet Union. In the same year, Vietnam, which had been providing crucial military assistance to the country in its continuing civil war with the Khmer Rouge, withdrew all of its troops. In April of that year the National Assembly embarked on the road to liberalization of the economy. The government began to develop more laissez-faire economic policies, while simultaneously undertaking a peace initiative.

Three aspects of the reform had a particularly dramatic impact on Phnom Penh's development. The first was the decollectivization of agriculture in 1989. Under the agricultural reforms the socialist agricultural policy of collective labor under solidarity groups was disbanded, and land title was given to peasant families. The continuing lack of adequate infrastructure in rural areas and the increase in employment opportunities in cities has led to increasing rural–urban migration. Due to the high degree of primacy in the urban system, this

migration has overwhelmingly been to Phnom Penh. Secondly, by decreasing the role of the state in the national economy, the reforms also had an impact on urban planning in Phnom Penh. The role of the state in planning for urban growth was weakened due to decreases in sources of state revenue, such as state ownership of industry and the state monopoly of foreign trade. The move towards austerity in government fiscal policy was backed by the International Monetary Fund (IMF). Later, the United Nations Transitional Authority in Cambodia (UNTAC) pursued a strategy of reducing government expenditure. The cumulative result of these policies was decreased government expenditure on infrastructure and social services. Finally, the reform which perhaps had the most far-reaching implications for urban development was a sub-decree that overnight changed urban residences from state property to private property. A booming real estate market soon developed as many people sought to supplement their incomes with the sale of their property. As many single-family dwelling contained multiple families, this resulted in considerable displacement.

In October of 1991, the Agreements on a Comprehensive Political Settlement of the Cambodian Conflict were signed in Paris, initiating one of the largest peacekeeping operations ever undertaken by the United Nations. For twenty-one months in 1992 and 1993 the United Nations implemented the United Nations Transitional Authority in Cambodia. The arrival of UNTAC in 1992 had several major impacts on Phnom Penh's development. First, it created a new market in the country for luxury goods and amenities. Imports increased six times in real terms during 1992 and 1993, and there was increasing investment from both domestic and foreign sources in leisure facilities. Second, it contributed to local employment both through direct employment and indirectly through the purchase of services in the country. Third, the influx of foreign capital in the form of investment and aid during the UNTAC period accelerated the process of commodification of land in Phnom Penh. Thus at the end of the UNTAC era in 1993, Cambodia had developed contacts with the global economy through trade, through the presence of international aid and lending institutions, through vastly increased transport and telecommunications links, and through investments (particularly in real estate and tourism facilities). In addition, the opening of the economy set the stage for Cambodia to be a location for FDI and tourism.

CAMBODIA IN THE EMERGING ASIAN REGIONAL SYSTEM: INVESTMENT AND TRADE

The changes that have taken place in Cambodia come at a time when Southeast Asia is one of the most rapidly industrializing regions in the world. In particular, the countries of the Association of Southeast Asian Nations (ASEAN) have experienced rapid economic growth as they have become a major target for FDI in manufacturing and services [. . .] Cambodia's dependence on trade with ASEAN states is particularly marked – ASEAN countries, particularly Thailand, Singapore and Malaysia, have increasingly viewed Cambodia as a potential market, as a source of raw materials, and as a target for investment in tourism-related services and, to a lesser extent, industry. Between 1993 and 1995, trade with ASEAN accounted for 59 per cent of total exports and 69 per cent of imports (IMF 1996). These numbers would likely rise if illegal exports (primarily of natural resources) and imports (of manufactured and luxury goods) were accounted for – smuggled goods were estimated to account for 40 per cent of consumer goods imports in 1995 (*Cambodia Daily*, 7 February 1995). Cambodia's major export is lumber, and the major imports are cigarettes, alcohol and petroleum.

Cambodia has seen a significant increase in foreign investment initiatives in recent years, the most important of which have involved ASEAN neighbors. These investments initially focused almost exclusively on resource extraction and hotels and tourism. The country has seen an increase in the number of international arrivals, from 17,000 in 1990 to 220,000 in 1995 (EIU 1996; Mullins 1999). The increase in the number of international arrivals has led to a major increase in investment in airport facilities and the formation of a national airline, which is a joint venture with a Malaysian company. Beginning largely in 1995 and 1996, an embryonic textile industry began to develop, centered for the most part in Phnom

Penh. While textiles accounted for only $3.5 million in exports in 1994, by 1996 there were 38 factories employing 16,000 workers, and exports totaled $33 million in the first six months of the year (EIU 1997). Another major area of development has been in the banking sector. In 1995 there were 29 registered banks in the country, including joint ventures with Thai, Malaysian and French banks. There has been increasing concern over the dubious practices of some of the domestic banks, many of which are suspected of being fronts for money laundering operations (*Far Eastern Economic Review*, 23 November 1995).

The prevalence of money laundering reveals another major economic and social development – Cambodia's increasing links with the global criminal economy. While improvements in transport and communications infrastructure have made it possible for international criminal organizations to move goods and coordinate activities in the country, Cambodia's technically deficient and often corrupt criminal justice system has had difficulty controlling their activities. While these criminal links are difficult to document, there is evidence that Cambodia is an increasingly important site in the global flows of narcotics, smuggled goods, prostitutes and illegal immigrants (*Phnom Penh Post*, 10–23 January 1997).

In sum, while Phnom Penh is not an important location of industrial production, the country has come to play a particular role in the Southeast Asian regional economy – as a source of increasingly scarce raw materials, as a playground for the region's wealthy elite, and as a major point in the region's criminal economy. The city's medium-term economic future is likely to be shaped by this role, although there is a possibility of continued industrial growth in the textile sector.

SPATIAL AND SOCIAL TRANSFORMATION IN PHNOM PENH

The combination of rapid urban growth, economic adjustment and the influx of cash from abroad has brought about a rapid growth in income disparity within Phnom Penh. Many of the new migrants to the city have taken up low-paying jobs in the construction industry, as pedicab and motorcycle taxi drivers, as petty sellers, and in other low-paying service occupations. Another factor in the high rates of urban poverty is the situation of civil servants – the government has resisted pressure from the IMF to cut the number of civil servants, yet has been unable to increase wages, which currently represent only a fraction of the cost of living for a family in the city. Many civil servants have taken second and third jobs. Finally, there exists a considerable gender imbalance in Phnom Penh due to the years of war, and 29.4 per cent of households in Phnom Penh are female-headed (Royal Government of Cambodia 1994). Single mothers are at extreme risk of poverty due to their responsibilities in the household, their generally lower level of education than men, and the undervalorization of female labor. These women are often forced to take on multiple sources of employment in order to subsist. The visibility of women working as petty sellers on the streets and in markets is testimony to this. In sum, the service sector has become the core of the urban economy. Service employment accounted for an estimated 74 per cent of employment in the city in 1994, and 84 per cent of all non-agricultural employment (ibid.).

Another dramatic and highly visible impact of the changes that have taken place in Phnom Penh in recent years is the proliferation of squatter settlements throughout the city. In 1989, squatter settlements were virtually unknown in the city, as the government had followed a policy of allowing newcomers to the city to settle relatively freely in unoccupied buildings and on vacant land left behind by the Khmer Rouge. Yet by 1994, an estimated 120,000 people, or approximately 12–15 per cent of the city's population, were living in squatter encampments (Urban Sector Group 1994). A particularly notable feature of Phnom Penh's squatter settlements is the high proportion of residents who have moved to the settlements from elsewhere in Phnom Penh. A survey conducted by the author in 1995 indicated that some one-third of the squatters in the city lived in Phnom Penh immediately before moving to the squatter settlement. This indicates that dislocation due to the rising cost of land and housing, caused by the demand for land as a source of investment and the increased cost of housing due to the booming market for wealthy Khmer and expatriate business people and aid

workers, is a major cause of squatting. The continued housing crisis and the likelihood of rapid urban growth in the future will most likely mean growth of the illegal housing sub-market into the foreseeable future.

The growth of squatter settlements has occurred in contrast to an obvious increase in wealth in the city. A remarkable construction boom is underway, as opulent nightclubs, restaurants, bars, hotels and casinos catering to a new Khmer elite, foreign investors, tourists and aid workers have appeared throughout the city. This has transformed the streetscape, as the main boulevards that once were quiet thoroughfares are now lined with flashing lights and thronging with people. Major boulevards are also beginning to experience congestion, as the number of privately owned cars and motorcycles has increased dramatically. Likewise, many of the city's residential buildings are newly constructed or have recently been renovated.

Thus the social and political changes that have taken place in the last several years are manifest in the spatial changes in the city form. The city has taken on a more cosmopolitan feel while emerging social problems related to the growing income gap are increasingly visible.

CONCLUSION

This chapter has attempted to use the case of Phnom Penh, Cambodia to demonstrate some ways in which globalization has impacted on cities in least developed countries. The intent is not to assert that cities in other LDCs can be expected to experience identical forms of urban development. The intent, rather, is to provide an example of how the particular attributes of the global economy impact on cities in LDCs through certain points of contact – FDI, improved telecommunications and transport links, links with regional economies, and the increased influence of international aid and lending organizations.

In the particular case of Cambodia, globalization has had political, social and economic impacts that have affected Phnom Penh. Politically, increased links with the global economy, and the increased influence of foreign and domestic capital in policy decision-making, have meant that access to political power is increasingly a function of access to sources of wealth. In addition, increased links with market economies in the region have exposed the country to political influence from these countries. Socially and economically, links to the global economy have meant different things to different people. For those who have been able to take advantage of the opportunities presented by the changes that have occurred – mostly people with access to resources and education – links to the global economy have meant a chance to join Phnom Penh's emerging middle and upper classes. For many others – particularly low level public sector workers, those with little education and female-headed households – these changes have often had neutral or negative effects on quality of life. The emergence of large squatter settlements in the city, and the growth of the low-income service sector, indicates to some degree the types of social issues the country is likely to confront in the new socio-economic order. Finally, globalization has also had cultural impacts, as the country is increasingly marketed as a tourist destination.

The degree of the impact of globalization on a city, and the form this impact will take, will differ significantly based on the regional context and historical circumstances. This complexity and variation in experience is not adequately represented by analyses that focus on the exclusion of least developed countries in the process of globalization. Theorists would do well to abandon their assumptions regarding the exclusion of these countries, and begin to look at ways in which these countries are integrated into and impacted by global economic forces. Such an approach holds greater promise in the effort to identify the potential problems and prospects faced by least developed countries in the global era.

REFERENCES FROM THE READING

Broad, R. and Landi, C. (1996) Whither the North–South gap, *Third World Quarterly*, 17, 1, 7–17.

Castells, M. (1996) *The Rise of the Network Society*. Cambridge, MA: Blackwell.

EIU (Economist Intelligence Unit) (1996) *Economist Intelligence Unit: Cambodia Country Report*, 4th Quarter. London: EIU.

EIU (Economist Intelligence Unit) (1997) *Economist Intelligence Unit: Cambodia Country Report*, 1st Quarter. London: EIU.

FitzGerald, E.V.K. (1993) The economic dimension of social development and the peace process in Cambodia. In P. Utting (ed.) *Between Hope and Insecurity: The Social Consequences of the Cambodian Peace Process*. Geneva: UNRISD (United Nations Research Institute for Social Development).

Friedmann, J. (1995) Where we stand: a decade of world city research. In P.L. Knox and P.J. Taylor (eds) *World Cities in a World-System*. New York: Cambridge University Press, 21–47.

Halfani, M. (1996) Marginality and dynamism: prospects for the Sub-Saharan city. In M. Cohen and B. Ruble (eds) *Preparing for the Urban Future*. Washington, DC: Woodrow Wilson Center Press.

International Monetary Fund (IMF) (1996) *Direction of Trade Statistics Yearbook*. Washington, DC: IMF.

Mullins, P. (1999) Tourism in Southeast Asia. In S.S. Fainstein and D. Judd (eds) *Places to Play: The Remaking of Cities for Tourists*. New Haven, CT: Yale University Press.

Royal Government of Cambodia (1994) *Report on the Socioeconomic Survey – 1993/1994 Cambodia (First Round)*. Phnom Penh: National Institute of Statistics, Ministry of Planning.

Urban Sector Group (1994) *Twelve Month Program Proposal*. Phnom Penh, Cambodia.

26

"Global and World Cities: A View from off the Map"

from *International Journal of Urban and Regional Research* (2002)

Jennifer Robinson

Editors' introduction

Jennifer Robinson is a South African urbanist based in the UK, where she currently teaches in Geography at The Open University. Robinson has written a large number of articles and book chapters on colonial and postcolonial urban development, with particular reference to South African cities. Robinson's 1996 book, *The Power of Apartheid*, explored the relationship between state power and space in South African cities, and she has also written extensively on feminist theory and politics. Most recently, Robinson has published a number of influential critiques of contemporary urban theory, including the reading here, from the point of view of critical postcolonial theory.

According to Robinson, the dominant approaches to global city theory have normalized the distinctive sociospatial features and developmental trajectories of North American and Western European cities: the latter have been presented as the paradigmatic "model" in terms of which all other cities are to be interpreted, regardless of their particular locations or histories. Within such a framework, cities located "off the map" – that is, in the developing countries of the global South – are almost invariably said to be lacking the characteristics that would qualify them as genuinely "global" cities. Much like David Simon (Reading 24) and Gavin Shatkin (Reading 25), Robinson is concerned to analyze the contextually specific patterns of globally induced urban restructuring that have been unfolding in cities located outside the developed capitalist North. However, whereas Simon and Shatkin suggest various ways in which extant approaches to global city theory might be modified or expanded in order to take such cities into account, Robinson aims her critique at the epistemological core of the theory itself. For Robinson, the entire conceptual apparatus of global city theory is problematic insofar as it is grounded upon basically static, decontextualized categorizations and typologies. Without denying the importance of global economic inequalities or the global urban hierarchy, Robinson suggests that a significant conceptual reorientation is required to grasp the distinctiveness of cities in the developing world. To this end, Robinson proposes a number of methodological innovations. First, the question of a city's "relevance" to the global economy needs to be explored not only with reference to its role as a basing point for transnational corporate activities, but also on the basis of a broader range of possible global–local linkages (see also Reading 25 by Shatkin). Second, based on work by Amin and Graham (1997) among others, Robinson mobilizes the notion of the "ordinary city," which she views as a conceptual basis on which diverse forms of urbanism, and the broad range of connections that link cities to global processes, might be more adequately appreciated and theorized. Robinson concludes her critique by underscoring its policy implications. For Robinson, the goal of becoming a global city is seriously unrealistic for most urban centers in the developing world. Other models of "successful" urban development are required, she argues, so that policy makers might harness the socioeconomic potential of developing cities while also alleviating their most pressing social, infrastructural and environmental problems.

There are a large number of cities around the world which do not register on intellectual maps that chart the rise and fall of global and world cities. They don't fall into either of these categories, and they probably never will – but many managers of these cities would like them to. Some of these cities find themselves interpreted instead through the lens of developmentalism, an approach which broadly understands these places to be lacking in the qualities of city-ness, and which is concerned to improve capacities of governance, service provision and productivity. Such an approach supports some of the more alarmist responses to mega-cities, which are more commonly identified in poorer countries. But for many smaller cities, even the category mega-city is irrelevant. My concerns in this chapter extend beyond the poor fit of these popular categories, though. I would like to suggest that these widely circulating approaches to contemporary urbanization – global and world cities, together with the persistent use of the category "third-world city" – impose substantial limitations on imagining or planning the futures of cities around the world.

One of the consequences of the unreflexive use of these categories is that understandings of city-ness have come to rest on the (usually unstated) experiences of a relatively small group of (mostly western) cities, and cities outside of the West are assessed in terms of this pre-given standard of (world) city-ness, or urban economic dynamism. This chapter explores the extent to which more recent global and world city approaches, although enthusiastic about tracking transnational processes, have nonetheless reproduced this long-standing division within urban studies [. . .]

I do this by reflecting on some fashionable approaches to cities from a position off their maps. Of course, the cities I am concerned with are most emphatically on the map of a broad range of diverse global political, economic and cultural connections, but this is frequently discounted and certainly never explored within these theoretical approaches. There is a need to construct an alternative urban theory which reflects the experiences of a much wider range of cities. This will involve disrupting the narrow vision of a (still) somewhat imperialist approach to cities, which has been reinforced by the strident economism in accounts of global and world cities. Elements of urban theory have become transfixed with the apparent success and dynamism of certain stylish sectors of the global economy, despite (and perhaps because of) their circumscribed geographical purchase and most unappealing consequences. These studies have been valuable, and offer great insight into the limited part of the world and economy that they study. My suggestion, though, is that these insights could be incorporated in a broader and less ambitious approach to cities around the world, an approach without categories and more inclusive of the diversity of experience in ordinary cities [. . .]

GLOBAL AND WORLD CITIES

In considering the dynamics of the world economy in relation to cities, a structural analysis of a small range of economic processes with a certain "global" reach has tended to crowd out an attentiveness within urban studies to the place and effect of individual cities and the diversity of wider connections which shape them (King 1995). Although status within the world city hierarchy has traditionally been based on a range of criteria, including national standing, location of state and interstate agencies and cultural functions, the primary determination of status in this framework is economic. The world cities approach assumes that cities occupy similar placings with similar capacity to progress up or fall down the ranks. The country categorizations of core, periphery and semi-periphery in world-systems theory have been transferred to the analysis of cities, and overlain on an extant but outdated vocabulary of categorizations (such as first/third world) within the field of urban studies. On this basis, from the dizzy heights of the diagrammer, certain significant cities are identified, labelled, processed and placed in a hierarchy, with very little attentiveness to the diverse experiences of that city, or even to extant literature about that place. The danger here is that out-of-date, unsuitable or unreliable data, and possibly a lack of familiarity with some of the regions being considered, can lead to the production of maps which are simply inaccurate. These images of the world (of important) cities have been used again and again to illustrate the perspective of world cities theorists.

A view of the world of cities thus emerges where millions of people and hundreds of cities are dropped off the map of much research in urban studies, to service one particular and very restricted view of significance or (ir)relevance to certain sections of the global economy. This methodology also reveals an analytical tension between assessing the characteristics and potential of cities on the basis of the processes which matter as viewed from within their diverse dynamic social and economic worlds, or on the basis of criteria determined by the external theoretical construct of the world or global economy. Although aiming to emphasize connections and not attributes, a limited range of cities still end up categorized in boxes or in diagrammatic maps, and assigned a place in relation to *a priori* analytical hierarchies.

The discursive effectiveness of the global city hypothesis depends on the pithy identification of the "global city" – a category of cities which are claimed to be powerful in terms of the global economy. If the global city were labelled as just another example of an industrial district (perhaps it should rather be called: new industrial districts of transnational management and control), it might not have attracted the attention it did.

FILLING IN THE VOIDS: OFF THE WORLD CITIES MAP

[. . .] I certainly appreciate that the focus of global and more recent world cities work is on a limited set of economic activities, which are assuming an increasingly transnational form, and in which relatively few cities can hope to participate. But it is the leap from this very restricted and clearly defined economic analysis, to claims regarding the success and power of these few cities, their overall categorization on this restricted basis, and the implied broader structural irrelevance of all other cities, which is of concern. These theoretical claims and categorizing moves are both inaccurate and harmful to the fortunes of cities defined "off the map."

The "end of the third world" is perhaps an accurate assessment of changes over the last three to four decades in places like Hong Kong, Singapore, Taiwan, South Korea and even Malaysia, and the appearance of these major urban centers in rosters of first- and second-order global cities reflects this. But in parts of the world where global cities have not been identified – the "voids" of world and global city approaches – the experience of many countries and cities has been much more uneven. For many, the 1980s and 1990s have been long decades of little growth and growing inequality. It is, however, inaccurate to caricature even the poorest regions as excluded from the global economy or doomed to occupy a slow zone of the world economy.

It is hard to disagree that some countries and cities have lost many of the trading and investment links that characterized an earlier era of global economic relations. A country like Zambia, for example, now one of the most heavily indebted nations in the world, and certainly one of the poorest, has seen the value of its primary export, copper, plummet on the world market since the 1970s. Its position within an older international division of labor is no longer economically viable, and it has yet to find a successful path for future economic growth. En route it has suffered the consequences of one of the World Bank/IMF's most ruthless Structural Adjustment Programmes. However, Zambia is also one of the most urbanized countries on the African continent, and its capital city, Lusaka, is a testimony to the modernist dreams of both the former colonial powers and the post-independence government. Today, with over 70 per cent of the population in Lusaka dependent on earnings from the informal sector, the once bright economic and social future of this city must feel itself like a dream.

Lusaka is certainly not a player in the new global economy. But copper is still exported, as are agricultural goods, and despite the huge lack of foreign currency (and sometimes because of it) all sorts of links and connections to the global economy persist. From the World Bank, to aid agencies, international political organizations, and trade in secondhand clothing and other goods and services, Lusaka is still constituted and reproduced through its relations with other parts of the country, other cities, and other parts of the region and globe. The city continues to perform its functions of national and regional centrality in relation to political and financial services, and operates as a significant market (and occasionally production site) for goods and services from across the country and the world.

It is one thing, though, to agree that global links are changing, some are being cut, and that power relations, inequalities and poverty shape the quality of those links. It is quite another to suggest that poor cities and countries are irrelevant to the global economy. When looked at from the point of view of these places which are allegedly "off the map," the global economy is of enormous significance in shaping the fortunes of cities around the world. For many poor, "structurally irrelevant" cities, the significance of flows of ideas, practices and resources beyond and into the city concerned from around the world stands in stark contrast to these claims of irrelevance [see Reading 25 by Shatkin].

And to pursue a more polemical line, mineral resources crucial to the global economy are drawn from some of the poorest countries of the world, where financiers and transnational firms negotiate with warlords, corrupt governments and local armies to keep profits, production and exports flowing. Widening the compass of analysis might help to encourage a more critical edge to the global and world cities literature. Moreover, it is precisely through avoiding "risky" investments (and pursuing vastly exploitative and violent forms of extraction instead) in the poorest countries and cities in the world that the western financial "mode of production" is able to aim to secure the stable shareholder returns which maintain post-Fordist finance-based economies. To the extent that they are absent from this aspect of the global economy, these places may well be central to sustaining it [. . .]

Most importantly, the particular "global economy" which is being used as the ground and

foundation for identifying both place in hierarchy and relevant social and economic processes, is only one of many forms of global and transnational economic connection. The criteria for global significance might well look very different were the map-makers to relocate themselves and review significant transnational networks in a place like Jakarta, or Kuala Lumpur, where ties to Islamic forms of global economic and political activity might result in a very different list of powerful cities. Similarly, the transnational activities of agencies like the World Bank and the IMF who drive the circulation of knowledge (World Bank 2000) and the disciplining power to recover old bank and continuing bi-lateral and multi-lateral debt from the poorest countries in the world would draw another crucial graph of global financial and economic connections shaping (or devastating) city life [. . .]

ORDINARY CITIES

A diverse range of links with places around the world are a persistent feature of cities. They can work for or against cities everywhere and are constantly being negotiated and renegotiated. To aim to be a "global city" in the formulaic sense may well be the ruin of most cities. Policy-makers need to be offered alternative ways of imagining cities, their differences and their possible futures – neither seeking a global status nor simply reducing the problem of improving city life to the promotion of "development." In developmentalist perspectives cities in poor countries are often seen as non-cities, as lacking in city-ness, as objects of (western) intervention. Ordinary cities, on the other hand, are understood to be diverse, creative, modern and distinctive, with the possibility to imagine (within the not inconsiderable constraints of contestations and uneven power relations) their own futures and distinctive forms of city-ness (Amin and Graham 1997).

Categorizing cities and carving up the realm of urban studies has had substantial effects on how cities around the world are understood and has played a role in limiting the scope of imagination about possible futures for cities. This is as true for cities declared "global" as for those which have fallen off the map of urban studies. The global cities hypothesis has described cities like New York and London as "dual cities," with the global functions drawing in not only a highly professional and well-paid skilled labor force, but also relying on an unskilled, very poorly-paid and often immigrant workforce to service the global companies. These two extremes by no means capture the range of employment opportunities or social circumstances in these cities. It is possible that these cities, allegedly at the top of the global hierarchy, could also benefit from being imagined as "ordinary." The multiplicity of economic, social and cultural networks which make up these cities could then be drawn on to imagine possible paths to improving living conditions and enhancing economic growth across the whole city.

A more cosmopolitan urban theory might be more accurate or helpful in understanding the world; it might also be more resourceful and creative in its output. But interrogating these categorizations of cities and theoretical divisions within urban studies matters primarily, I think, because they limit our potential to contribute to envisioning possible city futures. And given the gloomy prognoses for growth in poor cities within the context of the contemporary global economy, creative thinking is certainly needed!

From the viewpoint of global and world cities approaches, poor localities, and many cities which do not qualify for global or world city status, are caught within a very limited set of views of urban development: between finding a way to fit into globalization, emulating the apparent successes of a small range of cities; and embarking on developmentalist initiatives to redress poverty, maintain infrastructure and ensure basic service delivery. Neither the costly imperative to go global, nor developmentalist interventions which build towards a certain vision of city-ness and which focus attention on the failures of cities, are very rich resources for city planners and managers who turn to scholars for analytical insight and assessment of experiences elsewhere. It is my opinion that urban studies needs to decolonize its imagination about city-ness, and about the possibilities for and limits to what cities can become, if it is to sustain its relevance to the key urban challenges of the twenty-first century. My suggestion is that "ordinary-city" approaches offer a potentially more fertile ground for meeting these challenges.

THE POLICY IMPERATIVE: THE POLITICAL CASE FOR ORDINARY CITIES

If cities are not to remain inconsequential, marginalized and impoverished, or to trade economic growth for expansion in population, the hierarchies and categories of extant urban theory implicitly encourage them to aim for the top! Global city as a concept becomes a regulating fiction. It offers an authorized image of city success (so people can buy into it) which also establishes an end point of development for ambitious cities. There are demands, from Istanbul to Mumbai, to be global. But, calculated attempts at world or global city formation can have devastating consequences for most people in the city, especially the poorest, in terms of service provision, equality of access and redistribution (Robins and Askoy 1996; Firman 1999). Global and world city approaches encourage an emphasis on promoting economic relations with a global reach, and prioritizing certain prominent sectors of the global economy for development and investment. Alternatively, the policy advice is for cities to assume and work towards achieving their allocated "place" within the hierarchy of world cities.

Most cities in poorer countries would find it hard to reasonably aspire to offering a home for the global economy's command and control functions. More feasible for many poorer cities is to focus on some of the other "global functions" that Sassen (1994) associates with global cities. These include promoting attractive "global" tourist environments, even though these have nothing of the locational dynamics of command and control global city functions. Disconnected from the concentration of arts and culture associated with employment of highly skilled professionals in global cities, the impulse to become global in purely tourist terms can place a city at the opposite end of power relations in the global economy, while substantially undermining provision of basic services to local people. In addition, Export Processing Zones may be "global" in the sense that they are "transnational spaces within a national territory" (Sassen 1994: 1), but they too involve placing the city concerned in a relatively powerless position within the global economy, which is unlikely to be the city's best option for future growth and development. These are not places from where the global economy is controlled: they are at quite the other end of the command and control continuum of global city functions. More than that, the reasons for co-location would not involve being able to conduct face-to-face meetings to foster trust and cooperation in an innovative environment. Rather, they are to ensure participation in the relaxation of labor and environmental laws which are on offer in that prescribed area of the city. Cities and national governments often have to pay a high price to attract these kinds of activities to their territory. Valorizing "global" economic activities as a path to city success – often the conclusion of a policy reversioning of world cities theory – can have adverse consequences for local economies.

This is a familiar story, but one which scholars are more likely to blame on others – capitalists, elite urban managers – than on their own analyses, which are seldom the object of such reflection. It is when attention slips from economic process to a sloppy use of categorization that I think the most damaging effects of the world and global city hypotheses emerge. Categorizing a group of cities as "global" on the basis of these small concentrated areas of transnational management and coordination activity within them is metonymic in that it has associated entire cities with the success and power of a small area within them. In the process a valid line of analysis has reproduced a very familiar hierarchization of cities, setting certain cities at the top of the hierarchy to become the aspiration of city managers around the world. This has happened just as a burgeoning postcolonial literature became available to critique earlier categorizations of cities into western and third world, a categorization which had emphasized difference and deviation from the norm as bases for analysis and which had established certain (western) cities as the standard towards which all cities should aspire. Instead of pursuing the postcolonial critique, urban studies has replicated this earlier division by accepting the categories of world/global city as analytically robust and popularizing them in intellectual and policy circles. Global cities have become the aspiration of many cities around the world; sprawling and poor megacities the dangerous abyss into which they might fall should they lack the redeeming (civilizing) qualities of city-ness found elsewhere. This may not have been the intention of urban theorists, but

ideas have a habit of circulating beyond our control. It is my contention that urban theory should be encouraged to search for alternative formulations of city-ness which don't rest upon these categories and which draw their inspiration from a much wider range of urban contexts.

The political need for a new generation of urban theoretical initiatives is apparent. How can the overlapping and multiple networks highlighted in the ordinary city approaches be drawn on to inspire alternative models of development, which see the connections, rather than conflict, between informal and formal economies? Approaches which explore links between the diversity of economic activities in any (ordinary) city, and which emphasize the general creative potential of cities, are crucial, rather than those which encourage policy-makers to support one (global) sector to the detriment of others (see, for example, Benjamin 2000; Simone 2001).

CONCLUSION

The academic field of urban studies ought to be able to contribute its resources more effectively to the creative imagining of possible city futures around the world. One step in this direction would be to break free of the categorizing imperative, and to reconsider approaches which are at best irrelevant and at worst harmful to poor cities around the world. I have suggested that in place of world, global, mega-, Asian, African, former Socialist, European, third-world etc. cities, urban studies embark on a cosmopolitan project of understanding ordinary cities.

A second step must be to decolonize the field of urban studies. Theoretical reflections should at least be extremely clear about their limited purchase and, even better, extend the geographical range of empirical resources and scholarly insight for theorizing beyond the West and western-dominated forms of globalization. This has been initiated in a restricted form, through the transnational emphasis of global and world cities approaches, and the growing interest in globalization within a developmentalist frame. But a more cosmopolitan empirical basis for understanding what cities are, and how they function, is essential to the future relevance of the field of urban studies. In an age when most people now live in cities, and most of this urban population is in poor countries, irrelevance is a very real possibility for a field whose wellsprings of authorized theoretical innovation remain firmly fixated on the West and its successful satellites and partners.

This is not to insist that every study consider everywhere. But there is considerable scope for the spatial trajectories of theoretical imaginings to come closer to the spatiality of cities themselves, which are constituted on the basis of ideas, resources and practices drawn from a variety of places – not infinite, but diverse – beyond their physical borders. The conditions of incorporation, though, are crucial. Firstly, simply mobilizing evidence of difference and possibly deviation within the frame of dominant theory is not enough. Consideration needs to be given to the difference the diversity of cities makes to theory (not simply noting the difference that they are). How are theoretical approaches changed by considering different cities and different contexts, by adopting a more cosmopolitan approach? And secondly, as with cities themselves, power relations and their geographies cannot be avoided. If a cosmopolitan urban theory is to emerge, scholars in privileged western environments will need to find responsible and ethical ways to engage with, learn from and promote the ideas of intellectuals in less privileged places. This is not a call to western writers to appropriate other places for continued western intellectual advantage. It is a plea to acknowledge the intellectual creativity of scholars and urban managers in a wider range of urban contexts.

This will involve a critical analysis of the field's own complicity in propagating certain limited views of cities, and thereby undermining the potential to creatively imagine a range of alternative urban futures. It will require more cosmopolitan trajectories for the sources and resources of urban theory. Much innovative work is already being undertaken by scholars and policy-makers around the world, who have had to grapple with the multiplicity, diversity and ordinariness of their cities for some time. Ordinary cities are themselves enabling new kinds of urban imaginaries to emerge – it is time urban studies caught up. More than that, I would suggest that as a community of scholars we have a responsibility to let cities be ordinary.

REFERENCES FROM THE READING

Amin, A. and Graham, S. (1997) The ordinary city, *Transactions of the Institute of British Geographers*, 22, 411–429.

Benjamin, S. (2000) Governance, economic settings and poverty in Bangalore, *Environment and Urbanisation*, 12, 1, 35–56.

Firman, T. (1999) From "Global City" to "City of Crisis": Jakarta Metropolitan Region under economic turmoil, *Habitat International*, 23, 4, 447–466.

King, A. (1995) Re-presenting world cities: cultural theory/social practice. In P.L. Knox and P.J. Taylor (eds) *World Cities in a World-System*. New York: Cambridge University Press, 215–231.

Robins, K. and Askoy, A. (1996) Istanbul between civilisation and discontent, *City*, 5–6, 6–33.

Robinson, J. (1996) *The Power of Apartheid: State, Power and Space in South African Cities.* London: Butterworth-Heinemann.

Sassen, S. (1994) *Cities in a World Economy*. London: Sage.

Simone, A. (2001) Straddling the divides: remaking associational life in the informal African city, *International Journal of Urban and Regional Research*, 25, 1, 102–117.

World Bank (2000) *Cities in Transition: World Bank Urban and Local Government Strategy*. Washington, DC: World Bank.

F O U R

27

"Globalization and the Corporate Geography of Cities in the Less-developed World"

from *Annals of the Association of American Geographers* (2002)

Richard Grant and Jan Nijman

Editors' introduction

In this reading, a longer version of which was originally published in a major US geography journal, Richard Grant and Jan Nijman consider the spatial impacts of global economic restructuring in two cities in the developing world, Accra, Ghana and Mumbai, India. Grant and Nijman both teach in the Department of Geography at the University of Miami, and have written widely on the interplay between urban development and globalization. Grant is a human geographer who has conducted many years of research in Accra and has published extensively on globalization and urban sociospatial polarization. Nijman is a comparative urbanist and political geographer who has studied urban regions such as Miami, Mumbai and Amsterdam; his writings include major scholarly articles on globalization, urban restructuring and urban theory.

Like other contributors to Part Four, Grant and Nijman are sharply critical of the dominant approaches to global city theory due to their "western bias" and their pervasive failure to consider cities positioned beneath the upper tiers of the global city hierarchy. On the basis of a concise critique of global city theory, Grant and Nijman propose a new framework for the study of cities in the developing world in global perspective. The authors introduce a phase-model of urban development under colonial and postcolonial conditions, and they offer a number of generalizations regarding the evolution of urban spatial structures within each phase. In the main section of the article, Grant and Nijman present a comparative case study of Accra and Mumbai, two cities in former British colonies that have experienced significant sociospatial transformations in the wake of economic liberalization policies since the early 1980s. While Grant and Nijman devote considerable attention to the colonial and postcolonial histories of these cities, they focus their analysis primarily upon the post-1980s, "global" era, when foreign direct investment significantly increased. Building upon an original survey of foreign companies in each city, the authors explore three key empirical issues: the scope of transnational corporate activity, the geographical location patterns of transnational corporate activity, and the relation between transnational corporate location patterns and the urban geographies of domestic firms. In each city, their analysis illustrates the intensive agglomeration tendencies of transnational corporate activities and the rapid expansion of producer and financial services industries during the post-liberalization period. In addition, Grant and Nijman indicate the differential ways in which each city's major business distincts are connected to the global economy. Accra and Mumbai each contain at least one densely concentrated zone in which the vast majority of foreign companies are located, along with a number of additional business districts that host a mixture of foreign and domestic firms.

In sum, Grant and Nijman's reading provides an innovative contribution to the comparative investigation of globalization and urban development. Their analysis convincingly demonstrates that global firms are significantly transforming the sociospatial fabric of cities in the developing world. At the same time, Grant and Nijman's work sets forth a comparative research agenda on the study of sociospatial transformation in other cities of the global South. Due to space limitations, all bibliographical references have been removed from this chapter; however, readers are encouraged to consult the original article for the authors' extensive bibliography (see also the editors' introduction to Part Four).

■ ■ ■ ■ ■ ■

One of the ironies of the academic debate on globalization is its Western bias. Much of the theorizing and empirical research is based on the experiences of the United States, West Europe, and other countries in the core of the world economy. This is also true for most scholarly work on cities and globalization in geography and other social sciences. Overall, the globalization debate is not nearly as "global" as it probably should be.

In this chapter, we explore the changing urban geographies of cities in the less-developed world in the context of economic globalization.[1] We argue that the internal spatial structure of such cities can be understood in terms of their evolving roles in the wider-world political economy. We conceptualize the broader spatial context of cities in terms of a political economy because (sub)national government policies determine, to a large degree, the exposure of the urban arena to global economic forces. Cities differ in the ways they are linked to the external economy. This differentiation is as much a function of the idiosyncratic features of the city as a place and location as it is of developments in the global economy. In this study, we acknowledge the former, but we concentrate on the latter.

GLOBALIZATION AND CITIES IN THE LESS-DEVELOPED WORLD

The world city literature is mostly aimed at cities at the top of the urban hierarchy. In the process, cities in the lower strata of the global hierarchy – particularly cities in the less-developed world – are neglected. Hence, the literature often advances claims of universal validity without proper theoretical consideration of and without substantial empirical research in cities in the less-developed world. By contrast, the conceptual framework used in this chapter may be summarized with the following series of propositions:

A. With regard to the role of the city in the global political economy:

1. The historically evolving structures of the global political economy imply a variety of evolving roles of different cities. Fundamentally, cities maintain different positions in the spatial organization of power in the global political economy. This was quite obvious in the contrasting roles of London versus Accra or Bombay during colonial times. In the present organization of the global political economy, a world city hierarchy exists in which cities have more or less specialized functions with varying geographic reach. The role assumed by a city in the global political economy is conditioned by, among other things, the policies of the state in which it is located.

2. Many cities in the less-developed world have moved through four historical phases: the pre-colonial, the colonial, the national, and the global phase. These phases reflect fundamental changes in the nature and extent of the global political economy and differential external linkages of the city.

3. Cities in the less-developed world that have a gateway function – typically, large port cities – are particularly exposed to the global political economy. For such cities, the global phase is particularly significant.

B. With regard to ramifications of these global involvements for the city's economic geography:

1. The internal spatial organization of gateway cities in the less-developed world is in part a reflection of the city's role in the global political economy.

2. The precolonial phase and the national phase are characterized by relative insulation from the global economy. At these times, urban form was in large measure determined in local and national contexts.

3. In contrast, the colonial phase and the global phase are characterized by a relatively high degree of connectivity to the global economy, with a powerful imprint on the urban landscape. This imprint is essentially reflected in a spatially delineated foreign corporate presence.

4. For our purposes, the main difference between the global phase and the colonial phase lies in the political context of the linkages to the global economy. During colonial times, the city's economic linkages were by and large dictated by the colonial power, and economic relations were heavily biased to the "mother country." In addition, colonial governments exercised tight control over urban planning and land use in the city. During the global phase, the city's economic linkages are directly related to liberalization policies by national governments. The global connections are more intense and more diverse than in any previous phase. Further, urban form and land uses are less stringently regulated and are to a significant degree influenced by market forces.

We should note that, in our overall view of the evolution of cities, we do not privilege global or external influences over local determinants. Indeed, conceptually, the global and the local only make sense in relation to each other.

Our investigation concentrates on two cities: Accra in Ghana and Mumbai in India (in 1995 Bombay was officially changed to Mumbai; we use both names, depending on the historical period). A parallel study of urban change in two different cities in different settings helps to move beyond description and allows some degree of generalization. Our analysis focuses on the changing foreign corporate presence in the two cities as one important measure of economic globalization and the effects on the cities' overall corporate geographies. The research is based on extensive fieldwork in Accra and Mumbai between 1998 and 2000 and on assembled datasets on foreign and domestic corporations – their year of establishment, main activity, location, size, and so on. We find that there have been profound changes in recent times that were neither random nor accidental. We explain the main trends in Accra and Mumbai in terms of the changing linkages to the global economy since the implementation of liberalization policies.

The choice of Accra and Mumbai as case studies is based in our theoretical framework. The two cities are quite different in terms of size and in terms of their cultural settings, but they share a number of important characteristics in terms of their experiences in the global political economy. First, both cities are located in regions that are among the poorest of the world. Second, they have a similar historical experience as colonial port cities in the British empire. Third, both have been primary economic centers since the end of colonialism in, respectively, Ghana and India. Fourth, both cities – and the wider national economies of which they are part – have been subjected to large-scale national and subnational deregulation and liberalization policies since the mid-1980s. Finally, both cities function as major gateway cities, and as such they are significantly exposed to forces of globalization.

Accra and Bombay served as important colonial port cities under British rule. The mode of integration at that time was one-dimensional and dictated by Westminster. As a legacy of the colonial past, English is the main language used in commerce and government in both cities. After independence (India in 1947 and Ghana in 1957), national development policies and ideological perspectives in the two countries were quite similar. Nkrumah and Nehru, respectively, articulated a nonaligned position in world affairs and promoted national economic policies of self-reliance. In a sense, as we shall see later, Accra and Bombay were "nationalized" during these times, and the cities were integrated in the national economies. Generally, these policies are now widely acknowledged to have failed: according to most development indicators, Ghana and India consistently belonged to the category of the world's poorest and least-developed countries.

In a major change of course, Ghana and India witnessed a shift toward liberalization in the mid-1980s. The economic consequences of the reforms have been considerable in both countries, yet, as our ongoing research indicates, they have also been highly uneven at the regional level, in terms of the urban–rural divide, and among cities in the urban hierarchy. Gateway cities such as Accra and Mumbai have been disproportionately affected in terms of rapid increases in international trade, foreign investment, and the presence of foreign corporations.

In what follows, we focus on the changing economic geographies at a smaller scale, within the metropolitan areas of Accra and Mumbai. On the basis of our conceptual framework, and in view of the exploratory nature of this investigation, our empirical analysis of the changing corporate geographies of Accra and Mumbai is driven by three questions:

1. To what extent has the presence of foreign corporate activity increased across various economic sectors since the implementation of liberalization policies?
2. What has been the effect of the growing foreign presence on the cities' economic geographies, in terms of functional specialization and concentration?
3. Are there significant differences between foreign and domestic companies in terms of their activities and corporate locational behaviors? In other words, is there a difference in behavior that makes foreign corporate activity stand out in the urban landscape? More generally, is it relevant to speak about foreign and domestic spaces?

DATA AND SURVEY METHODS

Our data collection on foreign corporations consisted of three steps, implemented between 1998 and 2000. First, we identified and mined all available listings of foreign companies and their addresses from local sources. Second, we combined these listings to produce a single master list of foreign companies, with addresses for each city. Finally, we conducted a survey among all these companies. The resulting master lists for Accra and Mumbai contained, respectively, 655 and 611 companies, the largest ever compiled. The final response rate to our surveys in both cities was around 50 per cent.

Defining a foreign, or multinational, company is no easy task. Considerable controversy exists in the literature regarding the definition and measurement of "foreign control." For a long time, the national governments of Ghana and India have set certain conditions for the approval of foreign corporate activities, and in many sectors only joint ventures with limited foreign share-holding were allowed. Over time, the regulations have changed and have generally become less strict. However, many of the original foreign company listings mentioned above did not necessarily take government criteria into account; at any rate, most listings did not include a precise definition of the foreign company. Thus, included companies have variable foreign share-holdings: they may be subsidiaries (fully foreign-owned), liaison-offices with representative roles only, or they may be joint ventures that involve two or more companies (foreign and domestic) where the degree of foreign control is altogether unknown or unclear.

In the survey we targeted the headquarters of each company. The "location" of a company may not be confined to the address of the headquarters. Many of the larger companies have their headquarters and managerial staff in a prime business area, while back-office work and factories are located elsewhere in the city or beyond. Our mapping of the economic geographies of Accra and Mumbai is confined to company headquarters and should therefore be understood as a geography of corporate command and control. A second survey was conducted in the summer of 2000 to test hypotheses about the degrees of connectivity of different business districts to the global economy. It asked respondents to indicate the relative importance of global connections of their company (international phone calls, postal mail, electronic mail, international business trips, and so on). The second survey was confined to domestic companies to check the importance of location within the city as a predictor of global connections irrespective of ownership. In the empirical analysis below, we interpret the changing economic geographies of Accra and Mumbai in a historical context.

HISTORICAL BACKGROUND: ACCRA AND MUMBAI DURING COLONIAL AND NATIONAL TIMES

During the colonial era, both Accra and Bombay were spatially organized around the ports. They functioned as central nodes in the trade networks between their hinterlands and England. They were strategically located to link rail-lines and shipping routes. The docks, warehouses, and railway terminals highlight these functions of trade, storage, and distribution. Adjacent to the port area lay a well-defined European business district that functioned as the designated location for foreign companies. Most economic activities in the European commercial area involved trade, distribution, transport, banking, and insurance. Zoning and building codes were strictly enforced to maintain an orderly European character and atmosphere in the district. Traditional markets or bazaars were located in a business district in the Native Town, sometimes referred to as "colored town." This area comprised a mix of commercial and residential land-uses. Much commercial activity involved trade of agricultural produce and crafts, small-scale industry, and retailing. Importantly, the European Town and the Native Town were physically separated. In the case of Bombay, the barrier took the form of the Esplanade, a green area that served recreational purposes and that had been designed by the Portuguese in the sixteenth century. In Accra, as well, an open green separated the European and Native Towns.

In sum, these colonial landscapes exhibited high levels of segregation of foreign and native commercial and residential activities. The economic geographies of both cities also displayed high levels of functional specialization and concentration. It should be noted, though, that these patterns were not always stable throughout the colonial period and that they slowly eroded in the latter part of the nineteenth century. As participation in the international economy increased in the early twentieth century and population pressure built up, the boundaries between the different areas gradually blurred.

More drastic transformations proceeded in the wake of independence in each country. Apart from rapid overall population growth in Accra and Bombay, which resulted in significant geographical expansion, we should note four changes in the cities' spatial configurations.

1. The foreign corporate and residential presence declined in relative terms. The elimination of legislation that discouraged native enterprises led to rapid growth of domestic companies, which were free to locate around the city.
2. The former European central business districts (CBDs) of Bombay and Accra were at once de-Europeanized and nationalized, politically and economically. The national government took over administrative and military functions in the area. New large domestic companies favored a location in this area, leading to a steadily growing corporate density and a large majority of domestically controlled companies. In addition, the area was nationalized in a symbolic sense with the location of the newly established central banks, stock market (in Mumbai), and state-controlled companies.
3. The former native CBD became increasingly characterized by small-scale businesses, as larger companies moved to the emerging national CBD (the former European CBD). As a result of massive rural–urban migration in the postindependence years, the density and congestion in this area kept increasing. In the absence of the kinds of strict zoning policies present in the colonial era and as a result of the spillover of excessive growth in the former native town, the boundary between the former native CBD and the former European CBD became increasingly blurred.
4. The end of colonial segregationist policies, in combination with the rise of a national entrepreneurial middle class, meant that previously demarcated residential foreign spaces became diluted. Upscale residential neighborhoods that had been inaccessible to most natives were rapidly "nationalized."

By the early 1980s, on the eve of major economic reforms in Ghana and India, the economic geographies of Accra and Bombay showed a relatively modest foreign presence. More specifically, the foreign presence was not a salient spatial feature of the urban landscape. Compared to the colonial period, the spatial patterns were less punctuated and more diffuse, and boundaries between different areas had become increasingly blurred. This, then, was the situation in Accra and Bombay when liberalization policies opened the gates to the forces of globalization.

THE GLOBAL ERA: LIBERALIZATION AND THE INFLUX OF FOREIGN COMPANIES

In Accra and Bombay, the single most important increase in foreign corporate activity in the twentieth century occurred in direct relation to liberalization policies initiated in the 1980s. With the onset of liberalization in the mid-1980s (1983 in Ghana; 1985 in India), the arrival of new foreign companies started to accelerate to unprecedented levels. Liberalization and deregulation encompassed a range of measures and new types of legislation, the most important of which – for our purpose – related to foreign corporate ownership. In India, such laws have been relaxed several times since 1985, the reforms of 1991 being the most significant. Presently, foreign equity shares of 51 per cent are allowed on a routine basis in most economic sectors, and an increasingly lenient Indian government grants larger shares – up to 100 per cent in a range of industries. In Ghana, 100 per cent foreign equity is permitted in all sectors. There, the eventual liberalization of the mining and financial sector facilitated the recent influx of foreign companies. The reforms have had considerable consequences. In Mumbai, more than half of all foreign companies that are currently active were established after 1985; more than a third entered the city after 1991. In Accra, over 80 per cent of all foreign companies currently active were established since the initiation of reforms in 1983 (Figure 1).

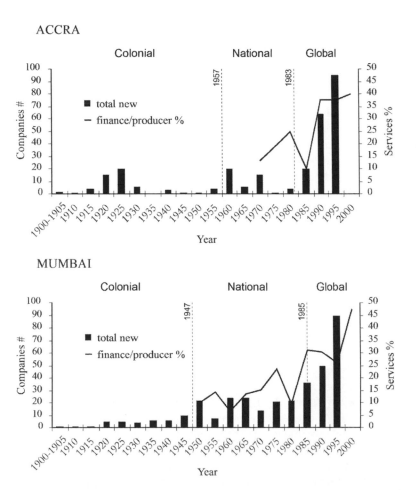

Figure 1 The number of newly established foreign companies and the share of companies in financial and producer services in Acra and Mumbai in the twentieth century.

New foreign companies in both cities are increasingly concentrated in finance and producer services. Finance companies comprise banks and investment brokers, while producer services include businesses in consulting, real estate, insurance, accounting, advertising, courier services, and so on. Of the foreign companies established in Mumbai after 1985, 32 per cent are in finance or producer services, as compared to 23 per cent of all domestic companies established after 1985. In Accra, of all foreign companies established since 1983, 16 per cent are in finance or producer services, as compared to 9 per cent of all domestic companies established since 1983. It is clear that trends of economic restructuring in both cities entail a decline in the primary and secondary sectors and growth of the services sector, with an emphasis on producer services and finance. As a consequence of liberalization policies, both Accra's and Mumbai's present integrations in the global economy are without historical precedent. Never before has foreign corporate activity been at these levels, nor have the global connections been as widespread.

THE CORPORATE GEOGRAPHIES OF ACCRA AND MUMBAI

Our analysis focuses on the metropolitan areas of Greater Accra and Greater Mumbai. Our study uses planning areas (Accra) and postal code areas (Mumbai) as the spatial units of analysis. In areas where a consistent spatial clustering of values existed among several contiguous areas, we decided to group them together into larger, more or less homogenous business districts.

Greater Accra consists of the "city" of Accra, the suburbs along the major motorways, and the Tema and Ga districts. The highways and the Ring Road in Accra are the key elements of its economic geography; motor transportation (taxis, buses and "tro-tros," or bush taxis) is the only way to commute within the city. The port of Tema has been developed since 1962, when Accra Harbor was abandoned in favor of Tema's deeper and more sheltered harbor. Since then, an industrial area, warehouse and storage facilities, and extensive housing for workers have been developed. The area of the present "city" covers a much larger area than that of old colonial Accra.

Figure 2 shows the high concentration of business activity inside Accra's Ring Road. We can identify a few areas of concentration.

First, Ussher Town, the traditional CBD, corresponds rather closely to the old European Town in colonial times. It is the most densely concentrated area in terms of corporate activities, containing many high-rise buildings that are in close proximity to the ministries. It contains the second highest concentration of Ghanaian-controlled companies of any business district in Accra and a small number of long-established foreign companies. Second, the area labeled "Central Accra" includes Adabraka, Tudu, and Asylum Down. This area corresponds to the old Native Town from the colonial era. It contains Makola Market, the largest market and most crowded commercial area in the city and the foci of most trips within it. The bustling market spills over onto the walkways and roadways, leading to acute congestion. The area has a mix of corporate and residential functions, a visible presence of firms of Lebanese and Syrian descent, and a maze of side streets and back alleys. This business district has the largest share of Accra's domestic companies, though they are generally of small size. The number of foreign companies in this district is not great. The newest business district in Accra is an area that stretches from Osu along Cantonments Road to the Ring Road. Corporate activities are concentrated in a ribbon development along the main thoroughfares. The area has modern low-rise buildings with off-street parking and is very different from the older parts of the city.

Greater Mumbai comprises the peninsula bound by the Arabian Sea to the west, Thane Creek to the east, and Vasai Creek and Ulhas River to the north. It is connected to the mainland to the north and northwest. The area's dimensions are about 50 kilometers from north to south and an average of 10 kilometers from west to east. The current population of Greater Mumbai is around 12 million people. The average population density is an overwhelming 24,000 people per square kilometer. The Mumbai Metropolitan region forms part of the larger economic region in Western Ma-harashtra that is sometimes referred to as the Mumbai–Pune corridor, one of the most industrialized, most urbanized, and most productive

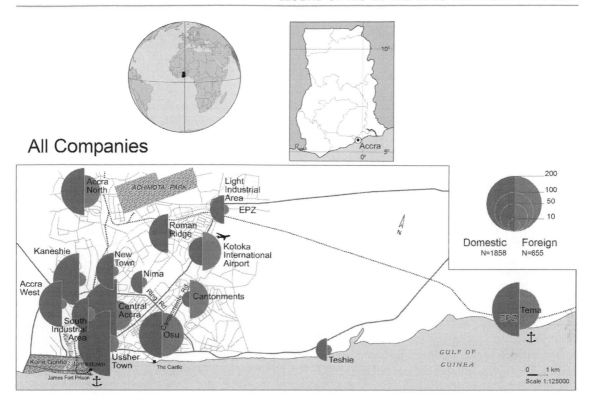

Figure 2 Business activity in Accra

regions of India. Greater Mumbai itself consists of the "city," located south of Mahim Creek, and the suburbs, which have gradually expanded northwards alongside the tracks of the Western and Central Railways and onto the mainland. The railways are a key element of Mumbai's economic geography and are said to account for five million passenger rides every day of the week. The geographical constraints of the peninsula put a premium on space and have historically influenced land values and land use in the city. A steep gradient in land values exists from the south to the north. In the mid-1990s, the influx of foreign corporations contributed to an extreme escalation of land values, making Mumbai, for a time, the most expensive city in the world. The area of the present "city" corresponds with that of colonial Bombay in past times, but it includes some additional land in the southwest that was reclaimed from the sea in the 1950s. The "city" contains the port, the railway terminals, governmental functions, and the region's main business districts. The northern part of the "city" contains the old textile industries (most of

which are derelict by now) centered in Lower Parel, and on its very northern edge what is known as the world's largest slum, Dharavi.

Figure 3 presents the general foreign and domestic corporate geographies of Mumbai, showing the generally high concentration of business activity in South Mumbai.

The Fort area corresponds rather closely to the old European Town in colonial times. It has the highest concentration of Indian-controlled companies of any business district in Mumbai, as well as a sizeable share of foreign companies. The area labeled on the map as Kalbadevi includes Khetwadi, Mandvi, Central Bombay, and Masjid and corresponds roughly to the old Native Town from colonial times. This part of the city continues to exude a distinctly bazaar atmosphere. It is the most crowded area in Mumbai, with over 100,000 people per square kilometer (the highest anywhere in India in neighborhoods this size). It contains a mix of corporate and residential functions and various ethnic neighborhoods. Much of the trade and retailing takes place in the open air in front of the

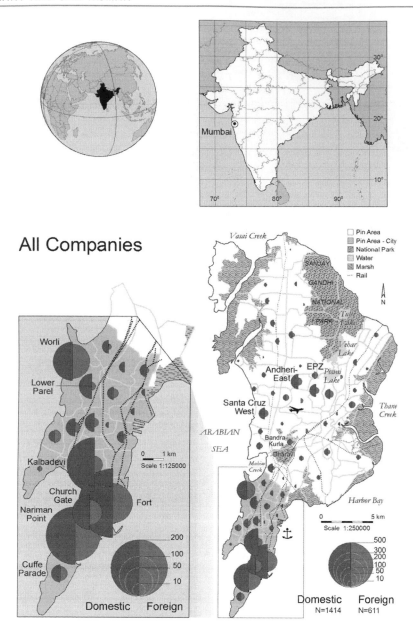

Figure 3 Foreign and domestic corporate geographies of Mumbai

small shops on the narrow streets and sidewalks. This business district has a large share of Mumbai's domestic companies, though they are generally of small size. There are few foreign companies in this district.

The newest – and arguably most prominent – business district of Mumbai is Nariman Point. This land was reclaimed from the Arabian Sea in the 1950s. It is a small sliver of land located west of

the Maidans and across from the Fort and south of Churchgate, measuring slightly less than a square kilometer (about one-fifth the size of the Fort area as defined above). Land values here are higher than anywhere else. Most of the construction dates from the 1960s and 1970s. The modern high-rise and broad avenues of Nariman Point stand in sharp contrast with the colonial architecture of the Fort and the bazaar atmosphere of

Kalbadevi. As Figure 3 shows, Nariman Point ranks second in terms of its importance for domestic companies, and it is by far the preferred location for foreign companies.

THE FASTEST-GROWING CORPORATE SECTOR: FINANCE AND PRODUCER SERVICES

Finance and producer services are the fastest-growing sectors of the urban economy in both cities, particularly among foreign companies. Figures 4 and 5 show the locational patterns in Accra and Mumbai of companies in finance and producer services only.

In Accra, this sector involves companies specializing in communication, real estate, advertising, and consulting. In Mumbai, all of these are important as well, but banking and finance companies are the most prominent. In both cities, finance and producer services are among the most spatially concentrated economic sectors.

In Accra, domestic companies in finance and producer services are most concentrated in Ussher Town and Central Accra. The largest share of foreign companies, on the other hand, is found in the districts of Osu, Cantonments, and the area around the airport. Over time, foreign producer services have concentrated more and more in the airport area and have moved to suburban locations

along the main thoroughfares that connect the airport area and the central city. Producer services as a sector are in a formative stage in Accra. It is worth noting that the vast majority of producer-service companies are small operations, and that many of the business and communication centers that specifically target businesses also have to take on consumer clients in order to turn a profit. The financial-services sector is much smaller compared to producer services. Although some new foreign companies have been established in the financial sector, it is still dominated by older companies, many of them established during the colonial era.

Many of the new foreign companies in Accra appear to be highly cognizant of the physical and social business environments, and they value propinquity to other foreign companies. This is especially the case for finance and producer services, since other multinational operations comprise their clientele. Their locational choices are limited in Accra, however. The land market in Accra consists of a combination of two or more systems of land supply (indigenous, illegal, and free-market) with considerable bureaucratic controls. The relaxation of traditional property notions in parts of the city has resulted in the operation of a free market in property that is geographically differentiated. This has allowed considerable land speculation in the Osu, Cantonments, and airport areas, where land is more freely available as well

Finance and Producer Services

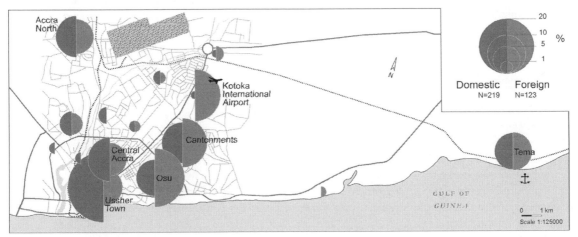

Figure 4 Finance and producer services in Accra

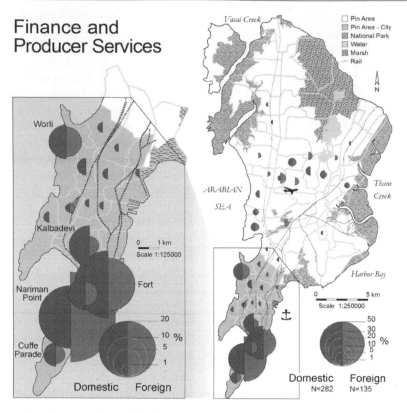

Figure 5 Finance and producer services in Mumbai

as favored. Land values in these areas are among the highest in the city. These trends have been reinforced by some national security land-use policies whereby property developers are prevented from developing the nearby vacant area adjacent to the Castle, so recently arrived foreign companies have located further away from the traditional central business district. Domestic companies are often more price-sensitive to land values, and they have tended to locate in less expensive areas such as Central and North Accra. However, domestic companies are also beginning to locate in greater numbers in areas of the city with concentrations of large multinational companies and foreign-producer service companies. Thus, although foreign and domestic companies have significantly different locational patterns, these differences have been decreasing in recent years.

In Mumbai, finance and producer services are strongly concentrated in the southern tip of the peninsula, more so than is any other sector of the urban economy. It should be remembered at this point that Mumbai is the unquestioned financial center of India, with large numbers of domestic and foreign banks, the largest stock exchange in the country, and a huge number of investment firms. The financial sector has attracted a large number of related producer services, especially accounting and consulting. The two main business areas are the Fort and Nariman Point, but the relative importance of these two areas for domestic and foreign companies is not the same: the Fort has the largest share of domestic companies, while Nariman Point houses an overwhelming share of foreign companies. As in Accra, the contrast is increasing over time.

Since 1986, Nariman Point has become the area of massive concentration of foreign companies in finance and producer services, while the appeal of the Fort has declined. Nariman Point's share of foreign companies in this sector has grown to twice its share of domestic companies, whereas the Fort's share of domestic companies has grown to three times its share of foreign companies in this sector. The area around Kalbadevi has about 10 per cent of domestic companies and only 1 per cent of foreign companies in finance or producer services. Overall, in Mumbai there has been only very modest suburbanization of activity in this sector. Figure 5 shows a slight increase of activity in the belt north of the airport. This shift has mostly involved companies in technological consulting and in the kinds of service activities that rely heavily on airport facilities (e.g., courier services). Many of the new foreign companies in Mumbai appear highly sensitive to the physical (and social) business environment, and they value proximity to other foreign companies. The latter is especially important in finance and producer services, since other multinational companies make up a significant share of their clientele. Often, they choose to work with large international (but Western) real-estate companies who "understand" their needs in their search for the appropriate location. Once the different characters of Nariman Point, the Fort, and Kalbadevi started to take shape, this accelerated the selectivity on part of newly arriving companies. This explains the sharpening distinction between the Fort and Nariman Point.

THE RISE OF THE GLOBAL CBD

The foregoing analysis suggests the emergence of distinct business districts in the two cities. In both cities, we can identify three CBDs that represent the main clusters of corporate command and control. Within each city, these CBDs have different historical origins, are different in terms of the nature of economic activity, and – most importantly – seem distinctly different in terms of their importance for foreign and domestic companies.

Central Accra in Accra and the Kalbadevi area in Mumbai are comparable, as they overlap in large part with the old Native Town from colonial times. A small foreign company presence and a mixture of residential functions with small trade businesses – especially crafts – and retailing characterize these areas. Ussher Town in Accra and the Fort area in Mumbai are comparable, as they correspond fairly closely with the European Town from colonial times, areas that were "nationalized" after independence. Their residential function is limited; they house many large domestic companies in the service sector and feature a substantial number of foreign companies. Finally,

Osu/Cantonments in Accra and Nariman Point in Mumbai are similar, as they are relatively newly developed business areas. They are dominated by finance or producer services, and they contain the largest share of foreign companies in the two cities.

These business districts are differentially linked to the global economy. This is not only due to the presence or absence of foreign companies: domestic companies, too, are differentially linked to the global economy. In the second survey, we found that the location of domestically controlled companies within the city serves as a predictor of the geographic reach of a company and of its global connectedness. Randomly selected domestic companies in the three CBDs were questioned about their global connections in terms of phone calls, postal mail, business travel, and e-mail. The results indicated that internationally oriented companies have a location preference that is different from more nationally or locally oriented companies. The domestic companies that are most linked to the global economy tend to locate in business districts with a disproportionately large share of foreign companies. Thus, domestic companies in Osu/Cantonments are more "global" than companies in Ussher Town. The latter, in turn, are more "global" than companies in Central Accra. In Mumbai, domestic companies in Nariman Point are more "global" than companies in Fort and much more "global" than companies in the area around Kalbadevi (Figure 6).

CONCLUSIONS

This study's main findings may be summarized in three points. First, our empirical research shows the reorganization of the space-economy of the two cities. On the ground, these cities have changed structurally in the past twenty years or so. Second, this change has not been random or accidental, but is tied very closely to the new roles of the two cities in the global political economy. Third, the study furthers theoretical understanding of global–local linkages. It does so especially by qualifying *place* (Mumbai, Accra) in historical and geographical context.

The existing literature has not yet caught up with these new realities. The study shows that the

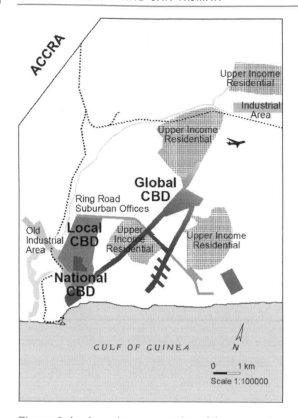
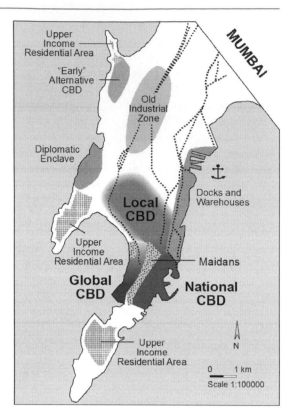

Figure 6 A schematic representation of the economic geographies of Accra and Mumbai during the global phase

economic geography of present-day cities in the less-developed world is fundamentally different from that of "Third World cities" *and* from globalizing cities in the West. Old notions of Third World cities do not appreciate the profound effects of relatively recent liberalization schemes and ensuing processes of globalization. Many cities in the less-developed world are changing rapidly to accommodate or cope with the influx of foreign investments. Moreover, the present spatial restructuring of cities in the less-developed world is fundamentally different from globalizing cities in the West. We found that, like cities in the West, Mumbai and Accra have experienced rapid growth of finance and producer services, but these two cities have also witnessed processes of corporate segregation based on domestic and foreign ownership that are not seen in the West. We view this as a reflection, in general terms, of the spatially fragmented integration of societies in the less-developed world at large in the global political economy.

We think that the experiences of Mumbai and Accra are shared by other cities in the less-developed world (especially gateway cities), each in their own context. In other words, we think that many other cities in the less-developed world with a comparable past and with currently high exposure to the global economy are likely to experience more or less similar spatial processes. Cities that come to mind include Chennai, Jakarta, Mombasa, and Lagos. It is hoped that the foregoing analysis will stimulate more interest in cities in the less-developed world. Research and theory on global urbanism in these cities is at a preliminary stage of exploration, given the rapid changes that are taking place. More is needed to assess the general applicability of our findings. We also need to go beyond these corporate patterns and relate them to evolving *human* geographies of cities in the less-developed world in the context of globalization. We hope that we have provided a basis for such future endeavors.

NOTE

1 We purposely employ the term "cities in the less-developed world." Alternative terms such as "Third-World city" and "postcolonial city" are more problematic. The notion of "Third-World city" has been debunked in the development studies literatures and elsewhere and hardly needs further discussion. The term "postcolonial city" is trendy but remains ill defined. More pertinently, the concept "postcolonial," as a historical category, does not acknowledge the significance of our distinction in this article between the national and global phases in urban evolution.

F
O
U
R

28

"São Paulo: Outsourcing and Downgrading of Labor in a Globalizing City"

Simone Buechler

Editors' introduction

In our final reading in Part Four, Simone Buechler develops an ethnographic analysis of working conditions in the squatter settlements of São Paulo, Brazil. A former PhD student of Saskia Sassen (see Readings 9 and 10) at Columbia University, Buechler is an expert on urban development in Brazil, where she spent several years engaged in fieldwork. She has worked as a consultant for the National Academy of Sciences on the informal sector in São Paulo and at the United Nations Development Fund for Women (UNIFEM). Buechler is currently Assistant Professor of Metropolitan Studies at New York University.

Buechler's contribution explores the ways in which economic globalization is transforming urban labor markets in São Paulo. Like Ross and Trachte (Reading 12), Buechler is concerned to investigate the effects of global city formation upon working conditions within low-income communities. As Buechler demonstrates, São Paulo is a significant global city "beyond the West" due to its role as a headquarters location for major transnational corporations operating in Latin America. However, Buechler focuses primarily upon São Paulo's industrial sector, which has been significantly restructured since the mid-1980s in the face of intensifying international competition. Buechler examines the massive expansion of informal labor conditions and the generalized downgrading of working conditions in São Paulo on a number of different levels. First, she summarizes the macroeconomic and spatial shifts that have led to the downgrading of urban labor markets in São Paulo. Second, she builds upon her interviews with workers in the squatter settlements of São Paulo in order to characterize the everyday experience of engaging in outsourced work. Buechler documents this experience in graphic terms. Her sweatshop interviewees describe their extremely difficult working conditions with reference to domineering and abusive bosses; a brutal work routine; long working hours; inadequate pay; chronic job insecurity; and a lack of employment benefits. Buechler's interview data thus provide us with a disturbingly direct impression of globalization "on the ground," a perspective that usefully complements the more theoretical and structural analyses presented in other contributions to Part Four.

INTRODUCTION

During my 1998 interview with participants in a union adult education program in the Metropolitan Region of São Paulo, "Nelba" exclaimed:

I entered into a firm in 1979 and stayed 10 years and today . . . you do not stay a year, that is, if you enter [at all]. These days everything is contracted [out]. It is by contract through an agency – two months, three and goodbye! You have to leave.[1]

"Maria José" added,

And today they are crushing us . . . The boss is crushing many of the employees, because there are many people who are unemployed. I am suffering a lot in my work . . . I am persisting because I have a small son . . . I am going through humiliation after humiliation and staying quiet. I am telling you the truth. It is because I need [the job]. It is not that I do not have skills. I know how to do many things. But [the situation] is so bad that if one goes out to sell clothing, one does not sell anymore and when one does sell, then one is not paid. They do not have money to pay. If you make snacks [people do not pay you]. [I know], because I already lived doing that. I made snacks to sell [on the street]. I already had a bar. I already did many things. I already worked eight years as a self-employed worker, but today it is impossible . . . The other day my boss said she would hit the employees in the face.

The response of a garment union leader to Maria José's story was:

It is true. It is difficult. There is unemployment. There is change. But if she lowers her head and accepts everything in the way that she is accepting it, she will return to being a slave. And before, black people like myself were slaves. Today, the slaves are all those who depend upon employment.

An investment banker located in New York, but working with Brazil, reacted to the complaints of my informants in the slums by exclaiming, "The party is over." It seemed that he was also worrying whether the party was over for him and his investors.

Nelba, Maria José and the union president live and work in São Paulo, Brazil, the fourth most populated metropolitan region in the world, after Tokyo, Mumbai and Mexico City, encompassing an area of over 8,000 square kilometers with approximately 17.8 million inhabitants in 2000 (IBGE 2001). It is a city of stark contrasts. Avenida Paulista, a large modern avenue, boasts the latest communication towers, international financial institutions such as Citibank and BankBoston, and McDonald's on almost every block. And new parts of Avenidas Faria Lima and Berrini sprout new office buildings for the overflow. Meanwhile, one-third of the population of the municipality of Diadema, where Nelba and Maria José reside, live in squatter settlements (called *favelas* in Portuguese). Stores such as Lacoste, Yachting Gear and Au Bon Pain abound to service the global elites. Well-equipped residences surrounded by high gates for the businessmen, with names such as Park Avenue, are located a block away from the stores and office buildings. The citadel-type apartment buildings allow the global elites to more easily separate themselves from the poor majority. Communities such as "Favela Leste," one of the squatter settlements in which I did my fieldwork, house those who service, clean and guard those buildings and produce material goods for the multinationals. Favela Leste is located in Diadema, one of the hardest hit municipalities in terms of industrial closings and loss of industrial jobs in part a result of the opening of the market, considered a crucial step for Brazil's entrance into the global economy. These spaces are where, as Sassen (1998) argues, the work takes place of producing and reproducing the organization and management of a global production system and a global marketplace for finance. They are the interface between the local and global.

We generally take for granted that investment bankers and other corporate elites are global actors. However, most of us are less likely to consider working people such as Nelba and Maria José as global actors, simply because they do not have the power that we associate with those who are directly involved in the management of the global economy. Against such assumptions, I suggest that São Paulo's financial districts should not be seen as the only parts of the city that are connected to the global economy. Indeed, even though they may have little direct contact with individuals in other countries, the populations of low-income communities also form an integral part of globalizing cities. They are enmeshed in the global economy because they are impacted by it and because they sustain it through their work – for example, by producing cheap goods for the low-income service workers. At the same time, they also may be excluded spatially from the centers of power (although they may spend their days and nights working in exclusive apartments right off Avenida Paulista); often politically (although the current president was once a factory worker); and economically in the sense that they are excluded from the profits and wealth that they produce for the owners of capital and for the global economy. They are also often excluded from the formal labor market and the social benefits attached to it.

This chapter focuses on the deteriorating employment situation for low-income women workers in São Paulo. The degradation of work can be seen within both the so-called informal and formal sectors. Employment often is unstable, poorly remunerated, and without benefits. Increasingly, production has been outsourced to smaller firms, cooperatives in name only, sweatshops and homeworkers. With the rise of competition from other unemployed salaried workers and unemployed customers, small-scale commerce has also become more precarious.

The field data for this reading were gathered in the Metropolitan Region of São Paulo during research in 1996, 1998, 2000 and 2003. The research was conducted primarily in two squatter settlements and one other low-income neighborhood with plots of land illegally subdivided by real estate speculators. In addition to interviewing low-income women in these communities, I interviewed students in union-organized adult primary education classes for unemployed people, street vendors, Bolivian sweatshop workers and owners, union activists, managers in industries operating close to the communities, local and national government officials, and investment bankers working in New York. Many of these voices, particularly of women workers, are missing in much of the scholarship on economic globalization and the city.

GLOBALIZATION AND CITIES BEYOND THE WEST

São Paulo has been labeled by some global city theorists as a global city of the second order (Friedman 1986), a Beta global city (Taylor 2003), a city with global city functions or in the mid-range of the global hierarchy (Sassen 2002), a world city beyond the west (Gugler 2004) or as an emerging global city (Olds and Yeung 2004). Such designations are based on the fact that São Paulo houses 6 of the 20 largest foreign-owned companies in Latin America, the headquarters of 36 out of 40 of the largest commercial foreign banks in the country and 63 per cent of the headquarters of the largest 100 foreign companies in Brazil (Schiffer 2002: 217). From 1993 to 2000, foreign direct investment (FDI) inflows have increased substantially from US$1,294 million in 1993 to US$32,779 million in 2000, but then, with increasing economic instability, decreased to US$16,566 million in 2002. Most of the increase can be attributed to the privatization process and foreign capital acquisitions of domestic companies and banks. FDI outflows from Brazil (investment by firms headquartered in Brazil flowing to other countries) also increased substantially from US$1.7 billion in 1997 to US$2.482 billion in 2002 with a large increase in flows to other Mercosur countries (UNCTAD 1999, 2003).

I prefer to use the phrase "globalizing city" for São Paulo in order to focus on globalization as a process with many dimensions. In contrast, the term global city has been used more narrowly to focus on a city's placement in a hierarchy defined primarily by certain types of financial and service firm networks. Such a narrow focus unnecessarily limits comparisons with other so-called global cities in both developed and developing countries. In particular, it fails to address the spread of neoliberal ideology through institutions like the International Monetary Fund and the pressure to borrow more capital, leading to indebtedness and to an exploitative labor market. The global city literature also often fails to take into consideration that São Paulo is the focal point for the Mercosur regional bloc, the Southern Cone regional trade agreement, that has resulted in its becoming a recipient of major national investment in regional infrastructure for transportation and digital networks. This characteristic of São Paulo differentiates it from many other globalizing cities. However, there are similar patterns in terms of the degradation of labor in both global and globalizing cities.

São Paulo's economy is undergoing a type of restructuring that many global cities in advanced industrial countries have experienced and that many globalizing cities are currently experiencing. This restructuring pattern includes the decline of the industrial sector and growth of the service sector, large-scale unemployment at least in the early stages of transformation, and the weakening of labor with the casualization of work relations. Like New York, London and Tokyo, industrial employment has decreased considerably over the years, but São Paulo continues to have an important industrial sector directly employing 15.38 per cent in the Metropolitan Region of São Paulo (MRSP) in 2003 (Seade/Dieese PED 1985–2003). Since the 1950s, São Paulo's industrial sector has been strongly connected to international capital, but it was not until the beginning of the 1990s that President Collor opened Brazil's market (Alves 2004). However, unlike many of the other Third World cities that play a role in the global economy, São Paulo's role is not limited to acting as a cheap producer for multinational companies because of the large domestic market. From 1985 to 2003, a period when the population of economically active individuals increased by 40 per cent or by over 3.87 million people, there was a 23 per cent decrease (338,916) in the number of workers employed in the industrial sector (Seade/Dieese PED 1985–2003). In addition, within the industrial sector the percentage of unregistered salaried workers and self-employed working for themselves and for firms almost doubled in 10 years from 1988 to 1998 in the Municipality of São Paulo. Including all sectors, the informal sector grew from approximately 33 per cent of the employed population in 1988 to 42 per cent in 1998 (and almost 50 per cent if we include all domestic workers). This trend continued in the five-year period from 1998 to 2003. This increase was primarily in unregistered salaried workers in larger firms of six or more employees (Seade/Dieese PED 1985–2003). Many of the industrial firms that have stayed in São Paulo have engaged in subcontracting, lowering wages and benefits through hiring their workers illegally, and hiring temporary workers. As the numbers of unregistered workers have soared,

so has the unemployment rate. The average annual rate of open and hidden unemployment increased from 18.2 per cent in 1998 to 19.3 per cent in 1999 and 19.9 per cent in 2003 in the MRSP.

The degradation of work in São Paulo is associated directly and indirectly with economic globalization. The opening of the markets creating the environment for an increase in foreign trade and investment and increasing global competition has helped produce an unstable labor market. Plants have closed because of global competition leading to unemployment. Unemployment has led to the increased vulnerability of labor to exploitation. Global competition has put downward pressure on prices leading at least to the perceived need and justification for outsourcing to reduce costs. In turn, the outsourcing of production can lead to sweatshop conditions with no benefits. The search for an even larger market has led to precariously employing residents of poorer neighborhoods as unregistered vendors of cosmetics, yogurt, and clothing in their communities. Precarization is not merely an effect of advanced capitalism, but an integral part of it. However, economic globalization cannot be seen as a completely uncontrollable external force. Rather, global forces are actively embraced, resisted, or transformed by national and local forces and actors.

THE DEGRADATION OF SALARIED WORK – OUTSOURCING PRODUCTION

There has been a resurgence of homework as a way to compete in the global market and gain higher profits. Although not yet a large part of the labor market, new kinds of homework seem to be surfacing such as threading strings through clothing tags, assembling parts of lipstick holders and pasting handles on shopping bags. The homework in all three communities studied had in common low remuneration per piece (or per thousand), instability and no benefits. In order to make any money, workers had to self-exploit. The factories transfer the insecurities of the market to the workers.

Although homeworkers often saw their work as exploitative, many of the women were happy for even these limited opportunities to earn money. "Márcia" and her daughters put 15,000 lipstick holders together a day, often working until midnight, but making only R$15 or approximately US$13 in 1998. A manager at the cosmetic packaging company, one of the suppliers for Avon, who outsourced part of its production to a small family firm which, in turn, outsourced production to women including Márcia, described the practice of outsourcing as the company's "social role" and as a way to deal with fluctuating demand.

"Josefa," an unemployed factory worker, was a homeworker in "Favela Sul" who for three months had been folding and/or gluing tags, putting strings through holes in the tags and tying the strings for beauty products, clothing and lingerie. She had a 15-month-old daughter at the time of the interview on 27 April 1998. With only an eighth grade education, but with hopes to continue when her child is older, Josefa struggles to earn a living. Her husband only occasionally brought in some income. She earned on average R$2 per thousand tags and less if they only had to be folded. Working more than eight hours, she was able to do approximately 1,000 to 1,500 tags a day depending upon her housework and the type. When her sister helped her, they could do 2,500 tickets together. She could therefore earn on average only R$10 a week or R$40 a month (or the equivalent of approximately US$35 before the steep devaluation of the currency a few months later). Josefa was on the fourth rung of a long production chain. The principal clients were multinationals such as Levi Strauss and Pierre Cardin. These companies hired graphics companies to produce clothing tags. The graphics companies outsourced part of the production to a woman, "Cica," who lived in a low-income neighborhood next door to Favela Sul. In turn, Cica outsourced part of the work to women like Josefa. Although Cica also put the strings through the tags herself, she usually had around 15 women working for her. The graphics firm paid a set rate per piece, but Cica took part of the money to compensate for her additional work retrieving the tags and string from the factory, counting the tags, putting them together, and returning them to the factory.

By 2000, most graphic companies had started using machines to do the work that Josefa was engaged in, and the particular graphics company that Cica and Josefa had been working for had closed. In 2001, Josefa moved to the interior of the state where her husband had found work.

Factories have left the Metropolitan Region of São Paulo to avoid strong labor unions and search for cheaper labor.

There is a long association between homework and the garment industry, but according to the president of one of the unions, in the last few years the use of homeworkers has accelerated in the industry. The trade union leader argued that there would be 160,000 instead of the 83,000 workers registered if all the workers in the sector were registered. "Where are the rest?" she queried.

[They are producing] in the manner I told you [working as outsourced workers]. And they are not only the Bolivians [sweatshop workers]. They include [workers for] large enterprises; let me make that clear: with a manager or courier carrying the sewing machine to their homes, so that they can work in that way.

In the center of São Paulo, unregistered Bolivian workers labor in garment sweatshops owned primarily by Koreans and more recently by other Bolivians. They work for long hours from 8:00 a.m to midnight during the week and half a day on Saturdays, earning often only little more than the minimum wage, but receiving lodging. Talking about a visit to a sweatshop, the president of the garment union commented:

[O]ne exploits the other. It is a prison, understand? There is a place where the food is locked up and so is your passport. It is guarded . . . It is a lie that they can leave . . . It isn't the truth.

In 1993, there were reported to be 2,500 Korean commercial establishments in São Paulo, of which 90 per cent were garment workshops responsible for the production of a minimum of 7 million pieces per month. Fetching an average price of five dollars, they represented a movement of $30 million per month (*O Estado de São Paulo*, 9 August 1993). The Koreans went directly to Brazil in the 1960s as part of a group emigration process, and at least individually in the beginning of the 1970s, but then started to go to Bolivia especially in the 1970s due to changes in immigration policies. With the economic crisis in Bolivia, especially in the 1980s, they moved to Brazil and started contracting Bolivian workers directly or indirectly. In turn, some of the Bolivian workers have been able to save enough after a few years to open up their own workshops and hire their compatriots to work for them. Here we have a complex interconnection with economic globalization: the perceived need by both Korean and Bolivian workshop owners to produce garments in sweatshops because of global competition and various types of transnational migration as a result of economic crises in part stemming from the debt crisis and structural adjustment policies in the case of Bolivia and war in the case of South Korea in the 1960s.

Outsourcing of work is not limited to the production process, but extends to the retail trade. Employing women to sell cosmetics, lingerie and other clothing, yoghurt, and other items from door to door or out of their homes both cuts labor costs in the area of sales and opens new markets in peripheral communities. These women are forced to assume the risk when goods are ordered and not purchased, and at the same time denied the right to be registered as formal workers.

THE PRECARIZATION OF SELF-EMPLOYMENT

Occupations in the so-called informal sector that perhaps are only indirectly linked to "formal" firms, those in petty commerce such as street vending or the ownership of stores in the slums have also become precarized because of the high wage unemployment that has led to the lack of salaried income coming into the household to support the enterprise, the inability of clients to pay, growing competition among workers in the neighborhoods and streets, and detrimental government policies.

As González de la Rocha (1997) argues for Mexico and is also true for São Paulo, there has been an erosion not only of the ability of the poor to find wage employment, but also in "their capacity to participate in 'alternative' occupations and self-provisioning activities in a kind of perverse process of cumulative disadvantages." From the 1950s until the early 1990s there was both a diversity of income sources and of occupations in households. As was the case in urban households in Mexico, it seems that low-income households in São Paulo also relied on at least one person

gaining wages in the formal labor market. This appeared to be true as late as 1996 when I began my fieldwork, when each family at least attempted to have one person in a "formal" job to receive a basket of basic food items negotiated as an integral part of compensation by some of the unions, bring in a steady income that could be used to support other activities as well, and gain access to health insurance. Today such a source of a steady income from a wage earner who can provide the necessary income to provide the startup capital and the continued support of the enterprise when needed is largely absent. Initial capital for micro enterprises and commerce has mainly come from an unemployment insurance fund called "Fundo de Garantia" which all registered waged workers contribute to while working. However, after this money runs out there often is little money left to "grow" the enterprise and support it during times of low sales.

The case of "Izabel" who runs a local store and bar is typical. She told me that she is unable to buy merchandise for her store/bar because her husband has not been able to earn much lately in construction and because there are many who are indebted to her. She is therefore unable to make a real profit. Similarly, "Sara" was not capable of buying material to make the bags that she sold in the market because her husband was unemployed. During this time all the money she earned from this and other small activities went into family expenses and therefore she could not use any of the earned income to put back into the enterprise even to keep the enterprise going.

As male and female factory workers lose their jobs, especially low-income women (unemployed factory workers or wives of unemployed factory workers), are opening up bars, bazaars, and grocery stores as well as providing services such as hair dressing, day care, catering, and dress-making out of their homes in the peripheral neighborhoods and squatter settlements. For example, in Favela Leste, with 3,000 families, there are approximately 66 bars and bars cum grocery stores. In 1996, there were only two grocery stores and two bars in Favela Sul, but by 1998 that number had increased to some seven grocery stores and bars selling food, one bakery, three hot dog stands, one store for ice cream and school supplies, and one used clothing store which also sells ice cream.

Discussing the growing competition, Lúcia, a grocery store owner, said:

> Before there was only me here . . . [N]ow, there are a lot. But it is like I said, everyone in the world wants to earn at your side, right? . . . Most of things that I sell, the others have also.

The number of street vendors has also increased dramatically. The estimates of how many street vendors exist in the Municipality of São Paulo in 1998 ranged from 20,000 to 70,000. As late as 1996, street vending was a fairly lucrative activity, but in 1998, with increasing competition and the new policy to rid certain areas of the city of street vendors, it had become more precarious. One of the vendors in Largo de Concordia responded to the complaint that street vendors made the city ugly and not modern:

> What about the poor children . . . that you pass sleeping on the sidewalk with everything (they have) at their side . . . Does none of this make the city ugly? Is only the street vendor ugly?

However, the harassment of street vendors because they are seen as nuisances and not "modern" is an old phenomenon, documented as early as the 1700s. I would therefore argue that while there has been a recent precarization of the work of the street vendors, it is a cyclical problem whereby when the rate of unemployment is particularly high, the measures against the street vendors take on a certain severity.

CONCLUSION

Globalizing cities do not consist of isolated islands of economic activity that are enmeshed in the global system within an unchanging sea of excluded workers. Josefa, the homeworker, is ultimately connected with Levi Strauss and Pierre Cardin, as she slaves away at putting strings through clothing tags. Márcia, the homeworker who puts together lipstick holders, is ultimately connected to Avon, another multinational corporation. And Nelba and Maria José are unemployed because of industrial closings due to strategies of capital seeking the cheapest labor. Korean and Bolivian garment sweatshop owners compete with Chinese imports, making it more advantageous for them to hire cheap Bolivian workers. Street vendors are both the unemployed and the conveyors for imported cheap goods. In conclusion, then, street vending, homework, and work in a multinational or international investment bank should not be viewed as activities that belong to different urban economic spheres, but as enmeshed, multicultural components of a single global economy.

NOTE

1   The names of my informants and communities studied have been changed in order to ensure anonymity.

REFERENCES FROM THE READING

Alves, M.H.M. (2004) São Paulo: the political and socioeconomic transformations wrought by the New Labor Movement in the city and beyond. In J. Gugler (ed.) *World Cities beyond the West*. Cambridge: Cambridge University Press, 299–327.

O Estado de São Paulo (1993) Select articles from August 9.

Friedmann, J. (1986) The world city hypothesis, *Development and Change*, 17, 1, 69–83.

González de la Rocha, M. (1997) The erosion of a survival model. Unpublished paper, Workshop Gender, Poverty and Well-being, United Nations Research Institute for Social Development, Trivandrum, Kerala, November 24–27, 1997.

Gugler, J. (2004) Introduction. In J. Gugler (ed.) *World Cities beyond the West*. Cambridge: Cambridge University Press, 1–24.

IBGE (Instituto Brasileiro de Geografia e Estatística) (2001) Census 2000 results.

Olds, K. and Yeung, H.W-C. (2004) Pathways to global city formation, *Review of International Political Economy*, 11, 3, 489–521.

Sassen, S. (1998) *Globalization and its Discontents*. New York: The New Press.

Sassen, S. (2002) Introduction: locating cities on global circuits. In S. Sassen (ed.) *Global Networks, Linked Cities*. New York and London: Routledge, 1–36.

Schiffer, S.R. (2002) São Paulo: articulating a cross-border region. In S. Sassen (ed.) *Global Networks, Linked Cities*. New York and London: Routledge, 209–236.

Seade/Dieese PED (Fundação Sistema Estadual de Análise de Dados – Departamento Intersindical de Estatísticas e Estudos Socioeconômicos) (1985–2003) *Pesquisa de Emprego e Desemprego na Região Metropolitana de SP-PED*.

Taylor, P.J. (2003) *World City Network: A Global Urban Analysis*. New York: Routledge.

UNCTAD (1999, 2003) *World Investment Report*. New York and Geneva: United Nations.

PART FIVE

Contested cities: state restructuring, local politics and civil society

Plate 22 Frankfurt (Roger Keil)

INTRODUCTION TO PART FIVE

In their original formulation, Friedmann and Wolff recognized the highly political nature of global city formation. In their contribution to this Reader (see Reading 6), they repeatedly describe the inherent contradictions of the world city as a source of conflict and struggle. However, neither these authors nor most other early contributors to world cities research engaged systematically with the politics of the global city. Although Friedmann and Wolff insisted on the strategic importance of sociopolitical struggles within global city-regions, the question of how political institutions structure, and are in turn structured by, processes of global city formation, was left largely open during the decade following the publication of their classic article in 1982.

In one particularly vivid passage, Friedmann and Wolff (Reading 6) evoke the image of the "citadel" and the "ghetto" as a metaphor for describing emergent patterns of sociospatial inequality in the global city. This metaphor powerfully illustrates the internal cleavages within the world city social fabric, which are said to be expressed spatially in the opposition between the gleaming office towers of the new downtowns and the impoverished residential quarters and production centers of the increasingly internationalized urban proletariat. In this view, the ghetto's inhabitants are isolated like a "virus" by the corporate and political elites (Friedmann and Wolff 1982: 325); this marginalized urban space is then in turn subdivided into ethnic and racial enclaves, which seem entirely separate from both each other and the power structure of the "citadel." Concomitantly, in this conception, the political sphere of the world city appears to be differentiated clearly into two different entities – one in which political and economic control capacities are concentrated; and one that is dominated by a politics of social reproduction and everyday survival. Meanwhile, national and local state institutions are seen as being unable to manage the proliferating regulatory problems associated with global city formation. Traditional welfarist policies are increasingly abandoned in favor of diverse boosterist strategies intended to attract and maintain investment by transnational corporations. The political conflicts which subsequently ensue within global cities are characterized, for Friedmann and Wolff, by a local/global dialectic that pits globally mobile corporations ("economic space") against diverse, territorially circumscribed interests associated with "life space." This intense struggle between life space and economic space assumes a concrete form in proliferating battles over livelihood, diversity and land use planning throughout the global city region (see Newman and Thornley 2005).

Friedmann and Wolff's initial formulation was bold, provocative and challenging to prevalent understandings of urban politics, but it ultimately proved unsatisfactory to later generations of researchers who proceeded to qualify and fine-tune this key aspect of the world city hypothesis. For, as Friedmann himself pointed out in subsequent publications, the emergence of a new global–local relationship could not be grasped adequately through metaphors of confrontation between two seemingly static poles (life space/economic space; local/global; ghetto/citadel); it needed to be examined, instead, as a process of mutual definition and as a result of material power relations (Beauregard 1995; Swyngedouw 1997). Moreover, not only did locally embedded communities take up the struggle against global capital, but also they transformed themselves, both politically and institutionally, in and through such struggles. In this sense, the citadel/ghetto metaphor could not adequately illuminate the dynamics of sociopolitical contestation and institutional transformation in global city-regions; it also left open the question of which political structures and agents actually produce, define and continually reorganize the global city.

As subsequent research on such issues has demonstrated, the sociospatial restructuring of global cities has contributed to the articulation of new political claims related to citizenship, identity, socioeconomic inequality and everyday life. In this context, various urban social movements for economic and environmental justice began to characterize themselves as being active political participants in the internationalization of the city (Keil 1998). The local politics of global capitalism thus experienced a qualitative redefinition as the discourses of "globalization" and "global city formation" were actively appropriated by oppositional social movements struggling to promote grassroots empowerment and various forms of sociospatial justice. Under these conditions, diverse class factions and territorial communities of the global city clashed directly with globally oriented forces such as transnational corporations, real estate developers and the boosterist local state apparatus. In these struggles, civil society-based agents (trade unions, neighborhood organizations, environmental groups, and so forth) must be viewed as actively shaping the politico-institutional dynamics of global city formation.

Secondary global cities such as Toronto, Frankfurt, Los Angeles and Zurich have been the focus of much research on the specifically political aspects of global city formation (see, for instance, Prigge 1988; Davis 1990; Keil and Lieser 1992; Keil and Ronneberger 1993, 1994; Hitz et al. 1994; Keil 1998; Keil and Ronneberger 2000; Kipfer and Keil 2002). This literature has studied, among other trends, the consolidation of boosterist local growth machines that are concerned to position the city strategically within supranational circuits of capital, generally through the consolidation of global command and control functions. Another key strand of this literature focuses on the role of urban social movements in global cities as an expression of the potential for oppositional mobilization and revolt against the capitalist everyday (Kipfer 1998; Schmid 1998; see also Reading 19 by Schmid; Ronneberger 2002). Others have examined the interplay between the consolidation of global cities and various ongoing transformations of state power at national, regional and local scales (Reading 30 by Brenner; Taylor 1995; Sassen 1998). Additionally, scholars working on such issues have examined the challenges of urban governance in cities that are attempting, simultaneously, to articulate themselves to the world economy and maintain internal sociospatial cohesion. In this explicitly political tradition of global cities research, a variety of intellectual approaches are applied – including Marxist and liberal theories of urban politics and urban political economy, such as urban regime theory and regulation theory (Lauria 1997; Jonas and Wilson 1999), neo-Gramscian state theory (Jessop et al. 1999; Kipfer and Keil 2002), Lefebvrian state and urban theory (Schmid 1998; Kipfer 2002) and local state theory (Kirby 1993; Keil 2003).

The urban politics of diversity – and exclusion

In light of the obvious demographic internationalization/transnationalization of the global city, much has been written about the politics of diversity. Mike Davis was one of the first authors who recognized the explosive transformations of world city polities that were unfolding due to accelerated immigration and growing ethno-cultural diversity. Nonetheless, in the mid-1980s, Davis noted that in Los Angeles the politics of white middle-class homeowners still eclipsed the "sleeping dragon" of the immigrant working class (Davis 1987: 86). The sleeping dragon, of course, awoke in 1992 with a loud roar spitting fire and ash, as the police beating of African American Rodney King and the subsequent acquittal of involved police officers triggered the Los Angeles Uprising, the most costly and most deadly of its kind in twentieth century America. This violent rupture in the fabric of multiculturalism was both an assertion of those groups – African Americans and Latinos in particular – who were shut out from the wealth and splendor of the global city, and a major setback to the "project world city" that had been espoused by the elites of the region (Keil 1998). Later, in the 1990s, the migrant communities and people of color

INTRODUCTION TO PART FIVE

who were just beginning to emerge as political actors in Davis' (1987) depiction of Los Angeles became central to the politics of the global city. Immigration was now recast as a form of "transnationalization" by authors such as Saskia Sassen and Michael Peter Smith (see Readings 43 and 46).

Many writers have analyzed the politics of ethnicity in an affirmative light, in which diversity is viewed as a politico-cultural challenge that could be confronted through multiculturalism. Leonie Sandercock, who had painted an enthusiastic and relatively hopeful picture of cosmopolitan politics in her book *Towards Cosmopolis* (1998; see also Reading 36), subsequently elaborated a more pessimistic, critical picture in her more recent sequel, subtitled *Mongrel Cities in the 21st Century* (2003). In this book, Sandercock attempted to move "beyond dreaming *cosmopolis*, to the practical challenges of creating it" (Sandercock 2003: xiv). Other commentators, such as Arjun Appadurai (2000), have been even more explicit in depicting the potentially racist, exclusionary and regressive reactions of certain segments of the global city polity to the internationalization of their city. Appadurai develops a chilling interpretation of global city formation as a process of what he terms "decosmopolitanization" in India's prime global city, Mumbai. Appadurai suggests that globalization may contribute to the exclusionary "ethnicization" of formerly cosmopolitan cities. Appadurai illustrates this point by examining a variety of culturally particularistic political consequences that have ensued in post-liberalization Mumbai. While many commentators have suggested that polities become more tolerant, cosmopolitan, transparent and democratic under conditions of globalization, Appadurai's incisive observations tell a radically different story. As Mumbai's economy is (re)opened to transnational capital investment through liberalization policies (see also Reading 27 by Grant and Nijman), its homegrown cosmopolitanism, which allowed diverse groups to live together in relative harmony, is increasingly replaced by a repressive, exclusionary and parochial politics of ethnicity, religion and "race." Appadurai's analysis thus directs attention to what might be termed the "dark side" of global city formation, in which the pressures and tensions induced by rapid socioeconomic change engender "militant particularisms" (Harvey 2000) based upon regressive, exclusionary and often violent political mobilizations.

An urban politics of citizenship and civil society

Linked to the politics of diversity and exclusion are the politics of citizenship, a realm in which citizens make claims on the state based on the principle of popular sovereignty (Isin and Wood 1999; Isin 2000). In the field of global cities research, the question of citizenship has only recently begun to generate theoretical debate and empirical research. This strand of research explores the question of whether new rights claims and new demands for state action are being engendered in globalizing urban spaces. While most theories of urban citizenship are rooted in the experiences of the western world, liberal democracy, the Fordist political economy and the postwar welfare state, there is also now a growing recognition that distinctive forms of urban citizenship have emerged in the contemporary era of global capitalism (Holston and Appadurai 2003: 304).

The resurgence of urban citizenship has been intertwined with diverse transformations of state space under contemporary conditions, which are increasingly relativizing the entrenched role of nationalized forms of economic regulation and political regulation (Brenner et al. 2003). In his brief contribution to this part (Reading 29), Warren Magnusson addresses such issues by arguing that inherited notions of nation-state sovereignty and nationalized models of state–society relations are no longer adequate. Magnusson perceptively argues:

> The concept of the global city invites us to abandon a number of old distinctions: between the local ("the city") and the global; between the economic, the social, the cultural and the political; and between the static ("structures," "systems," "space") and the dynamic ("movements," "time").
>
> (Magnusson 2000: 295)

Magnusson emphatically calls for a "politicizing" of the global city which he considers both "the venue and the product of our own struggles to become what we would like to be, and in the end there is no alternative but to take responsibility for what we have created" (2000: 304).

The challenge from below: contestation in the global city

Most literature on global cities deals with the structural conditions of urban change. It has been tempting to make quick and perhaps overstated generalizations regarding the relationships between the emergent social disparities in global cities and their political consequences. Yet, immiseration alone has rarely led to the formation or success of social movements. This is no different in the case of urban social movements in global cities. As the tremendous income inequalities, environmental injustices and political power differentials that plague most global cities become apparent to researchers, activists and communities alike, this awareness seldom prompts social movement activity or political outrage. Rather, it is only through a combination of lived community economic crisis (felt by ever more people in times of both prosperity and economic downturn) and a host of cultural and discursive events, the consolidation of movement milieux, tireless activism on the side of experienced individuals in well-organized groups, and a generally favorable political conjuncture that social movements in cities may become visible.

One strand of this literature has focused on the "rise of civil society." In his contribution (Reading 31), Mike Douglass examines the restructuring of state–civil society relationships in major cities throughout East Asia. In globalizing cities throughout the world, the role of civil society based organizations and networks has been greatly enhanced, as national and local states retreat from service delivery and corporate downsizing generates heightened unemployment and undermines job security. Under these conditions, civil society based institutions have become important economic and social stabilizers positioned at the margins of the neoliberalized political economy. This shift from the (local) state to (local) civil society as a basis for economic regulation has had profound implications for the governance of global city formation as functions such as welfare provision, labor market regulation, and even policing are increasingly "farmed out" to organizations in the non-profit sector and into non-governmental organizations "which, in one way or another, address the vast social needs of the city" (Friedmann 1998: 20; see also Eick 2004; Eick et al. 2004). Where, in general terms, it is correct to acknowledge "the reemergence of civil society as the collective actor in the construction of our cities and regions, in search of the good life" (Friedmann 1998: 21), it is equally crucial to add – and we are certain Friedmann would agree – that contestation is the name of the game in much global city politics. This creates subaltern counterpublics (Fraser 1997; Desfor and Keil 2004), an insurgent civil society in global cities (Keil 1998). And, as Janet Conway (2004: 247) explains, the praxis of social movements "points to cities as key sites for the new movement activism, especially in the North, and the possibility of locally rooted, multiscale politics in the globalized spaces of the world city" (see also Hamel et al. 2000; Köhler and Wissen 2003; and see Reading 35 by Mayer).

The politics of building the global city

In addition to studies of globalized elite politics and transnationalized popular contestation, scholars of global city governance have also begun to explore the political economy of the built environment and the real estate sector. A worldwide trend of associating globalized and spectacularized built forms (highrises, spectacular cultural buildings, corporate and political headquarters) with economic, cultural and demographic globalization has dramatically reshaped downtowns and suburban peripheries from Shanghai to Los Angeles, from Mumbai to Cape Town, and from Almati to São Paulo (Bodnár 2001; Olds 2001; King 2004; Sklair 2005; see also Reading 40 by Lehrer). Consequently, a distinct politics of the built environment has been emerging in globalizing cities. Real estate is one of the most prominent industries in these metropolitan regions: their success in the world economy is at once measured by, and visibly expressed through, the skyline of office towers and similar landmarks of economic success such as airports, ports and festival districts (King 2004). In her contribution to Part Five, Anne Haila explores the political maneuvers and struggles through which such spectacularized urban spaces are constructed in globalizing city regions (see Reading 33). We return to the politics of the built environment in a number of subsequent readings (see Reading 38 by King, Reading 40 by Lehrer and Reading 44 by Marcuse; see also Reading 23 by King).

Global cities and the ecological question

In a world increasingly connected financially and informationally, urbanization also runs up against certain ecological limits (Keil 1995, 1998; Derudder 2003; Terlouw 2003). This state of affairs has opened up a number of path-breaking discussions of sustainability and environmental justice in global cities (Davis 1998; Pezzoli 1998; Gandy 2002; Desfor and Keil 2004). From this "urban political ecology" point of view, global city formation, like the process of urbanization more generally, is significantly transforming humans' relationship to nature, leading in turn to a variety of deeply rooted governance problems – pertaining, for instance, to the provision of public goods (water, electricity) and the conditions of everyday social reproduction (housing, pollution, public health, transportation) in cities around the world. In Reading 32, Timothy Luke explores some of the ways in which global cities are being affected by, and are in turn influencing, emergent patterns of planetary interdependence and ecological sustainability.

The regionalization of global city politics

Most world city researchers have conceived urban *regions* as the relevant scale on which the city's global functions are realized. Airports, residences, jobs and educational facilities are not confined to the centralized, downtown financial districts of these cities, but can be found positioned in strategic locations throughout the metropolitan region (Sassen 2000). This means that the politics of the global city are, in practice, actually metropolitan or regional politics. In their study of metropolitanization in the Frankfurt region (Reading 34), Keil and Ronneberger illustrate this point by demonstrating that the politics of the global city continuously transcends the downtown city core, which has traditionally been the focal point for political studies of global cities. These dynamics of regionalization have also underpinned recent struggles over metropolitan institutional reform, regional economic governance, cross-border cooperation and inter-urban networking in many western European, North American and East Asian global city-regions (Friedmann 1997; Sassen 2001; Scott 2001; Brenner 2004; and see Reading 45 by Scott).

Taken together, the readings included in Part Five reflect some of the diverse ways in which global cities researchers have attempted to decipher the many dimensions of political life, institutional restructuring and sociopolitical mobilization in globalizing cities. These selections are intended at once to provide readers with an accessible survey of some of the main lines of research that have been elaborated in this research field and to demarcate a starting point for further inquiry into the problem of "governing complexity" (Keil 2003) within globalizing city-regions around the world.

References and suggestions for further reading

Appadurai, A. (2000) Spectral housing and urban cleansing: notes on millennial Mumbai, *Public Culture*, 12, 3, 627–651.

Beauregard, R. (1995) Theorizing the global–local connection. In P.L. Knox and P.J. Taylor (eds) *World Cities in a World-system*. New York: Cambridge University Press, 232–248.

Bodnàr, J. (2001) *Fin de Millénnaire Budapest*. Minneapolis, MN: University of Minnesota Press.

Brenner, N. (2004) *New State Spaces*. Oxford: Oxford University Press.

Brenner, N., Jessop, B., Jones, M. and McLeod, G. (eds) (2003) *State/Space: A Reader*. Boston, MA: Blackwell.

Conway, J.M. (2004) *Identity, Place, Knowledge*. Halifax, NS: Fernwood.

Davis, M. (1987) *Chinatown*, Part Two?, *New Left Review*, 164, 65–86.

Davis, M. (1990) *City of Quartz*. London: Verso.

Davis, M. (1998) *Ecology of Fear*. New York: Metropolitan.

Derudder, B. (2003) Beyond the state: mapping the semi-periphery through urban networks, *Capitalism, Nature, Socialism*, 14, 4, 91–120.

Desfor, G. and Keil, R. (2004) *Nature and the City*. Tucson, AZ: University of Arizona Press.

Eick, V. (2004) From SOLIDARI*City* to *Segregatio*TOWN. In INURA (eds) *The Contested Metropolis*. Basel, Boston, MA and Berlin: Birkhäuser, 53–64.

Eick, V., Grell, B., Mayer, M. and Sambale, J. (2004) *Nonprofit-Organisationen und die Transformation lokaler Beschäftigungspolitik*. Münster: Westfälisches Dampfboot.

Erie, S.P. (2004) *Globalizing L.A.: Trade, Infrastructure, and Regional Development*. Stanford, CA: Stanford University Press.

Fraser, N. (1997) *Justice Interruptus*. New York and London: Routledge.

Friedmann, J. (1997) World city futures: the role of urban and regional policies in the Asia-Pacific region. Occasional Paper no. 56, Hong Kong Institute of Asia-Pacific Studies, The Chinese University of Hong Kong, Shatin, New Territories, Hong Kong.

Friedman, J. (1998) 'The new political economy of planning: the rise of civil society. In M. Douglass and J. Friedmann (eds) *Cities for Citizens: Planning and the Rise of Civil Society in a Global Age*. Chichester: John Wiley, 19–35.

Friedmann, J. and Wolff, G. (1982) World city formation: an agenda for research and action, *International Journal of Urban and Regional Research*, 6, 3, 309–343.

Gandy, M. (2002) *Concrete and Clay*. Boston, MA: MIT Press.

Gottlieb, R., Vallianatos, M., Freer, R.M. and Dreier, P. (2005) *The Next Los Angeles: The Struggle for a Livable City*. Berkeley, CA: University of California Press.

Hamel, P., Lustiger-Thaler, H. and Mayer, M. (eds) (2000) *Urban Movements in a Globalizing World*. London and New York: Routledge.

Harvey, D. (2000) *Spaces of Hope*. Berkeley and Los Angeles: University of California Press.

Hitz, H., Schmid, C. and Wolff, R. (1994) Headquarter economy and city-belt: urbanization in Zurich, *Environment and Planning D: Society and Space*, 12, 2, 167–185.

Holston, J. and Appadurai, A. (2003) Cities and citizenship. In N. Brenner, B. Jessop, M. Jones and G. McLeod (eds) *State/Space: A Reader*. Boston, MA: Blackwell, 296–308.

Isin, E.F. (ed.) (2000) *Democracy, Citizenship and the Global City*. London and New York: Routledge.

Isin, E.F. and Wood, P. (1999) *Citizenship and Identity*. London: Sage.

Jessop, B., Peck, J. and Tickell, A. (1999) Retooling the machine: economic crisis, state restructuring, and urban politics. In A.E.G. Jonas and D. Wilson (eds) *The Urban Growth Machine: Critical Perspectives Two Decades Later*. Albany, NY: State University of New York Press, 141–159.

Jonas, A.E.G. and Wilson, D. (eds) (1999) *The Urban Growth Machine: Critical Perspectives Two Decades Later*. Albany, NY: State University of New York Press.

Keil, R. (1995) The environmental problematic in world cities. In P.L. Knox and P.J. Taylor (eds) *World Cities in a World-system*. Cambridge: Cambridge University Press, 280–297.

Keil, R. (1998) *Los Angeles: Globalization, Urbanization and Social Struggles*. Chichester: John Wiley & Sons.

Keil, R. (2003) Globalization makes states. In N. Brenner, B. Jessop, M. Jones and G. McLeod (eds) *State/Space: A Reader*. Boston, MA: Blackwell, 278–295.

Keil, R. and Lieser, P. (1992) Frankfurt: global city – local politics, *Comparative Urban and Community Research: An Annual Review*, 4, 39–69.

Keil, R. and Ronneberger, K. (1993) "Riding the tiger of modernization": reform politics in Frankfurt, *Capitalism, Nature, Socialism*, 4, 2, 19–50.

Keil, R. and Ronneberger, K. (1994) Going up the country: internationalization and urbanization on Frankfurt's northern fringe, *Environment and Planning D: Society and Space*, 12, 2, 137–166.

Keil, R. and Ronneberger, K. (2000) The globalization of Frankfurt am Main: core, periphery and social conflict. In P. Marcuse and R. van Kempen (eds) *Globalizing Cities*. Oxford: Blackwell, 228–248.

Keil, R., Wekerle, G.R. and Bell, D.V.J. (eds) (1996) *Local Places in the Age of the Global City*. Montreal: Black Rose.

King, A.D. (2004) *Spaces of Global Cultures*. London and New York: Routledge.

Kipfer, S. (1998) Urban politics in the 1990s: notes on Toronto. In INURA (eds) *Possible Urban Worlds*. Basel, Boston, MA and Berlin: Birkhäuser, 172–179.

Kipfer, S. (2002) Urbanization, everyday life and the survival of capitalism: Lefebvre, Gramsci and the problematic of hegemony, *Capitalism, Nature, Socialism*, 13, 2, 117–150.

Kipfer, S. and Keil, R. (2002) Toronto, Inc.? Planning the competitive city in Toronto, *Antipode*, 34, 2, 227–264.

Kirby, A. (1993) *Power/Resistance: Local Politics and the Chaotic State*. Bloomington and Indianapolis, IN: Indiana University Press.

Knox, P.L. and Taylor, P.J. (eds) (1995) *World Cities in a World-system*. Cambridge and New York: Cambridge University Press.

Köhler, B. and Wissen, M. (2003) Glocalizing protest: urban conflicts and the global social movements, *International Journal of Urban and Regional Research*, 27, 4, 942–951.

Lauria, M. (ed.) (1997) *Reconstructing Urban Regime Theory*. Thousand Oaks, CA: Sage.

Lefebvre, H. (1968) *Le Droit à la ville*. Paris: Anthropos.

Magnusson, W. (1996) *The Search for Political Space*. Toronto: University of Toronto Press.

Magnusson, W. (2000) Politicizing the global city. In E.F. Isin (ed.) *Democracy, Citizenship and the Global City*. London and New York: Routledge, 289–306.

Newman, P. and Thornley, A. (2005) *Planning World Cities*. Basingstoke: Palgrave Macmillan.

Olds, K. (2001) *Globalization and Urban Change*. New York and Oxford: Oxford University Press.

Pezzoli, K. (1998) *Human Settlements and Planning for Ecological Sustainability: The Case of Mexico City*. Cambridge, MA: MIT Press.

Prigge, W. (1988) Mythos Metropole: Wallmann lessen. In W. Prigge and H-P. Schwarz (eds) *Das NEUE FRANKFURT: Städtebau und Architektur im Modernizierungsprozeß 1925–1988*. Frankfurt: Vervuert, 209–240.

Ronneberger, K. (2002) Contours and convolutions of everydayness: on the reception of Henri Lefebvre in the Federal Republic of Germany, *Capitalism, Nature, Socialism*, 13, 2, 42–57.

Sandercock, L. (1998) *Towards Cosmopolis: Planning for Multicultural Cities*. Chichester: John Wiley & Sons.

Sandercock, L. (2003) *Cosmopolis II: Mongrel Cities in the 21st Century*. London and New York: Continuum.

Sassen, S. (1998) *Globalization and its Discontents*. New York: The New Press.

Sassen, S. (2000) *Cities in a World Economy*, 2nd edition. Thousand Oaks, CA: Pine Forge Press.

Sassen, S. (2001) Global cities and global city-regions: a comparison. In A.J. Scott (ed.) *Global City-Regions*. New York: Oxford University Press, 78–95.

Schmid, C. (1998) The dialectics of urbanization in Zurich. In INURA (eds) *Possible Urban Worlds*. Basel, Boston, MA and Berlin: Birkhäuser, 216–225.

Scott, A.J. (ed.) (2001) *Global City-Regions*. New York: Oxford University Press.

Sklair, L. (2005) The transnational capitalist class and contemporary architecture in globalizing cities, *International Journal of Urban and Regional Research* 29(3), 485–500.

Swyngedouw, E. (1997) Neither global nor local: "glocalization" and the politics of scale. In K. Cox (ed.) *Spaces of Globalization*. New York: Guilford, 137–166.

Taylor, P.J. (1995) World cities and territorial states: the rise and fall of their mutuality. In P.L. Knox and P.J. Taylor (eds) *World Cities in a World-system*. New York: Cambridge University Press, 48–62.

Terlouw, K. (2003) Semi-peripheral developments: from world systems to regions, *Capitalism, Nature, Socialism*, 14, 4, 71–90.

29

Prologue

"The Global City as World Order"

from *The Search for Political Space* (1996)

Warren Magnusson

One of the distinctive features of the city as a mode of order and domination is that it is not governed from a single centre. The principle of sovereignty does not work effectively within the civic domain. Of course, sovereigns often try to exercise control *over* cities, but they can rarely if ever work *through* cities. A city is in large degree a self-organizing system produced by a variety of cultural, social, and economic enterprises. It is where people come to do things outside the domain of sovereignty, in relative freedom from the dictates of church and state. The medieval proverb that "city air makes free" refers to more than the liberation of serfs, who could gain the status of free persons after a year and a day. It also alludes to the possibility for new enterprises that escape the dead hand of established authority. Such enterprises are not, in principle, contained within the territory of a particular city: they reach out to the surrounding countryside and to other cities in the world beyond. Urbanity in its fullest sense implies deterritorialized relations between people in different parts of the world. In this sense, the city is not fixed to a particular place the way a village is. Urbanity implies a kind of nomadism: a presence within a space of flows that connect and reconnect different places in the world. Obviously, the market structures many of these flows, and the logic of the market tends to determine which cities will expand and which will contract. However, the social and cultural flows that occur within urban

space are not reducible to the logic of the market. They have more complex origins and many autonomous effects.

Present-day municipalities are lineal descendants of the early medieval corporations that were designed to contain and control urban development. Whether constituted by Royal Charter or formed from below by civic insurgents, the municipal corporations were intended to fix economic activity in particular places and to give urban life the form that people believed would be most rational. However, the municipal corporations were generally unable to contain what they were supposed to manage. This was partly because they lacked sovereignty, but more fundamentally because the activities that typified a city burst the bounds of any particular place. London could not be kept in its square mile, Paris could not be confined to the Ile de la Cité, and even Manhattan was not big enough for New York. This physical overflow was a sign of economic, social, and cultural spillage of a much more profound character. By the seventeenth century the whole world was in London's domain, and, although London ultimately became the leading world-city, it was by no means the only city that could boast of a global or near-global reach. Moreover, the leading cities were all linked to one another in patterns of dizzying complexity. None was "independent" of the others, nor sought to be. Although there were governments of a sort in particular cities – sometimes even national governments with ostensibly sovereign authority – these governments exercised only a shadow control over their urban domains. Urbanism itself – the system of cities – was under no one's direct control, followed no one's orders in its development, and could not be managed from any single center. On the contrary, as Braudel and others have reminded us, sovereigns of one sort or another have always depended heavily on the urban system that they pretend to govern. A productive and dynamic urbanism will produce the surpluses necessary to pay soldiers and make arms. It will also generate the ideas and the functionaries necessary for effective government or imperial expansion. Sovereignties will be sustained or overwhelmed by urban dynamism.

Arguably, the true heirs of the medieval cities are not the municipalities but the multinational corporations of the contemporary world. The corporate form as exemplified by IBM and ICI is a late mutation of the municipal corporation of medieval times, and is not connected (except nominally) to a particular place. It is slimmed down for the pursuit of profit, and projected into a truly global space. As organizations for economic enterprise, municipalities are largely anachronistic, because they are tied to particular territories and burdened with tasks of government. The free corporations of the present day occupy deterritorialized spaces that cannot be mapped onto the world like countries or provinces. Increasingly, they function within a cyberspace that is characterized by instantaneous electronic communication. This cyberspace is not stable, and in fact the major actors within it are constantly changing the systems of communication and control for their own purposes. Thus, the space within which these organizations act is largely the product of their own activities. To keep up with the most innovative actors is enormously difficult, and the evolution of cyberspace is not governed from any single center. In this respect, the emergent cyberspace is typical of the spaces created by urban activity.

There is, of course, more to urban space than cyberspace. Urbanism is characterized by the continual production and reproduction of spaces of habitation, work, recreation, cultural expression, and so on. There is a dynamism to these processes that again defies any static representation. A building is simply a momentary expression of people's ideas about the way their activities need to be spatialized. Fixed as it is, a building is reformed in use until it becomes almost unrecognizable to its original founders. The physical form of the city as a whole is even more plastic. Once we take into account the city's relationship to the countryside and to other cities, it becomes apparent that urban space is a dynamic presence in the world as a whole. People are within urban space in their airplanes going from airport to airport, and in their cars speeding along the motorways. The airport café in Honolulu, which serves passengers on their way from San Francisco to Sydney, is a part of the space of all three cities – and many others. Similarly, the restaurants in the Black Forest, the beaches of Mauritius, and the mountains of British Columbia are extensions of the recreational space of urbanites in many parts of the world. Thanks to the means of transportation and communication that have been developed over the last thirty to thirty-five years, prosperous urbanites can inhabit any and all parts of the world in the routine course of their lives. It is a sign of parochialism – or poverty – if one fails to inhabit the whole world.

We might think of the wider habitat as a global "hyperspace," within which the "cyberspace" of the computer nets is a particular but important domain. Airports, motorways, offices, hotels, and boutiques are other elements in the most privileged domain of this hyperspace. Access to that domain is carefully controlled. In fact, cities are marked by exceedingly complicated strategies of territorialization. Among the obvious signs of this are the urban fortresses of the sort that were first established in the late medieval Italian cities and later replicated in many forms. These are not the fortresses of kings and bishops, but of burghers who seek security within the turbulent, expansive and absorptive space of the city. When people venture out from these fortresses, they carry with them their personal security systems. Much public and private enterprise is directed toward securing privileged people's routes and places of work, recreation, shopping, and cultural expression.

Whole cities (for example, Paris) and whole regions of the tropics are being re-formed to make them comfortable spaces for the public life of the prosperous. Beside, beneath, and often co-present with these spaces are other, more constricted spaces that provide for the less prosperous. Homeless people live in every crack and cranny, having been swept from the places of privilege with ruthless efficiency. The zones of exclusion and inclusion are subtly layered and dynamically articulated, so that they register in the consciousness only in the enactment (and often not even then). Nevertheless, these half-understood urban zones are generally of much greater significance in people's lives than the boundaries between states.

Although the world functions as a single city, it is not a "global village," as Marshall McLuhan once suggested. It lacks the fixity, community, and intimacy that is part of the image of the village. In fact, the global city is inherently complex, dynamic, and socially differentiated. It is at once expansive and inclusive, and it is more obviously marked by separations and exclusions than by the intimate relations of communal solidarity. The world as a city has many ethnic enclaves, rich neighborhoods and poor, a multinucleated central business district, suburban office centers, shopping precincts, and recreational complexes, overcrowded systems of public transportation, and vast slums on its fringes that flow over and through the better-ordered and more prosperous districts. The government of this whole is ineffective at best and nonexistent at worst – just as it is within particular metropolitan areas. And, just as in particular cities, there is intense competition among the different jurisdictions, intermittent co-operation among them, and a vague sense that the whole could be much better ordered. If it is held together as a whole, it is mainly by economic transactions, facilitated by a common physical infrastructure. There is some sense that there is a common environment to be maintained, and that violence has to be curbed; and there is something of a shared culture that is dependent on common media of communication. However, there is no effective sovereignty, and it is by no means clear that people would want such a mode of political organization if they could have it.

To assert their sovereignty over municipal authorities, states have had to strip those authorities of most of the powers they would need for effective governance. A self-governing city would have the power to regulate its own economy and determine its own foreign trade policy. It would have to control the flow of arms in and out of its domain, and break the power of the armed gangs that defy its authority. Even to deal with questions of public health, it would have to project its authority far beyond its immediate boundaries. In a sense, it would have to follow its particular connections throughout the world. Thus, a municipal government that was determined to protect the interests of its own citizens would need a kind of world reach and freedom of action that is clearly inconsistent with state sovereignty. On the other hand, without such powers, municipal governments are condemned to a sort of observer status within the cities they are supposed to govern. Ironically, this is the status to which most if not all national governments are presently being reduced. With respect to economic organization, public health, social services, and even "security," sovereignty seems increasingly like a "show" that offers a comforting illusion of national control over national destinies.

30

"Global Cities, 'Glocal' States: Global City Formation and State Territorial Restructuring in Contemporary Europe"

from *Review of International Political Economy* (1998)

Neil Brenner

Editors' introduction

Our second reading on the theme of global city governance is drawn from one of the editors' own articles, written in the mid-1990s while he was a graduate student in the Department of Geography at UCLA and subsequently published in an interdisciplinary journal of international political economy. Brenner's intellectual starting point is the observation that most global city theorists have postulated, either implicitly or explicitly, a declining role for national states in the governance of economic life. This assumption of "state decline" has in turn led scholars either to focus entirely on *local* scales of regulation within global city economies and/or to bracket the ways in which (reconstituted) national state institutions have, in many cases, actively facilitated the process of global city formation. Against these intellectual tendencies, Brenner argues that state institutions at national, regional and local scales have been instrumental in promoting the development of globally interlinked cities, both in the western European context (his central empirical focus) and beyond. This claim, which has been developed by several other urbanists since the mid-1990s (see, for instance, Reading 20 by Hill and Kim and Reading 48 by Olds and Yeung), leads Brenner to explore, more specifically, the ways in which state institutions themselves have been reorganized since the mid-1970s, in significant measure through their role in promoting urban restructuring. Brenner focuses upon the process of state "rescaling" in which established hierarchies of intergovernmental relations and political-economic regulation are being recalibrated so as to enable new forms of urban governance. While a significant part of Brenner's original article explored diverse empirical cases of global city formation and state rescaling in western Europe, the selection here is devoted primarily to an elaboration of these arguments on a more abstract, conceptual level. In the context of this Reader, one of the contributions of Brenner's work is to explore the interplay between global city formation and various ongoing transformations of statehood – including patterns of state spatial organization and changing modes of state intervention.

World city theory has been deployed extensively in studies of the role of major cities such as New York, London and Tokyo as global financial centers and as headquarters locations for transnational corporations (TNCs). While the theory's usefulness in such research has been convincingly demonstrated, I believe that the central agenda of world city theory is best conceived more broadly, as an attempt to analyze the changing geographies of global capitalism in the late 20th century. From this point of view, the project of world cities research is not merely to classify cities within world-scale central place hierarchies, but, as Friedmann (1986: 69) has proposed, to analyze the "spatial organization of the new international division of labor." The key feature of this newly emergent configuration of world capitalism is that cities – or, more precisely, large-scale urbanized regions – rather than national territorial economies are its most fundamental geographical units. These urban regions are said to be arranged hierarchically on a global scale according to their specific modes of integration into the world economy.

But how is this emergent global urban hierarchy articulated with the geographies of national state territories? Insofar as world city theory is directly concerned with the "contradictory relations between production in an era of global management and the political determination of territorial interests" (Friedmann 1986: 69), an analysis of changing relations between world cities and national state spaces is arguably one of its most central theoretical and empirical tasks. Yet, in practice, the methodological challenge of analyzing the changing linkages between different spatial scales has not been systematically confronted by world cities researchers.

Much of world cities research has been composed of studies that focus primarily upon a single scale, generally either the urban or the global. Whereas research on the socioeconomic geography of world cities has focused predominantly on the urban scale, studies of changing urban hierarchies have focused largely on the global scale. To the extent that the national state has been thematized, it has usually been understood with reference to its local/municipal institutions or as a relatively static background structure. Indeed, like many other prevalent approaches to the study of globalization, the bulk of world cities research during the 1980s

and early 1990s was premised upon the underlying assumption that intensified globalization entails an erosion of national state territoriality (see, for instance, Friedmann and Wolff 1982). This conception of globalization as a process of state decline has led world cities researchers to focus on the global scale, the urban scale and their changing interconnections while systematically neglecting the role of nationally configured political-economic dynamics. The privileging of the global/local dualism among world cities researchers has also been grounded upon what might be termed a zero-sum conception of spatial scales in which the global, the national and the urban scales are viewed as being mutually exclusive – what one gains, the other loses – rather than as being intrinsically related, co-evolving layers of territorial organization.

I argue here, by contrast, that national states are being rescaled and redefined in conjunction with processes of global city formation rather than being eroded. The resultant, rescaled configurations of national state space have come to figure centrally in mediating the processes of geoeconomic integration and urban-regional restructuring. While it is evident that the current round of geoeconomic restructuring has undermined Fordist-Keynesian forms of state regulation and economic governance, the narrative of state decline conflates the ongoing reconfiguration of the national state space with a withering away of national state power as such. Current transformations of state power may indeed herald the partial erosion of central state regulatory control over global flows of capital, commodities and labor-power, and they have also clearly entailed new fiscal and legitimation problems for national governments. Despite this, however, national states arguably remain central institutional matrices of political power and crucial geographical infrastructures for capital accumulation (Jessop 1994; Panitch 1994). More generally, as I argue in this article, national states have figured crucially in catalyzing and mediating the process of global city formation.

RESCALING URBAN SYSTEMS, RESCALED STATE SPACES

Cities are at once basing points for capital accumulation (nodes in global flows) and organizational-administrative levels of national states (coordinates of state territorial power). First, as *nodes in global flows*, cities operate as loci of industrial production, as centers of command and control over inter-urban, interstate and global circuits of capital, and as sites of exchange within local, regional, national and global markets. Second, as *coordinates of state territorial power*, cities are regulatory-institutional levels within each national state's intergovernmental hierarchy. The term coordinate is intended to connote the embeddedness of cities within the national state's organizational matrix. These coordinates may be interlinked through various means, from legal-constitutional regulations, financial interdependencies, administrative divisions of labor and hierarchies of command to informal regulatory arrangements.

During the Fordist-Keynesian period (circa 1950 to 1970), these two dimensions of urbanization were spatially coextensive within the boundaries of the national territorial state. As nodes of accumulation, cities were enframed within the same territorial grids that underpinned the national economy. The cities of the older industrialized world served as the engines of Fordist mass production and as the urban infrastructure of a global economic system compartmentalized into nationalized territorial matrices. It was widely assumed that the industrialization of urban cores would generate a propulsive dynamic of growth that would in turn lead to the industrialization of the state's internal peripheries, and thereby counteract the problem of uneven geographical development. Likewise, as coordinates of state territorial power, Fordist-Keynesian regional and local regulatory institutions functioned primarily as transmission belts of central state socioeconomic policies (Mayer 1994). Their goal was above all to promote growth and to redistribute its effects on a national scale. To this end, redistributive regional policies were widely introduced to promote industrialization within each state's internal peripheries (Albrechts and Swyngedouw 1989).

Since the 1970s, however, these nationalized geographies of urbanization and state regulation have been profoundly reconfigured. The crisis of global Fordism was expressed in a specifically geographical form, above all through the contradiction between the national scale of state regulation and the globalizing thrust of postwar capital accumulation (Peck and Tickell 1994). Consequently, since the global economic crises of the early 1970s, the scales on which the Fordist-Keynesian political-economic order was organized – national regulation of the wage relation; international regulation of currency and trade – have been significantly reconfigured. While the deregulation of financial markets and the global credit system since the collapse of the Bretton Woods system in 1973 has undermined the viability of nationally organized demand management policies, the increasing globalization of financial flows has diminished the ability of national states to insulate themselves from the world economy as quasi-autarchic national economic spaces (Agnew and Corbridge 1995). The intensification of global interspatial competition among cities and regions has also compromised traditional national industrial policies and led regional and local states to assume increasingly direct roles in promoting capital accumulation on subnational scales.

The central geographical consequence of these intertwined shifts has been to destabilize the most elemental building block of the postwar geoeconomic and geopolitical order – the autocentric national economy. Despite this, however, cities and national states have continued to operate as fundamental forms of territorialization for capital, even though this role is no longer tied primarily to the nationally configured patterns of urbanization and to nationally centralized strategies of economic governance. Since the crisis of North Atlantic Fordism in the early 1970s, new subnational and supranational patterns of urbanization and state regulation have been consolidated throughout the older industrialized world. Our task in the present context is to examine more closely the geographical-institutional interface between the rescaling of urbanization and the remaking of state spatiality.

First, as world cities researchers have indicated at length, the contemporary rescaling of urbanization must be viewed as a multidimensional reorganization of entrenched national urban systems in close conjunction with the consolidation of new world-scale urban hierarchies. To illustrate this ongoing rescaling of the urbanization process, Figure 1 depicts the ways in which the European urban hierarchy has been reconfigured since the crisis of the Fordist-Keynesian regime during the early 1970s.

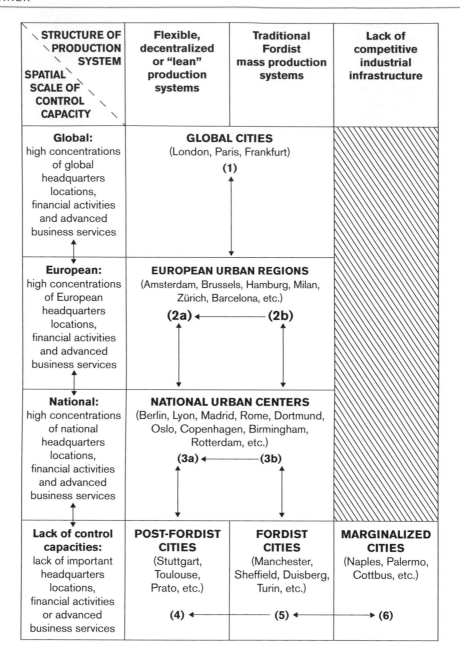

Figure 1 The changing European urban hierarchy

Source: derived from Krätke (1995: 141).

This schematic representation of the contemporary European city-system (derived from Krätke 1995: 140–141) focuses upon the first dimension of urbanization, the role of cities as nodes of capital accumulation. Krätke's model describes contemporary transformations of the European urban hierarchy with reference to two criteria – the industrial structure of the city's productive base

(Fordist vs. post-Fordist) and the spatial scale of its command and control functions (global, European, national, regional, non-existent). The arrows in the figure indicate various possible changes in position among cities within the European urban hierarchy; and various cities have been listed to exemplify each of these levels. As the figure indicates, global city formation has entailed the emergence of a new global urban hierarchy, defined through the increasing scale of urban command and control functions, of inter-urban exchange relations and of inter-urban competition. As nodes of accumulation, therefore, cities are no longer enclosed within relatively autocentric national economies, but have been embedded more directly within transnational urban hierarchies and inter-urban networks. Although the cities currently positioned at the apex of the global, European, North American and East Asian urban hierarchies present the most dramatic evidence of this transformation, their newly acquired positions within the global urban system are indicative of a more general rescaling of urbanization processes across the world economy.

Most crucially here, the current wave of global spatial restructuring has also had important implications for the role of cities as coordinates of state territorial power. Despite its neglect of nationally scaled processes, the methodology of world cities research provides a useful starting point for investigating recent rescalings of state spatiality. Much like the place-based territorial infrastructures of global cities, I would argue, post-Keynesian state institutions can be viewed as crucial *forms of reterritorialization* for capital in the current period. Whereas the centralized, bureaucratized states of the Fordist-Keynesian era converged around the national scale as their predominant organizational-regulatory locus, the national states of the post-Keynesian, neoliberal era have been restructured substantially to provide capital with many of its most essential territorial preconditions and collective goods at other spatial scales, including both the supranational and the subnational.

Soja's (1992) concept of the exopolis provides a strikingly appropriate image for describing the transformed spatial form of these rescaled state apparatuses. Like the exopolis, the spatial expression of post-Fordist forms of capitalist industrialization in which inherited urban spaces have been turned "inside-out" and "outside-in" (Soja 1992), the geographies of post-Keynesian state institutions are polymorphic, multitiered and decentered; and they are likewise being simultaneously turned inside out and outside in – *inside out* insofar as they attempt to promote the global structural competitiveness of their major cities and regions; and *outside in* insofar as supranational agencies and international agreements have come to play more direct roles in structuring their "internal" political

spaces. This rescaling of state space is rearticulating inherited political geographies in ways that are deprivileging nationally organized regulatory arrangements while ceding new roles both to supranational and to subnational institutional forms. Thus understood, state institutions retain a critical role as forms of territorialization for capital, but this role is no longer premised upon an isomorphic territorial correspondence between state institutions, urban systems and circuits of capital accumulation centered around national state boundaries.

This rescaling of state power has not only reshuffled entrenched political geographies and administrative hierarchies, but also been associated with a profound transformation of the relationship between states, capital and territory. Territorial organization has long operated as a force of production under capitalism through its natural goods, its supplies of fixed capital and labor-power, its technological-institutional infrastructures and other place-specific externalities and collective goods. The state has arguably played a crucial role in the production, regulation and reproduction of these socio-territorial and productive ensembles throughout the long-run history of capitalism. During the Fordist-Keynesian period, most older industrial states deployed indirect forms of regulatory intervention oriented towards the reproduction of labor-power, industrial relocation and the promotion of collective consumption. Although the collapse of the Fordist-Keynesian regulatory regime has undermined the monolithic unity of national states as territorially self-enclosed containers of socioeconomic activities, this development has also arguably *intensified* the state's role in the territorialization of capital (Brenner 2004).

The rescaled state institutions have come to play essential roles in the production, coordination and maintenance of the customized, place-specific configurations of socioeconomic organization upon which global competitive advantages today increasingly depend, both in global city-regions and in other major capitalist cities as well. For, in contrast to the various incentive-based and indirect policies of the Fordist-Keynesian era, contemporary post-Keynesian modes of state intervention have entailed a more direct, unmediated involvement of state institutions in the territorialization of capital. Faced with the apparently increased mobility of capital, commodities and labor-power across national borders, post-Keynesian state apparatuses are orienting themselves above all towards the provision of *immobile* factors of production – that is, towards those externalities associated with capital's moment of territorialized fixity within major cities and city-regions. From public–private partnerships, labor-retraining programs, science parks, conference centers, waterfront redevelopment schemes, technology transfer projects, information-sharing networks, venture capital programs and market research projects to large-scale investments in technopoles, innovation programs, enterprise zones and free trade areas, an immense range of state-organized economic development policies are being mobilized in order to enhance the territorially specific productive capacities of strategically delineated economic spaces. The overarching goal of these state strategies is to secure new locational advantages in international economic competition through the construction of territorially rooted immobile assets. In the current period, many if not all of the socially produced features of territorial competitiveness – such as human capital resources, cost efficiency, product quality, turnover time, flexibility and innovative capacities – have become central concerns of state institutions, at various spatial scales, in their governance of economic development at a range of spatial scales. And, even when such territorial assets are not directly produced by the state, a rapidly growing number of state agencies have become directly or indirectly engaged in financing, monitoring, coordinating and maintaining them.

More generally, by adopting new strategies of economic promotion and place-marketing, national state institutions have also come to play central roles in marketing their own territories (or strategic sites therein) as locational products on the world market. Under these conditions, the role of state institutions in economic governance is no longer merely to reproduce localized production complexes, but continually to restore, enhance, intensify and restructure their capacities as territorially specific productive forces. These developments lead Swyngedouw (1992: 431) to conclude that "the role of the state is actually becoming more, rather than less, important in developing the productive powers of territory and in producing new spatial configurations." The goal of creating place-specific or geographically immobilized

competitive advantages may be pursued through both deregulatory *and* reregulatory political strategies. The balance between the latter is frequently a matter of intense sociopolitical conflict.

It is in this context, I would argue, that the enhanced role of subnational institutional forms in contemporary processes of socioeconomic governance is to be understood. It is above all through their role in securing, promoting, maintaining and advertising any number of place-specific conditions for capital investment that local and regional states, in particular, are gaining structural significance within each national state's internal administrative hierarchy (Mayer 1994). Indeed, the process of state rescaling can be viewed in significant measure as a concerted political strategy through which political-economic elites at various levels of state power are attempting to propel major cities and regions upwards within the urban hierarchy depicted in Figure 1. Throughout Europe, local, regional and national governments are mobilizing diverse state strategies intended at once to revalorize decaying industrial sites, to promote industrial growth in globally competitive sectors and to acquire command and control functions in the world economy by providing various territorial preconditions for transnational capital, including transportation and communications links, office space, labor-power and other place-specific externalities (Hall and Hubbard 1996).

Figure 2 summarizes the ways in which the relations between urbanization patterns and forms of state territorial organization have been reconfigured since the Fordist-Keynesian period, highlighting at once the globalization of the world economy, the rescaling of state space, and the ramifications of these shifts for both dimensions of urbanization. As nodes of accumulation, global cities are embedded within flows of capital that no longer overlap coextensively with national economic space. As coordinates of state territorial power, global cities are strategic targets for rescaled state strategies oriented towards the continual enhancement of territorially specific competitive advantages and productive forces. In this sense, global cities are simultaneously spaces of global accumulation and coordinates of rescaled state spaces. The governance of contemporary urbanization patterns therefore entails not only the construction of "new industrial spaces"

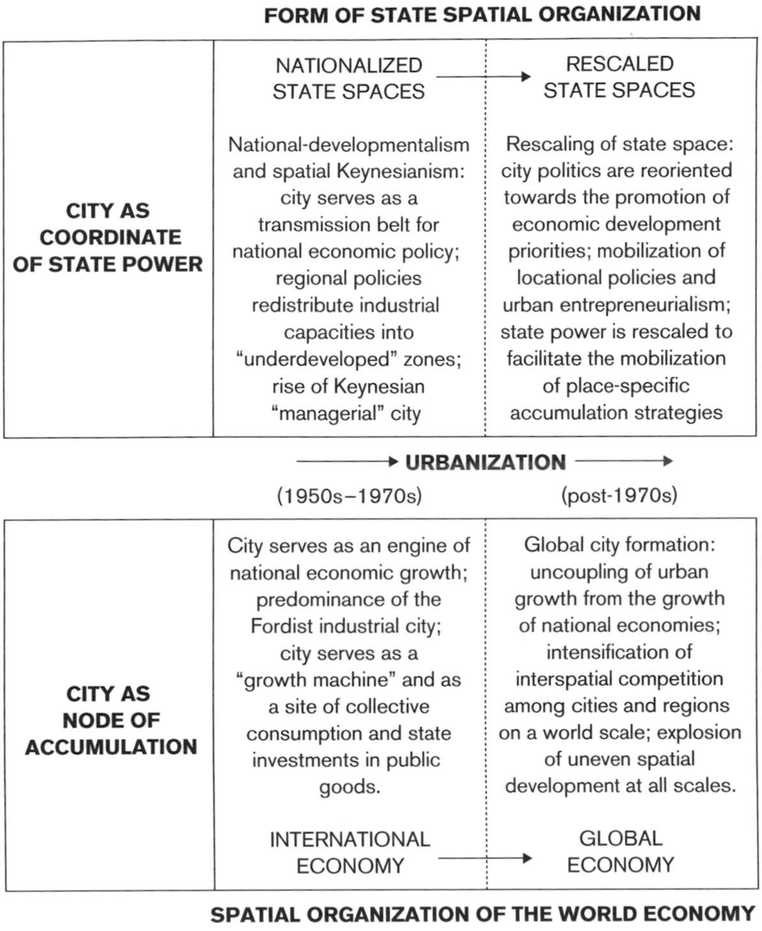

FORM OF STATE SPATIAL ORGANIZATION

	NATIONALIZED STATE SPACES	RESCALED STATE SPACES
CITY AS COORDINATE OF STATE POWER	National-developmentalism and spatial Keynesianism: city serves as a transmission belt for national economic policy; regional policies redistribute industrial capacities into "underdeveloped" zones; rise of Keynesian "managerial" city	Rescaling of state space: city politics are reoriented towards the promotion of economic development priorities; mobilization of locational policies and urban entrepreneurialism; state power is rescaled to facilitate the mobilization of place-specific accumulation strategies

→ **URBANIZATION** →

(1950s–1970s)   (post-1970s)

CITY AS NODE OF ACCUMULATION	City serves as an engine of national economic growth; predominance of the Fordist industrial city; city serves as a "growth machine" and as a site of collective consumption and state investments in public goods.	Global city formation: uncoupling of urban growth from the growth of national economies; intensification of interspatial competition among cities and regions on a world scale; explosion of uneven spatial development at all scales.
	INTERNATIONAL ECONOMY →	GLOBAL ECONOMY

SPATIAL ORGANIZATION OF THE WORLD ECONOMY

Figure 2 Urbanization, state forms and the world economy, 1950–2000

(Scott 1988) for post-Fordist forms of industrialization but, just as crucially, the consolidation of what might be termed *new state spaces* to enhance each state's capacity to mobilize the productive force of urban and regional spaces and to regulate the sociopolitical contradictions induced by such projects (Brenner 2004).

In the late 20th century, the state's own spatial and scalar configuration has become an important locational weapon in the interspatial competition between cities, regions and national states in the world economy. Under these conditions, a new "politics of scale" (Smith 1992) has emerged in which the scalar geographies of state power have become a direct object of sociopolitical contestation. If, as Friedmann and Wolff (1982: 312) have proposed, "world cities lie at the junction between the global economy and the territorial nation-state," then it seems appropriate to view the political-regulatory institutions of world city-regions as geographical arenas in which this new politics of scale are fought out with particular intensity.

CONCLUSION: THE URBAN QUESTION AS A SCALE QUESTION

Amidst the confusing and contradictory geographies of contemporary globalization, world cities represent a particularly complex "superimposition and interpenetration" (Lefebvre 1991: 88) of social, political and economic spaces. Because urban regions occupy the contradictory interface between the world economy and the territorial state, they are embedded within a multiplicity of political-economic processes organized upon a range of superimposed geographical scales. The resultant politics of scale within the political institutions of major urban regions can be construed as a sequence of groping, trial-and-error strategies to manage these intensely conflictual forces through the continual construction, deconstruction and reconstruction of relatively stabilized configurations of territorial organization. The rescaling of urbanization leads to a concomitant rescaling of the state through which, simultaneously, urban and regional spaces are mobilized as productive forces and social relations are circumscribed within new political boundaries and scalar hierarchies. These rescaled configurations of state power in turn transform the everyday social conditions under which the urbanization process unfolds. Whether these disjointed strategies of reterritorialization within European cities might eventually establish new spatial and scalar fixes for sustained capitalist growth on any geographical scale is a matter that can only be resolved through the politics of scale itself, through ongoing struggles for hegemonic control over the form, trajectory and territorial organization of the urbanization process.

REFERENCES FROM THE READING

Agnew, J. and Corbridge, S. (1995) *Mastering Space*. New York: Routledge.

Albrechts, L. and Swyngedouw, E. (1989) The challenges for regional policy under a flexible regime of accumulation. In L. Albrechts and Regional Studies Association (eds) *Regional Policy at the Crossroads*. London: Jessica Kingsley, 67–89.

Brenner, N. (2004) *New State Spaces*. New York: Oxford University Press.

Friedmann, J. (1986) The world city hypothesis, *Development and Change*, 17, 69–83.

Friedmann, J. and Wolff, G. (1982) World city formation: an agenda for research and action, *International Journal of Urban and Regional Research*, 6, 309–344.

Hall, T. and Hubbard, P. (1996) The entrepreneurial city, *Progress in Human Geography*, 20, 2, 153–174.

Jessop, B. (1994) Post-Fordism and the State. In A. Amin (ed.) *Post-Fordism: A Reader*. Cambridge, MA: Blackwell, 251–279.

Krätke, S. (1995) *Stadt, Raum, Ökonomie*. Basel: Birkhäuser.

Lefebvre, H. (1991) *The Production of Space*, trans. D. Nicholson-Smith. Oxford and Cambridge, MA: Blackwell.

Mayer, M. (1994) Post-Fordist city politics. In A. Amin (ed.) *Post-Fordism: A Reader*. Cambridge, MA: Blackwell, 316–337.

Panitch, L. (1994) Globalization and the state. In R. Miliband and L. Panitch (eds) *Socialist Register 1994*. London: Merlin Press, 60–93.

Peck, J. and Tickell, A. (1994) Searching for a new institutional fix. In A. Amin (ed.) *Post-Fordism: A Reader*. Cambridge, MA: Blackwell, 280–315.

Scott, A.J. (1988) *New Industrial Spaces*. London: Pion.

Smith, N. (1992) Geography, difference and the politics of scale. In J. Doherty, E. Graham and M. Malek (eds) *Postmodernism and the Social Sciences*. New York: St. Martin's Press, 57–79.

Soja, E. (1992) Inside Exopolis. In M. Sorkin (ed.) *Variations on a Theme Park*. New York: Noonday Press, 94–122.

Swyngedouw, E. (1992) Territorial organization and the space/technology nexus, *Transactions, Institute of British Geographers*, 17, 417–433.

31

"World City Formation on the Asia-Pacific Rim: Poverty, 'Everyday' Forms of Civil Society and Environmental Management"

from M. Douglass and J. Friedmann (eds), *Cities for Citizens: Planning and the Rise of Civil Society in a Global Age* (1998)

Mike Douglass

Editors' introduction

Mike Douglass is the Director of the Globalization Research Center and a Professor and former Chair of the Department of Urban and Regional Planning at University of Hawaii, Manoa. Douglass is one of the pioneers of world city research and has been one of the most consistent observers of urbanization processes on the Pacific Rim and in Pacific Island nations. His work on urban development and planning in cities of the South is widely recognized as a leading contribution to the field. In this riveting text, which is an abbreviated version of a chapter for an influential book he co-edited with John Friedmann (Douglass and Friedmann 1998), Douglass examines the interrelationships of world city formation and social and environmental struggles in East Asian cities. Isolating three themes – globalization, localization and community empowerment – Douglass examines the new forms of community politics that are emerging in the rapidly changing metropolises of East Asia. Focusing on diverse state–civil society relations in various East Asian developmental states, Douglass concludes that the project of empowering the urban poor remains extraordinarily difficult in the context of economic globalization and entrenched political authoritarianism. Nevertheless, Douglass places high hopes in the progressive potential of democratic, civil society based activism and makes a strong argument for including the politics of the poor within the narrative of global city politics.

INTRODUCTION

Three major themes run through the discussion of the processes of global–local interaction along the Asian arc of the Pacific Rim, with particular emphasis on the ways in which world city formation is part of a process of emergent social struggles around the built and natural environment.

The first is that the forces impinging on urban restructuring in Pacific Asia generate and reveal heightening tension between, on one hand, the (re)positioning of cities in a global system of cities and, on the other, citizen mobilization and demands for substantial improvements in daily life space. While linking up to and gaining higher positions in a system of world cities calls for increasing investments in mega-projects and the forced mobility of large numbers of people, the "discovery of civil society" associated with the rise of the urban middle class, organized labor, voluntary organizations and heightened political action from all quarters of society, including the poor, is making the implementation of such projects problematic as control over space becomes increasingly contested.

A second theme is that globalization is also a process of localization. Contrary to much literature on the tendency for capitalist penetration and Western cultural imperialism to lead towards the "rendering of the world as a single place" (King 1989: 5), variations in sociocultural, political and economic institutions continue to emerge to profoundly affect the ways in which global impulses are amalgamated into real historical settings. The array of possibilities for social mobilization and the enlargement of democratic spaces to create alternative development paths is greater than much of received world systems theories or mainstream economic theory allow. No matter how successful or unsuccessful, the future of Pacific Asia societies and their cities will be determined as much by localized socio-cultural and political processes as by global imperatives.

The third theme is that efforts by the poor to take command over community space are integral to a more embracing process of self-empowerment (Friedmann 1992). This includes not only securing land for housing, but also investing in efforts to gain access to and manage environmental infrastructure and resources. Again, contrary to conventional wisdom that the poor are "too busy being poor" to care about their habitat or the environment, research in low-income communities in Asia consistently shows that substantial amounts of household allocations of time, labor and resources are devoted to environmental considerations even among the very poor.

WORLD CITY FORMATION ON THE PACIFIC RIM

The appearance of "world city" as the new shibboleth of global achievement has not been missed by governments in Pacific Asia. The increasing strength of labor in these economies has shifted their comparative advantage away from labor-intensive manufacturing and towards higher-order production and service industries, including global information and control functions. The intensive economic restructuring necessitates a parallel urban restructuring, with major cities in competition with each other to capture key global functions. Those gaining top positions would be on the cutting edge of high technologically driven production and producer services and would enjoy a position of power unprecedented in their history. There is little doubt that a major contributor to both the accelerated growth and the rising tensions over the built environment in Pacific Asian metropolitan regions is what Friedmann has summarized as world city formation.

STRUGGLES OVER THE BUILT AND NATURAL ENVIRONMENT

The enthusiasm by Asian governments for achieving world city status stands in contrast to the imagery presented by Western writers on the subject, most of which argue that the processes attending world city formation result in profound ethnic, racial and other social divides. Although a powerful vortex of global accumulation, the world city is seen as an arena of deepening social and political crisis. Direct confrontation with the requirements of world city formation occurs daily in the metropolitan centers of Pacific Asian countries in the form of struggles over slum demolitions, evictions of the poor, destruction of older petty capitalist business neighborhoods, conversion of rich agricultural lands to urban zones, loss of open spaces, and longer commuting distances for the hapless wage worker. In the case of Bangkok, for example, by the mid-1980s there were more than 1,000 slum areas with a total population ranging up to 1.5 million (Kaothien and Rachatatanun 1991). Internationalization and the building boom during the same period saw high-rise commercial development displace more than 11,000 low-income housing units – more than 100,000 people – within a 10-mile radius from the center from 1984 to 1988 (Padco-LIF 1990). Many of these households were compelled to move to the metropolitan fringe. Most, however, chose to relocate in other slum areas in the inner city where job prospects are not only better but the petty economy of slums and neighborhood self-help relations are well established.

Low-income households that are able to maintain a hold on land and housing in the metropolis face another threat: severe urban environmental deterioration that has its greatest impacts in and around slum and squatter settlements. Not infrequently, the poor are able to stay in the city because certain quarters are so environmentally unsafe that private housing or commercial developers are not interested in them. These locations include sites adjacent to polluting industries, such as the remnants of the famous Diamond Hill slum in Hong Kong located below and downstream from textile dyeing factories (Chan et al. 1994), along heavily polluted urban canals and waterways, on steep slopes that easily collapse during the rainy season, in low-lying areas subject to heavy flooding, along railroad tracks and, on a smaller scale, underneath bridges. These patterns support the thesis that the increasing vulnerability of Third World cities to natural environmental hazards is part of a circular process of residential occupation of marginal urban environments by households that, under land markets that increasingly favor the wealthy, are unable to secure land at less vulnerable sites (Main and Williams 1994).

But the poor are not the principal source of urban environmental deterioration in Asia. The crowding of low-income households into environmentally poor areas is taking place in great urban regions that are experiencing widespread environmental deterioration. Untreated sewage and industrial effluents have left major waterways – including the magnificent Chao Phrya river running through Bangkok – unable to support life; breathing the air has become hazardous to health; and heavy metals and other pollutants are destroying coastal ecologies and fishing industries. In Jakarta, urban expansion into upland areas is resulting in a loss of ground cover that intensifies both flooding and drought in the city and its hinterland. In this city,

which has one of the world's highest levels of suspended particulate matter in the air, recent World Bank estimates are that infrastructure needed to begin to meet environmental standards would currently require annual investments of US$1 billion (World Bank 1993). In Hong Kong per capita municipal wastes continue to rise and water quality in Victoria Harbor, already very bad, continues to deteriorate steadily despite government White Papers and campaigns to clean up the environment.

The open competition by governments and corporate actors for world city status discloses the nature of choices which privilege economic growth over providing livable habitats for all citizens at the cost of environmental sustainability. Given the high visibility of this conflict in the day-to-day urban experience, the relative absence of explicit treatment of the built and natural environment in world city literature has meant that critical linkages between other features of Friedmann's world city hypothesis and with his subsequent writings on poverty and alternative development are difficult to make. Specifically, if the relationships between poverty and world city formation are to be explored fully in terms of relations of power, the question of decision making and control over the formation of the urban habitat needs to be included in the analysis.

POVERTY AS (DIS)EMPOWERMENT: LESSONS ON ENVIRONMENTAL MANAGEMENT FROM ASIAN SLUMS

Although variations are significant, if poverty is seen as a condition of low levels of social, economic and political power, the globally attached processes of urban spatial restructuring and environmental deterioration have been a major contributor to it. Gaining access to land, housing, basic infrastructure and environmental resources occupy a substantial portion of the time and energies of low-income urban households throughout Asia. As earlier observed by Castells (1983: 312), the urbanization process in a capitalist world entails a commodification of the city itself that disrupts communities and cultures, leads to unbalanced patterns of growth and creates chronic crises in housing, services and collective consumption. While the more affluent classes respond with ever higher levels of consumption, often in fortified neighborhoods and air-conditioned cars that insulate them from the deteriorating social and physical scenes around them, the poor are crowded into environmentally degraded areas that are systematically denied basic infrastructure and services. Although some forms of environmental deterioration, such as air pollution, cannot totally be avoided even by the rich, they nonetheless affect the lives of the poor more severely.

CIVIL SOCIETY, THE STATE AND EMPOWERMENT

A great deal of energy is spent by the poor just on maintaining a place in the city and managing daily routines. Household divisions of labor, reciprocity among neighbors, community leadership, NGOs and government assistance can all help in reducing the time burden on poor households in carrying out these life-supporting routines. But even where those burdens seem overwhelming, events can bring to communities an extraordinary sense of urgency that is shared by enough households to lead to organized resistance and, further, mobilization for change. Galvanizing resistance is itself a complex phenomenon that, while emerging from civil society, invariably confronts the state either as a potential ally against landlords, for example, or as the targeted cause of the problem. As it broadens its inclusion, mobilization of citizens also entails raising the "moral high ground" to subsume a wide range of interests under a banner of righteousness (Apter and Sawa 1984). This capacity, too, is highly contingent on shared cultural and moral values that can be drawn on to overcome class and other social cleavages.

In Pacific Asia the struggles engaging social energies arise in a milieu of accelerated urbanization and globally integrated local economies that have been managed by highly interventionist states. Social mobilization as a form of empowerment has not been achieved by unending and entrenched anti-statist resistance, but rather by longer-term achievement of democratic reforms allowing for political association in civil society to be included in a territorially defined political community. Participation in political, social and economic affairs is the goal that translates social power into political power through, ultimately, collaboration with the state. Achieving this goal is, however, seen as a process of struggle that principally swells up from the grassroots with the support of mediating organizations also emerging from civil society. A central task of the (democratic) state is to sustain the territorial basis for civil society's claim for inclusion in the formation of political community.

While much has been written about regulation of the (international) economy by the "developmental state" in Asia, its more recent role has been to institute processes of political empowerment through the establishment of an inclusive democracy. Evidence from Asia, while mixed, has shown remarkable advances in this direction over the past decade, giving cautious support to the thesis that social mobilization can lead to fundamental political reform. Yet it can also lead, as in the case of Burma, to brutal suppression or, as in the case of Singapore, may not appear in any substantial way at all. Moreover, without progress in the other dimensions of empowerment, constitutional provisions for elected governments may fail to provide for the type of level playing field of inclusiveness. In the Philippines, for example, the strong patron–client relations rooted in rural landlordism and concentration of land ownership in the hands of powerful families reaches into the state and deeply erodes the potential of elected governments to carry out popular economic and social reforms. This "Latin America in Asia" syndrome has left this country, which after the Second World War was thought to be the brightest star in Pacific Asia, far behind other market-oriented countries of the region.

This brings to the fore the contrasting differences in state–civil society–economy relations in Asia. Table 1 presents a summary glimpse of some of the key contrasts among the Asian NIEs, a group of societies that are all too often lumped together as a single "four tigers" or "neo-Confucianist" development model, but are in fact markedly different in culture as well as in state–society relations (Berger and Hsiao 1990; Douglass 1994).

Table 1 suggests that the struggle for inclusiveness through activating civil society involves many different strategies and tactics, depending on particular constellations of relations in a given setting.

Table 1 Variations in state–civil society–corporate economy relations in East Asia, early 1990s

Dimension	Hong Kong	Korea	Singapore	Taiwan
Domestic capital	"Accommodationist," no explicit industrial strategy, indirect subsidies to firms via housing and infrastructure. Small-scale firms dominate	"Corporatist state" allied with *chaebol* – large-scale oligopolies created/regulated via nationalization of banks	"Extroverted corporatist," extreme reliance on TNCs and biased against domestic firms	"Entrepreneurial state": state-owned enterprises used to promote major sectors. Domestic industry dominated by small-scale firms
Transnational capital (TNCs)	Same as above	Least reliance on direct foreign investment (FDI); use of military alliance with US to leverage investors	Virtually exclusive reliance on TNCs	High reliance on TNC FDI, but selectively used to avoid direct competition with domestic firms
Labor	Indirect control via housing, price controls and management of international labor migration. Small-scale firms dampen worker organization	Direct support of proletarianization and direct suppression of labor via use of police power. *Chaebol* and "Fordist" factory	Largely indirect via housing, international migration control, wage controls. Unions largely eliminated in 1960s	Domestic controls via prohibition of strikes, limited unionization under KMT Party. Small-scale firms inhibit worker organization
Civil society	Colonial, managed by public welfare and other agencies	Authoritarian use of police power to suppress social movements	Overt use of state power to regulate behavior and muzzle press in the name of "communitarianism"	Authoritarian use of police power to suppress anti-KMT movements by indigenous peoples

FIVE

A key distinguishing feature among the countries of the region is the degree to which the state allows collective association to emerge from civil society, which has a decisive effect on community-level strategies across the gamut of concerns. Where, for example, community power structures are replaced by the state or otherwise inhibited from emerging from within communities, community mobilization often moves away from self-provisioning efforts to forms of routinizing requests for state assistance, as in Hong Kong, or to street protests and political agitation, as in Korea.

In some instances, such as in Indonesia, the state permeates urban communities through networks of elected and appointed officials reaching all the way down to the neighborhood level in municipalities. Here the lines between community leadership and government officialdom are blurred. This has a tendency to foster a mini-bureaucratic process of implementing government programs in a "soft," top-down fashion. Neither spontaneous organization outside of official lines nor the establishment of non-government organizations is encouraged or facilitated by this system. In Bangkok, relations between slum communities and government are perhaps among the least defined and, as a consequence, are much less routine or predictable. While community leadership is recognized when it appears, it is not a part of an official apparatus of the Indonesian type. The result is that government attention to the more than 1,000 slum areas of Bangkok is sporadic and partial, often on an issue-by-issue rather than a programmatic basis. At the same time, the absence of a strong state presence in the community may allow for more authentic grassroots organizations and leadership.

These differing experiences show that currently fashionable policy of forging "partnerships" between state and community not only has many possible configurations, but also is fraught with lack of conceptual clarity about such key concepts as participation, citizen rights and the political versus technical role of non-governmental organizations. If a common pattern can be identified, it would be that extensive state bureaucratic involvement leads to increased reliance on the state to provide community infrastructure and services. Contrary to the commonly presented view that the poor are somehow naturally inclined to depend on the state for investments for collective consumption, case studies show that such dependence, where it does occur, is a product of a longer history of state–community and, on a larger scale, state–civil society relations that could have produced much more active community involvement.

The renewed interest in decentralization, democracy and citizen participation prevalent throughout Asia represents a potentially major shift in state–community relationships in cities in almost all settings. This has already been partially reflected in more open attitudes by governments toward communities, nongovernmental organizations and political association outside of the state. Whatever concrete manifestations these changes may have will, of course, continue to differ among the various national and local contexts in Asia, but most governments have already come a long way from the anti-slum policies and eradication drives of the 1950s and 1960s. It is now more common for governments implicitly to recognize the existence of squatters and, to varying degrees, enter into dialogue about redressing the concerns of poorer communities and provide some forms of government assistance to at least a select number of locations. Whether they will now take the next step toward validating the legitimacy of poor people to reside in the city by recognizing their rights to have access to land and housing, to organize and select leaders, and to become equal counterparts in political and planning affairs is one of the most important issues of the current decade.

As it reflects on the plight of poor people, the general evidence from this ongoing mobilization of households and communities from within civil society is that accountability of government, even through relatively weak and indirect forms of democracy, brings positive (if limited) attention to slum and squatter settlements. Where government officials must stand for election in poor sections of the city, palpable evidence of representing the voice of the people must be given. In Bangkok, for example, local elections led to promises of piped water and electricity to a squatter settlement that were later realized. This is not to say that democracy, particularly in its limited and often token and co-optive forms, is a panacea for eliminating poverty or improving the environment. Regimes in power still tend to be located within a range of authoritarian, paternalistic and non-democratic modes of governance. The ideal of authentic state–community partnerships that is currently being put forth in development plans throughout Asia is, in most instances, still in search of a real-world application. Even within the realm of democratic action, much also depends on class relations and how the emerging urban middle class in Asia will either support or move against slum and squatter settlements.

Much may also depend on whether the poor can tag their political agendas on to what is predominantly a middle-class movement for political liberalization. This is one of the most troubling features of democratization drives in situations of widening social inequalities that, in part, follow from the economic growth processes of Asia's miracle economies. In key aspects of the construction of (new) urban space, such as in housing, the implicit agenda of the middle class to live in posh neighborhoods near commercial skylines moves against the quest for empowerment of the poor.

To many the bridge across classes is manifested in the flowering of what are summarized as non-governmental organizations (NGOs), which are generally staffed by members of the middle class and disaffected elites and take on social justice and, currently, sustainable development concerns. Recognizing that even in the best circumstances, poor households and their communities have clear limitations on what they can accomplish on their own, the general proposition placing hope on NGOs is that, without some form of outside non-governmental support, sustaining community self-empowerment efforts will encounter severe, often insurmountable difficulties. As one dimension of collective association from within civil society, the NGO movement has generated a vast literature on their expected and real roles, typologies and orientations. There are NGOs that are genuinely involved in selfless endeavor for the poor, but there are also ones that are thin disguises for charlatans to exploit the poor. Some are completely autonomous from the state, while others are de facto arms of the state.

Generalizations about NGO–community relations must therefore be treated with care. Although the ideal type of NGO might be the "empowering" type described by Lee (1992) and identified by Friedmann, it can also be allowed that circumstances will dictate which approaches are viable in a giving setting. In several countries, for example, the emergence of NGOs is still severely limited by governments that remain entrenched in top-down, non-participatory approaches to development planning. The role of NGOs is also necessarily limited to small-scale, charitable activities rather than open advocacy of, for example, squatters' rights in these situations. In other settings, such as in the Philippines and Thailand, NGOs have appeared in almost astounding numbers and play a much wider variety of roles. As yet there are, however, few NGOs in Pacific Asia that express interest in urban environmental problems in low-income communities. What can be called the environmental NGOs still tend to focus on global and rural environmental issues, and when they do focus on cities, they do so at the urban rather than community level. Community-based NGOs, on the other hand, tend to focus either on immediate crises such as housing rights and eviction threats, or on health and education programs, and rarely focus on either income-generating or environmental management questions. Thus even in cities where NGOs are highly visible, the appearance of mediating organizations concerned with livelihood-environment questions is still rare.

CONCLUSIONS: THE GLOBAL–LOCAL NEXUS OF CIVIL SOCIETY AND EMPOWERMENT

The outstanding question raised by linking world city formation with an alternative development is whether the urban poor can gain power in the global–local nexus of economy and polity. While disempowering relations are the subject of their resistance, on a day-to-day basis the resistance is often muted, taking the form of non-compliance, avoidance and subtle disregard for rules established from above. It also takes the form of self-exploitation, of expending household energies to sustain the basic conditions for the reproduction of their own labor beyond levels made possible by selling labor power in the market alone.

Occasionally, resistance is galvanized into moments of collective action and, more rarely, urban social movements that span ethnic and class divides. As witnessed by the shifting patterns of direct foreign investments in response to political upheavals and political reforms, such moments reach outwards to the global scale as well. There is, however, no overriding imperative that global–local interaction will either deliver the poor from poverty or condemn them to conditions of immiseration. While the resolution of this question is a dynamic one that involves layers of interaction moving from the individual and household to the global scale, it fundamentally rests on the mobilization of the poor themselves. In the longer term, for the rise of civil society in East and Southeast Asia to bring an end to poverty, it must also transform adversarial relations between the state and the poor into collaborative ones. A change from confrontation to accommodation may seem improbable in the light of current political realities; yet events such as democratization taking root in the authoritarian states of Asia, which seemed equally unimaginable only a decade ago, suggest more possibilities for such accommodation than ever before.

REFERENCES FROM THE READING

Apter, D. and Sawa, N. (1984) *Against the State: Politics and Social Protest in Japan.* Cambridge, MA: Harvard University Press.

Berger, P. and Hsiao, H.M. (eds) (1990) *In Search of an East Asian Development Model.* London: Transactions.

Castells, M. (1983) *The City and the Grass Roots.* Berkeley, CA: University of California Press.

Chan, C., Chang, F. and Cheung, R. (1994) Dynamics of community participation in environmental management in low-income communities in Hong Kong, *Asian Journal of Environmental Management*, 2, 1, 11–16.

Douglass, M. (1994) The "developmental state" and the Asian newly industrialized economies, *Environment and Planning*, 26, 543–566.

Douglass, M. and Friedmann, J. (eds) (1998) *Cities for Citizens.* New York: John Wiley & Sons.

Friedmann, J. (1992) *Empowerment: The Politics of Alternative Development.* Oxford: Blackwell.

Kaothien, U. and Rachatatanun, W. (1991) *Urban Poverty in Thailand.* Bangkok: Government of Thailand, National Economic and Social Development Board.

King, A.D. (1989) Colonialism, urbanism and the capitalist world economy, *International Journal of Urban and Regional Research*, 13, 1, 1–18.

Lee, Y.F. (1992) Urban community-based environmental management. Paper for Supercities International Conference on the Environment, San Francisco State University, October.

Main, H. and Williams, S.W. (eds) (1994) *Environment and Housing in Third World Cities.* New York: John Wiley & Sons.

Padco-LIF (1990) *Bangkok Land Market Assessment.* Bangkok: NESDB/TDRI.

World Bank (1993) *Indonesia: Urban Public Infrastructure Services.* Washington, DC: World Bank.

32

"'Global Cities' vs. 'global cities': Rethinking Contemporary Urbanism as Public Ecology"

from *Studies in Political Economy* (2003)

Timothy Luke

Editors' introduction

Timothy Luke is a University Distinguished Professor in Political Science at Virginia Polytechnic Institute and State University in Blacksburg, Virginia. He is one of the leading radical political theorists in North America. The interface of Luke's work with the topic of Part Five is *politics*, particularly *environmental politics* or *urban political ecology*. In this innovative text, which is an excerpt from a collection of articles on political ecology published by the Canadian journal *Studies in Political Economy*, Luke discusses the importance of ecological processes in the dynamics of global urbanization. Counterposing the limited number of "Global Cities" (as broadly identified by many authors in this volume) to the overall process of urban restructuring within "global cities" around the world, Luke directs our attention to the overall process of globalized urbanization. Reminiscent of Lefebvre's notion of "urban society" (see Reading 50) and Friedmann's (2002) concept of "complete urbanization," Luke's insistence on the pervasiveness of the urbanization process worldwide is a reminder that globalization does not just occur in a select number of distinct places. Pointing towards the intricate links between sociotechnological as well as bio-metabolic processes in the current phase of urbanization, Luke demands that we explore the political ecology of these relationships. Criticizing the tendency for urban ecological issues to be channeled into "subpolitical" processes of "private ecologies," Luke argues for a democratically open urban political process that is capable of dealing with the global and local inequalities of the current phase of globalized urbanization.

A POINT OF DEPARTURE

Human beings always have proven destructive to their natural environments. Until the twentieth century, however, this damage was either limited and local or it was more broadly widespread in only a handful of large conurbations centered in the imperial economies of the planet's northern hemisphere. Today, however, the inhabitants of hundreds of large cities all over the world are relentlessly reshaping the traditional and modern economies of every continent as they exert global and local demands in "glocalized" spaces for energy, foodstuff, information, labor, and material through world markets. Hence, this analysis responds to the call for paying more attention "to urban ecologies

Plate 23 Green London (Ute Lehrer)

and the policies developed around them as part of the formation of world cities" (Keil 1995: 293) by exploring the environmental impact of generalized urbanism or "global cities."

Because of this unchecked proliferation of such citified spaces in the twentieth century, it is no longer as clear that "Nature" is what surrounds humans in cities. Instead, one must ask if the world-wide webs of energy, information, material, and population exchange flowing between cities around the world now are infiltrating Nature so completely that this new artificial ecology will undercut entirely the survival of human and non-human beings?

Plainly, there are a handful of major metropoles, like London, Frankfurt, Hong Kong, Tokyo or New York, where the command, control, communication, and intelligence functions of transnational commerce are highly concentrated. Many researchers have investigated the peculiar qualities of urban life in these Global Cities, and they are, in many ways, the limit cases of global urbanism. In many other ways, however, focusing upon such extraordinary Global Cities misses another qualitatively different transformation unfolding behind the quantitative proliferation of urbanized living in all "global cities." While the work of Global Cities leads into the spread of "global cities," it is at the latter sites, rather than the former, where the rising level of a globalized urbanization is overwhelming the Earth's natural ecology to the point of threatening the sustainability of the entire planet's human and nonhuman life.

Today's "global cities," then, are entirely new built environments tied to several complex layers of technological systems whose logistical grids are knit into other networks for the production, consumption, circulation, and accumulation of commodities. Along with sewer, water, and street systems, cities are embedded in electricity, coal, natural gas, petroleum, and metals markets in addition to timber, livestock, fish, crop, and land markets. All of these links are needed simply to supply food, water, energy, products, and services to their residents. Thus, "global cities" leave very destructive environmental footprints as their inhabitants reach out into markets around the world for material inputs to survive, but the transactions of this new political ecology also are the root causes of global ecological decline.

In 1900, only 10 per cent of the world's 1.6 billion people lived in cities. During 2000, just over 50 per cent of the world's 6 billion people lived in cities. And, by 2050, 67 per cent of a projected population of 10 billion people supposedly will live in cities. Today's premier Global Cities, plainly, are intriguing, but the more ominous numbers posed by all "global cities" taken together are far more threatening. Urbanism on this scale is creating a set of contested regions where command and insubordination, control and resistance, communication and confusion, and intelligence and incomprehension must all be rejigged daily as transnational commerce dumps an ever-accelerating turnover of goods and services into the global economy. With over 50 per cent of humanity now residing in urban areas, the quantitatively-growing logistical pull of all global cities together constitutes a new political ecology whose demands have acquired such operational mass that they are qualitatively more distinct and interesting.

Cities do have ecologies, and the ecological impact of all global cities as a system of biopolitics is building up into this wholly new built and unbuilt environment. The collective of global cities begin to constitute a "world of near complete internationalized urbanization" (Keil 1996: 42). Consequently, the larger ecology implied by the aggregate collectives of global cities is "an array of urban ecologies: 'environments' in the plural" that must address "global populations, globalized everyday practices, and internationally diversified gender relations, as well as images and uses of nature" (ibid.).

Many discussions of Global Cities, then, typically approach them in "metageographical" terms, trying to gauge how much they fully subsist in a space of flows or still persist in the space of territories. A cartography of states driven by Westphalian logic of territorialized sovereignty still occludes the charts and maps that tracking flows would require. Many traditional efforts at Global Cities analysis, as the Loughborough University Globalization and World Cities (GaWC) group illustrates, maps mostly these intercity linkages to see how Global Cities provide the high-level command, control, and communication services needed by global economy through inter-city linkages among themselves. This work on Global Cities and their interconnected networks is inter-

esting, but these static views of the links, magnitudes of interaction, and primacies in various industries among Global Cities is not the whole story. All of these links taken together now also constitute a vast megalogistical collective whose aggregate ecological effects are redirecting the world's built and unbuilt environments. The "global cities" approach must ask how this new "organizational architecture for cross-border flows" affects all local and global ecologies, fixed territorial sites, and streams of commodity circulation (Sassen 2002).

The ecologies between, beneath or behind the Global Cities' organizational architectures are increasingly public, but subpolitical; largely artificial environments, but rooted in many layered unbuilt ecologies; globally flow-based, but locally frozen in particular territorialized material sites and spaces. Global Cities do make possible a new metageography, but this spatial frame has its own metaecological realities that become manifest in the logistics of all the many global cities being infiltrated and influenced by the high-end command, control, and communication dictates of a few Global Cities.

Rather than focusing upon that handful of Global Cities which serve as the core nodes in networks for global capitalism, we need to ask instead about the collective impact of all "global cities." As a planetary system of material production and consumption, these built environments constitute much of the world-wide webs of logistical flows which swamp over the conventional boundaries between the human and the natural with a new biopolitics of urbanism. Here critical environmentalism must fuse the concerns of public health with the goals of sustainable ecology in public ecology. Decisions that are made, and patterns that become fixed, in a subpolitical fashion must be identified, addressed, and corrected in a more political register. Public ecology is a strategy for opening these discussions and effecting these changes.

Global Cities are the usual suspects in the line-up of world cities. They are, typically, presented in a fairly conventional manner. They are seen as limited in number, tightly interconnected in function, located at the center or semiperiphery of the global economy, and formed by abstract forces. A global cities perspective tries instead to address how ecological sustainability, municipal politics, and global citification interconnect in "local social struggles that try to keep damaging consequences of globalization at bay" (Keil 1996: 38).

In this respect, Keil is quite right, "the world city is a place where the global ecological crisis manifests itself concretely" (Keil 1995: 282). Looking at "global cities" takes this insight, and then combines it with Friedmann's and Wolff's recognition that "world cities are the control centers of the global economy" by looking at world cities as control centers as well as controlled centered sites. Moving from a perspective that counts and measures all Global Cities, like the work done by the GaWC project, to one that gauges the overall impact of the general citified formations marked by "global cities" is much more important. Here ecology does not stand outside of, and apart from, urbanism, which permits a critical analysis that emphasizes "the social nature of nature and the natural basis of society," and leads to a point where "finding a strategy to solve ecological problems leads potentially to a democratization of society, economy, and the state" (Keil 1995: 285).

Focusing upon the "metageography" of Global Cities, and studying these urban forms as core nodes in commercial networks, is highly useful for understanding how they shape and steer the world's megalogistical systems as a "private ecology." Yet, the work of these five, ten or twenty Global Cities now has led to over a half of the world's human population living in urban settlements. Global Cities are small in number, but "global cities" are many in number. The impact of the carrying charges against the Earth's ecological carrying capacity for these hundreds of settlements should force everyone to see that privatized ecology of Global Cities as, in fact, a highly public ecology, which must be repoliticized, resocialized, and relocalized by environmental activists in many everyday struggles. The costs of allowing large corporate formations to privatize urban ecologies are unacceptable, and the metaeconomic questions about how humans should live with all other nonhuman beings and things must be addressed by public ecology (Harvey 2000). Public ecology can, of course, be discussed apart from "global cities," but the environmental challenges posed by sustaining the logistical grids that Global Cities have propounded for "global cities," as well as global towns and countrysides, provide a critical point of departure for this discussion.

URBANISM AS LOGISTICS

As the art of moving war *matériel* and quartering troops, logistics is about organizing and sustaining supply chains, but it also suggests the bigger issues of lodging, accommodation, and shelter for people as well as moving whatever materials are needed to sustain those activities. In many ways, cities essentially are concretions of logistics past, articulations for logistics present, and speculations about logistics future. And understanding these enduring elements in the creation of urban civilization is captured best by reconsidering the ecological links they create between, with, and after human beings and nonhuman things. From small changes in the daily traffic of *matériel* in human collectives, major ecological outcomes occur later, in succession to, and after such modifications in the movement of things and people.

Immense logistical spaces, then, are always carved out beyond, beneath or behind the flows of urban existence. They help produce the permanent quarters of urban space, which fixes the conditions for quartering of city residents. These spaces also materially concretize all the arts and sciences of the broader civilization underpinning global cities. Here one sees the complex codes, collectives, and commodities of global commerce creating products and by-products out of global society's logistical exchanges, which are all dramatically recontouring the world's economies and environments around a highly privatized transnational, but still mostly transurban, trade. Cities remain pivotal sites at which the everyday exchanges between built and unbuilt environments occur, but they also are where much of what is regarded as international relations between different spatially-divided economies, governments, and societies transpire.

Clearly, any single Global City is a particular site where larger forces burrow into a given place in specific, but also varied, ways, localizing the global in some determinate fashion (Keil 1998). Yet, all of these determinate formations in the aggregate now also add up in the collective megalogistics of global cities which a handful of Global Cities has made much more possible.

The grids of global cities simultaneously are works in the present for what is hoped to be greater future logistical efficiency as well as past products of what was once believed to be efficient logistical greatness. Citification has led to rich civilizations, but those cultural advancements typically were highly localized, rarely permanent, and still subject to decay and collapse. Only with the advent of global capitalism and industrial production over the past five hundred years have cities become much more than huge agricultural villages. Medieval London held fewer than 60,000 people, and its core area was only about 700 acres. By 1800, it had nearly 1 million people, and a large network of roads, horse-drawn public transport, and latterly railways were needed to move people and things within an urbanized region composed of many scores of square miles. Until 1800, the cities were by and large not unlike they were in A.D. 800 or A.D. 80 Only about 2.5 per cent of the world's population lived in cities in 1800, but this quadrupled to 10 per cent in 1900, and then quintupled again to 50 per cent by 2000. In 1800 only two cities in the world – London and Edo – held a million people, in 1900 ten did, but in 2000 almost 300 did. The world's urban population, in turn, grew from around 225 million in 1900 to 3 billion in 2000. While large cities covered about 0.1 per cent of the world's land in 1900, this figure grew ten-fold by 2000 to about 1.0 per cent.

To comprehend fully the destructive demands of today's transnational urbanism, one must accept how globalization is operating now in 2005. This acceptance is important if one hopes to understand how fully the reticulations of power and knowledge work in most locales through what Baudrillard has identified as "the system of objects" in culture, urbanism, and globalization on a local, national or global level. All of these terms, however, are quite mutable in their meanings, and they constantly are evolving everyday in new objectifications of the systems at play in objects – capitalism, nationalism, technology, urbanization – within globalization (Baudrillard 1981, 1996).

FINDING PUBLIC ECOLOGY IN POLITICAL ECOLOGY

Articulating the many ambiguous interconnections between urbanization and the environment requires one first to come to terms with privatized ecologies. Only then can globalization and the nature of modern urbanism, the boundaries of the political and the subpolitical, the nature of personal and public health, and the social formations that link health and the environment be connected as a public ecology. The recurring motif that emerges from this reconsideration is inequality, so to cope with globalism's ambiguities this discussion speaks in favor of funding a new normative discourse, or a public ecology, to guide critical thinking and political activity to improve life in both the unbuilt and built environment for human beings as well as nonhuman life. While recognizing that the concept of "public" brings a great deal of baggage with it, this notion still provides a useful point of departure for rethinking our collective ecological future in terms of sustainability, equality and justice.

While there perhaps is no single hegemonic power today, the dilemmas posed by preserving the health of the world environment suggest there is this hegemonic form of globality at work. Moreover, many globalists are willing to push certain conditions of consumption to advance globalization into where it does not yet exist, even though there are many resistances. Most political rulers as well as most corporate managers are all working quite openly to perfect this new private ecology. It is being built to bring many more global goods and services to consumers as a part of, first, their on-going programs to advance globalization, second, as an implicit sign of their globality, and, third, as a marker, complicitly of their shared submission if only for now, to globalism.

The privatization of collective ecological goods, as it is celebrated in the quest for greater performativity, is not advancing everyone's welfare. On the contrary, these practices are leading down paths that often are producing greater and greater malfare. Not only is the earth's "natural ecology" being degraded, but so too is the "social ecology" being neglected. Corporate visions of private ecology often discount healthy built environments, health care systems, and health centered lifestyles to advance global growth; hence, these utilities are not being maintained or not being developed at all. Likewise, a provision of potable water, edible food, safe housing, efficient sewerage, reliable hospitals, and effective medical care as mandatory features of many built environments has never been done.

There are severe inequalities at work today in global affairs. Some are very old, and well known.

Some are quite old, and only now being recognized. Some are new, and just now being felt. Most of them, however, can be tied back to unequal levels of access, power, status, and wealth, which are becoming so quantitatively unbalanced on a global scale that they are turning into something qualitatively different. The analytical tools in both global studies and environmental studies are perhaps not adequate to the tasks of interpreting what is now unfolding. Instead, too many of these existing tools occlude what needs to be analyzed, who needs to be criticized, and what must be done to overcome these trends toward powerlessness and inequality.

All too often, global studies is relegated to the realm of "Society" and its analysis is assigned to only the cultural and social sciences, while environmental studies are shuttled off to the domain of "Nature" and its consideration is given over exclusively to the biological and physical sciences. To really get at what is happening today, however, we need to focus on hybridities of Nature/Society at sites which intermix the natural and the social, like the "built environment," "natural history," or "political ecology" in privatized ecology. These amalgams of Nature/Society are what sustain and/or degrade overall levels of health and environmental quality for both human and non-humans, and they materially manifest themselves in patterns of urban settlement, industrial ecology, and natural economy.

A public ecology must fuse the administrative concerns of civic public health with the activist engagements of a critical political ecology (see the Public Ecology Project at http:// www.cnr.vt.edu/publicecology). Public ecology should mix the insights of life science, physical science, social science, applied humanities, and public policy into a cohesive conceptual whole. Public ecology should work at the local and global level to develop "pre-pollutant" or "noncontaminant" approaches to environmental problems by using political pressures to work back up the commodity chain to lessen ecological damage by mobilizing solutions drawn from collaborative management, green engineering, industrial ecology or vernacular design. Public ecology must show how private ecology has turned the built and unbuilt environments into a formation that is one and of a piece, not two and wholly separable.

CONCLUSION: RESISTING INEQUALITY

What surrounds one in Dallas is not what surrounds one in Delhi, but those different surroundings have high economic, political, and social costs inside and outside of both environments. A persistent feature of all global societies today are toxic wastes. Like weather, water, and wildlife, such waste is to be found everywhere in the planetary environment, making this by-product a new fixed characteristic of the Earth's ecology as it is being transformed by modern agricultural, industrial, and technological development. Nonetheless, many mechanisms in the world's political economy permit Dallas more than Delhi to dump more toxic wastes outside specific locales, boost their concentrations beyond permissible thresholds, raise exposures so intensively as to threaten health, and disperse effects indiscriminately across space and time. These irrationalities in the private ecology of global cities come from in a subpolitical realm, but they now are negatively affecting every political system on a global scale as transnational environmental problems. All of this, in turn, exposes the key metaeconomic issues raised by the metageography of Global Cities.

In the realm of the subpolitical, ordinary processes of democratic legitimation fail, because modern industrial revolutions with all of their profitable products and toxic by-products are highly technified economic actions. Each always "remains shielded from the demands of democratic legitimation by its own character" inasmuch as "it is neither politics nor non-politics, but a third entity: economically guided action in pursuit of interests." Because of property rights and expert prerogatives, most occupants of this planetary subpolis have yet to realize fully how "the structuring of the future takes place indirectly and unrecognizably in research laboratories and executive suites, not in parliament or in political parties. Everyone else – even the most responsible and best informed people in politics and science – more or less lives off the crumbs of information that fall from the tables of technological sub-politics." This elaborate subpolis evolves in the reified dictates of industrial ecologies, whose machinic metabolism, in turn, entails the planned and unintended destruction of nonhuman and human lives in many different environments (Beck 1992: 222–223).

Beck worries about how to face this modernity as he recognizes how fully "the possibilities for social change from the collaboration of research, technology, and science accumulate" in new loci of social order and disorder when real power and knowledge "migrates from the domain of politics to that of subpolitics" (ibid.). In the subpolis, activities that often may begin at an individual level as a rational plan combine at a collective level into the irrational, unintended, and unanticipated. It is difficult to resist these outcomes inasmuch as the workings of modern technics and markets are "institutionalized as 'progress,' but remain subject to the dictates of business, science, and technology, for whom democratic procedures are invalid" (ibid.).

The subpolis shapes, and then is itself shaped, in the global market's imbrication of the polis for humans and the subpolis of things. Modernity becomes an inegalitarian mechanism whereby the few who know-how and own-how maintain domination over the many who do not know-how or own-how. The illusion of progress through greater education and broader opportunity, in fact, always belies grittier realities of exploitative avarice fostered by growing disinformation and greater dispossession. Consequently, the subpolitically-structured inequalities in global cities need to be more closely policed in public policy and political practice to correct the inequalities of overall health and environmental quality behind today's economic crises and political contradictions. The notion of a public ecology offers a set of values and practices to push such decisions out of the subpolitical domain.

Environmentalism, urban studies, and public health are among some of the last remaining discourses available to provide some ethical consideration or political reflection regarding these issues. Private ecology degrades the overall civic life of society as the privileged millions still benefit from the international misery of billions. We can-not continue on this track if the Earth's ecologies are ever to be mended. Here one must not only talk about the iterated ecological links of the top ten Global Cities. Instead, we must consider the imposition of productive material extraction systems and material consummative discharge grids by all the settlements occupied by over half the world's human population. Ecology here becomes inescapably public, and thereby it becomes political. A public ecology can begin to develop a more formalized discourse about the unecological conduct and anti-environmental practice of living in global cities, which will help determine what must be done.

REFERENCES FROM THE READING

Baudrillard, J. (1981) *For a Critique of the Political Economy of the Sign.* St. Louis, MO: Telos Press.

Baudrillard, J. (1996) *The System of Objects.* London: Verso.

Beck, U. (1992) *The Risk Society: Towards a New Modernity.* London: Sage.

Friedmann, J. (2002) *The Prospect of Cities.* Minneapolis, MN: University of Minnesota Press.

Harvey, D. (2000) *Spaces of Hope.* Berkeley, CA: University of California Press.

Keil, R. (1995) The environmental problematics in world cities. In P.L. Knox and P.J. Taylor (eds) *World Cities in a World-system.* Cambridge: Cambridge University Press.

Keil, R. (1996) World city formation, local politics, and sustainability. In R. Keil, G.R. Wekerle and D.V. Bell (eds) *Local Places in the Age of the Global City.* Montreal: Black Rose Books.

Keil, R. (1998) *Los Angeles, Globalization, Urbanization, and Social Struggles.* Chichester and New York: John Wiley & Sons.

Sassen, S. (2002) Introduction. In *Global Networks, Linked Cities.* London: Routledge.

33

"The Neglected Builder of Global Cities"

from O. Källtorp et al. (eds), *Cities in Transformation – Transformation in Cities* (1997)

Anne Haila

Editors' introduction

Anne Haila is a Professor of Urban Studies in the Department of Social Policy, University of Helsinki. In recent years, she has focused her research on the role of state intervention in the regulation of urban property markets. As the epitomic global city researcher, Haila has lived in and conducted research on several global cities, including Singapore, Helsinki, Hong Kong, New York and Los Angeles. Building upon a conceptual critique of the global cities literature (which, unfortunately, could not be included here), Haila is concerned to examine "the politics of the global city," which she detects in most large urban centers around the world. In her eyes, it is the political economy of the *building* of the global city which needs more extensive research. Haila's particular interest lies in the politics of real estate. By identifying four areas of global convergence in the real estate development in global cities – global actors concerned to internationalize the city, their spatial strategies, their decision-making environment, and the landscape they subsequently produce – Haila lays out a map for "real" global city politics in globalizing cities throughout the world economy.

THE POLITICS OF THE GLOBAL CITY

Rather than defining global cities in terms of a local response to global economic forces or in terms of cities as necessary conditions for the new economic order, I suggest an alternative approach to the solution of problems concerning the scope and focus of global city research. This is to focus on the process through which cities try to achieve the status of the global city. Instead of global cities, therefore, I prefer to talk about the politics of the global city. My claim is that there has emerged a new type of urban politics that has gained popularity not only in world cities like New York, London and Tokyo, but also in small and middle-size cities. Similarities in urban politics are what characterize modern cities. If one defines "globalness" in this way, even small and middle-size cities can manifest the trends identified in New York, London and Tokyo.

The content of the politics of the global city may be outlined thus: a belief that attracting investments, especially foreign and real estate investments, represents a way out of the present recession. In order to attract investments cities develop a strategy in which the creation of the image of the city has an important role. An important medium to make the image known is the media. One consequence of this image promotion and use of the media is that local politics come to focus more on so-called "big issues." This makes the politics of the global city symbolic politics in which rhetoric has an important role.

THE NOVELTY OF THE POLITICS OF THE GLOBAL CITY

There is nothing new in local politics attempting to attract investments, industries and inhabitants. Wherein, then, lies the novelty of the politics of the global city? The first novelty is the increased importance of real estate investments and the increased influence of real estate investors. Lefebvre (1972) has suggested that the real estate sector has supplanted the industrial sector in importance. Baudrillard (1972) has claimed that the laws of production have become obsolete, and that consumption, symbolic exchange, simulations and signs instead are what is important. The claims by Lefebvre and Baudrillard are connected, not only in arguing for a decrease in the importance of the productive and industrial sectors, but also in making an implicit connection between the real estate sector and consumption: real estate capital has found an unexploited area of consumption, where it increases not only its own profits but also consumption through development of places of consumption.

If the politics of the global city of the 1990s are compared to the growth policies of the 1960s, which also attempted to attract investments – see, for example, Molotch (1983) – they differ in the sense that while the primary aim of the growth policies of the 1960s was to expand the productive sector and attract industries, in the 1990s real estate investments play a more important role than before, and a remarkable portion of investments solicited and received consists of construction and speculative investments. Granting planning permission for shopping centers and sports halls and investing in entertainment complexes and facilities in science parks are elements of the urban politics of the 1990s, together with donating industrial sites and giving tax breaks, the elements of the old

industrial politics. Harvey (1989) has used the term "entrepreneurialism," as distinct from "managerialism," to refer to that type of urban governance that favors entrepreneurs at the expense of the inhabitants. The politics of the global city also favors entrepreneurs, but even more real estate developers and investors.

The second novel aspect of the politics of the global city is the symbolic aspect. In explaining why local decision-makers can make decisions about development and land use and redistribute material benefits, Molotch made use of the distinction drawn by Edelman (1964) between two kinds of politics. "First, there is the 'symbolic politics,' which comprises the 'big issues' of public morality and the symbolic reforms featured in the headlines and editorials of the daily press. The other politics is the process through which goods and services actually come to be distributed in society . . . this is the politics that determines who, *in material terms*, gets what, where, and how . . . This is the kind of politics we must talk about at the local level" (Edelman 1964; cit. Molotch 1983: 348). The politics of the global city of the 1990s resembles symbolic politics more than the local growth politics of the 1960s. One reason for this change in local politics is the appearance of new influential agents in urban development: global real estate investors and developers. The image of the city advertised in the media, the symbolic reforms featured in the press, and the conformity to moral principles, indexed, for example, by crime rates and propensity to riots, are important indicators for global real estate investors considering the possibility of investing in the city. Therefore these are elements in the politics of the global city.

Symbols are important for real estate investors. Decisions to invest in real estate are based on estimates concerning future rents and value increases. Because of the unpredictable nature of the property market, these estimates are based on beliefs rather than facts. Beauregard (1993) has recently argued that in the United States there exists a discourse of urban decline that has motivated and assured people and investors, assisting them in the selection of place of residence and investments. This discourse is more important in motivating investors than facts. The politics of the global city are like this discourse in the sense that they affect the beliefs of global investors and reduce the calculated risk by giving a promise that the city will follow the familiar rules of the real estate game.

A comparison between Los Angeles and Singapore will illustrate the importance of conformity to moral principles. In the 1980s, Los Angeles was a city favored by Japanese investors. The riots in 1992 changed this situation. A property market where an asset can be demolished in one night is not a good market.

Singapore, a small city-state without hinterland, has been very successful in its global city politics. This success can be measured by the figures showing high economic growth, and by the fact that several international companies have selected Singapore as their location. For example, more than 70 Finnish companies have located in Singapore. The presence of some of these companies is surprising – for example companies operating in the forest industry. Why did these companies select Singapore, and not the neighboring Malaysia or Indonesia where their customers and markets are, and that are much cheaper than Singapore? The Singapore Government and the Economic Development Board (EDB) have a role to play in this success. EDB provides many high-class business services, such as up-to-date information about companies in Singapore in CD ROM form, and the city has plenty of high tech office space in intelligent buildings. In addition to this material infrastructure there is another factor that favors Singapore: the conformity to moral principles, Confucian values that promote economic growth.

The third novelty is the use of other cities, not as a model for new ideas nor as a lesson to learn from their experience, but as legitimation for certain measures. In Finland, we used to take England and Sweden as a model when we wanted to reform our planning law, and our laws were influenced by the laws in those countries. In the 1990s, Helsinki wants to beat Prague and become the European cultural capital in the year 2000. Helsinki wants to beat Berlin and become the gateway to Russia, and beat Stockholm as the conference city of Scandinavia. London with its "dirty streets and no Disneyland" is compared to Paris with its clean streets, Disneyland and mayor, in order to point out that London can lose out to Paris and that London needs a mayor (Hebbert

1992). Singapore's rival is Hongkong, and for example, high real estate prices in Singapore are justified by saying that in Hongkong prices are even higher.

The novelty of the global city politics is, however, relative. Boosterism and "cities as growth machines" are old practices in the United States. But in Finland and in other European countries this type of politics is new. In Europe, because of the feudal heritage of landed property (where the public domain plays a more important role) and the less commodified nature of landed property, the politics that create wealth and growth through land development and real estate speculation are new politics.

NEGLECTED ASPECT: REAL ESTATE

It is surprising that real estate has not gained popularity as a topic in urban studies. The global city literature has repeated the neglect commonly seen in urban studies. This is unfortunate because global real estate investment flows could have provided the explanatory link connecting cities, thereby clarifying the role played by control which remains unexplained in the global city literature.

Literally the term "global" refers to something that is worldwide. In very general terms, one way to legitimize the use of the term global is to argue that in recent years cities around the world – even formerly closed cities like those in China – have become open and dependent on each other. What is the nature of this dependency, and what is the link that connects cities worldwide? Rather than talking about control, or world cities as locations for international headquarters and important financial institutions, or as places where important decisions are made, I prefer to discuss another link that connects cities. This is real estate capital that has become more mobile: globally operating developers and real estate investors create a network in which developments in one city are dependent on events in another city. In what follows I will discuss four trends in contemporary urban development: those concerning actors, their strategies, their decision-making environment, and the landscape produced. Together, they show the interconnected nature of real estate investment flows and the politics of the global city.

Plate 24 Chongqing: the latest city in the world seeks the global image (Anne Haila)

FOUR TRENDS IN URBAN DEVELOPMENT

(1) The operation of global actors to create an international image for the city

Zukin (1992), in her comparison of London with New York, remarks upon the striking similarities between the landscapes in these cities, despite all the differences with respect to tradition, culture, and planning. She explains these similarities by referring to the fact that "the same worldly superstars, including developers, architects and private-sector financial institutions" design the landscape in all global cities. "A city that aims to be a world financial center makes deals with Olympia & York and Kumagai Gumi, welcomes Citibank and Dai-Ichi Kangyo, and transplants Cesar Pelli as well as Skidmore Owings and Merrill and Kohn Pederson Fox" (Zukin 1992: 203–204).

Modernism in architecture was an international movement. The politics of the global city is even more international. Jet set architects invited to global cities transplant their designs without consideration for locality, and replicate similar forms worldwide. The strange familiarity of global cities demonstrates the irony that while postmodern architecture pretends to celebrate diversity, it actually promotes uniformity.

In Helsinki, the American architect Steven Holl designs a museum of contemporary art. In Kuala Lumpur, Cesar Pelli, an Argentine whose firm is based in New Haven, designs the world's tallest tower, Petronas Towers. The building of Petronas Towers is an example of contemporary international development projects: the construction companies include Japanese, South Korean and Australian firms and most of the laborers are Indonesian, Bangladeshi and Filipino (reported in *Straits Times*, Singapore, February 9, 1995). Petronas Towers (450m) has been designed with other towers in the world in mind: the Sears Tower in Chicago (443m), the World Trade Center in New York (405m) and the Empire State Building (380m).

(2) The spread of similar methods of finance and construction

Broadgate in London has been welcomed as a novel type of construction, introducing a new term, "groundscraper" (Williams 1992). In Europe, where smallness of scale, grids, and details consistent with each other are the norm, groundscrapers that fill the whole block and consist of a variety of uses and styles are the new type of construction.

British urban scholars have pointed out that not only the product – the building – but also the methods by which Broadgate is produced, are new. Stuart Lipton, the chief executive of Stanhope Properties, imported the US concept of shell and core construction: the developer builds the frame and cladding and runs the main services into cores in the building; the client completes the fit out of the internal spaces. With energy, and sheer force of personality, Lipton set a fantastic space at Broadgate, achieving speeds of construction that were substantially faster than anywhere else in the world (Williams 1992: 255).

Broadgate is, perhaps, only the beginning. Edwards in his study of London's King's Cross argues that developers Rosehaugh and Stanhope have made remarkable innovations in finance and construction methods, following American practice. In finance, the new method is to marshall short- and medium-term syndicated bank loans on a huge scale and raise loans through partly-owned subsidiary companies whose debts are not declared in companies' balance sheets (Edwards 1992: 171).

(3) Foreign investments and the integration of markets

In the 1980s the Japanese began to invest heavily in the U.S. real estate market. Among the reasons behind this investment flow were the strong yen, the economic surplus in Japan, the change in Japanese monetary policy, low interest rates, the willingness of Japanese banks to provide low-cost loans, the scarcity of land for sale in Japan, high Japanese real estate prices and relatively low prices in the United States. The Japanese were not the first group of investors to invest in the U.S. property market. Although British, Dutch and Canadians were already there, the Japanese sparked off a rush to invest abroad, which was reinforced by real estate economists who recommended the global diversification of real estate investment.

Although the bids foreign investors make are dependent on local price levels, their capacity to bid is determined by their home market. In the situation where there is a surplus produced in the home market and the price level is higher than in the foreign market (reasons for investing abroad, as in the case of the Japanese), this capacity is higher than that of local bidders. When the Japanese landed in the New York property market, *The Economist* calculated that Japanese investors were willing to pay up to 5 per cent above the prevailing market values. The effect of this globally mobile real estate capital is to create a situation in which urban development and price levels in one city are dependent on urban development and price levels in other cities. This is one way in which cities can control each other.

(4) Buildings as signs

Investing abroad usually involves greater risks than investing in a local market, so investors need to be persuaded to invest in foreign markets. The politics of the global city signal that the point of view of the real estate investor is acknowledged in the city and can provide this kind of persuasion. The physical signs of the appeal of such politics are written in the urban landscape.

In the 1990s, buildings are not just frameworks for activities (like in the time of the productive capitalism), nor objects of investment (as in the casino economy and the speculative boom of the late 1980s), but are built and purchased for another reason. I will call this third reason, besides use value and exchange value, sign value. Three types of buildings as signs are trophy buildings, exclusive buildings and image buildings.

Trophy buildings give name, fame and prestige to the owner. An example is the deal carried out by Mitsubishi Estate Co. (the real estate division of Mitsubishi Conglomerate) in New York in 1989 in buying a 51 per cent interest in Rockefeller Group (the owners of New York's Rockefeller Center). The purpose of exclusive buildings is to segregate and exclude, for example a shopping mall that monitors customers with its electronic eye, thereby excluding those without purchasing power. Exclusive buildings have a role to play in segmenting the markets and guaranteeing monopoly prices.

The purpose of image buildings is to create a favourable image for the city in order to attract foreign investors. Image buildings are landmark buildings that give the city the appearance of the global city, like Petronas Towers in Kuala Lumpur. Such visible signs, which serve to persuade global investors and inspire their confidence, are especially important in the rising East Asia that, after all, still follows non-western rules. Although an accounting firm or a legal firm in the Western world can locate in a loft building in a gentrified neighborhood, in the East the presence of such firms in old Chinese shop houses deep in non-tourist Chinatowns would hardly inspire much confidence. "The rise of East Asia as an economic powerhouse has been accompanied by unprecedented building projects that seem to be bringing more and more 'North American' architecture to the exotic East. The tallest buildings outside of North America are now in Singapore and Hong Kong, but Japan, Malaysia, and Korea are not far behind" (Ford 1994: 271).

REFERENCES FROM THE READING

Baudrillard, J. (1972) *For a Critique of the Political Economy of the Sign*. St. Louis, MO: Telos Press.

Beauregard, R. (1993) *Voices of Decline*. Oxford and Cambridge, MA: Blackwell.

Edelman, M.J. (1964) *The Symbolic Uses of Politics*. Urbana, IL: University of Illinois Press.

Edwards, M. (1992) A microcosm: redevelopment proposals at King's Cross. In A. Thornley (ed.) *The Crisis of London*. London: Routledge.

Ford, L.R. (1994) *Cities and Buildings*. Baltimore, MD and London: Johns Hopkins University Press.

Harvey, D. (1989) From managerialism to entrepreneurialism: the transformation in urban governance in late capitalism, *Geografiska Annaler*, 71, 1, 3–18.

Hebbert, M. (1992) Governing the capital. In A. Thornley (ed.) *The Crisis of London*. London: Routledge.

Lefebvre, H. (1972) *Die Revolution der Städte*. Munich: Munich Syndicat.

Molotch, H. (1983) The city as a growth machine: toward a political economy of place. In M. Baldassare (ed.) *Cities and Urban Living*. New York: Columbia University Press.

Williams, S. (1992) The coming of the ground-scrapers. In L. Budd and S. Whimster (eds) *Global Finance and Urban Living*. London and New York: Routledge.

Zukin, S. (1992) The city as a landscape of power: London and New York as global financial capitals. In L. Budd and S. Whimster (eds) *Global Finance and Urban Living*. London and New York: Routledge.

F
I
V
E

34

"The Globalization of Frankfurt am Main: Core, Periphery and Social Conflict"

from P. Marcuse and R. van Kempen (eds), *Globalizing Cities: A New Spatial Order?* (2000)

Roger Keil and Klaus Ronneberger

Editors' introduction

In this chapter by Roger Keil, a Professor of Environmental Studies at York University in Toronto, and Klaus Ronneberger, an independent scholar based in Frankfurt am Main, global city formation is examined as both product and producer of local political conflict. Veterans of urbanist debates and struggles in Germany's prime global city, Keil and Ronneberger reflect on the ways in which global restructuring in the downtown city core and suburban peripheries of Frankfurt has been influenced by the political strategies of local politicians, social movements and other collective actors. Reading 34, which is a selection from a longer contribution to a book on "globalizing cities" (Marcuse and van Kempen 2000), is based on research that was originally undertaken in the early 1990s, when the authors were working in the GreenBelt planning project for the City of Frankfurt. Based on a hybrid methodology that combines political-economic analysis, cultural anthropology and participant observation, the authors worked closely with communities bordering the GreenBelt, which covers about one-third of the city's area. With a special interest in spatial conflicts that were emerging due to the expansion of the global city into the region and in conjunction with the specific plans related to the GreenBelt, Keil and Ronneberger mapped global city formation in Frankfurt as a landscape of sociospatial contestation. They explore, in particular, the fault lines that have emerged around the process of glocalization: the elite project of internationalization and transnationalization of Frankfurt's banking, services and transportation sectors sat uncomfortably with all manner of social movements, political groups and "local" communities, who politicized land use decisions, employment and social welfare issues and cultural struggles in all parts of the city. Inasmuch as the expansion of the global city beyond the Frankfurt city core also entailed a rescaling of political projects into the broader Rhine-Main region, new forms of sociopolitical mobilization – of both progressive and reactionary varieties – began to emerge in the spatial periphery, the newly urbanized exurbs on the metropolitan fringe. The main points to be taken from this reading are that (local) politics matters in the formation of world cities and that the sociospatial conflicts generated in the transnationalization process are regulated through an urban politics that frequently reaches beyond the municipal boundaries of the city core.

Plate 25 Conservative campaign advertising for the World City 1989 (Roger Keil)

INTRODUCTION

In this paper, we examine the ways in which globally induced socio-spatial restructuring has occurred both in the core and in the periphery of Frankfurt, and how social conflict influenced this process: while we recognize the dynamics and power of globalization and restructuring, we will show how local politics and social conflicts shape the transformation we describe. We also argue that in this process the very social (class, ethnicity and gender) and spatial (neighborhood, city, region, countryside) categories with which we commonly understand the urban region are changed as well in the process. Frankfurt's world city formation really began in the late 1960s. Local subsidiaries of internationally important growth sectors and new middle class residential populations started to evict existing economic and residential uses in the city center. The combination of the political regime of Mayor Walter Wallmann after 1977 and changing global conditions of capital accumulation in the 1980s repositioned Frankfurt in the

global interurban competition. During the 1980s, Frankfurt was transformed from a national economic core into a classical world city, a hub of the transnationally organized world economy. This has been expressed functionally and symbolically by the continuing growth of the downtown skyline of office towers, home to the headquarter functions of German and international financial capital. Clustered around and catering to the needs of the financial sector, one finds the globalized super-structure of the fairgrounds, the airport and a diversified sector of business services ranging from insurance companies to advertising and soft-ware development.

SPATIAL DIMENSIONS OF FRANKFURT'S RESTRUCTURING

The narrative of restructuring and globalization in Frankfurt is mostly one of growth and expansion – both in discursive construction and in material terms. We can roughly differentiate two growth dynamics: the continuing vertical and horizontal expansion of the core and a new type of self-sustained peripheral expansion. At first glance, growth in Frankfurt appears as a derivative of the centralized global city functions with the financial industry as their flagship. Yet, in recent years, significant parts of the regional world city economy have been established in the peripheries of Frankfurt-Rhein-Main. An increasingly self-sustaining horizontal growth in a para-urban zone beyond a 10 km radius and inside a 100 km radius around the center is linked to the advent of the global city in the countryside, i.e. the establishment of economic functions linked to the global economy in the urban periphery *without* mediation through the core. The periphery is becoming the prime location for important sectors of the post-Fordist economy (like software production, logistics etc.). The concentration of office towers in the downtown often eclipses the tendency of the "real" economic center of Frankfurt to shift away from the downtown into the forest on the fringe of the city: Frankfurt airport, the second largest in Europe, has developed into the crucial modem of the global flows of people, goods and information. The largest employer in the region, the airport places aggressive demands on space in the area. While the axis airport–downtown remains the single most important spatial connection in Frankfurt-Rhein-Main, airport-related activities are being diffused into the region.

The spatial form of the periphery is produced by a contradictory dynamic. The periphery has not just been colonized by the core but appears as the product of exchange processes both of core and periphery and of various subcenters on the fringe. One can speak of an oscillating growth dynamics between the urban core and the urban periphery. Growth in the downtown triggers expansion in the periphery; in turn, the newly emerging polynuclear nodes in the region engender higher density in the central city. Frankfurt is not just a hub of international capital but also the central place of an economically diversified European growth region with its homemade demands on the citadel. World city formation in Frankfurt shifted from a central focus to a regional focus. During the 1980s, conservative urban governments pushed centralized urban development based on the "citadelization" of the core of Frankfurt and neglected both individual

communities in the city proper and the surrounding region. Since the 1990s, some of the attention has been redirected towards the neighborhoods on the one hand and towards the region on the other hand. Whereas local actors in the periphery fought the world city's encroachment on their semi-rural neighborhoods, the metropolitanization of the *Umland* proceeded apace exactly because of the new attention given to formerly peripheral areas in the evolving multi-centered, nodal, flexible and globalized urban region.

POWER AND SPATIALITY

We are starting to find, in the periphery, a socio-spatial structure equally complex as (or even more complex than) in the inner city. This means that our traditional understanding of urbanity and urban civil society is insufficient. The difference of core and periphery implies a competition of different factions of urban elites fighting over territorial control of the city or its important parts, in order to put their specific stamp on the structured coherence of this particular urban region. The first line of power is redrawn horizontally mostly within the urban elite and the middle classes. During the 1980s' Wallmann-regime in Frankfurt, neo-urban (progressive and conservative) elites defined urban phenomena from the point of view of the core and continued to regard the periphery as a compensatory space (recreation, elite housing) and a container of problems (social housing megaprojects). By reclaiming the urban in its cosmopolitan world city incarnation as a positive force, these strata served as translators of global dynamics into local space. To the neo-urbans, urban oriented middle classes, who are still the most influential (popular) faction of the political class in Frankfurt, the city they recognize and use has swivelled to a set of highly selected partial spaces wherein the periphery has a minor role. The model behind the resurgence of "urbanity" remains the 19th-century modern (European) metropolis.

In the periphery, and mostly with the newly arrived former urbanites who we will call neo-rurals, concepts of "urban villages" have taken hold which borrow from images reaching back to pre-industrial urban form. The world view of the

Plate 26 Frankfurt: activist map (Roger Keil)

neo-rurals is marked by a peculiar contradiction. A large part of the peripheral population is fairly dependent on the economic power of the metropolis for their professional success and income, and many of them have been contributors to the urbanization of the periphery. Yet, it is particularly in the suburban areas, where the social and ecological consequences of the "growth machine" are starting to be felt, that various forms of "slow growth" and "no growth" movements spring up. The suburban actors, often economically strong and politically astute individuals and groups with professional backgrounds and histories of activism, distance themselves from both the city and the region by urging to concentrate urban ills in the inner city or by advocating to concentrate problem areas outside of their neighborhood elsewhere in the region. While they downplay the fact that they consume the cultural and commercial amenities of the central city, they would not want to miss the growing skyline as the symbolic representation of the power of the region and an aesthetical backdrop to their semi-rural life-styles.

SOCIAL STRUGGLES AND SPATIAL CHANGE: CORE AND PERIPHERY IN TURMOIL

But the suburban middle classes are not alone in their secret admiration of the globality represented by the Frankfurt skyline. Dany Cohn-Bendit, a leader of the French student rebellion in Paris 1968, was Frankfurt's municipal councillor for multicultural affairs in the 1990s. In an interview in 1996, he observed: "My route (on bicycle) leads me past the university and down the Bockenheimer Landstraße. Here, in the early 1970s, we fought to stop the destruction of the lovely patrician homes – and lost. In their place stand postmodern office complexes, the symbols of Frankfurt's position as the financial center of Germany and Europe. My nostalgia for the noble old houses – which in our youthful zeal we occupied but whose leveling we failed to halt – is tempered with the pleasure at the sight of Frankfurt's new skyline. (But please don't tell anyone: my friends here would be shocked!)" (Cohn-Bendit

1996: 17–18). As the erstwhile squatters and radicals of the 1970s adjusted their political values and esthetics and moved into positions of power locally and beyond, a new political landscape started to emerge in the global city of Frankfurt. But the "cooptation" of some previously critical groups of activists into the fold of the global city has created new and sometimes surprising actor constellations. If globalization leads to a newly partitioned city, it does so only through the filter of localized political processes. The structural changes in the Frankfurt region over the past decade have been influenced, resisted, produced and furthered by social and political activities of local agents. The emergence of a new urban form as described above is, in fact, connected to a distinctively novel sphere of social and political agency. During the past forty years, growth periods in Frankfurt have always been linked to forms of local resistance against the disadvantages of such growth for the resident population. Urban social conflicts usually erupted where the new phase of expansion manifested itself most visibly. At the end of the 1960s, when the downtown began to spread into the residential areas closest to the inner city, a culture of resistance developed in the largely middle class Westend. It resulted in a violent housing struggle (*Häuserkampf*) and a preservation ordinance (*Erhaltungssatzung*). In the early 1980s, when the expansion of the Frankfurt airport was pursued as the most obvious sign of Frankfurt's world city ambitions, the region saw its most severe social disturbances and struggles of post World War 2 history.

The displacement of residential population from the inner city neighborhoods and the conversion of apartments into office space were the main issues of urban discourse in the 1970s. The governing Social Democratic Party was intent on controlling emerging conflicts. This implied a more repressive stance of the local state. At the same time, however, more comprehensive and democratic planning processes were instituted. Increasing political repression against the radical squatters alongside more open and participatory planning processes made urban politics in Frankfurt during that era a contradictory affair. It eventually led to an irresolvable crisis of legitimacy for the Social Democratic government, a weakening of its powers vis-à-vis both the right-wing

opposition and popular resistance. This phase of heavy social struggles ended with the end of a phase of radical opposition against the expansion of Fordist mono-structures and the dissolution of local Social Democratic hegemony by 1977.

During the 1980s, a host of inner city groups were continuing the struggle against the destruction of housing and community spaces by the expansion of the office economy. In both rhetoric and action, these groups were rehashing the anti-speculation-politics of the 1970s housing movement. Often based in the remnants of the 1970s left wing radical milieu of communes and collectives, this traditional movement of resistance to the global city gained political support in the 1980s mostly from two groups. Both the fundamentalist wing of the local green party and working class activists from the trade union movement, from the far left of the SPD and from the German Communist Party (DKP) operated as the cadres of neighborhood-based struggles. In the case of the Campanile office tower project across the street from the central train station, a combination of political activism, legal rulings and a slump in the real estate market led to a surprise success of neighborhood political activism in keeping a major symbol of global economic forces out of the community. This project would have threatened the integrity of this neighborhood with a majority immigrant population. At the same time, however, in other parts of the city, this type of neighborhood-based defensive activism against gentrification and conversion of housing into office space, based on left wing ideologies and working class neighborhood cultures, became increasingly irrelevant politically and was eclipsed by other forms of social protest. Throughout the end of the 1980s and into the 1990s, squatting as a form of defensive appropriation of urban living space and as a politics of resistance to gentrification became less relevant in the political discourse as a whole and even more isolated in the neighborhood than it had been in earlier years.

Throughout the 1970s and into the 1980s, opposition to growth was fuelled by left-wing analyses of the urban process and integration of the urban struggle into the general context of anti-capitalist politics in the city as well as by petty bourgeois groups in defense of inner city residential areas. In contrast to the eclectic 1970s combination of revolutionary rhetoric on the one hand and steadfast defense of property values and living space on the other hand, resistance to growth has now become a populist endeavor which does not lend itself to simple ideological classification. The cultural hegemony of modernity has given way to a contradictory and fragmented reality. Correspondingly, social agents in local movements are critical both of traditional forms of habitus and of specific consequences of modernization. Actors now often combined conservative, anti-modern ideology with an acceptance of technological or economic modernization without much effort. Local resistance to growth has, therefore, become increasingly hard to classify in terms of its material or ideological background.

Materially, social and political agents are faced with a new political space created by peripheral urbanization. In Frankfurt, this new emphasis on peripheral spaces as the real centers of growth turns areas which had until recently been agricultural and rural into complex modules of world city formation. Spaces in the geographical periphery of the city become crystals of current urbanization where contradictory growth dynamics occur in a variety of forms: zones of affluence mix with marginalized housing areas; industrial development happens adjacent to GreenBelt-planning; new housing construction both creates the need for and competes spatially with new infrastructures such as transportation corridors. This new thrust of urbanization was in turn met with protest and political resistance where it had or threatened to have its major effects. After the "discourse of the metropolis" had been preeminent throughout most of the 1980s, the impacts of world city expansion now began to be noticed far beyond the citadel of the financial core, in the exurban spaces of the agglomeration. What growth meant was put up for discussion by new social groups who had mostly not been considered politically relevant during previous periods of expansion: long time village dwellers, farmers, suburbanites. This complicated the very meaning of growth in public discourse. Class remains the most important qualifier in articulating residents' relationship to growth – both business and residential. The experience of social class (here understood in terms of income and property ownership) is embedded in a set of cultural conditions. Spatiality, ethnicity, gender, urban versus rural

identity are some of these conditions. They influence the way in which social collectives and individuals attach themselves to questions of expansion in place-specific and peculiar ways that can only be understood through detailed empirical analysis. While such attachments and articulations follow certain general rules – like those elaborated upon by the political economy literature in the United States (Logan and Molotch 1987) – they often take on unpredictable and often surprising forms.

In the western suburb of Sossenheim, for example, the encroachment of expansion triggered by and shaped by world city growth (economic and demographic) has led to a unique constellation of social and political struggles articulated in complex ways with novel and traditional social identities. Long time, German residents of housing estates feel threatened by both immigration and the expansion of the metropolitan character of their neighborhood. They tend to perceive most changes as encroachments on their life worlds and local identities. These identities have a core of class positions but are strongly influenced by the location of social groups on other socio-political faultlines: old versus new, native versus immigrant, periphery versus city, property owner versus tenant, and so forth.

In the North of Frankfurt, the single most important peripheral conflict erupted when the redgreen municipal government decided to move the city's slaughterhouse from the banks of the Main River to a suburban location in the Northern district of Nieder-Eschbach. The old site at the river was to be used for a major housing project which was part of the City's ambitious plan to use the axis of the River as one of the main organizing principles of Frankfurt's urban space. In addition to angry protests against the slaughterhouse project, the City also met with fierce resistance by mostly conservative citizens over major housing projects on the urban periphery.

The configuration of protest in the North of Frankfurt was characterized by a populist particularism which links the desire for local identity and the disdain for centralized urban planning on the one hand with aggressively protectionist status politics and the exclusion of marginal segments of society on the other hand. This political brew has led to a brand of populist regionalism and has created a challenge both to traditional left wing

Plate 27 Frankfurt: northern suburb (Roger Keil)

movement politics and to the established conservative mainstream in city politics. In any case, the new movements represent a clear departure from the notion that the social movements in Frankfurt were hegemonized by the left wing or green milieu alone. Rather, as was expressed by the near total eclipse of Green Party politicians in the uprising of the North, the new populist milieu, while having connections to earlier periods of struggles, social institutions like trade unions or environmental groups and established parties, has developed an original and fairly independent stance (which is not to say that their ideas were new and without influence from outside developments). Far from being the lunatic fringe of Frankfurt's suburban political space or "uppity garden elves" as they were viewed by many of the traditional neo-urbanist political class in the center of Frankfurt, the new social agents are politically savvy, professionally educated middle class activists whose populist agenda has anchored them deeply in the social space of the periphery.

CONCLUSION

The new spatial model has been sketched above with only the most important new lines of vertical and horizontal power shifts, but it clearly puts new demands on the regional mode of regulation which remains stuck in the logic of core–periphery. This is true both for conflicts pertaining to issues of spatial restructuring inside the municipality of Frankfurt and for intraregional affairs involving a number of local states. Politicians have hesitated to realize that Frankfurt by no means is autonomous but "the center, the inner city" of the Rhein-Main Region. But in the mid-1990s, the global restructuring of Frankfurt-Rhein-Main is articulated mostly in a discourse of regionalization. This discourse mirrors the aspects of the current urban crisis: the metropolitanization of the region, "fattening" of the urban periphery and social peripheralization of the core. Metropolitanization means the dissolution of the city into the region. The regionalist discourse builds on but increasingly replaces the discourse of urbanity which during the 1980s had focused on the development of the core. Regionalizing the urban crisis also means that conflicts are now spread all over the territory of Frankfurt and its environs. In contrast to previous images of core and periphery, the region is now increasingly being viewed as a complex urban landscape from which a new consciousness, a new politics, a new economic regime of accumulation as well as a new regional mode of regulation emerge.

The new geography of industrial and commercial development engenders a new structure of urban conflict. Generally, the importance of struggles over urban space has increased. Both core and periphery are in turmoil. Lines of social and spatial distinction and separation are redrawn in these conflicts. The very meaning of social status, territorial community, and urbanity is redefined as new spatial practices and representations evolve. Political and social practices of exclusion, especially racism, redraw the lines of racialization and prejudice and lead to new forms of discrimination. Policies and reformist concepts like multiculturalism gain currency and redefine identities of social groups. In the central city, socially marginalized people, remaining liberal middle class groups, and residual forces from older urban social movements wage a defensive battle against the takeover of the built environment by global capital, mostly symbolized by more office towers. On the edge of the city, farmers and suburbanites have embarked on a campaign against the threat to their livelihoods, privileges and lifestyles posed by what they call "urbanization."

While the causes of urban growth in Frankfurt appear in rather abstract terms at first sight (glob-

alization, restructuring), it has become clear that the real globalized flows of capital and people moving into and out of the city are very concretely guided, fought over and facilitated by local actors. Local actors rarely initiate growth and decline but they do have to deal with the consequences of each new period of creative destruction. For the time being, competing centralized neo-urban notions and peripheral neo-rural notions of regionalism are engaged in a struggle of middle class hegemony in an entirely globalized urban region. Competing concepts of regional hegemony include politically articulated ideas and practices of exclusion, segregation and separation. Frankfurt's immigrant communities have started to express their identities and demands on the political world of the global city in their own terms and with the help of policy reforms at the municipal level. Our analysis of Frankfurt shows that spatial restructuring cannot be understood as a purely statistical event in which social categories (class, ethnicity, gender) or spatial categories (neighborhood, city, region, countryside) remain unchanged. Social and spatial restructuring rather must be seen as a complex process resulting from and producing social struggles and the articulation and disarticulation of different spatial images of local and global agents in the urban. As of yet, no clearly delineated regional mode of regulation has emerged. In the absence of a viable progressive alternative both to unrestricted access of global capital to every nook and cranny of the region and to restrictive conservative-populist tendencies, the search for a new regionalism based on communitarian democracy remains a main task.

REFERENCES FROM THE READING

Cohn-Bendit, D. (1996) Frankfurt: my homeland Babylon, *Enroute*, April, 17–18.

Logan, J.R. and Molotch, H.L. (1987) *Urban Fortunes*. Berkeley and Los Angeles: University of California Press.

Marcuse, P. and van Kempen, R. (eds) (2000) *Globalizing Cities*. Oxford and Malden, MA: Blackwell.

35

"Urban Social Movements in an Era of Globalization"

from P. Hamel, H. Lustiger-Thaler and M. Mayer (eds), *Urban Movements in a Globalizing World* (2000)

Margit Mayer

Editors' introduction

Margit Mayer is a Professor of Political Science at the Free University Berlin. She has also taught at many American and Canadian universities. Her work in the United States and Germany has provided her with a unique perspective from which to explore the globalization of urban social movements on both sides of the Atlantic. Mayer is a leading social movement researcher who works predominantly in the New Social Movements tradition, which both acknowledges the wider socioeconomic contexts of movements and focuses on the identity changes that occur in and through social movement struggles. Mayer's recent work has focused on the interactions of urban social movements with local states, particularly in the spheres of homelessness, nonprofit organizations and employment policy. In this reading, she examines the intersection of urban social movement activity with processes of worldwide interurban competition. Mayer discusses the erosion of traditional welfare rights and the restructuring of urban governance under these circumstances. Mayer's analysis underscores the contradictions and conflicts associated with urban social movements, as well as the opportunities for social transformation they appear to open. In so doing, she indicates the key role of urban social movements in the process of global city formation.

This chapter discusses first the changes in local politics which recent urban research has identified as reactions to globalization practices, and then relates today's prevailing movements to these shifts in urban politics, in order to, finally, discuss strategic implications and options for local actors. The focus is mostly on major cities, which, regardless of historical tradition, geographical location, or general level of economic development, are increasingly tied into global flows and networks in very similar ways.

NEW TRENDS IN LOCAL POLITICS AND OPPOSITIONAL MOVEMENTS

Reviewing the literature on recent developments in urban governance, there is consensus that three trends in particular are significant and novel on the level of local politics:

- the new competitive forms of urban development,
- the erosion of traditional welfare rights,

and the expansion of the urban political system, also described as a shift from "government" to "governance."

Each of these trends has provoked or influenced new or existing urban movements, which I will present now in the context of their respective urban political setting.

The contemporary forms of urban growth and development consist primarily of the efforts of cities to upgrade their locality in the international competition for investors, advanced services, and mega-projects. Local political actors everywhere emphasize economic innovation, seek entrepreneurial culture, and implement labor market flexibility in order to counter the crisis of Fordism and to meet the intensified international competition. Other policy areas are increasingly subordinated to these economic priorities. The higher the position of a city within the global competitive structure of the new economy, the more important the role of advanced services in the central business district, and the more intense the restructuring of urban space. For global cities in particular, which compete as much for foreign investment and the economic megastructures of internationally oriented growth as they do for world class culture, everything from the production of the built environment to the priorities of the municipal budget has become subject to the function of the city as command center and its corresponding service industries. But other major cities have also seen the rebuilding and expansion of their downtowns into producer-oriented service centers. The intense tertiary development in the central business districts and the construction of huge new infrastructure projects have had undesired consequences for large parts of the resident population, because their effects have been gentrification and displacement, congestion and pollution, and often the loss of traditional amenities. While the city centers are being turned into luxury citadels, other neighborhoods are turned into preferential sites for unattractive functions, yet others are given up to abandonment. Further, the rebuilding of the central city has entailed an urban expansion into peripheral "green pastures."

Opposition movements have formed both in the cities and at their peripheries. They have either built on existent (latent) networks or organizations, or have sprung up anew, and they range from defensive and pragmatic efforts to save existing quality of life or privileges (which are sometimes progressive, environmentally conscious, and inclusive, but other times selfish, anti-immigrant, or racist) to highly politicized and militant struggles over whose city it is supposed to be (as in anti-gentrification struggles or movements against other growth policies).

Social movement research has produced most work on those mobilizations that are to protect the home environment – of too much traffic, too much development, or any other project which people don't like to have "in their own backyard." These often middle class based, quality-of-life movements frequently succeed in averting an unwanted facility (NIMBY [not in my backyard]), with the effect that then a poor or minority neighborhood is targeted.

Case studies show that such groups quickly become skilled at a variety of tactics and repertoires such as petition drives, political lobbying, street confrontations, and legal proceedings. Researchers tend to lament the fact that social justice orientations, which used to characterize the goals and practice of such citizens' initiatives during the 1970s, have been replaced by particularist interests and/or a defense of privileged conditions. But there are also case studies of local movements composed of working class and middle class participants mobilizing against highway construction plans, traffic congestion, or housing shortages, and, particularly in U.S. minority/working class communities, against polluting industries and hazardous facilities, with which they are disproportionately burdened. The action repertoire of such groups goes well beyond that of the defensive NIMBY movements: beside direct action (demonstrations, blockades, corporate campaigns) to put public pressure on polluting firms, they also undertake independent analysis of urban problems, and they demand representation on relevant decision-making boards.

Frequently, movements against urban growth policies and gentrification are directly triggered by what have become increasingly used instruments of big city politics. These include large, spectacular urban development projects, such as London's Docklands or Berlin's Potsdamer Platz; festivals such as the Olympics, World Expo, international garden shows or 1,000 year birthdays; or the attraction of mega-events, sports entertainment complexes, theme-enhanced urban entertainment centers – all of which depend on the packaging and sale of urban place images. The movements have attacked the detrimental side effects of and the lack of democratic participation inherent in these strategies of restructuring the city and of raising funds, and they criticize the spatial and temporal concentration of such development projects, as they prevent salutary effects for the city overall. The concentration on prestige projects tends to detract attention and finances from other urban problems and to restrict investments in other areas. Thus, protest campaigns against these forms and instruments of city marketing raise questions of democratic planning that urban elites concerned with intra-regional and international competitiveness like to downplay. Furthermore, they have the potential of bringing otherwise scattered local movement groups together in broad coalitions (as happened, for example, in the NOlympia Campaign in Berlin in 1991–1993). A leading actor in such campaigns are often radical, so-called autonomous movements, who consciously seize on the importance image politics have gained in the global competition of cities, and seek to devise image damaging actions to make their city less attractive to big investors and speculators, to creatively prevent the takeover of the city by "global capital."

The trend of the eroding local welfare state has been another trigger of the structural change in the profile of urban social movements. It has two elements. First, there is the dualization of labor markets, the expansion of precarious and informal jobs, and the shift in social policies, which produced a new marginality, the most visible manifestation of which are the tens of thousands of homeless inhabiting major cities. Other, less visible forms of social exclusion and new poverty also concentrate in urban areas, even if their causes are increasingly identified in global processes. Secondly, since the image of cities is playing such an important role in attracting supra-local investment, stern anti-homeless and anti-squatter policies have been drafted, and regular raids are carried out at the showcase plazas of all major cities. This kind of regulation of public space has been observed since the early 1990s, even in cities with "progressive"

governments, which have also adopted laws that prohibit people to sit or lie on sidewalks in business districts. In order to drive out beggars, homeless people, or "squeegee merchants" from the center of the cities (where they concentrate for a variety of reasons), these groups are being constructed as "dangerous classes" or "enemies of the state." Social policies have been abandoned in favor of punitive and repressive treatments.

In reaction to the combination of these trends, new poor people's movements have sprung up and actions by supporter groups and advocacy organizations, frequently also anti-racist initiatives. Research findings on the forms of self-organization by the new poor are most scarce, which is – among other things – due to the fact that most authors assume this population to be not just poor and without resources but also disempowered and passive. In fact, the resources of these groups consist primarily of their bodies and time, so that their protest activities tend to be episodic and spontaneous, local in nature, and disruptive in strategy. At best, their disruptive tactics block normal city government operations and threaten the legitimacy of local policies of exclusion. This was the case, for example, when the homeless in Paris defended their right to the city in a campaign around the slogan "droit au logement" ("The Right to Housing") during the mid-1990s. In general, though, such new poor people's movements face an increasingly recalcitrant and punitive state and it is only under rare conditions that their struggle against efforts to drive them out of the downtowns, their setting up of encampments, holding public forums and making demands on the city allows them to develop solidarity, political consciousness and organizational infrastructures – i.e. the elements of which social movement research assumes that they are preconditions for the emergence of mobilization. However, when resource-rich political advocacy groups dedicate themselves to the problems of the homeless or when professional activists make their resources available to such organizations, durable and effective mobilizations can be achieved.

Next to churches, political activist groups, and local coalitions, in Germany another relevant support network is provided by autonomous movements and anti-racist initiatives. These latter groups scandalize the production of new poverty

Plate 28 Amsterdam: squatter symbol (Roger Keil)

and homelessness while also mobilizing against their own eviction from squatted buildings and "liberated areas" of the city centers. In Berlin, where the city government has marked 14 so-called "danger zones," from where individuals looking "suspect" may be deported, newly formed initiatives stage protest demonstrations, provide legal aid, and put public pressure on the local government. In June 1997, a "Downtown Action Week" took place in 19 German and Swiss cities to create public awareness and pressure about this widespread practice of driving out the new marginality from the core areas.

Another new movement form has emerged in the context of housing need and new poverty, though its members do not see themselves as a "poor people's movement." The majority of the so-called "Wagenburgen," i.e. groups of people squatting on vacant land, living in trailers, circus wagons or other mobile structures, see their action as a form of resistance against the dominant pattern of political, social and economic relations in German cities. There are about 70 to 80 such sites in Germany; of the 15 in Berlin most have been in

the downtown area and thus threatened by eviction or already displaced to other locations. Their political orientations cover a wide spectrum: while some use the freedom this lifestyle allows them for political activism (such as sheltering refugees without legal status), others are content to explore alternative ways of living. But evictions and the threat of evictions have brought them together in campaigns to pressure city governments to tolerate the sites, delay construction, or provide other acceptable locations.

The new conditions on the labor market and the shift from social welfare to more punitive workfare policies have impacted on the urban movement scene in further ways. Not only have hundreds of new organizations sprung up, nonprofits run by and for the homeless, the unemployed, and the poor, but also the number and variety of institutions and projects "servicing" the marginalized has exploded, many of them function within municipal programs that harness the reform energy of community-based groups. Their labor seeks not just to "mend" the disintegration processes which traditional state activities cannot address, frequently they develop innovative strategies acknowledging the new divisions within the city.

Finally, the third novel trend in urban politics is that the local level of politics has gained renewed significance (and in the process has transformed itself), simply because the concrete supply-side conditions making for structural competitiveness can neither be provided by multinationals' strategies nor by uniform national policy (Mayer 1992). These conditions can only be identified and implemented at the local level of politics. Local politics, which seek to make local economies competitive in the world economy, are increasingly organized in partnership with an extended range of non-governmental stakeholders holding relevant resources of their own (such as private finances, local knowledge, community-based or locality-specific expertise). Besides investors and chambers of commerce, also education bodies, research centers and local unions have become such partners, as well as voluntary sector groups and associations, including former social movement organizations (Mayer 1994). This trend, which political science has described as shift from government to governance arrangements, means that the state's involvement, besides that of a plurality of other actors, is becoming less hierarchical and more moderating than directing (Jessop 1995).

The opening up of the urban political system to non-governmental stakeholders and the strategy of many municipalities to employ former social movement organizations in the development and implementation of (alternative) social services, cultural projects, housing and economic development has been a new and important force shaping the trajectory of urban movements since the 1980s. By including and funding third sector organizations the municipalities hope to achieve political vitalization as well as financial relief, though these goals frequently conflict with each other.

For the movements, this trend posits a particularly ambivalent opportunity structure, because it makes parts – but only parts – of the urban movement sector into "insiders." This development is rather more advanced in North America and Great Britain than on the European continent. Community-based and client-oriented groups now play a polyvalent function, in more and less developed forms, in and for cities all over North America and Western Europe. The establishment

of alternative renewal agents and sweat-equity programs and the funding of self-help and social service groups was in most places a long and contested process, but since the late 1980s municipal social and employment programs everywhere have been making use of the skills, knowledge and labor of such movement groups. Similarly, many cultural projects have become part of the "official" city, and youth and social centers play acknowledged roles in integrating "problem groups" and potential conflict. The participation of community and movement groups in different policy fields has, in other words, become routinized, especially in fields where both groups of the alternative sector and the political administration are keenly interested in solutions, such as urban renewal, drugs, immigrant integration, AIDS, or unemployment.

The bulk of the research focusing on these novel forms of institutionalization of social movements within the shifting relations of welfare systems and provision emphasizes the "contestatory character of their constituency" and the counterweight they pose to "conventional views of local economic planning" (Lustiger-Thaler and Maheu 1995: 162, 165). Whether in the economic development sector, the field of alternative services, or that of women's projects, the work of the groups is generally found to be an innovative and progressive challenge to public policy, as improving access to the local political system, and providing potentially more active citizenship.

Closer examination, particularly of the more recent developments, reveals, however, that these (former) movement organizations that have inserted themselves into the various municipal or foundation-sponsored funding programs play a rather complicated role within the urban movement scene. On the one hand, they enhance organization building and lend stability to the urban movement infrastructure and thus to the conditions for continuing mobilization. But on the other hand, the widening and the growing internal differentiation within the movement sector has led to new conflicts and antagonisms. The movement organizations now participating in the new governance arrangements are subject to the danger of institutional integration, "NGOization," and of pursuing "insider interests," and their own democratic substance is far from guaranteed. Especially since these organizations find themselves threatened by cuts and are faced with the reorientation of public sector programs toward labor market flexibilization, competition among them for funding has intensified, and the groups engage more in private lobbying strategies to secure jobs and finances instead of creating public pressure. Furthermore, some of the alternative renewal agents and community-based development organizations, who are busy developing low-income housing or training and employment opportunities for underprivileged groups, find themselves criticized and attacked by other movement actors who do not qualify for the waiting lists or who prefer squatting or other non-conventional forms of action.

While such attacks serve to illustrate new polarization tendencies and antagonisms within the movement sector, a series of indicators points toward an interpretation that the inclusion of movement groups in revitalization and other partnerships means, for many, that they become tied up with managing the housing and employment problems of groups whose exclusion by normal market mechanisms might otherwise begin to threaten the social cohesion of the city. This kind of instrumentalization of (former) movement groups thus harbors the real danger that their reform energy evaporates in the processing of urban disintegration tendencies or might even be used for the smooth implementation of state austerity policies.

But at the same time it is also the case that the increasing dependence of city governments on such (former) social movement organizations for processing the complex antagonisms within contemporary cities does also enhance the chances for tangible movement input. While this dependence is meanwhile institutionalized with the routinized cooperation between the local state and the former social movement organizations with regard to community economic development, client-based social services and women's centers, these new partnership relations are also beginning to influence interaction between the local state and movements described under the first two categories, i.e. no-growth and anti-poverty movements. The eroding local competence and dwindling resources which many city governments are suffering increases the pressure on the local political elites to negotiate and bargain with movement representatives within the channels and intermediary frameworks generated by the wave of routinization

of alternative movement labor in the context of municipal (employment or revitalization) programs. Thus, today's movements making a stand on the use value of the city, such as environmental and poor people's movements, now may also expect to profit from the new culture and institutions of non-hierarchical bargaining systems, forums, and round tables. It is true though that these new structures of governance are open to the less progressive, xenophobic, and anti-social movements as well.

THE ROLES AND OPTIONS FOR URBAN MOVEMENTS IN GLOBALIZING CITIES

The specific socio-spatial context which cities provide for social movements, as well as its consequences for the dynamic and development potentials of movements, needs to be further differentiated for a contemporary assessment. Due to the position of cities within globally and regionally restructuring hierarchies a variety of city types, with different conflict patterns, is emerging, which means that the homogeneous pattern of conflicts and movements of the Fordist era is dissolving. Metropolitan regions at the top of the global hierarchy develop particularly pronounced conflict patterns along the internationalization of their working classes and neighborhoods, their precarious labor relations (made use of especially by migrants), and their eroding municipal powers. At the same time, large metropoles facilitate the emergence of a critical mass, which is precondition for the building up of movement milieus and the construction of collective projects and identities. This is where movements against central city development in favor of global headquarters as well as new poor people's movements proliferate and may expect – because of the presence of handed down movement cultures and institutions (such as community organizing in North America and leftist political organizations in Western Europe) – both support and instrumentalization.

Old, deindustrializing cities on the other hand feature struggles over plant closures and new employment possibilities, and, depending on the profile with which the city seeks to reposition itself in the new urban hierarchy, more or less intense cooperation between the municipality and community groups. Cities trying to make their fortune as "innovation centers" frequently provoke environmental protest and slow growth movements with this strategy. Cities transforming themselves into module production places are particularly dependent on cheap and flexible labor and thus provide a difficult terrain for movements struggling for social citizenship rights and sustainable development. Such different movement activities would need to be analyzed in order to systematically explain the heterogeneous picture of urban movements in the 1990s.

Even though such differences among urban movement milieus are far from adequately researched, the stocktaking of some of the changes movements have undergone as presented in the preceding section allows some preliminary statements about their current role and possibilities for action. The argument here is that just like the movements of the phase of the 1970s and 1980s have contributed to shaping and changing the forms of governance as well as the structure of the city, the movements active in and around the city today play a role, if a contradictory one, in contributing to and challenging the shape and regulation of the city. While their practice with innovative urban repair and their inclusion in municipal governance structures may well feed into the search for (locally adequate) post-Fordist solutions and arrangements (making the movements appear "functional" and co-optable), their challenge of undemocratic and unecological urban development schemes may yet contribute to a more participatory and more sustainable first world model of city. In order to realize this potential, however, the new problems confronting urban movements of today have to be addressed head-on.

One of these new problems is the new antagonisms within the movement sector, which are also a product of the restructuring of the urban polity that has expanded and now includes some but not others in its governance arrangements. Besides this tendency to produce new forms of marginalization and new "losers," a second new problem demands attention, that is the evidence that the inclusion of movement groups in revitalization and other partnerships has meant, for many, that they become tied up with managing the housing, employment, or survival problems of groups whose exclusion by normal market mechanisms might otherwise begin to threaten the social cohesion of the city. And finally, a third problem arises with the pressures to entrepreneurialize the social and community work of these groups, as funding support for them is increasingly only available through workfare programs and microcredit arrangements. These structurally new constellations have to be acknowledged and their specific constraints – as well as the opportunities peculiar to them – have to be identified.

At the same time, the conquered positions and new institutional avenues described above offer opportunities that allow tackling the new problems. The growing role of local politics, even within global contexts, and the simultaneous inclusion of a variety of non-governmental stakeholders, including former movement organizations, into local politics have made new avenues available for those forces amongst the urban social movements that can seize them and that can tease out their ambivalence. But rather than doing so only for particular defensive spaces or individual threatened privileges, the challenge consists of making use of these avenues in the complex struggle for a democratic, sustainable and social city.

Some urban theorists see this struggle as one between global elites and local communities, reduced to the simple antagonism between distant powerful forces (such as global capital) and local victims "retrenched in their spaces that they try to control as their last stand against the macro-forces that shape their lives out of their reach" (Castells 1994). Such an idealized view of local movements would already have been problematic for the 1960s and 1970s, when the majority of urban movements was still part of a larger social struggle against broadening forms of domination. Today's local movements certainly cannot in their entirety be listed on the positive side of the ledger, since they are such highly differentiated products of recent shifts in urban politics (their proliferation as well as their fragmentation can be shown to result from the three trends described above). They themselves are contradictory and complex agents in the shaping of globalizing cities and have to deal with the new fragmentation within the movement sector as well as with massive social disintegration processes increasingly characteristic of urban life. The institutionalized, professionalized or entrepreneurial movements which now benefit from routinized cooperation with the local state, frequently want nothing to do with younger groups of squatters and cultural activists. Because of their preoccupations due to the new funding structures, they are often at quite a distance from the growing marginalized and disadvantaged social groups. But since the latter's organizations and forms of resistance do not automatically lead to mobilization or widespread support, it becomes crucial that those parts of the movement sector that enjoy some stability, access, resources and networks devote part of their struggle to creating a political and social climate where marginalized groups can become visible and express themselves.

Movement actors will thus need to acknowledge and make transparent their new dependencies (both on state and market) in order to identify the opportunities that exist under contemporary conditions. The new and difficult task consists in transforming the funds and the stability of the resource-rich movements into support for precarious movement groups. Existing opportunities, whether workfare programs or poverty initiatives, need to be seized and used to attack and to restrict marginalization and discrimination at the root of the new form of poor people's movements.

Urban movements need to politicize the new inequality and they can exploit the new access structures and the dependency of the new negotiation frameworks on local residents' input for this purpose.

REFERENCES FROM THE READING

Castells, M. (1994) European cities, the informational society, and the global economy, *New Left Review*, 204, 18–32.

Jessop, B. (1995) The regulation approach, governance and post-Fordism: alternative perspectives on economic and political change?, *Economy and Society*, 24, 3, 307–333.

Lustiger-Thaler, H. and Maheu, L. (1995) Social movements and the challenge of urban politics. In L. Maheu (ed.) *Social Movements and Social Classes*. London: Sage, 151–168.

Mayer, M. (1992) The shifting local political system in European cities. In M. Dunford and G. Kafkalas (eds) *Cities and Regions in the New Europe*. London: Belhaven.

Mayer, M. (1994) Post-Fordist city politics. In A. Amin (ed.) *Post-Fordism: A Reader*. Oxford: Blackwell, 316–337.

F
I
V
E

PART SIX

Representation, identity and culture in global cities: rethinking the local and the global

Plate 29 Potsdamer Platz, Berlin (Roger Keil)

INTRODUCTION TO PART SIX

Culture as practice, talk about culture, cultural politics, cultural production and products are all – and have always been – an integral part of the formation of world and global cities. British social theorist Raymond Williams (1976: 77) noted in the mid-1970s: "Culture is one of the two or three most complicated words in the English language." Culture is a multifaceted concept that includes notions of civilization, product, anthropological practice and so forth. Additionally, Williams distinguishes culture as *material* production and culture as a *signifying* or *symbolic* system. In the most colloquial and widespread usage of the term, "culture is music, literature, painting and sculpture, theater and film" (Williams 1976: 80). Additionally, and quite centrally for our discussion of globalizing cities, we can also identify culture as being closely related to forms of sociological difference – culture as ethnicity, for example.

The most direct and intuitive connection to culture that has been made in the literature on global cities has been in work on multiculturalism. Often without making use of the term "culture," Friedmann and Wolff and other early analysts of global city formation noted that the arrival of migrants from the global periphery enhanced social polarization within major urban regions (see Reading 6 by Friedmann and Wolff; see also Reading 12 by Ross and Trachte). Yet, the reality of cultural diversity in world cities remained fairly underexplored in most accounts throughout the 1980s and 1990s. The arrival of the "South" in the world cities of the North led many observers in the 1980s to declare that "the Third World has come home"; many argued, further, that the new diasporic populations were not coherently integrated into social formations that were frequently hostile to immigration from non-European countries (Amin 2002). Subsequently, as world cities emerged as more clearly recognizable hubs of global migration during the 1980s, their multicultural character attracted greater academic attention (see Davis 1987). As Saskia Sassen (1998) has noted more recently:

> The large Western city of today concentrates diversity. Its spaces are inscribed with the dominant corporate culture but also with a multiplicity of other cultures and identities. The slippage is evident: the dominant culture can encompass only part of the city. And while corporate power inscribes these cultures and identifies them with "otherness" thereby devaluing them, they are present everywhere.
>
> (Sassen 1998: xxxi)

In the late 1990s, Leonie Sandercock's two-volume project *Cosmopolis/Mongrel Cities* (1998, 2003) explored the transformation of everyday life within large, diverse cities through the lens of planning history and theory. As outlined symptomatically in Reading 36, Sandercock believes that the condition of multiculturalism is now quite typical of major global urban centers. Sandercock presents multiculturalism as her "self-declared social project for the mongrel cities of the 21st century," and suggests that its ideology contains a "multiplicity of meanings" (2003: 100–101). Drawing upon work by British cultural studies theorist Stuart Hall (2000), Sandercock proposes a multicultural perspective for the 21st century composed of nine premises emphasizing (1) the cultural embeddedness of humans; (2) the evolving and hybrid nature of culture; (3) the necessary relationality of cultures; (4) the politically contested character of multiculturalism; (5) the right to urban difference that lies at the core of multiculturalism; (6) the perpetual contestation of the right to difference against other human rights; (7) "an agonistic democratic politics that demands active citizenship"; (8) a sense of belonging that must be "based on a shared commitment to political community"; and (9) the need to address the "material and cultural conditions of 'recognition'" (Sandercock 2003: 102–103). Sandercock notes that "[t]he 21st century is indisputably the century of multicultural cities and societies" (2003: 105).

Other authors share Sandercock's view that the right to difference and the right to the city are central to an understanding of today's global cities, but adopt a more critical and less explicitly norm-ative approach. Toronto-based urbanists Kanishka Goonewardena and Stefan Kipfer, for example, are "suspicious of not only the inflated promises of bohemian 'creativity,' but also postmodern valorizations of aestheticized ethnicity: variations on the theme of liberal pluralism that furnish, wittingly or not, a much needed human face for bourgeois urbanism" (Goonewardena and Kipfer 2005). They argue against a

> debilitating disposition shared by otherwise quite different political positions: *culturalism*. For the major devotees, detractors and deconstructors of difference who are engaged in the debate on multiculturalism in Canada all tend to operate with fundamentally culturalist conceptions of identity, while maintaining a symptomatic silence on socio-economic divisions that are especially influential in the everyday life of big cities.
>
> (Goonewardena and Kipfer 2005)

In this manner, Goonewardena and Kipfer advocate an integration of studies of urban cultural dynamics and struggles with critical analyses of urban political-economic divisions.

Another scholar who has creatively linked the original world city hypothesis with an analysis of cul-tural dynamics has been Ulf Hannerz, who has systematically explored the transnationalization of urban cultural processes (see Reading 37). Building upon his extensive anthropological studies, Hannerz exam-ines global cities as places that are produced in the forcefields of local and global cultural interactions; he also explores the everyday processes of exchange and consumption that unfold on the streets of major cities. Analogously, Sharon Zukin has analyzed the dynamics of cultural contestation in the built environments of global cities (see Reading 16). Working with the notion of a "landscape of power," Zukin (1991) perceptively chronicled the emergence of a globalized post-industrial urban culture. For Zukin, the culture of cities can serve as a basis for transcending everyday urban life and also as a "powerful means of controlling cities" (Zukin 1995: 1). Consequently, as Zukin shows, urban cultures may serve as a method of "sorting out" different types of urban dwellers, be it in public space, private housing markets, streets or museums. At the extreme end of this development is the powerful role that cultures of difference play in gentrification (Smith 1996).

A departure from western modernist and postmodernist projects has occurred in "postcolonial stud-ies," which have also begun to influence global city theory. For instance, Anthony D. King builds upon the notion of postcolonialism, a complex and contested term which he defines as a combination of material practices and discursive formations that emerge "after the colonial" (King 2004: 45). Beyond this relatively straightforward use of the term, the notion of postcolonialism also denotes the type of knowledge produced in postcolonial studies, a distinct theoretical approach that has emerged in close conjunction with the migration of diaspora academics to metropolitan universities. Postcolonial studies is focused on questions of knowledge and power, "issues of agency, representation and especially, the representation of culture(s) under asymmetrical political and social conditions" (King 2004: 48; see also Reading 38 by King).

Cultural dynamics also figure crucially in the production of built environments, urban form and sociospatial divisions around the world, from the global cities of the West to the expanding metropolises of the global South. As King (Reading 38) shows, a culture of distinction associated with gentrified built environments has now reached the rapidly expanding exurbs of Indian and Chinese world cities such as Bombay/Mumbai and Shanghai. As King (1996) explains,

> The increasingly differentiated cultural discourse on the city is itself self-generating and part of the accumulation of cultural capital in the city. It is linked to the massive expansion of the cultural industries in (especially) the western metropolis: to publishing, education, the media, the symbolic and representational realms of advertising, and the "aestheticisation" of social life.
>
> (King 1996: 3)

In recent years, public and academic debates on urbanism have been strongly influenced by the notion of creativity. On the one hand, the consolidation of creative industries (such as media and cultural products) across the network of global cities has been explored systematically by Stefan Krätke (see Reading 39). As Krätke explains, various clusters and cliques of media production centers, so-called "global media fields," have emerged in close conjunction with the world city network (see also Scott 2000, 2005). At the same time, other scholars have argued that demographic diversity, tolerance (e.g., the "gay index"), overall creativity and an ability to attract talent represent as important bases for urban economic development (Florida 2002). The idea of "creative cities" frequently presupposes that particular types of urban images must be mobilized in order to project the identity of particular places into the global space of flows. This theme is explored at length in Reading 40 by Ute Lehrer. Much like Anne Haila in Reading 33, Lehrer argues that the "willing" of the global city in post-unification Berlin rests on a distinctively culturalist discourse of "spectacularization." Lehrer emphasizes the importance of the culture of the built environment and the discourse on urban design for the status and competitive strategies of global cities (see also King 2004).

In sum, as several contributors to Part Six argue, the global city must be viewed as a project of cultural hegemony in that it creates an ideological and material domain that is based upon the domination of western cultural formations. Indeed, in many ways, global cities research has itself been permeated by a form of cultural imperialism insofar as it has been grounded predominantly upon western cultural forms and intellectual paradigms (see Reading 26 by Robinson). In Reading 41, Nihal Perera's incisive observations on New York City and Colombo raise our awareness of this entrenched problem in global city research. Likewise, in Reading 42, Steven Flusty's recent project of "making the globe from the inside out" can be viewed as an effort to call into question some of the prevalent conceptions of globalization that have dominated much of global cities research. In reflexively grappling with these issues, the readings in Part Six also point towards new and productive ways of thinking about globalized urbanization that transcend conventional Eurocentric categories, assumptions and narratives.

References and suggestions for further reading

Amin, A. (2006) Ethnicity and the multicultural city: living with diversity, *Environment and Planning A*, 34, 959–980.

Appadurai, A. (1996) *Modernity at Large*. Minneapolis, MN: University of Minnesota Press.

Davis, M. (1987) *Chinatown, Part Two. New Left Review*, 164, 65–86.

Davis, M. (2000) *Magical Urbanism*, revised edition. London: Verso.

Florida, R. (2002) *The Rise of the Creative Class*. New York: Basic Books.

Flusty, S. (2004) *De-Coca-Colonization*. New York and London: Routledge.

Goonewardena, K. and Kipfer, S. (2005) Spaces of difference: reflections from Toronto on multiculturalism, bourgeois urbanism and the possibility of radical urban politics, *International Journal of Urban and Regional Research* 29(3), 670–678.

Hall, S. (2000) Conclusion: the multicultural question. In B. Hesse (ed.) *Un/settled Multiculturalisms*. London: Zed Books.

King, A.D. (ed.) (1996) *Re-Presenting the City*. London: Macmillan.

King, A.D. (2004) *Spaces of Global Cultures*. London and New York: Routledge.

Sandercock, L. (1998) *Towards Cosmopolis*. Chichester: John Wiley & Sons.

Sandercock, L. (2003) *Cosmopolis II*. London and New York: Continuum.

Sassen, S. (1998) *Globalization and its Discontents*. New York: The New Press.

Scott, A. (2000) *The Cultural Economy of Cities*. London: Sage.

Scott, A. (2005) *On Hollywood*. Princeton, NJ: Princeton University Press.

Smith, N. (1996) *The New Urban Frontier*. London and New York: Routledge.

Williams, R. (1976) *Keywords*. Glasgow: Fontana.

Zukin, S. (1991) *Landscapes of Power*. Berkeley and Los Angeles: University of California Press.

Zukin, S. (1995) *The Cultures of Cities*. Cambridge, MA and Oxford: Blackwell.

36

Prologue

"Towards Cosmopolis:
A Postmodern Agenda"

from *Towards Cosmopolis* (1998)

Leonie Sandercock

COSMOPOLIS: DREAM AND REALITY

According to *The American Heritage Dictionary of the English Language*, "cosmopolis" is "a large city inhabited by people from many different countries." Working with this definition, the world already has many cosmopolises – but the word itself tells us nothing of their qualities, or their quality of life, so I would prefer to call such cities cosmopolitan metropolises, or metropolises that are characterized by significant cultural (racial, ethnic, and sexual) diversity. We need to construct a normative cosmopolis, a Utopia if you like, but a Utopia with a difference, a postmodern Utopia to which I will not ascribe built form, and which I insist can never be realized, but must always be in the making.

A NEW WORLD DISORDER: FEAR OF "THE OTHER"

These are the three dominant socio-cultural forces of our time: the age of migration, the rise of post-colonial and indigenous peoples, and the emergence of a range of so-called minorities (women, gays, etc.), hitherto invisible/suppressed, as political actors. Linked with the destabilizing effects of global economic restructuring and integration, these new forces are literally changing the faces of cities and regions, which are becoming much more culturally diverse. This cultural diversity is generating processes of socio-spatial restructuring which are becoming a focal point in the managing of cities and regions, in their urban governance and planning. There would appear to be no way back to a static or homogeneous urban/regional culture and politics, despite the nostalgic appeals of parties of the right with their promises of a return to an earlier time when life was "simpler" – that is, when women knew that their place was in the home, and indigenous, ethnic, immigrant and other troublesome groups hadn't begun to make claims for their own space. Multi-ethnic, multi-racial, and multi-national populations are becoming a dominant characteristic of cities and regions across the globe, and this is causing a profound disturbance to the values, norms and expectations of many people. The multicultural city/region is perceived by many to be much more of a threat than an opportunity. The threat is many-layered. It is perceived as economic, as cultural, as religious, as psychological. It is a complicated experiencing of fear of the Other, alongside fear of losing one's job, fear of a whole way of life being eroded, fear of change itself. These fears are producing a new world disorder – rising levels of violence against those who are different, who don't belong, rising levels of racism, and increasingly repressive responses to the claims of insurgent citizenship. From the growing political strength of the right in France, to the United States' and Germany's recent repressive legislation against immigrants, to the bloodshed in West Kalimantan between indigenous dayaks and Javanese immigrants, or the attempt by ultra-orthodox Jews to drive secular Jews out of their neighborhoods and out of Jerusalem, these global processes of cultural differentiation and transformation and their spatial expression as conflicts over who can live where, and with what social and economic and cultural rights, is emerging as one of the most profound problems of the next century.

THE POLITICS OF DIFFERENCE

If cultural imperialism and systemic violence are features of contemporary global urban and regional changes, then a politics of difference is a prerequisite for confronting various oppressions (Young 1990). A politics of difference is a politics based on the identity, needs, and rights of specific groups who are victims of any of the faces of oppression discussed above. The emergence of numerous social movements in the past two decades – feminism, gay liberation, Black power, indigenous rights – embodies the practice of a politics of difference, perhaps the most important aspect of which is a discursive politics. By that I mean the effort to reclaim and revalorize the meaning of difference by asserting the positive qualities of the particular group and refusing to accept the dominant culture's definition of itself. In asserting gay pride, or Black is beautiful, there is a reversal of the devaluation of difference and an effort to overcome the internal colonization of selves by the dominant culture's definition.

The emergence of this identity politics has not gone unchallenged, by either left or right, and a number of concerns do need to be aired. One is the assumption of group unity and homogeneity that often seems to be implied, or stated as essential to identity politics. Clearly none of the social movements which have asserted a positive group identity is in fact a unity. This has been most systematically discussed in the women's movement, where the importance of attending to differences among women has now been well established. Another perceived problem with a politics of difference is the alleged impossibility of working on broader agendas of social, economic, and environmental justice so long as oppressed groups insist on only fighting for their group-specific concerns. But while this may have been a reasonably accurate interpretation of identity politics when it first emerged, it is evident in the late 1990s that many of these social movements are now engaged in broader coalition politics to achieve precisely the broader issues of social justice which are seen to transcend group interest. Further difficulties with identity politics may arise from the increasing "hybridity" of global populations, "the inbetween character of increasing numbers of people" (Bhabha 1994): for example, the Turk who has been living in Australia for thirty years and still identifying with Turkey, until he or she makes a trip "home," only to discover that he or she no longer feels "at home" there; or the third-generation "Asian origin" British person. The point here is that a politics of difference is not based in essentialist notions of identity but in situations, historical contexts, in which there are social relations of domination. Hybrid identities are just as vulnerable to stereotyping, vilification, and exclusion. A more serious charge against identity politics is that it, too, is oppressive, both to those inside the group in question (demanding allegiance to specific meanings and behaviours) and to those outside (who may be treated with contempt or indifference or exclusion). Anyone who has been involved in social movement politics has experienced these practices; but so too have they experienced the internal struggles against them, as social groups strive to embody a more democratic politics.

REFERENCES FROM THE READING

Bhabha, H. (1994) *The Location of Culture*. London: Routledge.

Young, I.M. (1990) *Justice and the Politics of Difference*. Princeton, NJ: Princeton University Press.

37

"The Cultural Role of World Cities"

from *Transnational Connections* (1996)

Ulf Hannerz

Editors' introduction

Ulf Hannerz is Professor of Social Anthropology at Stockholm University, Sweden, and has taught at several American, European and Australian universities. His research has been focused on urban anthropology, media anthropology and transnational cultural processes. In this reading, Hannerz explains how local cultures in world cities are subject to a transnationalized production process, which involves a specific set of actors. In the selection included here, Hannerz introduces us to four such groups – transnational corporate elites, "Third World" migrant populations, cultural producers/consumers, and tourists – who he believes play key roles in the production and consumption of world city cultures. For Hannerz, the interaction of these four groups and their cultural practices does not reflect a hierarchical relationship between "high" and "low" culture, but should be understood, rather, as the outcome of a co-production and co-consumption of cultural formations within largely shared urban spaces. Also, the inhabitants of world cities are never merely passive consumers of other peoples' cultural products. Rather, they are, at any given moment, active co-producers and "participant observers" in the process of cultural production. Hannerz's text is thus a provocative rethinking of both the traditional notion that culture is fixed within dominant societal institutions (such as museums, opera, theatre and so forth) and of the idea that sociocultural formations within world cities represent linear outcomes of abstract socioeconomic forces and hierarchical power relationships. Hannerz distinguishes between "market" relations of culture (in which there is some form of buying and selling of cultural products) and those that are "form-of-life" (which are grounded upon more or less informal interactions among people in the shared space of the world city). As Hannerz indicates, many such cultural production/consumption processes involve the street as a key site of exchange and consumption.

INTRODUCTION

World cities are places in themselves, and also nodes in networks; their cultural organization involves local as well as transnational relationships. We need to combine the various kinds of understandings we have concerning the internal characteristics of urban life in the world cities with those which pertain to their external linkages.

As a first step I want to identify four social categories of people who play major parts in the making of contemporary world cities. If New York, London, or Paris are not merely localized manifestations of American, British, or French culture, or even peculiarly urban versions of them, but something qualitatively different, it is in very large part due to the presence of these four categories. What they have in common is the fact that they are in one way or other transnational; the people involved are physically present in the world cities for some larger or smaller parts of their lives, but they also have strong ties to some other place in the world. They do not together make up the entire urban populations, and need not even constitute majorities within them. Nor do they altogether exhaust the possible ways of being transnational in the cities in question. Without these people, in one constellation or other, however, these cities would hardly have their global character.

The first of the categories is that of transnational business. To borrow Redfield and Singer's (1954) description, these are the "cities of the world-wide managerial and entrepreneurial class," nerve-centers of the world economy. Their main functions, as listed by Friedmann and Wolff [see Reading 6], are those of management, banking and finance, legal services, accounting, technical consulting, telecommunications and computing, international transportation, research, and higher education. Whatever part manufacturing may still have in the economies of world cities, for reasons which may now be passing into their history, it is not what most directly and dynamically involves them with a wider world. The second obvious transnational category in the world cities is made up of various Third World populations (in what are, in the cases I have in mind, First World locations). As a third category I would identify an undoubtedly considerably smaller number of people who yet tend to maintain a rather high profile in the world cities I am concerned with: people concerned with culture in a narrower sense, people somehow specializing in expressive activities. The fourth category, finally, consists of tourists – with their quick turnover not officially included in the populations of world cities at all, but always present in considerable numbers, and engaging with a remarkable intensity with the cities for as long as they are on the scene.

The people in these four categories are presumably actively engaged in the transnational flow of culture by being mobile themselves. Clearly their numbers have grown because of (and their forms of life in their transnational aspects have in no small part been shaped by) recent changes in the technology and economy of transportation. In the age of jet planes, people can move over great distances, back and forth almost in a shuttle fashion, in ways which our more timeworn understandings of migrants and migration hardly account for very satisfactorily.

Culture also moves without people moving, however, not least through the media. Having identified the four categories above, I must add that special relationships also tend to exist between media and world cities, in that the former are frequently both based in and somewhat preoccupied with the latter. There will be various opportunities to come back to the implications.

THE LOCAL SCENE AS SPECTACLE

The flow of meaning and meaningful form through contemporary societies can be seen as organized mostly within a few major kinds of organizational frames, and in their interrelations. These frames each have their own principles which animate cultural flow within them, their particular temporal and spatial implications, their different relationships to power and to material life. I will here be concerned largely with two of them, those which I term "market" (where people relate to each other in the cultural flow as buyer and seller, and meaning and meaningful form have been commoditized) and "form-of-life" (where cultural flow occurs simply between fellow human beings in their mingling with one another, in a free and reciprocal flow). We can view a large part of world city cultural process, I will argue, in both its local and its transnational facets, in terms of an interplay of cultural currents within and between these organizational frames. Some of it, quite conspicuously, is in the streets.

There is, in other words, what Georg Simmel pointed to in his essay on mental life in the metropolis, "the rapid crowding of changing images, the sharp discontinuity in the grasp of a single glance, and the unexpectedness of onrushing impressions" – a sense of spectacle. Obviously, this is something one gets in large part for nothing, simply by being present, through the reciprocities of the form-of-life frame. And we should be aware of the reciprocities here. To one degree or other, the spectacle of the world city is something people constitute mutually. Everybody is not merely an observer, but a participant observer, and the prominent features of the spectacle may depend on one's perspective. For the First World – western European, North American – visitor to (or resident in) London, Paris, or New York, it is perhaps the presence of the Third World populations that is most striking; to the accountant from the Dominican Republic, the sheer idiosyncrasy of an individual; to a Jamaican student, homosexuality being explicitly announced, or simply the diversity, in which from her point of view even the Quebec housewife or the Minneapolis health care inspector may count.

For the visitor from far away, people in a city may be all alike, and yet strikingly different from what he or she sees at home, and therefore still a spectacle. But in the world city this is hardly all there is to it; it is also a matter of internal diversity. One may argue that in at least two ways, such diversity finds a likely habitat in the world city, as in most large cities. On the one hand, in so far as conspicuously different styles of life are collective phenomena, subcultures carried by more or less cohesive groups, their members can together offer whatever moral, emotional, or intellectual support is needed to evolve and maintain them, whether they are rooted in the historical backgrounds, current circumstances, or simply the preferences of group members. And other things being equal, large cities simply have the critical masses for more such groups, creating diversity together, than do smaller places. On the other hand, large cities tend to be places where social relationships and personal reputations have to be achieved, where people may work on being personally distinctive because they find the alternative of anonymity unattractive. So some of the diversity is generated at the individual rather than at the group level. The fact that the world city population is of unusually varied origins naturally combines with both these more general tendencies in urban life to promote diversity yet more strongly.

Now people can certainly be quite varyingly appreciative of the spectacle of the world city local scene. To some, it is mostly a nuisance or a threat, an experience of symbolic violence. These are perhaps mostly people who are not in the world city because it is world city, who were even there before it became one in the current sense of the word; often indigenes who would prefer Paris, New York, or London to be just like any French, American, or British city, and for one thing certainly not a Third World frontier. Here is one of the paradoxes of world city cultural process: while those of us who are at the periphery or semi-periphery always sense the cultural influence of the center, that is the center-to-periphery flow, at the center itself, whatever passes for a native culture frequently seems to view itself as beleaguered or invaded by the local representations of the periphery.

The reaction to diversity on the part of these inhabitants of the world city may be to shield themselves from it as much as they can, by living in their very own neighbourhoods, or if affluent enough, in a house with a doorman, and traveling by taxi rather than subway. These are the people of the center wanting the periphery to go away from their doorstep, or at least to show up there only discreetly, to perform essential services. "You've got to insulate, insulate, insulate," as someone says in Tom Wolfe's New York novel, *The Bonfire of the Vanities* [1987]. Yet as that book suggests, they are still vulnerable. The diversity of the city can impinge on their lives suddenly and dramatically. The local media, for one thing, tend again and again to suggest this to them. Moreover, they often have difficulties telling symbolic and other violence apart.

At the other end of the scale with regard to stances toward world city spectacle it may well be that we find many of the most temporary, most voluntary visitors, such as those quoted above. Writing on "the semiotics of tourism," Jonathan Culler (1988: 154) draws on Roland Barthes as he points out that a tourist is apt to make of everything a sign of itself. Other people are more likely to "refunctionalize" a practice, to make it a more or less pure instance of use, to claim that the fur coat one wears is simply something that protects one from the cold. But the tourist is unconcerned with such alibis. The pub becomes a "typical British pub," a restaurant in the Quartier Latin becomes a Quartier Latin restaurant, even a thruway is not just an efficient way of getting between places but a typical thruway.

Indeed tourists, arriving with their guidebooks and cameras in search of signs, may not get around to refunctionalizing anything, for before they do so, they are already on their way home. But if they enjoy the juxtapositions of ongoing life around them, they are hardly alone in the world city in doing so. It seems likely that at least some members of several of the other categories of inhabitants of the world city have something of the same attitude. If they claim functional alibis for many of their own practices, they can still take much of everything else to be spectacle. And in so far as they are in the world city voluntarily, for one reason or other having to do with its world city character, they may hold that to be something of value.

One characteristic of the world city spectacle is its pervasiveness – if you are a tourist, you may trek from one more specific, authorized local sight/site (the Louvre, Buckingham Palace, Empire State Building) to another, while in between, the wonders of the street are continuous. But at the same time, the sense of spectacle is inseparable from the local setting. You cannot take it with you as such; only as memory and anecdote, or perhaps in piecemeal technical representation. And this inseparability of sense from place helps keep the world city a world city. People really have to make the journey to see, and hear, and smell it for themselves.

Probably it must be inserted here that the media need also be taken into account as contributing to the wealth of serendipitous experience in the world city. Because of their particularly high density and ubiquitous presence there (more television channels, more radio stations, more daily papers, more journals, newsstands everywhere), what is not really in its own streets may still impinge on one's consciousness almost as much as if it were.

THE CULTURAL MARKET-PLACE

Yet here we are slipping out of the form-of-life framework of cultural flow, through a vaguely defined boundary zone, into the market framework. World cities derive much of their importance from being cultural market-places. If all information is culture (and that is a matter of definitional taste), then the managerial elites are also in these market-places, even during business hours, since information is what they are primarily dealing in: minute, ephemeral units and their aggregates. The question

is there, in any case, of what kind of more durable, broadly patterned metalearning goes on in a form-of-life where there is such intense attention to information and its small shifts and differences. Quite possibly such everyday practices help form the intellectual and aesthetic consumer habits and thus the relationships of this group to the cultural market in a narrower sense.

In any case, we understand world cities to be cultural market-places primarily because of the presence, in large numbers, of the expressive specialists. On the one hand, that presence is constant. The expressive specialists remain there – but not least, move there – because of the availability of mentors and peers, to learn from with or without acknowledgment, in institutionalized or uninstitutionalized ways; of a highly elaborated structure of gatekeeping institutions; and of a mass of more or less affluent consumers.

On the other hand, the particular cultural commodities are just as continuously changing. It is from the world cities we tend to get the *dernier cri*, the latest in a wide spectrum of fashions, isms, avant-gardes and "movements" (which in this case are not much like what we otherwise think of as movements, but rather more stylistically and temporally marked clusters of cultural commodities and producers). Because of this durable generative capacity, attention to the world cities as cultural market-places cannot be allowed to slacken, at least among those concerned to be *au courant*.

What are the sources of this creativity? In part, the inherent tendency within the market framework to try to gain competitive advantage through innovation. In part also, no doubt, intentionality and will-power – the expressive specialists know well enough that they are in the world cities to produce new culture. In part, again, a concentration of ability – the world cities draw, surely not all, but presumably a greater than average proportion of, "the best and the brightest."

There is also, however, the fact of diversity. Cultural creativity seems to feed on it. Raymond Williams (1981: 83–84), for example, surveying avant-gardes in the decades around 1900, found that they usually grew in metropolitan settings (in which context he emphasized both the relative cultural autonomy and the internationalization of the metropolis), and that a high proportion of the contributors were immigrants, often from what were regarded as more provincial national cultures. It was in the encounter between modes of expression of different derivation, each now of doubtful relevance in its original form, that new kinds of consciousness and practice emerged which had a significance beyond the nation-state.

Probably Williams' analysis of the implications of diversity is in large part valid with regard to world city cultural process around the coming turn of century as well, only with a wider transnational scope yet. I would like to draw attention here, however, a little more specifically to the particular local potentialities of world city interrelations between the form-of-life and market frames of cultural flow.

The expressive specialists are most likely often among those who find intellectual and aesthetic stimulation in spectacle as a part of the general urban ambience. This is one kind of form-of-life/market connection, although one that tends to be rather vaguely defined. Yet there are more specific interrelations as well, involving a shift of items, or clusters of items, of meaning or meaningful form from one frame to the other.

I will sketch a rough sequential model. It applies, I believe, to such cultural careers as those of some ethnic music and some ethnic cuisines, but the same kind of process may be more widely recurrent. In a first phase, as it were, the items of meaning and meaningful form at issue flow fairly freely within some subcultural community, as long as the latter is large and cohesive enough to offer sufficient moral and other basically non-material support. People eat their home cooking and make music together. Whatever slight degree of specialization is involved among the members of the community with regard to the production of the items involved, these have not become commodities. They move largely within the internal matrix of personal relationships of the community, in its array of private settings.

In phase two, a higher degree of division of cultural labor is introduced within the community. Apparently the latter has also reached another kind of critical mass here, where it is profitable enough to commoditize subculturally distinctive items for consumption by community members, as an alternative to the free flow of the form-of-life framework. And the market, of course, tends always to be looking out for such opportunities for expansion. At this point, members of the subcul-

tural community may or may not make "signs" of these commodities, in line with what was said before – that is, they may see them as taken-for-granted items of their everyday routines, or they may be more acutely aware of, and appreciative of, their subcultural distinctiveness. In any case, the commodities tend to move here into more public arenas. Subcultural cuisines are on display in area food stores, carry-outs or restaurants; music is performed on stage in the ethnic quarter (by local talent or, if the subcultural community is the diasporic extension of a society which is mostly somewhere else, by the celebrities of the periphery, flying into the world city from home), or it is broadcast on the local ethnic radio station.

And so we reach stage three in the career of cultural commodities (where, of course, they were not yet commodities in stage one): having become more public, they are also more available to the constant scanning for novelties in the wider cultural market-place. The ethnic cuisine is discovered by people on an outing of gastronomic slumming; the music (Caribbean reggae in London in the 1970s, North African rai – not quite so successfully – in Paris in the 1980s) becomes ethnopop.

The sequential model must certainly be allowed to encompass a number of variations. As a subcultural, transnational community is established in the world city, it may already be into the second stage, and sharing it with the country of origin. Sometimes, on the other hand, stage two is skipped; items from the subcultural free flow economy pass directly into commoditization in the wider cultural market. In stage three, the appeal may not be quite so much to the wider cultural market, but rather, through some kind of elective affinity, with another subcultural community – the way the reggae music of Caribbean immigrants, for instance, appealed at first particularly to the adherents to certain British youth styles. Sometimes stages two and three coexist, as both the subcultural community and the wider cultural market are able to accept the cultural commodity in a single form; but sometimes the third stage involves further change, to render that commodity in a version more agreeable to existing tastes in the wider market. In music, a split may appear between "fusion music" and "roots music," but in so far as they remain in contact, and perhaps in a dialectical relationship, this may keep leading to more innovation, and more market crossovers.

FROM CENTER TO PERIPHERY

The model of how cultural items move, from within the internally varied form-of-life frameworks, by way of commoditization in segmented local cultural markets, into yet wider markets, is not necessarily relevant only with respect to ethnic items with some kind of Third World connection. In the American context, it has clearly long been applicable to the influence of Afro-American culture on American culture more generally; in part, but not only, in the world cities. In the British case, one could fit various youth cultures into the first stage of the sequential model and see what follows from there on. But the special point, in the world city context, of considering the cultures of transnational groups of Third World background in this light is that we can see how the center–periphery relationships in culture now, because of the makeup of world city populations and the structure of world city cultural markets, fairly often become, as we view them more completely, periphery–center–periphery relationships. The world cities are no doubt still frequently the points of origin of global cultural flow, but they also function as points of global cultural brokerage. Third World music, to return to perhaps the most obvious source of examples, may become world city music, and then world music.

Here, however, we are moving out from the local scene, and on to the field of transnational relationships between the world cities and their variously distant hinterlands. What are the channels through which culture flows here?

The actual physical mobility of the members of the four transnational social categories identified above must certainly be taken into account. Unless they are complete cultural chameleons, changing color depending on context, they presumably carry something from the world city back to wherever else they belong. The double vision I have suggested here, of the world city as both a place in itself and as a source of culture flowing out from it, which in its combination is most directly represented by the members of the various transnational groups, seems necessary to understand fully the contemporary cultural role of the world cities. We should understand that for those who do not go there themselves, the media again play a part in offering substitute world city spectacles. As the transnationally effective media are in large part at home in the world cities, and often preoccupied with portraying them to the world, hardly anybody can remain quite ignorant and unconcerned with them as places. And thus, every time we at the periphery come across one of those cultural commodities which can more readily be removed from the source, we remember the sense of spectacle, and note that "there is more where this came from."

REFERENCES FROM THE READING

Culler, J. (1988) *Framing the Sign*. Norman, OK: University of Oklahoma Press.

Friedmann, J. and Wolff, G. (1982) World city formation: an agenda for research and action, *International Journal of Urban and Regional Research*, 6, 69–83.

Redfield, R. and Singer, M. (1954) The cultural role of cities, *Economic Development and Cultural Change*, 3, 53–73.

Williams, R. (1981) *Culture*. London: Fontana.

Wolfe, T. (1987) *The Bonfire of the Vanities*. New York: Farrar, Straus.

38

"World Cities: Global? Postcolonial? Postimperial? Or Just the Result of Happenstance? Some Cultural Comments"

Anthony D. King

Editors' introduction

In his work on global cities, Anthony D. King has been particularly interested in the social production of building form, the relationships of colonialism and urbanism, social and spatial theory, postcolonial theory and criticism, as well as transnational cultures. In this reading, King powerfully summarizes and integrates these various facets of his impressive body of work on global cities. King begins with a brief discussion of the concept of a postcolonial critique, which he defines as being "essentially concerned with questions of agency, representation, and especially, the representation of cultures under the asymmetry of global political and social conditions." On this basis, King points to the neglect in global cities research of historical (e.g. imperial, colonial) and cultural (e.g. religious) considerations. After criticizing the present-oriented economism of this literature, King also calls into question the assumption that quantitative representations can adequately capture the lives of migrant populations in world cities. In the next section, King discusses the ongoing race among global cities to construct the tallest skyscraper. King contrasts this struggle for recognition, centrality and symbolic power in the sphere of downtown architecture with the reality of global suburbanization, which he views as an expression of the new forms of sociospatial exclusion that are associated with global city formation. Consequently, in the shadow of the official quest for cultural recognition, King detects a global culture of the *banlieusards*, the inhabitants of the suburbs. Finally, King emphasizes the colonial legacies that are embedded in the spatial structures of many postcolonial cities and suggests that such legacies have important effects upon contemporary power structures.

INTRODUCTION

Postcolonial theory and criticism has been slow in penetrating the social sciences, not least, the literature of urban studies. At the basis of the postcolonial critique are questions of knowledge and power. A postcolonial critique is essentially concerned with questions of agency, representation, and especially, the representation of cultures under the asymmetry of global political and social conditions. The critique of "Eurocentrism" is its basic task.

Postcolonial criticism most obviously assumes a knowledge of colonial histories in the contemporary world, not only those of major European imperial powers (France, Britain, Spain, Portugal, the Netherlands) but also of the USA, Russia and Japan. Recent interpretations of globalization have done much to correct the initial Eurocentric and ahistorical focus of that process, both by examining globalization from the viewpoint of countries (and religions, including Islam) outside the West, but also tracing earlier, Islamic phases of globalization, before European hegemony, as well as later ones, including, since 1950, that of "postcolonial globalization" (Hopkins 2001).

Despite these theoretical developments, work on the historical origins of what have come to be called "global cities," either individually or in the form of a network or system (Taylor 2004), is still woefully neglected. A glance at any good 1900 atlas would, however, show that the majority, if not all, of the 82 cities "cited in world city research" were at that date (and earlier) either capitals or major cities of European or Asian empires or, alternatively, cities established in colonial or one time colonial territories. (Here, I treat the "world cities" of North and South America, South Africa, Asia as well as Australia, as resulting essentially from European and Asian colonial expansion. And in relation to their indigenous peoples, as colonizing or imperial.) The networks linking these cities to their real or one time metropolitan core, whether by sea, over land, and eventually by air, telegraph, telephone, the press, by economic and social processes of migration, were synchronized most powerfully through various economic, social, political and cultural institutions, adapted and transformed to various degrees by local and indigenous agency. Of these institutions, language is crucially important.

Other powerful networks, sustained by the formal institutions of particular belief systems (Catholicism in South America and also elsewhere; varieties of Protestantism) which were significant influences in some locations but in other, more secularized societies, were less powerful, have nonetheless left sedimented values. The earlier networks of Islam, before European hegemony, were equally powerful in the Asian and Middle Eastern world, known for radically different conceptions of space and time, and also in relation to the cosmology of Hindu, Buddhist and other belief systems.

In the context of language, following the onset of European hegemony from the sixteenth century onwards, anglophone, francophone, hispanophone and lusaphone networks, transformed by national, transnational and regional identities, continue today to provide the basis for transnational processes in important spheres of public and commercial life. The knowledge regimes established through these languages, both in the one time metropole and its one time colonies, and especially resulting from the expansion of the historically anglophone, EuroAmerican cultural sphere (i.e. US hegemony), are primarily accountable for the fact that, in regard to the geographical sciences, for example (but also in many other areas of knowledge), what is frequently referred to as "international" knowledge (Gutierrez and Lopez-Nieva 2001) is, in fact, postcolonial (King 2004). How else can we explain why this book is published, and also being read, in ("international") English and not in Arabic, Mandarin Chinese or Hindi? And while religion has, with some minor exceptions effectively been ignored in the debate on global cities, it has taken the events of 9/11 to remind us that the vast majority of these 82 cities were in territories historically part of the Judeo-Christian ecumene (where, in Europe, political parties still describe themselves as "Christian Democrat" and in the Americas, religion is a significant factor in politics). Where this is not the case, as with Mumbai, Jakarta, Hong Kong, Kuala Lumpur, it is worth examining whether their putative world city credentials rest on institutions and practices developed, in conjunction with indigenous agency and knowledge, in their colonial past. It is also worth asking where the networks of Chinese diasporic capitalism fit into this picture. As for Cairo (another of the 82 "world cities" mentioned above), this was already described as the "mother (city) of the world," "the glory of Islam and the center of the world's commerce" as far back as the tenth century, well before the era of European hegemony (Abu-Lughod 1989: 149, 225). In brief, research with a cultural focus might provide an answer to one of the still unanswered questions in "global city" research: why so many major cities in the world are not, according to the criteria currently used to identify them, "global cities."

This does not, of course, deny the significance of the empirical findings in regard to which the present day concept of the "world" or "global city" has been (largely) defined, nor the discourses that circulate around them: the increasing concentration of producer services such as accountancy, advertising, banking, law, insurance and the rest, whether introduced from "the West" or emerging locally. What seems conceptually problematic, however, is the appropriation of the term "global city" to categorize a class of city based primarily on these criteria (Sassen 1991). It implies that as the *presence* of these producer services becomes increasingly universal in major cities worldwide the *effects* of these services will somehow be similar. Or to put it differently, that focusing on an apparent similarity manifested by the growth of these services – which have effects in relation to the realm of the world economy – is more important than focusing on the differences which may result from them in relation to the realm of global culture. The separation of both these realms is, of course, purely methodological. What I wish to argue here is that different questions demand not only different kinds of research but also different positions from which to start.

My argument about the imperial and colonial origins of many so-called world or global cities is not just a matter of "getting the record straight" or to add a missing history, even though these are legitimate tasks. The aim is rather to draw attention to the overly economistic nature of the criteria driving the "world city paradigm" and its framing within a narrowly restrictive framework of urban political economy. It is also to highlight the ahistorical and analytically feeble nature of the category, "*global* city." As others have pointed out, assumptions in much of the literature, including some of that on global cities, that "global urbanism can be regarded as a uniform or homogenous outgrowth from Europe and America, belatedly affecting Africa, Asia and South America" or that "cities in Africa, Asia or South America can be understood on the model of cities in Europe, Australia or the USA" (Bishop et al. 2003: 2) are seriously flawed. While the largely quantitative

data which characterizes much of recent global city research may tell us something about the organization of the contemporary world economy and the worldwide growth of contemporary capitalism, it fails to address the distinctive *cultural* forms of that economy and also the cultural characteristics of all cities, including postcolonial cities, not least those affecting, and in particular cases determining, the nature of contemporary economic and political activity. Here, I use the term cultural in its widest sense both to refer to questions of meaning, identity and representation as well as language, religion and ways of life.

Of major importance here has been the rapid growth in recent years of information technology enabled service employment between different parts of the one time anglophone empire, particularly between the USA, Canada, the UK and India, resulting in the outsourcing of tens of thousands of jobs from high cost to low cost states. The rapid growth of call center employment in India, which has created 336 call centers between 1997 and 2002 employing over 100,000 operatives and generating a $1.4 billion industry (King 2004: 152), is based on the existence of a highly articulate, *English-speaking*, under-employed graduate population, educated in an essentially postcolonial, western-oriented university system. It is rarely, if ever, acknowledged that, after the USA, India has the largest English-speaking population in the world. The development of this language and technology based sector not only has created extensive employment opportunities but also is related to the massive construction boom on the outskirts of cities such as Delhi, Bangalore and Hyderabad.

Recognizing these postcolonial cultural links is equally vital for addressing economic developments in "world cities" in the West. At one level in the economic and social hierarchy, virtually half the cabs of New York City are driven by English-speaking migrants from India, Pakistan and Bangladesh; at another level, some 10,000 highly educated graduates of India's six elite and highly selective Institutes of Technology occupy some of the top ranks of business, banking, IT development in the USA. We also need to recognize that the cultural politics of India include anti-English protests by "Hindutva" supporting nationalists.

MIGRATION AND ITS CULTURAL EFFECTS

In the discourse about the multiculturality of global cities, attention is often drawn to the presence of "other" nationalities and language groups. That one city has representatives from 102 nations or language groups, and another from 192, is somehow taken as a "sufficient and necessary" condition of its cosmopolitan character. This issue needs addressing both in more detail and with more sophistication.

Viewed from a demographic viewpoint, it might be useful here to imagine what an "ideal" or "utopian" global city would be, for example, one whose population was composed of representatives (say 1,000) of each nation-state, and in proportion to its total population, from each of the almost 200 nation-states in the world. Irrespective of the artificiality implied by this notion of the "nationally constructed subject," such an "international" city would in no way get close to the tens of thousands of ethnic and linguistic groups which make up these nation-states.

The relevance of suggesting this ideal – or perhaps absurd – model, however, is to demonstrate that, from a demographic viewpoint, no so-called world city can, or ever will, approximate towards it. This is because in all world cities (at the time of writing, at least), there is a numerically dominant population from the host society and in each, the proportion who are "foreign born" comes from a diverse range of cultures, ethnicities, religions and regions, and also as a result of very different historical circumstances. Moreover, the monolithic (and also xenophobic) category "foreign born" can, for all except legal purposes (though this is obviously a highly significant exception), be dismissed. From specific points of view, to assume that "foreign birth" provides commonality amongst a vast range of peoples, in either eighty or even thirty "world cities" in five continents, is as absurd as suggesting that "domestic birth" can be used to characterize, along cultural or ethnic lines, those born in the host society. Similarly, to suggest that the 15 per cent of the Paris population who are "foreign born" and coming primarily from North Africa (Algeria, Morocco, Tunisia), Armenia or Mauritius, the 28 per cent in New York, over half of whom are from the Caribbean and Central America, with significant proportions from Europe, South America, East Asia and South and Southeast Asia, and some 20 per cent "foreign born" in London, from South and Southeast Asia, Ireland, continental Europe, East, West and South Africa, the Caribbean, North America and Australasia, have somehow more in common than they have differences is equally problematic.

Clearly, the historical, cultural and political status and power (or lack of it) possessed by migrants from different countries, when relocated in the cities of another society, is highly variable and differentiated. Given that a large proportion, both in Europe and North America, are from "Third World," postcolonial societies, their colonial histories – as I discuss below – place different kinds of migrants in very different situations of power and lack of it, irrespective of their relation to the (economic) labor market.

Quite apart from their influence in different sectors of the economy, however, the influence of postcolonial subjects on the culture and politics (as well as cultural politics) of the dominant society can clearly be substantial. There is no better example than the significant impact which postcolonial criticism (largely developed by scholars such as Arjun Appadurai, Homi Bhabha, Dipesh Chakrabarty, Edward Said, Gayatri Spivak and others) has had on the epistemology of the western academy. In terms of cultural politics and theoretical critique, numbers are irrelevant. Salman Rushdie, after all, is only one person.

REPRESENTING AND SYMBOLIZING THE GLOBAL AND POSTCOLONIAL

How are the global and postcolonial represented and symbolized in the spaces of the world, global, or postcolonial city? Prior to September 11, 2001, a partial answer to that question might have been found in the global discourses prompted by the construction, at increasingly frequent intervals, of what is represented as "the world's tallest building." Roger Keil has suggested that the gigantic tower has become the most important symbolic product of the world economy (King 2004: 4), a sign used by states and cities both to challenge the existing economic order and to make their own claims to contemporary modernity. The twin-towered Petronas building in Kuala Lumpur, for instance, built in 1996, was some meters taller than the previous claimant, Chicago's Sears Tower. The event generated global publicity, including predictions about "the decline of the West" and "the coming of the Asian century." The logic behind the construction of the "world's tallest building" rests on the acceptance of a conceptualization of globalization as a process by which "the world becomes a single place" (Robertson 1992). The logic also assumes that creating a similitude, or joining a contest, with other states and cities whose conceptions of modernity it wishes to share will, ipso facto, place that state or city in the same reference group. Other cities assume that it will attract inward investment.

Two recent instances of the "world's tallest building" phenomenon include the completion of Taiwan's "Taipei 101" building in October 2003, the aim of which was said to be "to put Taipei on the global map" as well as "bring in foreign companies." This is also the case with the (elite) sponsors of other mega-projects in Asia which, like Taipei 101, aim to provide "world class" accommodation to attract financial and other producer services (Marshall 2002). Indonesia's proposed Jakarta Tower is "five meters taller than the Canadian National Tower in Toronto," currently the "world's tallest *tower*." This would be "like other famous structures in big countries" and, from the city governor's perspective but not necessarily that of the inhabitants, would "enhance the image of Jakarta as a metropolitan city."

What these few examples illustrate, however, is that, for both the producers as well as the public, there is not just one but rather many worlds and they do not necessarily coincide. Malaysia's Petronas Tower not only was a postcolonial gesture from Prime Minister Mahathir Mohammed but also, in incorporating Islamic motifs in its design, was both a signifier, as well as signified, of the world of Islam. The mammoth multistorey tower designed for London by British architect, Richard Rogers, to celebrate the millennium (though never built) prompted a leading paper, the *Guardian*, to suggest that, if constructed, it "would confirm the status of Britain as a Third World country." According to the paper's architectural correspondent, super highrise towers are "increasingly symptoms of 'second city syndrome'." After September 2001, the discourse in the United States about the "world's tallest building" has been decidedly subdued, and the contest, both metaphorically and literally, has been Shanghaied by particular representatives of the Chinese elite.

If the city center highrise tower is, for some at least, the prototypical sign of the global city and its postcolonial aspirants, equally important is the space and culture of the suburbs. Simultaneously both global and postcolonial, their transformation in many "world cities" is best expressed in Hopkins' term as postcolonial globalization (2001). In Paris, most of the postcolonial minorities (from Algeria, Tunisia, Morocco) are in the *banlieues* which, "in the 1990s, have become a byword for socially disadvantaged peripheral areas of French cities" (Hargreaves and McKinney 1997: 12). Structurally equivalent to British and American inner city areas, and often referred to as ghettoes, the *banlieues* provide a natural space in which to develop "a separarist cultural agenda marked by graffiti, music, dancing, and dress codes" with which the *banlieusards* (suburb-dwellers) reterritorialize the "anonymous housing projects" (ibid.). In Britain, the urban landscapes of the eastern, western and southern suburbs of the postcolonial/postimperial global city of London are regenerated and transformed by South Asian Muslims, amongst others, from other networks, who, though "united by belief, are nonetheless divided along national, ethnic and sectarian lines" (Nasser 2003: 9). In the terrain and terraced housing of Britain's "second city" of Birmingham, the largest group of the 80,000 South Asian Muslims are from Pakistan, and comprise 7 per cent of the city's population (Nasser 2003: 9). Though not sufficient in themselves, postcolonial histories are nonetheless central to any analysis of the multicultural nature of both suburbs as well as central areas of global cities in Canada, the USA, Australia, the Netherlands, France, Spain and many other countries in Europe as well as elsewhere in the world.

HERITAGE, CULTURAL IDENTITY AND SOCIO-SPATIAL EQUITY

In other postcolonial states around the world, what was once the "western" space of the colonial urban settlement – in Delhi, Karachi, Singapore, Jakarta, Shanghai, Accra, Cape Town, Kolkata and elsewhere – has, in the half century following independence, frequently become the preferred residential area for the indigenous elite, or more recently, under neo-liberal political regimes, been redeveloped to provide luxury housing in gated communities, transnational tourist hotels, shopping malls, and other spaces of contemporary capitalism. In these real or potential "global cities," one of the most urgent questions to address is whether the social and spatial polarization, frequently seen as the key characteristic of the contemporary "world" or "global city," is simply a continuation of the social, spatial, and racial divisions occupied by colonized and colonizer in the earlier layout of the city. Whether in their nomenclature, design, financing or use, new luxury housing developments – apartments, villas, highrise towers – from Delhi to Jakarta, reveal both the persistence of postcolonial cultural connections to the metropole as well as localized versions of transnational forms of consumption (King 2004).

In recent years, an increasing number of scholars, institutions and authorities have argued for the preservation of parts, sometimes all, of the architectural, urban design and planning forms of one-time colonial cities. In addition to reports by Unesco's International Committee on Monuments and Sites (ICOMOS) there are studies on cities ranging from New Delhi to Jakarta, Singapore to Karachi. According to these commentators, fifty years after independence, the symbolic significance of colonial buildings had lost its old political meaning. Younger citizens of some republics (and their postcolonial capitals) see colonial architecture and urban design as more important for generating revenue from tourism than for summoning memories of colonial oppression.

Yet few of such reports address buildings and spaces as social and political as well as aesthetic objects. The elite Dutch colonial suburb of Menteng, outside Jakarta, with its Art Deco houses and spacious tree-lined boulevards, continues, as in colonial days, to house the rich and powerful (including former President Suharto). In New Delhi, government ministers fight for the privilege of living in spacious colonial bungalows set in three and four acre compounds. Thousands of low paid clerical workers cram into over-packed buses to traverse the colonial wastes or either walk or cycle the interminable avenues in 110 degree summer temperatures.

Despite the many studies that have been made on the "divided city" of colonial times, there has not yet been a study which puts together the views of both sides. That is, the urgent need for a "decolonization" of "symbolic space" and its more equitable redevelopment, along with a sensitive, socially critical acknowledgment of its value as cultural heritage. Such studies would provide different perspectives on the real and potential postcolonial global city.

The more challenging question, however, is how to view what are called "world" and "global cities" not only from a world outside the West, such as Asia, but also from a time, such as the future, that is beyond the present.

ACKNOWLEDGMENTS

Many thanks to Roger Keil and Neil Brenner for suggestions and Abidin Kusno for comments on an earlier draft.

REFERENCES FROM THE READING

Abu-Lughod, J. (1989) *Before European Hegemony*. New York: Oxford University Press.

Bishop, R., Phillips, J. and Yeo, W.W. (eds) (2003) *Postcolonial Urbanism*. New York and London: Routledge

Gutierrez, J. and Lopez-Nieva, P. (2001) Are international journals of geography really international?, *Progress in Human Geography*, 25, 1, 53–70.

Hargreaves, A.G. and McKinney, M. (1997) *Postcolonial Cultures in France*. London and New York: Routledge.

Hopkins, A.G. (ed.) (2001) *Globalization in World History*. London: Pimlico.

King, A.D. (2004) *Spaces of Global Cultures*. London and New York: Routledge.

Marshall, R. (2002) *Emerging Urbanity: Global Urban Projects in the Pacific Rim*. London: Spon.

Nasser, N. (2003) The space of displacement: making Muslim South Asian place in British neighborhoods, *Traditional Dwellings and Settlements Review*, 15, 1, 7–21.

Robertson, R. (1992) *Globalization*. London, Thousand Oaks, CA and New Delhi: Sage.

Sassen, S. (1991) *The Global City*. Princeton, NJ: Princeton University Press.

Taylor, P.J. (2004) *World City Network*. London and New York: Routledge.

39

"'Global Media Cities': Major Nodes of Globalizing Culture and Media Industries"

Stefan Krätke

Editors' introduction

Stefan Krätke is Professor of Economic and Social Geography at the Europa-Universität Viadrina in Frankfurt (Oder). He has been one of the most productive and creative European global city researchers and has recently collaborated with the GaWC research group in Loughborough to explore the role of media industries in the global cities network. Krätke's work on European urban regions combines detailed economic analysis with sophisticated modes of sociospatial explanation. Krätke's interdisciplinary approach to urban and regional studies is evident in this reading, in which he examines the institutional and geographical dimensions of the global media industry, emphasizing in particular the production of cultural commodities. According to Krätke the products of global cities' cultural industries simultaneously serve corporate demand in the centers of the global economy and are widely consumed throughout the world economy. However, as Krätke points out, the global map of media cities is only partly congruent with the global map of global financial centers. Krätke's contribution also critically engages with recent debates on the role of the urban creative class − the cultural producers −in urban economic development.

This contribution examines the link between cities and culture from the point of view of the *production* of cultural goods and media products. The culture and media industry is a prime mover for globalization processes in the urban system, in which cultural production clusters act as local nodes in the global networks of the large media groups. The analysis of "global media cities" enables those locations to be identified, from which globalization in the spheres of culture and the media proceeds. Global city research has predominantly emphasized the role of advanced producer services in the development of the contemporary world city network. This reading emphasizes that for the process of globalization the

globally operating media firms are at least as influential as the global providers of corporate services, because they create a *cultural* market space of global dimensions, on the basis of which the specialized global service providers can ensure the practical management of global production and market networks.

MEDIA CITIES AND THE INSTITUTIONAL ORDER OF A GLOBALIZING CULTURE INDUSTRY

Media city is a term currently used to describe culture and media centers operating at very different geographical levels. They range from small-scale local urban clusters in the media industry to the cultural metropolises of the global urban system. An *up-to-date* examination of culture and cities ought to have the "commodification of culture" as a central theme, i.e. the worldwide assertion of the market economy in the form of the market-focused production of cultural commodities and the market-related self-stylization of individuals competing for positions in societies characterized by the all-embracing mediatization of social communication, consumption patterns and lifestyles. The culture and media industry embraces those branches of social activity that are determined to a large extent by creative work and the production and communication of symbolic meanings and images.

A main characteristic of the culture and media industry's geographical organization is the selective concentration of culture and media producers in a limited number of large cities and metropolises within the *global* urban system. The other characteristic is the formation of clusters *within* the boundaries of large cities, i.e. the local concentration of cultural production in particular urban districts, preferably in the inner city area (Krätke 2002a). The locational patterns of the culture and media industry in selected global cities such as Los Angeles and London reveal that cultural production tends to the formation of local agglomerations of specialized firms (Scott 2000). The second, most important feature of the present-day cultural industry's institutional order is the *globalization* of large cultural enterprises, which enables global media firms with their worldwide network of subsidiaries and branch offices to forge links between the urban clusters of cultural production. This supra-regional linkage of local media industry clusters lies at the heart of an emerging system of *global media cities* within the worldwide urban network.

Today's culture industry is a highly differentiated business incorporating diverse sectors that range from traditional artistic production to technology-intensive branches of the media industry. The products of these activities are of the utmost cultural importance in that they function as agents of information, influence and persuasion or as vehicles of entertainment or social self-portrayal. They include primarily the diverse branches of the entertainment and media industries, e.g. theatres and orchestras, music production, film production, television and radio productions, the printing and publishing trade, as well as design agencies and the advertising industry. The "image production" activities of the cultural economy (Scott 2000) in today's marketing society include not merely the product images created by advertising and design agencies, but also the lifestyle images communicated via the programme formats of the entertainment and media industries. There is considerable overlapping between the culture industry and the media sector, since a large part of cultural production is organized directly or indirectly as a special value chain within the media industry. The media industry acts as a focus for the commercialization of cultural production and is also to be found at the heart of the "culturalization of the economy," given that its market success is founded on the construction of images and extensive marketing activities that are supported by the media industry.

New communications technologies and the emergence of large multinational groups within the culture and media industry contribute to a global flow of cultural forms and products, whose reach, intensity, speed and diversity far exceed the cultural globalization processes of previous eras (cf. Held et al. 1999). The emphasis below will be on the *corporate* infrastructure that produces and distributes the content and products for a globalized cultural market. The present-day flows of cultural globalization proceed primarily from the industrialized nations of the West and their market-focused culture and media industry. However, the global firms in the culture and media industry are obliged to take account of specific tastes and cultural preferences in other countries. The market strategists employed by global media firms are well aware of the cultural variety and differentiation of their global audiences and customers, and have long given their products and programmes a "regional touch" with a view to stabilizing or enhancing their global market success. In other words, they have adapted their products and programmes to specific regional or national tastes and cultural preferences. This trend towards cultural market differentiation is at the same time a driving force for the organization of global production *networks* in the culture and media industries with "local" anchoring points in different regions and nation-states.

Today's cultural economy is characterized by a marked trend towards the globalization of corporate *organization*. The formation of huge media groups is accompanied by the creation of an increasingly *global network* of branch offices and subsidiaries. This global network of firms linked under the roof of a media group has its *local anchoring points* in those centers of the worldwide urban system that function as "cultural metropolises," i.e. as centers of cultural production. The global media groups are organizing the worldwide spread of media content and formats which are generated in the production centers of the global media industry, in particular in Los Angeles, New York, Paris, London, Munich and Berlin. The culture and media industry is a prime mover for globalization processes in the urban system, in which cultural production clusters act as local nodes in the global networks of large media groups.

The globalization *strategy* pursued by media firms is not geared, as is the case in many industrial groups, to the use of "cheap" labor and the like, but primarily to market development and extension through the establishment of a presence in all the major international centers of the media industry. Second, the strategy of media firms reveals a strong trend towards using creativity resources on a global scale. A presence in the leading centers of cultural production offers global media firms the chance to incorporate the latest fashion trends in the cultural industry as quickly as possible.

In a "global" media city there is an overlapping between the location networks of *several* global media firms. The *local and the global* firms in the culture and media industry are linked here in a development context that fosters the formation of an urban media cluster, whose international business relations are handled primarily via the global media firms that are present. The local media cluster in Potsdam/Babelsberg on the outskirts of Berlin might be taken as an example (Krätke 2002b) to show that cluster firms are not only closely networked within the local business area, but also integrated into the supra-regional location networks of *global* media firms. In the case of Babelsberg the *local* cluster firms are directly linked with the resident establishments of global media firms from Paris, London and New York. The establishment of a global network of business units that are integrated at the same time into the

local clusters of the culture industry enables the large media groups to tap place-specific creativity resources on a global scale.

GLOBAL MEDIA CITIES IN A WORLDWIDE URBAN NETWORK

Most of the studies on global cities and the economic and functional structures of the international urban system reveal a tendency to reduce the "high-ranking" global cities to their function as financial centers and centers of specialized corporate services. This reading highlights the particular role of the culture and media industry in the formation of a world city network, thus emphasizing the *variety* of economic activities which are involved in globalization processes.

There has been a severe data deficiency in the study of world cities, particularly for measuring inter-city relations. One solution, that pioneered through GaWC researches (Taylor 2003), has been to conceptualize the world city network as an *interlocking network* which then allows relations between cities to be measured through data collected on firms. An interlocking network has three levels: as well as the usual two levels comprising the nodal level (the cities) and the network level (all nodes and links, cities connected), there is a subnodal level comprising firms (corporate service firms, media firms, etc.). The latter "interlock" the cities to create a city network through their inter-city locational policies (Taylor 2003). Intra-firm flows of information, knowledge, instruction, ideas, plans and other business between offices/enterprise units are creating a world city network based upon the organizational patterns of global firms.

The requirement for an interlocking network analysis is a matrix of cities and firms showing which firms have establishments in which cities and the relative importance of the cities within a firm's organizational network. The research on the media industry's world geography started by identifying the most important urban nodes of the global media firms' locational network (Krätke 2003). The analysis covers the location networks of 33 global media firms with a total of 2,766 establishments. To qualify as "global" a media firm had to have a presence in at least three

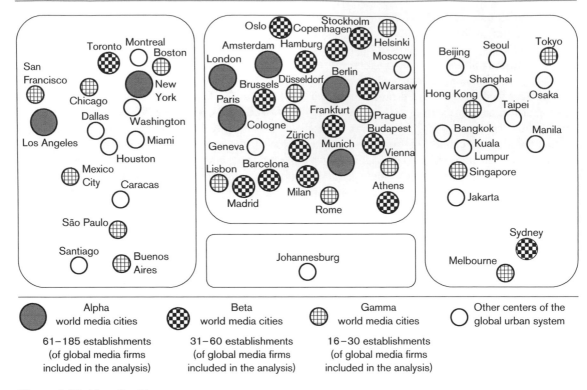

Alpha world media cities — 61–185 establishments (of global media firms included in the analysis)

Beta world media cities — 31–60 establishments (of global media firms included in the analysis)

Gamma world media cities — 16–30 establishments (of global media firms included in the analysis)

Other centers of the global urban system

Figure 1 World media cities

different national economic areas and at least two continents or "world regions." For the media firms included, the locations of all branch offices, subsidiaries and holdings were ascertained and entered in a list of 284 cities (distributed all over the world). The result is a relational data matrix of global media firms and cities, with the matrix cells indicating the number of establishments of a particular global firm in a particular city. These data can be used for a ranking of the cities based on the number of establishments of global media firms that are located in the city. By selecting certain threshold values it was possible to present a set of "global media cities" in the form of readily distinguishable groups (Krätke 2003). However, this relational data matrix can be used for more thorough analyses, e.g. for measuring the *connectivity* of media cities and identifying distinct *geographical configurations* of global media firms' locational strategies (Krätke and Taylor 2004). In the basic analysis, global media cities were divided into three groups: *alpha*, *beta* and *gamma* world media cities. An *alpha world media city* had to have a presence of more than 50 per cent of the global

players analysed here, and more than 60 *business units* from among the global media firms included had to be present. The characteristic feature of world media cities is that they can point to an *overlapping* of the location networks of a *significant number* of global (plus local) media firms. The results are presented here by a synthetic "world map" of the global media business network *nodes* (see Figure 1).

GLOBAL MEDIA CITIES AS CENTRAL NODES OF GLOBAL MEDIA FIRMS' LOCATIONAL NETWORKS

The analysis shows, first of all, a *markedly unequal distribution* of the establishments of global media firms over no more than a handful of cities. The 7 cities in the alpha group account for as much as 30 per cent of the 2,766 establishments of global media firms that were recorded, while the 15 cities in the beta group provide the location for a further 23 per cent of the registered establishments. Thus, over 50 per cent of the branch offices and subsidiary firms of global media groups are concentrated in 22 locations within the global urban system. If the gamma group is included, the share of the global media cities increases to 70 per cent of the establishments registered. The organizational units of the globalized media industry reveal a highly *selective* geographic concentration on a global scale.

Prominent among the *alpha world media cities* are New York, London, Paris and Los Angeles, which are ranked as "genuine" global cities in virtually every analysis of the global urban system. This research again stresses that global cities are to be characterized not only as centers of global corporate services, but also as major centers of cultural production and the media industry. However, among other cities that qualify as global media cities there are interesting deviations from the widely employed global city system: the alpha group of global media cities also includes *Munich, Berlin* and *Amsterdam*, three cities that in global city research which focuses on corporate services were ranked as ("third-rate") gamma world cities (Beaverstock et al. 1999). In the system of global *media* cities, by contrast, Munich, Berlin and Amsterdam are included in the top group. These cities have achieved a degree of integration into the location networks of global media firms that qualifies them as internationally outstanding centers of the culture and media industry. Thus we can conclude that the diversity of globalized activities leads to *multiple globalizations* within world city network formation.

Whereas the global city network constituted by advanced producer services has major nodes which are relatively evenly represented *in all the major regions of the world* (with the exception of Africa), the network of global *media* cities reveals a strongly uneven distribution in favour of the *European* economic area. Of a total of 39 world media cities, 25 are in Europe, 9 in the USA, Canada and Latin America, and 5 in Asia and Australia. What is remarkable is the polarity in the group of US global media cities: integration into global location networks of the media industry is concentrated on just *two outstanding centers* – New York and Los Angeles. *Europe*, on the other hand, has the largest number of media cities with a high global connectivity. The reason for this is cultural *diversity*, since the European economic area has a large number of different nation-states compared with the USA and a multitude of distinct "regional" cultures. This cultural market differentiation is the driving force for the global media firms' strategy to establish local anchoring points in different nation-states. The media cities network as a whole is a reflection of the locational system run by the *western* media industry, which concentrates primarily on North America and Europe. Large media groups with a transnational impact also exist in Asia (particularly in Mumbai and Hong Kong), but the cities of Asia are not incorporated to the same extent as cities in Europe and North America in the western-style globalized media industry.

GLOBAL MEDIA CITIES AS CENTERS OF CREATIVITY AND THE PRODUCTION OF LIFESTYLE IMAGES

To return to the point of departure – the comprehensive *merging of culture and market* – the relationship between urban development and the media industry shall be outlined. Global media cities are functioning as "lifestyle producers" which includes the *production of lifestyle images*. The current lifestyle producers in the culture and media industry are concentrated in leading media cities, from which they spread lifestyle images in the global urban network. In conjunction with the increasing mediatization of social communication and entertainment, the culture and media industry functions as a "trend machine" that picks up on the trends developing primarily in the leading media cities, exploits them commercially in the form of a packaging and repackaging of lifestyle elements, and transmits them worldwide as part of the phenomenon of globalization.

The production locations of lifestyle images are *urban clusters* of cultural production. The local concentration of culture and media activity in specific "districts," which tend to be situated in inner-city areas, is not determined solely by economic driving forces of cluster building. In cities such as New York, London, Berlin etc., culture and media firms prefer "sexy" inner-city locations in which living and working environments merge with leisure-time culture. The specific quality of urban life clearly becomes an attraction factor here. For corporate operators and employees in the media industry the local connection between working, living and leisure time activities is an attraction factor that is in harmony with their lifestyle. These people deliberately seek out locations in a "subcultural" urban district that they can use as an extended stage for self-portrayal during working hours and in their leisure time. In the local media clusters there is thus a direct link between certain lifestyles and urban forms of creative production activity, and thus a specific overlapping of the geographies of production and consumption.

The growth of culture and media industry clusters in selected urban areas is related to the fact that such cities have the sociocultural properties to become a prime location of the "creative class" in terms of Florida's (2002) concept. A concentration of the creative class attracts the music industry as well as other branches of cultural production and the media industry, and also a whole range of other knowledge-intensive activities (like the software industry, the life sciences sector etc.). Florida (2002) emphasizes the sociocultural properties which make a city like London, New York or Berlin particularly attractive as a place of living and working for the creative class: "Creative people . . . don't just cluster where the jobs are. They cluster in places that are centers of creativity and also where they like to live" (Florida 2002: 7). Thus lifestyle attributes of the creative class and a supportive sociocultural milieu are at the center of a city's attractiveness to the creative economy. Florida highlights the role of a

> social milieu that is open to all forms of creativity – artistic and cultural as well as technological and economic. This milieu provides the underlying eco-system or habitat in which the multidimensional forms of creativity take root and flourish. By supporting lifestyle and cultural institutions like a cutting-edge music scene or vibrant artistic community, for instance, it helps to attract and stimulate those who create in business and technology. It also facilitates cross-fertilization between and among these forms, as is evident through history in the rise of creative-content industries from publishing and music to film and video games. The social and cultural milieu also provides a mechanism for attracting new and different kinds of people

and facilitating the rapid transmission of knowledge and ideas.

> (Florida 2002: 55)

Moreover, the city *as a whole* can become an attraction factor for the media business in that the symbolic quality of the specific location is being incorporated into the products of the culture and media industry (Scott 2000). Hence production locations such as New York, Paris and Berlin are perceived in the sphere of the media as being "brand names" that draw attention to the attractive social and cultural qualities of the cities concerned. This includes, in particular, the perception of the respective city as a social space in which there is a *pronounced variety* of different social and cultural milieus. As regards the content and "design" of their products, media firms have to contend with rapidly changing trends. For that reason the media firms wish to be near the source of new trends that develop in certain metropoles such as New York, Paris and Berlin. A marked social and cultural variety and openness, therefore, represents a specific "cultural capital" of a city, which is highly attractive for the actors of the creative economy. On a local level, this cultural capital of a city might also be characterized as a specific "subcultural" capital of particular districts within the city. These thoughts support Florida's thesis that the metropoles' economic growth "is driven by the location choices of creative people – the holders of creative capital – who prefer places that are diverse, tolerant and open to new ideas" (Florida 2002: 223).

A flourishing creative and knowledge economy is based on place-specific sociocultural milieus which positively combine with the dynamics of cluster formation within the urban economic space. Creativity and talent thus depend on the dynamic interplay of economic, sociocultural and spatial factors, and might become a central basis of successful urban development in the future. However, with regard to the specific sociocultural base of the creative economy, the concentration of knowledge-intensive activity and creative forces within the urban system is *highly selective*, so that only a certain number of particular cities and metropoles (i.e. those with "attractive" sociocultural properties) can draw on the creative economy as a focus of their development strategy.

CONCLUSION

The present-day culture and media industry is characterized by the globalization of large media groups. Within urban media clusters, these global players interact with specialized local media firms and form at the same time a global network of branch offices and subsidiary firms, by means of which urban centers of cultural production are being connected with each other on a global scale. This locational strategy enables the global players to make use of worldwide distributed creativity resources. An analysis of the global media cities leads to the identification of the prime locational centers of the contemporary culture and media industry, from which globalization in the spheres of culture and the media proceeds. At the same time, major global media cities are locational centers of the "creative class," since they contain the specific sociocultural properties and lifestyle attributes which are most important for a city's attractiveness to the creative economy.

REFERENCES FROM THE READING

Beaverstock, J.V., Smith, R.G. and Taylor, P.J. (1999) A roster of world cities, *Cities*, 16, 6, 445–458.

Florida, R.L. (2002) *The Rise of the Creative Class*. New York: Basic Books.

Held, D., McGrew, A., Goldblatt, D. and Perraton, J. (1999) *Global Transformations*. Cambridge: Polity.

Krätke, S. (2002a) *Medienstadt*. Opladen: Leske & Budrich.

Krätke, S. (2002b) Network analysis of production clusters, the Potsdam/Babelsberg film industry as an example, *European Planning Studies*, 10, 1, 27–54.

Krätke, S. (2003) Global media cities in a worldwide urban network, *European Planning Studies*, 11, 6, 605–628.

Krätke, S. and Taylor, P.J. (2004) A world geography of global media cities, *European Planning Studies*, 12, 4, 459–477.

Scott, A.J. (2000) *The Cultural Economy of Cities*. London: Sage.

Taylor, P.J. (2003) *World City Network*. London and New York: Routledge.

SIX

40

"Willing the Global City: Berlin's Cultural Strategies of Inter-urban Competition after 1989"

Ute Lehrer

Editors' introduction

Ute Lehrer teaches Environmental Studies at York University in Toronto, Canada and has previously written extensively on architectural history and urban planning in western Europe and North America. The particular slice of this work she presents in this reading stems from a larger research project on the role of the built environment, the culture of construction and the "spectacularization" of the building process in the formation of global cities or "wannabe world cities." Lehrer's trademark style combines visual, planning and geographical analyses to forge a cultural ecology of the city. Lehrer mobilizes this combination of methodological strategies in order to explain the political and cultural strategies through which, following reunification, Berlin's local growth coalition embarked upon a deliberately aesthetics- and form-oriented debate on urban regeneration. By mobilizing approaches to redevelopment that were grounded upon stylized notions of history, geography and place, the local growth machine attempted to reposition the city (once again) as a global command center. While competition with Germany's other global cities (most notably Hamburg and Frankfurt) was sidelined in this discourse, comparisons with the classic metropoles of Europe – such as Paris and London in the West and Budapest, Vienna and Moscow in the East – served as backdrops to a localized *Kulturkampf* (cultural struggle) around the meaning of architecture, built form, symbolic place and historicity.

After the fall of the Berlin Wall in November 1989, the largest city in central Europe had suddenly lost its significance as the major switching station and place of confrontation of the Cold War. Previously assured of its national and international importance through the rivalry between the showcase of the West in the East and the capital of the richest Eastern bloc country, Berlin was now devoid of its mission both sides of the crumbling dividing line, which had physically and militarily separated the city's communities and by extension – symbolically – the world. Not surprisingly, the first reaction of decision- and image-makers in the newly unified city banked on globalization and Berlin's potential role in it as a process through which the city could grow together. When leaders within the government and the business community of Berlin promoted images of the future, their shared vision was that

Berlin would become a major player within the global economy, a world city, a service metropolis, a bridge between East and West, and the old/new capital city of the reunified Germany. The social construction of these images that would lead to a discursive formation over Berlin's world significance, represents a strategic attempt to position the city within the accelerated global interurban competition.

Fantasies of global city status were rampant in the early 1990s, when demographic growth projections predicted to turn the city of 3.46 million into 6 million inhabitants within two generations, when investments from global corporations would bring European-scale headquarters and general "buzz" would attract global cultural attention to Berlin, potentially reviving the capital's glory days of the Roaring Twenties. Pink Floyd's *The Wall* performed at The Wall in 1990, U2's *Zooropa* album and tour from 1993, a gigantic Love Parade techno-fest in 1995 with rapidly increasing numbers of dancing bodies in the central city to over 1 million in the following years, as well as Christo/Jeanne-Claude's packaging of the Reichstag, which was visited by 5 million spectators before its make-over were some indication that cultural attraction, at least at the scale of global mega-events, was working as planned. In addition, investors and developers at the central Potsdamer Platz construction site organized operatic performances and philharmonic concerts. Fireworks illuminated the sky over the construction site on a number of occasions, and arts projects at the Daimler-Benz site elevated the banal everyday place of a construction site to a temporarily spectacular space. The employment of star architects and the labelling of the mega-project at Potsdamer Platz as "Europe's largest construction site" all were part of Berlin's strategy to become recognized as a significant player on a global stage.

The dominant forces behind the redevelopment schemes of Berlin believed that a transformation of the built environment was one of the means to this end; hence, architecture was supposed to work as the catalyst for Berlin's quest for a new identity as global city. By the mid-1990s, the inner city had turned into a major construction site, with Potsdamer Platz as the most spectacular of Berlin's projects.

With these practices, Berlin had put itself on the map of global cultural spectacle once again. However, much to the disappointment of local boosters and investors, economic development and demographic expansion did not quite come along in the same way. And this despite concerted efforts by many among the city's decision-makers to "will" the global city and "channel" the global. Blueprints of the global city were scaled down to the realities of the capital city, which had become Berlin's official function from 1998 on.

In this reading, I will sketch this process of "willing" the global city through a range of explicit

Plate 30 Construction of Potsdamer Platz (Ute Lehrer)

cultural strategies by Berlin's growth regime. The underlying argument is that the building culture of a specific locality is changed by the dynamics of globalization where the building *process* – and not only the outcome – becomes part of a concerted marketing strategy (Lehrer 2002). I will specifically look at three aspects of this development in the context of redeveloping Potsdamer Platz: first, the phenomenon of the spectacularization of the building process; second, the projection of Berlin's globality as a place of centrality and culture via a selective and manipulative presentation of local history; and third, a brief discussion of Berlin's architecture debate as *Kulturkampf* (cultural struggle).

THE SPECTACULARIZATION OF THE BUILDING PROCESS

Individual buildings and whole building complexes are being used increasingly as a means of establishing a city on the map of world locations and destinations. In this spatial transformation of cities, three distinct expressions of the city's "symbolic politics" stand out: the trophy building, the mega-event, and the large-scale project (Lehrer 2003). The trophy building usually results from hiring a world-renowned architect as a certain guarantee to get recognition on a world scale. The second strategy for putting a city on the map of world locations is the mega-event. These temporally limited events serve as impetuses and legitimizing forces in the structural and physical redevelopment schemes of major parts of cities and their regions. The third symbolic strategy in the transformation of cities is a straightforward attempt to compete with other cities for symbolic preeminence in a global environment by emphasizing scale. Large-scale projects, however, play a dominant role in image production of cities not only because of their sheer size but also because of the impact they have on the urban, and therefore social, fabric.

The importance of the built environment's symbolic value in an advanced service economy has been the subject of extensive discussion since at least the mid-1980s (Harvey 1989; Zukin 1991; Debord 1994; Fainstein 1994; King 1996). What is new is that the process is being sold. In other words, the building process no longer involves just the conception of an idea and its realization; it is equally about the production of images between the project's inception and completion. In underscoring the significance of these developments, I call this new strategy for producing global images of place the "spectacularization of the building process." In this process, not only the scales and scopes of projects, but also the speed and the assertiveness with which images are produced, leave a physical imprint in the global production of images, long before the building is able to "speak" for itself.

Potsdamer Platz in Berlin is a prime example of such a spectacularization of the building process, one in which the full range of strategies have been deployed to draw attention to the construction site and its future. Each of the three modes of architectural image production – trophy building, large-scale project and mega-event – found its way into the redevelopment scheme of Potsdamer Platz. Promoted as Europe's largest construction site, Potsdamer Platz is certainly a large-scale project. It can be identified as a series of trophy buildings since world-renowned architects participated in its design. By turning the construction process into a major spectacle, the redevelopment of Potsdamer Platz had become also a mega-event.

The spectacularization strategy was based on the usage of superlatives, the creation of tangible objects, and the production of hands-on events on construction site. From the beginning, investors at Potsdamer Platz labelled the mega-project "Europe's largest construction site." Politicians, bureaucrats and the mainstream press were quick to adopt this affinity for superlatives. While one could easily have believed these hyperbolic claims – after all the project's scale and scope were huge – it was unclear on what grounds such a statement was based. Did it refer to the footprint of the construction site, to the scale of the construction activities, or to its budget? Many additional superlatives were exploited during the course of construction to describe and celebrate the site's remarkable scale: the number of cranes being used, the number of construction workers employed, and the ethnic and cultural diversity of the people engaged in the project. The investors declared even the overcoming of logistical hurdles and the unique means of removing debris as a "documentation" of the success of their "unusual" and "spectacular" building techniques.

Playing a crucial role in the spectacularization process, the so-called Info Box was the most

Plate 31 Info Box and Berlin Wall (Ute Lehrer)

successful object of image production at Potsdamer Platz. Shaped like an oversized version of those containers that usually appear next to construction sites, this temporary building alluded directly to its immediate surroundings. With its bright red metal facade and its elevated position, it worked simultaneously as a billboard and as a place of orientation within the messiness of the construction site. In fact it represented a microcosm of the building activities at Potsdamer Platz, since it was home to an exhibition that explained the scale and scope of the construction project as well as the site's historical significance. As a marketing strategy, the Info Box exhibition/building had two main objectives. On the one hand, Info Box created a concrete "place" in the middle of a wide-open space in Berlin's geographic center, one that became a point of attraction for tourists and Berliners alike. On the other hand, Info Box drew attention away from the nuisances associated with such a large-scale project. Moreover, it turned the site and its building process into the happening place for a New Berlin.

Another temporary but hugely successful strategy to produce images during the construction period was a personalized experience of construction sites at Potsdamer Platz and other places around the city. Introduced for the first time in the summer of 1996, *Schaustelle Berlin* (Show site Berlin – a play on words implying both stage and construction site) allowed Berliners and tourists alike to participate in guided tours and other innovative ways of sightseeing of buildings under construction during the summer months of June, July and August. Daimler-Benz announced that during *Schaustelle Berlin* as many as 50,000 people per day visited its construction site. The driving force behind this summer event was *Partner für Berlin*, a public–private agency between the Berlin Senate and about 120 national and international corporations, created in fall of 1994 with the mandate to promote Berlin as a location for good investment.

THE PROJECTION OF GLOBALITY: THE CULTURE OF THE IMAGE AS GLOBALIZATION

Despite some initial scepticism, within a few years of its erection the Potsdamer Platz redevelopment found wide acceptance among tourists and Berliners alike. In 2001, Alexander Smoltczyk argued, "No place in today's Berlin is as history-less as Potsdamer Platz. Perhaps this is why it is so successful" (Smoltczyk 2001: 60–61). To become "history-less," however, was a process that seems to have been produced intentionally. The paradox is that it was in fact history that was exploited as an argument by the supporters of Berlin's attempt at going global, however, a very selective history that to a good part avoided to draw attention to the problematic past of Berlin. Therefore, the treatment of history played a central role in the debate over redevelopment schemes in Berlin. Because Potsdamer Platz had become a non-place in most Berliners' mental maps after the erection of the Wall, the challenge was to reclaim the space and fill it with new meaning, meaning that would contribute to the projection of Berlin's globality as a place of centrality and culture.

Image production understands the built environment not only in physical and aesthetic terms but also as an outcome of socioeconomic relations and as discursive practices. This means that along with the production process of the "real-material" built environment, there is a production process of the imaginary, a social construction of a particular image and meaning. Images, as they are understood here, include three overlapping and communicating levels of visual, symbolic and metaphorical products and processes. The production of these images has to be understood as processes through which members of society make sense of their individual worlds and of each other's discursive and visual contributions to the general process of communication in society (Deutsche 1996). Therefore images should be treated as substantial elements in the three-pronged spatiality people encounter in cities (Lefebvre 1991).

Ridding Berlin of certain symbolic elements, while manipulating others and constructing new ones, was the preferred method of re-envisioning history in Berlin. While the new Potsdamer Platz can be seen as amnesic in terms of its treatment of history, certain historical details and allusions were used to enhance aspects of the project that were considered key to the reappropriation of space; others were crucially elided. The promotion of a new version of Potsdamer Platz's history accompanied its physical renovation, and took the form of a visual deconstruction and reconstruction of Berlin's material markers.

In the process of rewriting its history, Potsdamer Platz was described as the heart of Berlin. It is true that in its earlier incarnation, Potsdamer Platz was a major traffic intersection with five large streets, several streetcar lines and the site for the first traffic light in Europe, and with two nearby train stations constantly feeding people and goods into the area. However, when investors promoted Potsdamer Platz's cosmopolitanism during the Roaring Twenties, they assiduously avoided any mention of elements that would have linked the site to the Nazi period of the 1930s and 1940s, or to its own masters of image production – in particular Albert Speer with his master plan for Germania. The proximity of the Hitler Bunker, the Gestapo prison and the Ministry of Propaganda to the north of the redevelopment of Potsdamer Platz was hardly mentioned. Omitted was any inkling that today's Daimler-Benz site also housed the *Volksgerichtshof*, where about 13,000 death sentences were issued beginning in 1935 (Winter 2001: 24). There was little reference to the expropriation of Jewish-owned businesses in the area, such as the Wertheim department store on Leipziger Platz, during the Nazi period. The same silence prevailed about Daimler-Benz's World War II role in the production of weaponry and its forced labor camps. The official story of Potsdamer Platz even fails to mention the fact that on May 2, 1945, a delegation of the Nazi leadership initiated Germany's capitulation to the Allies at Potsdamer Brücke (Winter 2001: 24). This selective deployment of historical detail not only was part of the marketing strategy of the investors when Potsdamer Platz was under construction, but also has continued ever since.

In the first years after reunification, the area behind the Info Box became a safe haven for Berlin's historical relics, a bizarre open-air museum and dumping place for the remnants of the GDR period. In the end, however, those historical markers had to make room for the prospect of new development. The displacement of watchtowers and wall sections was also a symbolic act of historical annihilation. Berlin's officials consciously sought to free the city of most reminders of its partitioned past.

The dismantling of Info Box in 2000 can be interpreted as a similar kind of historical deconstruction. Because of its peculiar shape, its distinctive position on the construction site, and its function as the foremost symbol of the rebuilding of Berlin, the persistence of Info Box would have been a constant reminder of a transitional phase in Berlin's new history and identity as global city. Ironically, the space that Info Box occupied is replaced by the perfect octagon of Leipziger Platz – as if there never had been a World War II or two distinct German states.

While certain images were suppressed, others were manipulated. One of the very few buildings that had survived the bombing in 1944–1945 as well as the bulldozers of the 1960s and 1970s was the *Kaisersaal* (Emperor's Hall), which used to belong to the Grand Hotel Esplanade. Built in 1907 as a speculative investment, this hotel became one of Berlin's most exquisite places to be and to be seen. During World War II about three-quarters of the hotel was destroyed, and after the city was partitioned, the surviving halls, staircases and bathrooms were turned into a dance hall and theater. When construction began at the site that Sony had purchased for its European headquarters, this remnant of a formerly grandiose Berlin was in the way of the architect's plan. With great fanfare and superior media coverage, the Kaisersaal was relocated from its original site to about 70 meters to the west in March 1996. It was the most spectacular event on Potsdamer Platz and it demonstrated the literal deconstruction of a historic relic. This relocation represents a plain instance of the manipulation of Potsdamer Platz's material history: after being surgically separated and partly relocated, the remainder of the Kaisersaal became an enshrined part of the Sony Corporation where pieces of it were enclosed into a glass box.

In spite of a strong local building culture that favoured clear height limitations and stone facades, images that fit into the rhetoric of global city formation were strengthened and elaborated: highrise buildings, glass and metal facades, an enclosed inner city shopping center all fit this agenda. They all demonstrate the urge to create the image of a global city that is in sync with other global cities by using an architectural language as well as building materials that is global in its uniformity.

THE ARCHITECTURE DEBATE: *KULTURKAMPF* IN THE GLOBAL ARENA

Architecture plays a significant role within the public discourse on urban development and local politics in Germany. Understood as more than merely aesthetically or economically motivated, the discourse about architecture draws on previous time periods and their ideologies about architecture, including the time when architecture became one of the various strategies used by the Nazi regime to claim cultural superiority over other "races" (Frank 1985). Hence, the relevance that architecture plays within public discourse on city-building processes in Germany in general and in Berlin in particular often bewilders outsiders.

The roots of Berlin's current discourse on architecture and urban planning go back to the 1970s and 1980s, when locally specific strategies for urban redevelopment met with internationally celebrated architectural approaches. With the International Building Exhibition and at the height of a postmodern style in architecture, Berlin had become an open-air museum for buildings designed by world-renowned architects. Yet, at the same time, the city also could demonstrate the value of a more ecological approach to urban planning (the so-called *Behutsame Stadterneuerung*). With a lively squatter scene, the discursive realm of architecture and urban redevelopment was not left to the self-acclaimed experts, but instead became part of a general public discourse. Not surprisingly, therefore, all of the proposed visions of a reunified Berlin were heavily contested, both within the expert world as well as in the general public sphere.

While the discourse was about creating an identity through the built environment, and about the role of marketing this identity, it was also about a localized *Kulturkampf* (cultural battle) over the meaning of architecture. The dispute was articulated as one between the American city and the European city. There were those who wanted to use new projects in Berlin in order to connect with the global language of office towers made out of glass and steel. Opposing them were those who wished to save the European city by demanding various types of restrictions on building style and form. Whereas the former position was shared among the investors and some of the architects, the latter was favored by a local alliance composed of various city planners, politicians and architects. They saw the redevelopment of Potsdamer Platz as a defense of the European city; it was about streetscape, density, scale and urban pattern. This clash of ideologies appeared, at first glance, to be about stylistic considerations; in effect, however, it concerned the relationship between the construction of built environment and larger social, cultural and economic transformations. The dispute led the investors in Potsdamer Platz to urge the municipal government to reject the image of the European city as backwards and provincial. Of course, this was a slap in the face to a city that was trying hard to strengthen its position within the global economy through visual statements in the built environment. In the end, the redevelopment of Potsdamer Platz presents a compromise between the European and the American city: height limitations that were in reference to Milan throughout the development, but skyscrapers right at the tip of Potsdamer Platz as well as on the opposite side; predominant use of stone facades, but glass and steel for the Sony complex; a street plan that is relatively fine grained, but an entire section of a street is roofed off and turned into a shopping mall.

CONCLUSION

Long-established forms of boosterism have now evolved into coordinated efforts to turn cities into spectacles and the urban experience into image consumption. This is particularly true for large-scale projects, where it is difficult – for both the specialist and the non-specialist – to imagine the future shape of new built environments and to anticipate their impact on the urban fabric. What is new is that the building culture has changed in such a way that it is no longer about the end result but also about what images are created and disseminated during the building *process*. Therefore, not only buildings and the advertisement of their potential success contribute to locational competitiveness in a global marketplace but also the marketing of the building process, which draws attention to the real physical as well as messy and procedural aspect of global city formation.

REFERENCES FROM THE READING

Debord, G. (1994) *The Society of the Spectacle*. New York: Zone.

Deutsche, R. (1996) *Evictions*. Cambridge, MA: MIT Press.

Fainstein, S. (1994) *The City Builders*. Cambridge, MA: Blackwell.

Frank, H. (1985) *Faschistische Architekturen: Planen und Bauen in Europa. 1930 bis 1945*. Hamburg: Christians.

Harvey, D. (1989) *The Condition of Postmodernity*. Cambridge, MA: Blackwell.

King, A. (1996) Worlds in the city: Manhattan transfer and the ascendance of spectacular space, *Planning Perspectives*, 11, 97–114.

Lefebvre, H. (1991) *The Production of Space*. Oxford: Blackwell.

Lehrer, U. (2002) Image production and globalization: city-building processes at Potsdamer Platz, Berlin. Dissertation, UCLA.

Lehrer, U. (2003) The spectacularization of the building process: Berlin, Potsdamer Platz, *Genre: Forms of Discourse and Culture*, Fall/Winter, 383–404.

Smoltczyk, A. (2001) Auf neutralem Boden, *Der Spiegel*, 60–61.

Winter, F. (2001) Verpatzter Potsdamer Platz: Geschichtsverleugnung im Europäischen Städtebau, *Die Wochenzeitung*, June 14, 24.

Zukin, S. (1991) *Landscapes of Power*. Berkeley, CA: University of California Press.

41

"Exploring Colombo: The Relevance of a Knowledge of New York"

from A.D. King (ed.), *Re-Presenting the City: Ethnicity, Capital and Culture in the 21st-Century Metropolis* (1996)

Nihal Perera

Editors' introduction

Nihal Perera is Associate Professor of Urban Planning and Director of Asian Studies at Ball State University. He studied at the University of Sri Lanka, University College, London and MIT, and has a PhD from the State University of New York. Perera's contribution is a fascinating account of how experiences in two (at first glance) rather different global cities can interact in the construction of global city cultures. Through a concrete case study of urban cultural dynamics in Colombo and New York City, Perera takes the discussions of cultural diversity and postcolonialism found in previous contributions to Part Six one step further. These two urban centers were established at the same time, during the first global expansionary phase of European colonialism in the sixteenth and seventeenth centuries, and both were at one point urban centers in Dutch colonies. However, these cities' vastly different geopolitical locations (both geographically and in terms of their position in the global urban hierarchy) have made them quite different places during the past two hundred years. Concentrating on the interplay between culture, ethnicity and spatial organization, Perera suggests that the conception of cities (as opposed to the countryside) as places where identities are formed is a specifically western one. Perera shows how identities of people in Colombo continue to be formed by their relationships to the "external" rural surroundings of the city. Cultural production likewise assumes very different forms in New York and Colombo: in New York, a postcolonial history is not obviously visible in urban design or public culture, whereas in Colombo new forms of postcolonial cultural production have been taking shape. This explicitly postcolonial position, which decenters the perspective of world city theory and introduces a "view from the South," represents an important critique of many of the contributions to this Reader (see also Part Four). A deliberately pluralizing approach to the study of culture in the world city system, as suggested by Perera, now appears essential to the overarching project of world cities research.

INTRODUCTION

I shall explore the relevance of a knowledge of contemporary urban developments in Colombo, the former capital of Sri Lanka, for the understanding of New York. In doing so, I shall argue the following. One, New York is too unique to be representative of the large majority of major cities of the world, even so-called "world cities," if we are to understand the cities outside of the top of the hierarchy of the "World Cities." Two, even if a broader knowledge of New York and other major cities of the world-economy is useful, a lack of mediating concepts, local knowledge, and frameworks prevents us making use of such a knowledge to understand other cities.

Though a knowledge of New York may not be directly relevant for an understanding of developments in Colombo, I shall nevertheless begin by pointing out some similarities. New York and Colombo have at least four or five historic characteristics in common. Both were founded as colonial outposts by Europeans in non-European continents, north America and south Asia. In both cases, the early (if not the first) colonists included the Dutch, followed by the British. As important colonial port cities, both developed, along with Cape Town and Batavia (now Djakarta), as crucial nodes of the seventeenth-century Dutch imperial system: New York (as New Amsterdam, between 1626 and 1664, when it was taken over by the British), and Colombo from 1658 (when it was acquired by the Dutch from the Portuguese who had first established it in 1517) to 1796, when it was taken over by the British. Both were later developed as part of the British imperial system, New York for just over a century (1667–1776), Colombo for one and a half centuries (1796–1948). Finally, both are contemporary financial centers – one, long-established and in reference to the global economy; the other, aspiring and recent, with reference to the regional economy.

Here the correspondence between the histories of these two cities tends to fade. Unlike the Dutch in New York, what the Portuguese appropriated in Colombo was an established trading port of the Indian Ocean trade network, displacing the Marakkala people who had operated and inhabited it. The most significant difference is, however, in the population that subsequently gained control of and inhabited these cities, as well as the states to which they belonged. (Modern) Colombo, like New York an essentially foreign implantation, was appropriated by the indigenous population, the Ceylonese, in 1948, after 150 years of British political, cultural, and spatial control and 430 years of European colonial presence. Yet in New York, as also in the rest of the United States, Canada, Australia, New Zealand and elsewhere, the colonial situation continues, albeit not under the original British, Anglo-Saxon regime, but through a cultural, spatial, if not political conflation of the metropole and the colony. The city of New York, therefore, unlike Santa Fe in New Mexico, for example, not only lacks any visible indigenous political, cultural, or spatial presence in the city with an attachment to land or in the vicinity, but there is virtually no material or spatial evidence on the landscape of the once indigenous population. To that extent, New York City approximates a situation of total colonial control, with the elimination of the indigenes, a phenomenon that is reflected in the brute gridiron division of space dedicated to the buying and selling of land. As Sennett (1992) points out, this represented a straight relationship between capitalist economics and the grid of the city.

The elimination of the indigenous populations, cultures and languages also provided the linguistic space for the development of one important language – English. It is this which provides a common lingua franca for addressing, in this volume for example, issues concerning New York, Lagos, Mombasa, Colombo and other cities, by scholars of Argentinean, British, Canadian, Puerto Rican, Nigerian, Kenyan, and Sri Lankan origins, in a spatially, politically, and culturally "native free" environment. This common European, especially Anglo-Saxon, colonial past is equally important for the understanding and explanation of certain dominant social, cultural, and political phenomena in these cities, for example, the relevance of a "foreign," Western and European-oriented Colombo for the construction of an indigenous Other that is represented as "Sri Lankan" culture. In New York, failing the presence of a "genuine" indigenous Other, their position is occupied by the Puerto Rican (colonial) immigrants. As Columbus supposedly "discovered" the Americas for Europe, so the Portuguese in colonial Ceylon romanized the old name of Kolamba to Colombo in recognition of Columbus. The common European cultural space linking north America and south Asia is therefore reflected in nomenclature and language, among numerous other phenomena.

CULTURE, ETHNICITY, AND CLASS

For the remainder of this chapter, I concentrate on ethnicity, culture and the city. The main point I want to make here is that the discursive emphasis on cities, and increasingly on the larger ones, is especially a "Western" pre-occupation. The notion that the city is a "crucible of change" is a cultural product of western industrial societies. The situation of societies in the world outside Europe and North America, where the majority of people are more intimately tied with the so-called rural, is quite different, even though we recognize that the largest cities in the world are, indeed, increasingly in the "Third World."

As in almost every major city, Colombo represents a different ethnic and religious composition than the island as a whole (McGee 1971). For example, the Singhalese, who are about 77 per cent of the total population of the island, and more than 90 per cent in the surrounding areas, are only about 50 per cent in the city (see Table 1). It also accommodates some smaller ethnic groups that are not widespread in the island. Nominal and real religious affiliations of inhabitants enhances this contrast between the city and the state (see Table 2). As in other postcolonial or postimperial cities, Colombo too has ethnically defined "quarters," for example the Tamil quarter of Wellawatta.

Although Colombo shares a number of characteristics in common with New York, largely due to its colonial past, the relationship of ethnicity to territory is quite different. In Sri Lanka the attachment of ethnic and cultural groups to their territory of origin is still strong and there is considerable movement of people between the city and

Table 1 Ethnic composition of Colombo, Colombo District and Sri Lanka population (per cent)

	Colombo city	Colombo District	Sri Lanka
Singhalese	51.1	77.9	74.0
Muslim	18.8	8.3	7.1
Sri Lankan Tamil	17.2	9.8	12.6
Indian Tamil	6.6	1.3	5.5
Burghers	2.6	1.1	0.3
Malays	2.2	1.1	0.3
Others	1.5	0.5	0.2
	100	100	100

Source: *Census of Population and Housing 1981* (Dept of Census and Statistics, Colombo, 1982).

Table 2 Religious composition of Colombo, Colombo District and Sri Lanka population (per cent)

	Colombo city	Colombo District	Sri Lanka
Buddhist	43.2	70.8	69.3
Muslim	21.5	9.9	7.6
Christian	19.7	11.4	7.5
Hindu	15.4	7.6	15.5
Others	0.2	0.3	0.1
	100	100	100

Source: *Census of Population and Housing 1981* (Dept of Census and Statistics, Colombo, 1982).

ethnically specific rural areas, not least because distances are relatively small. In New York however, while there may be strong ties between ethnic groups and particular territories elsewhere in the world, the connection is more likely to be through memory. This makes the construction of ethnic identity in Colombo somewhat extrovert, as compared to New York which is largely introvert. By extrovert I refer to the outward orientation of particular cultural groups who maintain a continued physical relationship with their "homeland" outside the city and its culture, as a primary source for the construction and replenishment of their identities. Introvert, on the other hand, implies that the primary focus is in the reproduction of a culture and identity within the city but with no direct relation to the place from which the cultural group originally stems.

As in the cultural and spatial practices of Puerto Rican immigrants in New York (Sciorra 1996), most cultural groups in New York City construct their identities through the reproduction of a distant, sometimes completely unknown homeland and culture, often far from the place of origin. Not least because of distance and high costs of travel compared to the low income of immigrants, but even among the so-called middle classes, there is often little intention to return to their lands of origin. Hence people deploy a cultural past to inscribe a present, a difference in the city. Most inhabitants of Colombo, however, continue to retain strong ties with their gamas, or "home villages," to which they might return after retirement, loss of employment or illness. Their perception of Colombo is therefore largely as a place of temporary residence which, in some cases, is merely a desire, an image of "the good life." In these circumstances, there is little need or incentive to inscribe signs of their own ethnic identity.

The poorest strata of migrants to the city, who were largely "coolies" for the colonial community, and later "squatters" for urban planners, reproduce in Colombo the built forms and cultures with which they are familiar. They also, however, continue to have one foot in their home village. In Colombo, what are called "shanties" by urban experts are in fact the urban production of rural house forms, employing similar methods of construction but with urban materials. Despite aspirations to build a modern house in the city, they

are prevented from doing so by prevailing market conditions.

The aspirations of the middle-income strata, largely composed of state and private sector employees, are more ambivalent; mostly, they do not consider Colombo as their home. Middle-level government and private sector employees who occupy ethnically defined "quarters" also maintain a strong relationship with their home village, one expression of which is building a house there. While in the city, the identity of Colombo's residents is more a result of their employment and economic status, frequently in some form of government service; their cultural identities are produced in Colombo both as part of their class and status identities through the continuing relationship with their home villages. This points to one of the most basic differences between Colombo and New York, namely that the former is the nation's capital city – with all that this implies for the presence of administrators, politicians and bureaucrats, and their effect on the social characteristics of the city – and New York is not.

The complex relationship between culture, territory and the city is expressed in the challenges to Colombo's postcolonial centrality within Sri Lankan society and space. Colombo's role as the principal site of political negotiation has so far been a short-lived, postcolonial phenomenon. Despite the efforts of the 1977 government to reorganize Colombo's economic centrality within Sri Lanka, its political centrality has waned. This is understandable if we remember its colonial history as a capital was imposed from outside to serve the colonizers' interests and not those of Lanka itself. Although the government has built a new government complex at Kotte, the locus of political negotiation, the principal place where political organizations resolve regional and cultural differences within broader national policies, has shifted not to Kotte but to rural areas.

It was only during the first three decades after independence in 1948 that Colombo enjoyed a position as the national center in which political conflicts were resolved, constitutionally through voting and parliamentary debates, and by representatives sent to Colombo from electorates elsewhere. This was a particular form of decision-making developed in the "West," known as "democracy," institutionalized in Ceylon by the colonial government. The major political conflicts during colonial rule, including earlier revolts led by the Kandyan aristocracy and Buddhist clergy and later in the twentieth century socialist-led strikes and plantation uprisings, have largely taken place in rural areas. The major political conflicts of the last two decades, one of which calls for a separate Tamil state, Tamil Eelam, led by the Liberation Tigers of Tamil Eelam, and the rebellions led by Janata Vimukti Peramuna between 1971 and 1989, have also taken place in rural areas outside Colombo. Instead of using "democratic" means of debate and voting, both the Liberation Tigers and the Janata Vimukti Peramuna have resorted to armed struggle, inviting representatives of the state to their own territory to negotiate the outcome. They have thus devalued Colombo's role as the center of national politics and government. It is the exacerbation of this culture–land–people conflict that is apparent in these two struggles.

It is therefore quite pertinent to ask to whose domain does Colombo belong? Like New York, Colombo was a European implant, a secular and essentially commercial city. It was race and ethnicity which provided the principal criteria for social stratification and segregation in the city, colonial masters from the indigenous population as well as white officials from white business and working classes. Here, the British did not tolerate the close presence of other Europeans, the descendants of the former Dutch and Portuguese colonial communities, later identified as Burghers. In squeezing them out of the fort area – first constructed by the Portuguese and later modified by the Dutch in the mid-seventeenth century – the British effectively divided the city into three ethno-racially defined components. In the words of Hulugalle, "the Fort was chiefly occupied by British residents; the Pettah [area outside the fort] by the Dutch and the Portuguese; and [the 'outer Pettah'] by Singhalese, Tamils, and [Muslims]."

In the postcolonial city racial segregation has given way to increasing class stratification to which new ethno-linguistic and also caste divisions are being appended. By "class," I refer to social groups defined primarily along economic and occupational lines vis-à-vis cultural groups defined along ethnic, religious and caste ones, with their accompanying values. This emphasis on class is certainly an outcome of the British colonial system with its large bureaucratic and socio-spatial segregation as well as the hierarchic industry-based culture. Since the colonial division of labor was organized around ethnicity and race, social stratification was based on a combination of race, ethnicity, and class. The economic and political elite who moved into powerful positions as postcolonial administrators in this capital city, and occupied prestigious places such as the suburb of Cinnamon Gardens left by the colonials, had more of an elite outlook than an ethnic or religious one. They were the "old boys" of elite schools who shared the same culture, cultivating a value system which supported the view of the city as home, promoting secularization and undermining ethnic, religious and linguistic differences. By the time of independence, most of the Ceylonese political, economic and cultural elite were more at home in England and the colonial world of Colombo than in the Singhalese, Tamil or village-based world, that had once been their own. Hence Colombo is now largely the home of the power elites and capitalists, also executives, bureaucrats and traders – though this is hardly the producer service class of New York. Those who occupy the top rungs of an economically and administratively defined hierarchy of inhabitants of Colombo are the ones who can afford to buy or invest in expensive real estate and share the similar "introvert" values to those found in New York. That Colombo is still the "capital" and there is no wholly commercially-oriented city (as represented by the relation of Washington to New York) probably dampens the development of market values in the city.

Elite attitudes towards culture, ethnicity and religion have produced confusion, especially since independence made Colombo more directly a part of larger Sri Lanka, as well as one of its cultural centers, primarily of "Western" capitalist culture. Hence there is a desire among groups that are culturally and linguistically more close to the "West," to reproduce ethnic and religious identities acceptable to the Sri Lankan public. This is particularly the case with politicians who, for example, now wear national dress for important political events.

Colombo's political and economic centrality, and cultural supremacy over the crown colony of Ceylon, was a result of British colonialism. It resulted from the incorporation of the island into Colombo's sphere of domination. The independence of Ceylon has not only enabled indigenous religions to play a greater role in the society, but also revived historic ethno-religious centers – the Buddhists in Kandy and Anuradhapura and Hindus in Jaffna – producing a multicentered cultural arena. In this context new cabinet members made pilgrimages to the Temple of the Tooth Relic of Buddha at Kandy, and the first executive president, J.R. Jayawardena, following a former royal tradition chose to address the nation in 1978 from the *pattirippuwa*, the octagonal podium of the temple. Thus governments reinforced this multicentric cultural and spatial formation until the 1980s, even though Colombo still remained the principal center of accumulation and "Western" capitalist culture.

The former president, Ranasinghe Premadasa (1988–1993), who was also prime minister under Jayawardena, attempted to promote the Buddhist temple at Gangaramaya to a "national" level by, among other things, holding an annual procession with elephants, comparable to that held in Kandy. This is perhaps the latest and most profound example of mobilizing historic temples for political means in the vicinity of the municipal area. Attempts to advance Colombo to a significant ethno-religious center have not been limited to Singhalese-Buddhists, but can also be seen among Hindu-Tamil elites and the upper middle classes of Colombo. It is, therefore, the upper classes and politicians of all ethno-religious groups that have attempted to construct a more consciously "postcolonial" cultural role for Colombo. It is hardly necessary to point out that there is no comparable resurgence of indigenous cultural expression, either in New York or its vicinity.

Moreover Colombo, where different ethnic groups meet and conflicts arise, is seen by them as common territory. Although the wave of anti-Tamil violence of 1983 caused a large migration of Tamils out from Colombo and its vicinity to the north, as well as emigration out of Sri Lanka, the continuing battle between Tamil separatists and the military, and among militant groups in the north and east, has caused many Tamils to return to the city. Here, the former separatist groups, for example, the People's Liberation Organization of Tamil Eelam, also use what they view as the "neutral territory" of Colombo for political activity and organization.

For the most, the identities constructed in Colombo are not complete; they are only one of a number of other identities that exist in tension. As the majority of Tamils settled in Colombo seem not to favour a separate state, this would suggest that their economic and social identities are also important. Colombo is therefore much more a part of a larger single territory of Sri Lanka, a city that cannot be separated from the state in which it exists. Should there be a separate state of Tamil Eelam, this would oblige the Tamils of Colombo, like the transnational immigrants of New York, to provide a meaning for their lives and culture by referring to an unrelated land.

THE CULTURE INDUSTRY AND TOURISM

The cultural industries of the "global cities," particularly in New York, have developed, according to various scholars, mainly to serve the new professionals of the producer service sector as well as domestic and foreign tourists. In the newly independent states, a more general tourist industry can be seen as related. Although art markets and tourism are not new, they have expanded, especially from the 1970s. After the failure of policies directed at economic development, many Third World states developed tourism as a means of earning foreign exchange and balancing trade deficits.

Cultural industries, derived from a particular commodification of culture, have also emerged in Sri Lanka. The cultural artifacts exhibited for the consumption of mainly European and American tourists are located in the ancient metropolitical centers of, for example, Anuradhapura and Polonnaruwa, far away from Colombo which, in this respect, remains a somewhat alien port-city of less interest. This exemplifies the stereotypical colonial and postcolonial split site, of the so-called "traditional" and "modern" city. This is certainly a departure from an earlier British representation of Sri Lankan cultures through dead artifacts in museums, whether in Colombo or London. Here the Sri Lankan government has restored ancient capitals, historic and architecturally important buildings, as well as culturally significant landscapes in their original sites. Promoted as a foreign exchange earner in the 1970s, tourism has renewed an interest in the past, expressed in the establishment of an institution separate from the Archaeology Department, the Cultural Triangle, which refers to three former Lankan capitals, Anuradhapura, Polonnaruwa and Kandy. Significantly however, the production of this history and the organization of this particular culture industry takes place in Colombo, still reflecting its external origins.

CONCLUSION

I have drawn attention in this paper to the inadequacy of the available intellectual tools to address issues of postcolonial space, in this case, cities such as Colombo. It is clear that, apart from some commonalities, frameworks produced for the examination of New York are insufficient to make much sense of contemporary developments in Colombo, even though they do provide some significant insights. New York, as the leading economic center of the leading economic power of this century and with its continuing colonial situation, is too unique to be directly relevant for the study of many other cities of the world. Despite significant developments in the capacity to explore cities as part of a larger spatial system, particularly through the concepts of the global or world city, much of this theory is inadequate to address the economic restructuring of cities outside the capitalist core, and concerning aspects other than the economic.

Moreover, "cities of the world economy" is only one among many ways of representing systems of cities. Regional and national as well as cultural and political formations have come to play an increasingly significant role following the independence of a large number of states after the mid-twentieth century. This is even more so after the end of the Cold War and the demise of US hegemony. As the case of Colombo demonstrates, it is these different spatial, cultural, and political contexts that produce differences among cities. This is not to reject the importance of studying larger structures of cities and commonalities among those, which, in an increasingly integrated world, are increasingly visible. Nonetheless, this global integration has itself produced more and new differences in seemingly similar spaces. Addressing more carefully the changes taking place in cities outside the capitalist core – which is hardly a unified entity – and relating those to phenomena in that core, however, requires the mediation of more conceptually sophisticated frameworks and the pluralization of our urban perspectives. Developing analytical frameworks that are friendly to local perceptions and employing local vantage points of inquiry are crucial steps in this direction.

REFERENCES FROM THE READING

McGee, T.G. (1971) *The South-East Asian City: A Social Geography of the Primate Cities of South-East Asia*. New York: Praeger.

Sciorra, J. (1996) Return to the future: Puerto Rican vernacular architecture in New York City. In A.D. King (ed.) *Re-presenting the City*. New York: New York University Press, 60–92.

Sennett, R. (1992) *The Conscience of the Eye*. New York: Knopf.

SIX

42

"Culturing the World City: An Exhibition of the Global Present"

Steven Flusty

Editors' introduction

Steven Flusty teaches in the Department of Geography at York University in Toronto after having worked in Los Angeles for many years. His contribution to this volume is based on his book, *De-Coca-Colonization: Making the Globe from the Inside Out* (2004). Flusty's work is concerned with the everyday practices of global formation. In particular, Flusty approaches this issue by examining the production, distribution and consumption of cultural artifacts in the global urban system. Not satisfied with the standard, western-oriented global city discourse, Flusty follows four specific, and very different, artifacts as they traverse the cultural microcircuits of individual global cities and in the larger global space of what he calls "metapolis." Starting from this multifaceted approach to urban cultural artifacts, Flusty explores the web of networks within global cities through which diverse cultural practices collude to produce what he terms "the global present." The global city described by Flusty is a "fluidly demarcated global urban field upon which we all wrestle with the very definitions of alien and native, foreign and domestic, cosmopolitanism and locality." After entertaining the reader with a detailed gaze at three cultural objects situated within a virtual/imagined museum, Flusty concludes by holding up a mirror to all of us as active participants in the production of the global city. Flusty's turn to agency as an important aspect of world-cityness resonates with some of the other readings in Part Six, particularly with the postcolonial emphasis on agency, social movements and decentralized cultural production (see also Reading 46 by M.P. Smith). Flusty's explorations of the culture(s) of the world/metacity point towards a cultural politics based on the principle of "xenophilia," a radical differentialist mode of interaction in the spaces of the global city.

INTRODUCTION

Imagine yourself visiting an exhibition at some new museum or other. This should not prove too difficult a task, given the fecundity with which museums have been cropping up in eccentrically angled glass or fluidly excreted titanium across the postindustrialized world. The exhibition itself, likely sponsored by some consortium of corporations anxious to advertise their civic-virtuousness despite their lack of any plausible local or national identity, is regrettably a small one – a collection that consists of a mere four artifacts. These are, after all, times of fiduciary stringency. Nonetheless, the few objects on display (available as authentic replicas in the attached gift shop, and at cosponsoring department stores) should provide a serviceable impression of ongoing excavations into what might best be called the metapolitan moment.

CULTURED WORLD CITIES

Our first artifact is a small gold-tone and cloisonné lapel pin, circa 1990. It bears the slogan "Building a World City," surmounted by a handful of stylized cubes forming a skyline in symbolic shorthand (Figure 1). This pin, issued to then-executives of Los Angeles' Community Redevelopment Agency, celebrated L.A.'s long-sought ascension to "world class city" status through the of-a-piece installation of a high-rise central business district where none had existed a scant fifteen years prior. The skyscraper has become the universally agreed upon icon of world cityhood, a complex concatenation of material culture whereby, if you build them

Figure 1 CRA World City pin

in sufficient density, the world will come. And so they have been built in great numbers, from Los Angeles to Frankfurt to Shanghai, in every city moved to signify indisputably its emergent presence on the world stage. Not that this fetishization of tower-studded horizons is restricted to the urban apparatchiki: ask any child to draw a city and you will likely receive in return a picture of numerous, grid-bestudded rectangles all standing at attention. But within the symbolic system of the world city makers, the economystic arcana of location theory, urban entrepreneurialism and A-class office space are empowered by the mudras and mantras performed on stock-exchange floors to render the skyscraper not merely an iconic synecdoche inextricable from our collective cultural consciousness, but a powerful ritual instrument of practical magic. Erected and consecrated, it channels capital from on high to transubstantiate the city into a circuit for the electro-ethereal web of plutocratic global "flows." In the process, dispersed cities attain union with one another across vast distances to become a city of cities, a world city system, a metapolis predicated upon a common culture of cash and commodities in-transit – New York and London become NY-LON, ascendant within that supreme trinity of world citydom: New York/ London/Tokyo (Sassen 1991). Beneath these "alpha class" world cities, others take their rightful places in the beta or gamma classes arrayed along a great chain of municipal being determined with recourse to enumerations of each city's corporate head offices and producer-service firms (see Knox and Taylor 1995; Taylor 2003).

While our pin depicts only the architectonic lingam at the heart of this process, the skyscraper requires its attendants if it is to work its worlding magic. In the lay definition, world cities are places where the world's business is transacted (Hall 1966). But in their primordial genesis such cities were imperial metropolises, and along with their crown corporations and banking houses came sites where both the most sublime and grotesque of humanity's creations, stripped from empire's hinterlands, could be collected, admired and consumed. Commerce bedecked itself irrevocably in Culture, and to this day the contemporary world city is without a soul in the absence of the art museum and the concert hall – without the cultural capital, the intellectual capital at the helm of fiduciary capital will not come. Thus L.A.'s skyline arrived with a complement of two highly celebrated art museums and a completely remodeled third, Frankfurt's a museum row consisting of more than a dozen fresh-built museums on the banks of the Main, and franchises of Manhattan's Guggenheim proliferate across the face of the earth (Friedman 2003). It is Culture not just as commodity (an old story, that) but also as bait, prepackaged events and exhibitions that are shunted, for a fee, from one world city aspirant to another. A condition appropriately isomorphic to the metapolitan harmonization of settings in which these unitized Culture flows take up their temporary residence – the meltoid metallic curves of Frank Gehry's Guggenheim Bilbao being all but indistinguishable from those of his Disney Concert Hall in Los Angeles.

THE CULTURE OF THE WORLD CITY

Our second artifact is a boxful of LEGO ® bricks – multicolored, snap-together building blocks (Figure 2). This particular boxful is one selected from the LEGO collection of "World City" building sets that, when assembled according to the photograph on the box, creates a high-tech police surveillance truck accessorized with a three-wheeled motorcycle to apprehend fleeing suspects. The city of the child's imagination may be a simplified rectilinear skyline, but the LEGO pedagogy of play elaborates that vision with the detailed specifics of a world city culture – build-it-yourself high-speed passenger and cargo trains on the one hand, and on the other police helicopters, armored cars and surveillance vehicles; on the one hand mobility, on the other its delimitation and suppression.

These are not oppositions, though, but complements. Machines for moving the possessions and persons of those who have to rely upon machines that immobilize the dispossessed. And then some – for every chauffeured town-car there are legions of surveillance cameras arrayed along its route, ensuring unmolested passage across the world city and throughout the metapolis, from gated community to corner office suite, from Four Seasons Hotel New York to Four Seasons Hotel London to Four Seasons Hotel Tokyo at Marunouchi, from Parisian café to Phuket beach

resort. Along the way, each stop is an opportunity to acquire new and different tastes, sights, objects and experiences, to sample and edit and recombine them. A vast cosmopolitan hybridity engine, lubed and fueled for perpetual motion with the sacrificial blood, sweat and "ethnic food" of the immobilized – held back at border crossings and beyond the gates of guarded neighborhoods, in favelas and gecekondus and refugee encampments, kept in their place as assemblers in Export Production Zones and waiters in Club Meds and janitors in skyscrapers' washrooms or on trading room floors, fixed in shrunken places where the new is what's on TV now and the different are those who aren't quite right a few blocks over thataway.

The privilege of mobility plugs into cosmopolitanism and tabs together with ever-accelerating hybridization, while the immobilized slot in with localisms that snap tight with desperation (Friedmann 1994; Bauman 1998). This is the prescription for connecting the blocks of world city culture, the illustration of a sharply bifurcated complementarity emblazoned on front of the metapolis' box. But prescription does not entail subscription, instructions can be disregarded, and the blocks of the LEGO surveillance truck serve equally well to make a tuktuk, a "technical" bristling with RPGs a'blazing, or a dancing lowrider pickup truck. The building blocks of world city culture are similarly incorrigible – the immobilized prove mobile, privilege cleaves to parochialism, and hybridity is born of desperation. Consider, for example, the executive elite, flying business-class from business-class airport lounge to business-class airport lounge, from business-class hotel to business-class hotel, rarely compelled to speak an alien tongue, ingest an unfamiliar food or negotiate a foreign street. Now, compare with the West African taxi-driver negotiating a fare through the streets of London or Tokyo, a Michoacaña hotel-maid walking a Manhattan picket line, a sailor recruited from Luzon Island to tend the containerized leviathans that ply the shipping routes linking these three cities into one – all obliged to adapt their everyday worlds to that in which they find themselves subsisting, and it to theirs. Now, who among these is the cosmopolitan, and who the blinkered local? True, the circumstances of the refugee compelled across a border differ radically from those of the tourist who crosses by choice,

Figure 2 World City Lego © 2005 The LEGO Group. All rights reserved

but still, which is doing the real work of hauling other worlds into the world city?

Such are the problematics and potentials of the metapolis, a fluidly demarcated global urban field upon which we all wrestle with the very definitions of alien and native, foreign and domestic, cosmopolitanism and locality. And during such contests we kick up dislocalized localisms, new majorities and emboldened minorities, ever-shifting constellations of popular coalitions, and maybe . . . just maybe . . . a chance to reimagine the world city's fiduciary rationale not as an underlying truth but as just one of many strategic, and inherently cultural, agendas.

Figure 3 Tasbeh

WORLDS OF CITY CULTURES

Our third artifact is a tasbeh, a set of ninety-nine black wooden beads strung along a tasseled green cord to form an Islamic rosary (Figure 3). This component of the collection was made in Bangkok and acquired in Toronto, but could equally well have been purchased in New York, London or Tokyo. We could think of this tasbeh as proof of how the world's ex- (or neo-, if you prefer) colonial hinterlands penetrate to the global center, infusing their many worlds into the world city. But another interpretation is equally valid: in determinations of how the tasbeh is (or is not) to be employed for performing *dhikr* (personal prayer), New York, London and Tokyo are the recipient hinterlands of a very different global center – a system of world cities comprised of such places as al Madinah, Cairo and Karachi.

Of course, it might be argued that such locales cannot be world cities. They are not centers of corporate command and control, they are woefully undersupplied with skyscrapers, they do not attract armies of immigrant labor, they don't even have a Guggenheim. In a count of corporate head offices, al Madinah would not even appear as a delta class world city! But for many of those whose patterns of commonplace, symbolically charged material practices – whose culture – is more beholden to the Qur'an than to the capital markets, al Madinah is central in a very different world city system that relegates New York, London and Tokyo to beta or even gamma class status, at best.

The tasbeh is a reminder that while corporate head offices are readily countable, this in no way entails they are all that counts, and counting other cultural indicators yields some very other world cities organized into some very other world city systems indeed. Head offices, after all, are no less cultural artifacts than any other, components of the dynamically patterned practices within which capital and economies are embedded. And if we shift our vision to focus upon the material practices that circulate not conceptions of capital but of divinity, cities we never thought to notice before take pride of place: Vatican City, al Madinah, perhaps Salt Lake City and Dharamsala and, insofar as neoclassical economics presently constitutes the planet's preeminent theology, Chicago. Nor need we stray so far into the realm of the theological. In a world imbued with cinematic communications, for instance, Mumbai has long stood astride all others for sheer quantity of output while Tokyo, that paragon of world cityhood, constitutes a hinterland wherein anime and videogaming fruitfully miscegenate and multiply – here there be monsters!

The metapolis, then, is not simply a world city system but a system of world city systems, and at these systems' proliferating intersections divergent cities manifest within one another across wide distances – the culture of the arbitrageur embodied by the branch office of a Manhattan-based bank in Saudi Arabia invariably implies the presence of al Madinah's priestly culture in any number of masjids dotting the northeastern seaboard of the United States. Such ongoing cultural exchanges generate a landscape of interleaved world cities, one that systematizes differently depending upon how one looks, and what one looks for. Further, the disjunctions between these different systematizing views are not just an artifact of how we see, but also an impetus to how we act: consider, for instance, the currently escalating tension that indirectly pits the metaphysical logic of al Madinah against the fiduciary logic of NY-LON, tension that manifests at scales ranging from the geopolitical to the city block.

It may well be that being a place where the world's business is conducted determines world city-hood. The business of the world, however, takes many forms indeed – embedded as it is in wildly divergent patterns of widely differing practices that are simultaneously material and symbolic – and the world city necessarily follows suit. For some time now it has been commonplace to assume the world city as a deculturated economic formation, and to pursue from there such cities' more-or-less epiphenomenal cultural correlates. Meanwhile, the battle cry against these market-driven urban machines has been: "another world is possible." A crippling understatement, that, when the polyvalent cultural embeddedness of the metapolitan condition proclaims something simpler yet far more radical: "other worlds are." Rigorously universally scientifically objective determinations of world cityhood notwithstanding, there are far more kinds of world cities, organized into far more world city systems, than are dreamt of in our geostatistical algorithms.

WORLD CITY CULTIVATORS

Our fourth and final artifact is a mirror. Its provenance and details are unimportant. What matters is that you view yourself within it, cease to be an exhibition attendee, and become instead a participant.

We are free to depict the global circulation of cultural materiel, whether instruments of fiduciary capital, the exertions of migrant labors, cinematic genres or ritual implements, as flows. But conversely, we can describe them as discrete units comprised of those who send, receive and deploy them, who carry them from place to place and adapt them to new settings. Material practices exist and become meaningful only on account of their practitioners, and that means us. In innumerable and diverse ways,

sometimes intentionally but more usually without realizing it, we are the world city makers and the sites at which systems of world cities intersect. Whether an executive flying business class between New York and Tokyo, an immigrant laborer sending remittances back home to the outskirts of Morelia or Accra, or a Muslim doing *dhikr* in Toronto, we carry our worlds with us, refit them to the cities in which we find ourselves, and transmute the city as best we can to accommodate our worlds.

The aggregate of our practices is the culture of the world city, its ongoing hybridizations and metapolitan outcomes, and to act accordingly is to recapture just the slightest bit of command and control so long held aloft in those alpha class conurbations of head offices. The more of us who do so, the more control we recapture from on high. But failure to do so is acquiescence to a marketist multiculturalism, one in which privilege makes great show of tolerating all comers while zealously insulating itself against them, leaving those so excluded to eye one another with suspicion and fear.

Establishment of common cause amongst the divided and conquered would constitute the best response to such a dog-eat-dog metapolitan dystopia. A shame this seems no mean feat, given how the divisions in question are predicated upon the seemingly irreconcilable divergences of viscerally affective, meaning-laden material practices: *Kulturkampf* or, more specifically, the purported clash of separate and distinct "cultures." Cultures that only grow more entrenched when confronted by the imperatives of acculturation, the threats of Coca-Colonization and McDonaldization.

But while our worlds may remain divergent, in the world city they must also make their homes cheek by jowl, rub up against one another, swap bits and pieces, and propagate the hybrids that result. Interculturation reconciles irreconcilable worlds without sacrificing their irreconcilability. In the streets and the everyday, it mocks both the dictatorship of acculturation and the disingenuousness of multiculturalism, whether in forms as ephemeral as the appearance of soju cocktails in Persian restaurants or as radical as the Zapatista's anti-authoritarian tactics embodying on the streets of Quebec City or Genoa. In the world city it is everywhere, and it constitutes a small but mighty tool for re-Building a World City from the underside out . . . especially when fortified with a modest dose of xenophilia.

By xenophilia I do not mean the romanticization of some Other and the consumption of the othered's cultural forms, although this at least can constitute a first step – we have become too prone to underestimate the power of breaking bread with others on their own terms. Think instead of xenophilia as a driving thirst to openly engage, and be engaged by, that which is unfamiliar, a sensibility that regards difference as much more than just something that happens and needs to be grudgingly dealt with or, worse, defended against. Xenophilia reminds us how we too are different, prepares our psyches for deep relations with those who differ from us, and at the same time acts as a remedy to and an inoculation against our mistrust of otherness. So while our lived worlds, and our apprehended knowledges of how the world is, are necessarily too divergent to ever completely integrate, cultivating xenophilia is requisite to appreciating, respecting and even empathetically occupying each other's ineluctably partial perspectives (Haraway 1991). Through xenophilic engagement we develop the capacity to experience a semblance of diverse realities, interact dialogically with others who live those realities, and so negotiate across differential positions (Bakhtin 1988) amiably and with mutual affect. Such dialogical negotiations, in turn, are indispensable if radically diverse social actors are to take a stand against practices whereby power works to silence and disappear many, and order the metapolis as a whole, for the benefit of a few.

So, which shall it be? A place where difference divides, privilege is conserved, and the devil take the hindmost? Or a place where otherness engages, disparity is dismantled, and the production of a metapolitan culture becomes a common, conscious project? We culture the world city, so the choice is ours.

REFERENCES FROM THE READING

Bakhtin, M. (1988) *The Dialogic Imagination*. Austin, TX: University of Texas Press.

Bauman, Z. (1998) *Globalization*. Cambridge: Polity.

Flusty, S. (2004) *De-Coca-Colonization*. New York and London: Routledge.

Friedman, A. (2003) Build it and they will pay: a primer on Guggenomics, *The Baffler*, 1/15, 51–56.

Friedmann, J. (1994) *Cultural Identity and Global Process*. London: Sage.

Hall, P. (1966) *The World Cities*. New York: St. Martin's Press.

Haraway, D. (1991) *Simians, Cyborgs and Women*. New York: Routledge.

Knox, P.L. and Taylor, P.J. (eds) (1995) *World Cities in a World-System*. Cambridge: Cambridge University Press.

Sassen, S. (1991) *The Global City*. Princeton, NJ: Princeton University Press.

Taylor, P.J. (2003) *World City Network*. London and New York: Routledge.

PART SEVEN

Emerging issues in global cities research: refinements, critiques and new frontiers

Plate 32 Shanghai (Henry Yeung)

INTRODUCTION TO PART SEVEN

On July 13, 1988, a "ballet" performed by four caterpillars was the centerpiece of a groundbreaking ceremony for what remains Europe's second tallest skyscraper, the 257 meter Frankfurt MesseTurm on the German city's west end fairgrounds. The highrise, designed by German-born Chicago architect Helmut Jahn, under the direction of New York development Tishman Speyer and with money from Citibank, had both local and global significance. For Jahn, it was a landmark assignment in his native Germany; for Jerry Speyer, whose Jewish family had to flee Frankfurt during the Nazi years, it was a symbolic "homecoming" to a city which desperately tried to remake itself from the capital of Germany's post-World War II economic miracle into a node of the global economy. The redevelopment of the Frankfurt fairgrounds and the construction of the iconic MesseTurm supported Frankfurt's emergent role as a player in a world of seemingly unlimited economic possibility.

Two days before the groundbreaking, one of this Reader's editors had co-authored a full-page feature article in Frankfurt's highest circulation daily newspaper, the *Frankfurter Rundschau*. The article, entitled "The New Frankfurt – Between Citadel and Ghetto," was a critical commentary on the kind of global city formation the arrival of the MesseTurm signaled for West Germany's economic powerhouse (Keil and Lieser 1988). The piece made extensive interpretive use of work by John Friedmann and other critical urbanists, who at that time had just begun to elaborate the intellectual foundations of world city research. In particular, the newspaper article critically underscored the increasingly polarized socio-economic and sociospatial urban landscape that could be expected to emerge in the wake of Frankfurt's globalization; it thus drew extensively upon Friedmann's metaphorical opposition between the "citadel" and the "ghetto."

The article's public reception turned out to be peculiar. A few months after its initial publication, the authors were approached by the developer of the project, Tishman Speyer, with a request to permit the article's translation and subsequent use for advertising purposes among the company's corporate clients. Surprised and amused that the article could be viewed in such a positive light by a party that had been implicitly criticized within it (and due to the fact that the piece was in the public domain anyway), the authors agreed to have their work used in this way. But this was not the end of the story. A few years later, an anarchist organization linked to the most radical factions of the Frankfurt political spectrum held a weekend conference on the future of Frankfurt as a globalized city. The organizers used the 1988 newspaper article on posters announcing the event on walls and billboards around the city and as reading material for the conference.

What does this anecdote reveal? In our view, it shows, first and foremost, that research on global cities is intrinsically political. Like all forms of urban knowledge, representations of the global city are never neutral, but are always embedded within ongoing struggles to shape and reshape the everyday geographies of social, economic and political relations within cities. Consequently, an analysis that one group views as an endorsement of globalized urbanization may also be mobilized for diametrically opposed, critical purposes by other social forces. One of the more persistent criticisms that has been leveled at global city researchers is that their work serves to glorify the status of particular cities in worldwide interurban competition, and thus represents an uncritical affirmation of global neoliberalism. Relatedly,

Plate 33 Frankfurt MesseTurm (Roger Keil)

it has also been insinuated, at times, that research on global cities more or less embraces the policies of municipal boosters in search of distinction on the world stage. In our view, the misunderstanding that underlies these criticisms is based on a mistaken identification of the colloquial notion of the global/world city with the scholarly concept developed in the literatures we have reviewed in this volume. While the former is a descriptive, affirmative notion often used by municipal power brokers to draw attention to specific places (Boyle 1999; Short 1999), the latter is a polysemic analytical term that has been employed by critical urbanists concerned to decipher the globalizing dimensions of contemporary urbanization.

Still, some of the confusion around the notion of the global city may also be attributed to the substantive content of social science research on this topic. In some cases, such as Los Angeles, it would appear that the "hype" generated through studies of the purported "globality" of a particular place actually permits academic researchers to be enlisted, often unwittingly, as "mercenaries" into the camp of global city boosterism (Davis 1990; Gottdiener 2002). In this context, it is crucial to recall that Friedmann and Wolff's first foray into global cities research contained the programmatic subtitle "an agenda for research and *action*" (our emphasis). For Friedmann and many of his colleagues, the analysis and description of the global city was meant to be a first step in actively effecting positive, progressive and even radical social change. Thus, data on the formation of global urban hierarchies and on the intensification of sociospatial polarization within global cities were clearly understood as a call to arms for progressive planners. Their role, in Friedmann's view, was to mobilize new public policies designed to reduce the suffering of the global city's increasingly impoverished internationalized working classes and migrant populations and, more ambitiously still, to subject the apparently "deterritorialized" operations of transnational capital to localized, democratic political control. For others, of course, this call to action was interpreted as an imperative to establish the positive business climate and general investment conditions that were deemed necessary for world city formation. However, in an incisive intervention into the public policy debate in East Asian city states craving world city status in the 1990s, Friedmann reminded his audience:

[U]rban outcomes are to a considerable extent the result of *public policies*. They are, in part, what we choose them to be. The cities of the next century will thus be a result of planning in the broadest sense of that much abused term. This is not to fall into the naïve belief that all we need to do is to draw a pretty picture of the future, such as a master plan, or adopt wildly ambitious regulatory legislation as a template for future city growth . . . Instead of waxing enthusiastic about megaprojects – bridges, tunnels, airports, and the cold beauty of glass-enclosed skyscrapers – which so delight the heart of big-city mayors, I am talking about people, their habitat and quality of life, the claims of invisible migrant citizens and now, in yet another turn, the concept of civil society.

(Friedmann 1997: 2, 26; emphasis in the original)

Open questions and missing links: a provocation to further research

The readings we have assembled in this final part are a selection of more recent contributions to the debate on globalization and urbanization. All of them are, in some manner, critical of the global city research agenda and all are equally wary of the possibly contradictory political valences that may be associated with it, both in scholarly and policy contexts. In general terms, these contributions may be distinguished according to whether they advocate conceptual and methodological reformulations within the current parameters of world city theory, or whether, by contrast, they adopt a more fundamentally critical stance and propose a more incisive change of direction in this research field. Concomitantly, the critiques differ according to whether they are primarily social-scientific, and thus target the "research" element of Friedmann's original agenda, or whether they are more explicitly political in orientation, and thus strive to address the question of "action" in globalizing city-regions.

Reading 43 by Saskia Sassen as well as the subsequent contributions by Scott, Samers, and Olds and Yeung (Readings 45, 47 and 48) are situated most clearly in the methodological tradition of global city theory as originally developed by Friedmann and the other contributors to Parts One and Two of this Reader. Readings 47 and 48, in particular, advance a renewed and re-energized research agenda that combines the most sophisticated work in urban and regional studies with key insights from established work on global cities. Reading 45 by Scott reformulates and extends Friedmann's original contention that world city regions constitute the spatial foundations for a new configuration of global capitalism. One of Scott's key contributions is to analyze the political and institutional consequences of the resurgence of global city regions in a post-Westphalian world. In a closely related analysis, which also resonates with Reading 32 by Timothy Luke, Olds and Yeung (Reading 48) draw our attention to the variant pathways of global city formation that have crystallized in different zones of the world system. On this basis, they emphasize the diverse *processes* through which globalized urban spaces are produced in specific historical-geographical contexts. While critical of some aspects of the global cities literature, these authors attempt to strengthen and extend the main lines of world city research exemplified by previous work from Friedmann, Sassen and other major contributors to this literature. In Reading 47, Michael Samers considers analyses of the interplay between immigration and global city formation. Focusing on questions of causality and empirical evidence, Samers' critique points to a more general problematic in global cities research – the tendency to presuppose rather than to demonstrate causal linkages among diverse macro-level processes that happen to be articulated within globalizing cities. While sympathetic to Sassen's general approach to global city formation, Samers outlines a number of methodological and empirical challenges that, he believes, must be confronted more rigorously in order to pursue some of the basic research questions that have been posed by scholars of globalized urbanization.

Reading 44 by Peter Marcuse, which builds upon a number of book-length investigations (see Marcuse and van Kempen 2000, 2003), examines the *micro*-level of global city formation. Marcuse rejects the assertion that globalization replicates an identical pattern of sociospatial inequality within major urban centers. Instead, he presents a more empirically differentiated model of how and why processes of urban sociospatial polarization unfold, which is intended to be applicable to globalizing cities throughout the world system, and not only to those that have previously been characterized through the global city concept.

Two other readings in Part Seven are more explicitly critical of the methodology used in the bulk of global city research. Reading 46 is an excerpt from a major book-length publication by Michael Peter Smith, *Transnational Urbanism* (2001), which develops a fundamental critique of the structuralist bias of much global cities research. M.P. Smith's interpretation of globalized urbanization suggests that, despite the extensive empirical research of global city researchers, there can be no positive determination of a "real place" that could be legitimately described as a "global city." As he explains, "The global city is best thought of as a historical construct, not a place or 'object' consisting of essential properties that can be readily measured outside the process of meaning-making" (see Reading 46). For him, the structuralist bias of global city theory blinds researchers to the transnationalized forms of human agency that, in practice, actually construct and sustain the global city. While Michael Peter Smith appears to suggest that the project of global cities research should be abandoned altogether, Richard G. Smith takes a different approach in Reading 49. R.G. Smith's concern is to reconceptualize, on a fundamental level, the geography of the global urban system as a whole. To this end, he contrasts the hierarchical notions of global city relationships with a "topological" view, in which global cities are viewed as arenas and outcomes of diverse transnational flows and networks. Accordingly, R.G. Smith treats global cities as heterogeneous products of multiple, multiscalar and networked relations rather than as fixed and bounded containers embedded within a nested global hierarchy (see Reading 49).

The Reader ends with an excerpt from an author who was arguably one of the most important voices in twentieth-century urban thought, the French sociospatial theorist Henri Lefebvre (Reading 50). Lefebvre never wrote explicitly about global cities in the sense in which they have been examined in this book. Nonetheless, in one of his major works of urban theory, *The Urban Revolution* (2003 [originally published in 1969]), Lefebvre alludes to the globalization of urbanization in at least two ways that, in our view, have considerable significance for future work in this field. First, Lefebvre contends that we have now collectively entered into a phase of world history in which questions of human survival are tied, intrinsically, to processes of urban development. For Lefebvre, then, the world city has emerged not because certain types of places have become control centers for the global economy, but rather because a generalized process of worldwide urbanization is now unfolding. Lefebvre referred to this process as the worldwide expansion of "urban society." Second, in a prescient critique of Maoist revolutionary strategy, Lefebvre argues against the notion that there could, under contemporary conditions, be a rural alternative to global urbanization. Like other authors after him (Magnusson 1996; Friedmann 2002) Lefebvre insists that there is no "outside" to the world city. We all reside in it; therefore, one of the basic political questions of late modern capitalism is: what kind of urban world do we want to live in?

Conclusion

What can research on world cities/global cities teach us about the state of late modern social formations more generally? Beyond its significance to urban specialists, does research on global cities make a more general contribution to our understanding of contemporary social life? According to John Berger (2003: 13), we currently live in a globalized tyranny grounded upon a decentralized power structure "ranging from the 200 largest multinational corporations to the Pentagon"; for him, this imperial system rules in a style which is "interlocking yet diffuse, dictatorial yet anonymous" and its "aim is to delocalize the entire world." Global city research, in our view, offers us some bearings, some intellectual and political grounding, as we attempt to orient ourselves within this fundamentally disjointed, yet profoundly authoritarian, new world order. Whether or not this intellectual perspective can help open up possibilities for radical or progressive social change is ultimately a political question that can only be decided through ongoing social mobilizations and struggles. For Friedmann and Wolff, the different worlds of the citadel and the ghetto within the global city co-constituted one another; yet the inhabitants of the ghetto were said to be "isolated like a virus" from the hegemonic power structures. By contrast, in Reading 49 by R.G. Smith, the global city's rigid, fortress-like boundaries appear to be dissolved into fluid flows of actor-networks, where viruses seem to coexist quite easily alongside corporate leaders and bankers within the global urban system (for similar arguments see Swyngedouw 2004; Ali and Keil, 2006). The politics that spring from such radical openings in the world city fortress may generate unexpectedly trans-formative outcomes – for instance, new types of citizenship claims, new modes of political struggle, and a new globalized urban political ecology. The contributions to Part Seven begin to lead us down this path of possibilities.

References and suggestions for further reading

Ali, H. and Keil, R. (2006) Global cities and the spread of infectious disease: the case of Severe Acute Respiratory Syndrome (SARS) in Toronto, Canada, *Urban Studies*.

Berger, J. (2003) Where are we?, *Harper's Magazine*, March 13.

Boyle, M. (1999) Growth machines and propaganda projects. In A.E.G. Jonas and D. Wilson (eds) *The Urban Growth Machine: Critical Perspectives Two Decades Later*. Albany, NY: State University of New York Press, 55–70.

Davis, M. (1990) *City of Quartz*. London: Verso.

Friedmann, J. (1997) World city futures. Occasional Paper no. 56, Hong Kong Institute of Asia-Pacific Studies, The Chinese University of Hong Kong, Shatin, New Territories, Hong Kong.

Friedmann, J. (2002) *The Prospect of Cities*. Minneapolis, MN and London: University of Minnesota Press.

Gottdiener, M. (2002) *Understanding the City*, J. Eade and C. Mele (eds). Oxford and Malden, MA: Blackwell.

Keil, R. and Lieser, P. (1988) Das Neue Frankfurt – zwischen Zitadelle und Getto, *Frankfurter Rundschau*, July 11, 13.

Lefebvre, H. (2003 [1969]) *The Urban Revolution*. Minneapolis, MN: University of Minnesota Press.

Magnusson, W. (1996) *In Search of Political Space*. Toronto: University of Toronto Press.

Marcuse, P. and van Kempen, R. (eds) (2000) *Globalizing Cities*. Oxford: Blackwell.

Marcuse, P. and van Kempen, R. (eds) (2003) *Of States and Cities*. Oxford: Oxford University Press.

Short, J.R. (1999) Urban imagineers: boosterism and the representation of cities. In A.E.G. Jonas and D. Wilson (eds) *The Urban Growth Machine: Critical Perspectives Two Decades Later*. Albany, NY: State University of New York Press, 37–54.

Smith, M.P. (2001) *Transnational Urbanism*. Cambridge, MA: Blackwell.

Swyngedouw, E. (2004) *Social Power and the Urbanization of Water*. Oxford: Oxford University Press.

Prologue

"Whose City is it?"

from *Public Culture* (1996)

Saskia Sassen

The space constituted by the global grid of cities, a space with new economic and political potentialities, is perhaps one of the most strategic spaces for the formation of transnational identities and communities. This is a space that is both place centered in that it is embedded in particular and strategic locations; and it is transterritorial because it connects sites that are not geographically proximate yet are intensely connected to each other. An important question is whether it is also a space for a new politics, one going beyond the politics of culture and identity, though at least partly likely to be embedded in it.

Globalization is a process that generates contradictory spaces, characterized by contestation, internal differentiation, continuous border crossings. The global city is emblematic of this condition. Global cities concentrate a disproportionate share of global corporate power and are one of the key sites for its valorization. But they also concentrate a disproportionate share of the disadvantaged and are one of the key sites for their devalorization. This joint presence happens in a context where the globalization of the economy has grown sharply and cities have become increasingly strategic for global capital; and marginalized people have found their voice and are making claims on the city. This joint presence is further brought into focus by the increasing disparities between the two. The center now concentrates immense economic and political power, power that rests on the capability for global control and the capability to produce superprofits. And actors with little economic and traditional political power have become an increasingly strong presence through the new politics of culture and identity, and an emergent transnational politics embedded in the new geography of economic globalization. Both actors, increasingly transnational and in contestation, find in the city the strategic terrain for their operations. But it is hardly the terrain of a balanced playing field.

"Space in the Globalizing City"

Peter Marcuse

Editors' introduction

Peter Marcuse is a lawyer and Professor of Urban Planning at Columbia University. He has been one of the most influential writers on cities since the mid-1970s; his academic and political writings have been an inspiration to generations of planners and urban theorists. Among his major contributions are studies of housing and gentrification, planning theory, the sociospatial reordering of what he has termed the "quartered city," and sociospatial change on both sides of the Atlantic, with a particular focus on the United States and Germany. More recently, Marcuse has become one of the most recognized voices in the international debate on cities and globalization. While critical of the global cities approach in the narrow sense, Marcuse has systematically explored the specific effects of globalization on urban sociospatial structures. In two widely recognized publications with Dutch urban geographer Ronald van Kempen, Marcuse developed a sophisticated matrix for the investigation of sociospatial restructuring in globalizing cities (Marcuse and van Kempen 1999, 2002). Reading 44 is based on this work but also discusses new theoretical and empirical developments.

What difference has globalization made in the space within cities? More specifically, what are the characteristic internal structures of cities today, and how, if at all, are they different from what they were before globalization? What aspects of globalization account for what changes we find? Is the result a new Global City? If not, what does account for the spatial structure of cities today? These are frontier questions for anyone concerned about cities today, and suggests the research agenda this reading attempts to outline.

To start with, a definition of globalization is necessary. Then we pose in more detail the key questions that globalization raises about cities. We then go on to examine the actual patterns of space in cities, and changes in them, in the current period. We conclude by asking whether globalization has created a new form of global city, and what other forces may be involved in shaping cities today.

FORMAL DEFINITIONS

One seldom finds a formal definition of globalization in the literature. Very often, simply a listing of parallel developments is given: greater mobility of capital, greater mobility of labor, rapid development of communication technology, computerization, world-wide conveyor belt in manufacturing, shift from manufacturing to services, integration of all economies in a single market, permeability of borders, decline of the nation-state, homogenization of culture, and so on. That is not very satisfactory; it mixes causes and effects, fails to distinguish the inherent from the temporary or circumstantial, the determining from the result, the necessary from the variable. What almost all uses of the term have in common, however, is the dating: those changes, that set of events, that began about 1970. Sometimes the dating is referred to the "oil shock" of 1973, sometimes to the breakdown of the Bretton Woods arrangements, sometimes symbolically to events such as the destruction of the Pruitt-Igoe pubic housing development and the end of the dominance of the "modern" in architecture. So considering globalization to encompass those events that began or developed dramatically after about 1970 is a useful initial working definition.

For our purposes, however, a more substantive definition is useful. It may run as follows: globalization, in its really existing form, is the further strengthening and internationalization of capital using substantial advances in communications and transportation technology (see also Marcuse 1997, 2002). The definition is important for two reasons. First, it highlights two separate components of globalization, the social/economic and the technological, and suggests that the former determines the latter, not the opposite, as is often implicitly assumed. Second, it notes that there may be many forms of globalization, and we are concerned with its specific form in the world of the 1970s.

Since the two-part definition of globalization is critical for the discussion that follows, it is worth detailing a bit further. Really existing globalization is:

1. the qualitative leap in information and transportation technology since *c.*1970, permitting increased internationalization of information and physical production;
2. used by dominant social-economic-political groups to produce a further increase in the concentration of private economic power since *c.*1970 (both within the economic sphere and vis-à-vis government), permitting increased internationalization of control over economic and political processes;
3. with consequences for cross-border integration of production and investment, cultural homogeneity, United States dominance, inequality and polarization, environmental quality, popular movements, culture, etc. (and perhaps for the spatial structure of cities, the topic of this reading).

This is a definition of really existing globalization; it is not a definition of all possible forms of globalization. Specifically, one could imagine a form of globalization in which advances in technology, (1) above, were used, contrary to (2) above, to improve the standard of living of all, promote world-wide democracy, and reduce inequality of opportunity, with results quite the opposite of (3) above. When the World Social Forum adopts the slogan, "Another World is Possible," it does not mean a world without globalization, but a world in which globalization takes a quite different form from its really existing model.

I would hypothesize that a detailed examination of the spatial patterns in cities we witness today would show that they are in fact significantly, but only partially, linked to really existing globalization; that the patterns affect all cities, not only global cities; that, to the extent they are linked to really existing globalization, they are linked as much to its second aspect as to its first; and that thus the specific form they have taken is not an inevitable product of globalization, but rather of its specific current form. And I would finally hypothesize that the forces shaping contemporary cities are not simply the products of the current form of globalization, but that they, like really existing globalization itself, are the results of more deeply embedded forces of longer standing in our economies and political structures.

THE FOUR KEY QUESTIONS FOR RESEARCH

As to the spatial patterns we witness in cities today, we might ask the following questions:

1. Is the pattern new in the present period of globalization? Has it produced a new spatial model of a city?
2. To the extent it is new, to what aspect of globalization is it attributable? To oversimplify, are they the result of technological change or of social/economic/political change?
3. Is the pattern an aspect of all cities, all globalizing cities, or only global cities?
4. To what extent is the pattern not new, and, for what is not new, what are its underlying determinants?

This reading proposes only, at the end, a very partial answer to some of these questions, but first suggests what the spatial patterns are that need examination, with the suggestion that the detailed answers are the appropriate agenda for further work for those interested in the production of better spaces for all people within which to live.

CONTEMPORARY SPATIAL PATTERNS

What in fact are the spatial patterns we witness in cities today, and to what extent have they changed in the present period of globalization? Let us begin with the conception of a quartered city (Marcuse 1989). It may be seen both in the spatial arrangement of residential life and in the spatial arrangement of business activities. Is the fact that cities today, at least in the advanced industrialized economies of the West, are not "dual," but more like "quartered," cities, new (Marcuse 1991)? The answer becomes clear if we recapitulate the argument about the patterns of the contemporary city briefly.

The residential cities

Within the city of today we see a set of sometimes overlapping but quite different residential cities:

- The luxury city, with luxury housing, not really part of the city but in enclaves or isolated buildings within it, occupied by the top of the economic, social, and political hierarchy.
- The gentrified city, the city of winners, occupied by the professional-managerial-technical groups, whether yuppie or muppie without children.
- The suburban city, sometimes single-family housing in the outer city, other times apartments near the center, occupied by skilled workers, mid-range professionals, upper civil servants.
- The tenement city, sometimes cheaper single-family areas, most often rentals, occupied by lower-paid workers, blue and white collar, and generally (although less in the United States) including substantial social housing.
- The abandoned city, the city of the victims, the end result of trickle-down, left for the poor, the unemployed, the excluded, where in the United States home-less housing for the homeless is most frequently located.

The economic cities

These felt divisions in the residential city are roughly paralleled by divisions in the economic city:

- The city of controlling decisions include a network of high-rise offices, brownstones or older mansions in prestigious locations, but are essentially locationally not circumscribed; it includes yachts for some, the back seats of stretch limousines for others, airplanes and scattered residences for still others.
- The city of advanced services, of professional offices tightly clustered in downtowns, with many ancillary services internalized in high-rise office towers, heavily enmeshed in a wide and technologically advanced communicative network.
- The city of direct production, including not only manufacturing but also the production of advanced services, in Saskia Sassen's phrase, government offices, the back offices of major firms, whether adjacent to their front offices or not, located in clusters and with significant agglomerations but in varied locations within a metropolitan area, sometimes, indeed, outside of the central city itself.
- The city of unskilled work and the informal economy, small-scale manufacturing, ware-housing, sweatshops, technically unskilled consumer services, immigrant industries, closely intertwined with the cities of production and advanced services and thus located near them, but separately and in scattered clusters, locations often determined in part by economic relations, in part by the patterns of the residential city; spatially, the overlap with the city of advanced services is substantial, for the service economy produces both high and low end jobs in close proximity to each other: janitors in executive offices, etc.
- The residual city, the city of the less legal portions of the informal economy, the city of storage where otherwise undesired (NIMBY) facilities are located, generally congruent with the abandoned residential city.

These patterns are spatial, but they are not rigid, in the old sense in which Burgess and Parks tried to describe city structure. And their spatial pattern varies widely from city to city, country to country;

SEVEN

Los Angeles, for instance, has a pattern I have described as fluid, separations as of oil and water together with walled enclaves, rather than the more clearly bounded and more homogeneous quarters of New York City. But the spatial patterns are always there, if differing in intensity and sharpness. Initial work on the 1990 census in the United States has demonstrated that fact.

The built environments

Paralleling the divisions within the residential and the economic cities, are different concrete forms of the built environment. Again, they may be listed in outline form:

- The citadel, as the assemblage of residential and commercial space used by the upper classes, separated from the rest of the city – indeed, withdrawn from it, as Ray Pahl has argued – often gated and always secured, increasingly in the form of mega-projects.
- The older city neighborhoods, near the core, rehabilitated, "regenerated," gentrified by the professional and managerial groups whose role in the cities is growing both quantitatively and qualitatively.
- The edge cities, in-between cities, ex-urban centers, in which the older forms of suburbia are combined with jobs and all of the cultural and recreational and commercial necessities of life formerly assumed to be found only in the urban core.
- The diluted and manipulated areas of social and working class housing, in which mixes of incomes and population groups are designed to attenuate any coalescence of "lower-class" based behaviors or mobilizations.
- The ethnic enclaves, in which immigrants and the lowest-paid workers integrated into the mainstream economy are able to find the possibilities of life supportable by their jobs towards the bottom of the economic structures of the city.
- The excluded ghettos, the areas to which those at the very bottom of the economic ladder, and not needed in the dominant economy, are confined, with reduced public services and neglected physical surroundings.

The soft locations

Within these divisions there is a particular set of soft locations in which the process of change is particularly striking, and which appear at first blush to be linked to the processes of globalization and post-Fordist economic changes that were likely to have a particular impact.[1] Recapitulating but adding to that list:

- waterfronts
- currently centrally located manufacturing
- brown fields (formerly industrial sites)
- central city office and residential locations
- central city amusement locations and tourist sites
- concentrations of social housing
- residential locations on the fringe of central business districts
- ethnic areas of concentration
- suburbs
- historic structures
- public spaces.

The new aspects

What is new about these patterns? It is, it seems to me, an under-debated question, but an important one. A preliminary answer would include at least the following:

- The growth in the size of the gentrified city and the shrinking of others: expanding gentrification.
- The growth in the size of the abandoned city: increasing ghettoization.
- The dynamic nature of the quarters, in which each grows only at the expense of the others: tensions among quarters, with displacement as the mechanism of expansion, and both walling in and walling out more common.
- The growing importance of the identity of the quarter for the lives of most residents: the defensive use of space, including the intensity of turf allegiance.
- The shift in location and structure of certain types of economic activity, such as manufacturing, finance, consumer services.
- The role of government, not only acceding to but also promoting the quartering of the city in the private interest, fortifying both the gentrified

and the abandoned city: the subsuming of the public interest under the private, the dominance of the neo-liberal agenda.

But in each case (and more might be added), these changes represent trends manifest well before the period of globalization, and are represented in cities in various relationships to globalization (see Marcuse 2002; Marcuse and van Kempen 2002).

THE APPLICATION OF THE RESEARCH AGENDA AT THE CITY SCALE

As to each of these spatial patterns of the contemporary city, the four key questions as to their nature and causes raised at the beginning of this reading may be addressed. To give only a few examples:

■ *Waterfronts.* The contemporary pattern is certainly the abandonment of shipping uses at old port facilities and the construction of giant facilities for much larger container-oriented shipping. That appears a result of new technologies of transportation. Is it (or is it not?) also the result of a practice in which the expense of the necessary new harbor facilities is publicly paid, but the profit from its use private?[2] Is the essential motivation for the change the replacement of labor by capital, and if so is this use of technological possibilities dictated by social priorities or the priorities of capital? Is the change in transportation technology greater than previous changes? Is it limited to a few global cities, or is it general?

■ *Ghettoization, polarization, walling and the construction of citadels.* The pattern is a widening gap between the rich and the poor, linked to the gap between the well educated and the ill educated. Is the existence and impact of that gap a product of the greater education advances in technology demand of workers? Or is it a result of the social/political strength of the haves to skew public investment in education in their favor, and to maintain grossly unequal wage rates for different kinds of economic activities? Is the division of cities represented by these phenomena something new, or is there historical continuity in them?

■ *Brownfields.* Is their abandonment a result of the technological obsolescence of the manufacturing processes that had taken place there, or of the skewed responsibility for the externalities of production, which permits abandonment without cleaning up, and zoning that follows rather than directs the economic uses of land? Are major changes in land use a new phenomenon historically? Are they limited to a handful of cities, or are they general?

■ *Megaprojects and skyscrapers.* Is the use of ever more sophisticated building technology to build higher and larger simply the replacement of better technology replacing outdated technology, or is it determined by the symbolic desire to express power and the social desire to insulate, withdraw, and secure?

■ *Gentrification.* Is gentrification, now at the global scale (Smith 2002), the result of increasing technologically enabled prosperity, or of the replacement of industrial by service activities in city centers, or is it the result of shifts in the balance of political power that have permitted the imposition of neo-liberal policies for the determination of land use? What have been the historical determinants of changes in the social uses of the built environment?

The first three of our four key questions may be specified in similar fashion to every one of the changes in city spatial structure outlined above. The fourth question is at a different scale.

THE RESEARCH AGENDA AT THE SCALE OF GLOBAL STRUCTURES: THE DETERMINANTS OF THE STRUCTURE OF GLOBALIZING CITIES

To look at the determinants of the spatial structure of cities in broader historical perspective, we need first to determine whether what we are examining is a set of phenomena confined to a few cities, or whether it is a general characteristic of cities in the contemporary period. It is a question that deserves considerable attention, and at least the broad outlines of the answer are evolving.

The difficulty arises from the misuse of the concept of the global city. Its original meaning called attention to a hierarchical pattern among cities, based on their position in a network of cities increasingly linked by newly developed modes of communication and transportation, later with a focus on financial networks among them (Friedmann and Wolff 1982). The term has however been very widely used in a related but essentially different sense, by city boosters that would like to see their city also designated "global," meaning modern, up-to-date, "in," real parts of a global network, at the cutting edge of globalization. It is in this usage then often followed by a call for a certain arrangement of space within the city, for instance extra-high skyscrapers, as appropriate to a "global" city, differentiating the "global" city from "non-global" cities.

But the patterns described in this reading which are characteristic of the so-called global cities are by no means confined to them. Every spatial pattern found in New York City, London and Tokyo can also be found in Cleveland, Vancouver, Detroit, Stuttgart, Accra or Calcutta. Some aspects characterizing the key Global Cities may be further developed in cities much further down the hierarchy; even among the Global Cities, there are major differences in urban form (Abu-Lughod 1999). If anything, we may speak of "globalizing cities," cities – virtually every city in the world – that are impacted by the processes of globalization, that reflect tendencies that are common to all cities linked to each other in a global pattern (Marcuse and van Kempen 1999).

Is globalization, really existing globalization as here defined, then the dominant, determining factor shaping all globalizing cities, explaining all the patterns described above?

And the answer suggested here is: no. The changes described here are only in part, and ultimately only in small part, the results of the really existing globalization of the post-1970s. Every one of the changes has a long history; every one is influenced by multiple factors; every one can be linked to more fundamental causes than those newly arrived with globalization.

Beyond that, if one looks more narrowly at the impacts on space that can be traced to globalization, the patterns outlined above, another suspicion arises, but one as to which there is as yet no clear empirical evidence. It may well be that even putting all these influences of globalization together, they

affect only a minority of a city's population, only the lesser part of a city's spaces. Granted that, to some degree, everything that happens in or to a city affects everyone and every place within that city – granted that no man is an island, nor no woman either, nor no spaces either except in the narrowest geographical sense – the question is how important that effect is. The impact of globalization on the outer boroughs on New York City can be easily seen: the airports, the change in job opportunities, the back offices, the immigrant quarters, the gated communities. But how many of the outer boroughs, residents are significantly affected by what is new in these areas (and of course much is not new, and while internationally linked pre-existed the globalization of the period after 1970), and how many continue their lives and lifestyles much as before, if not entirely the same? The question is worth exploration.

And in any event the various aspects of city spatial structure described in this reading have all been with us for several centuries, in some cases even longer. They are aspects perhaps more striking, perhaps more extensive, perhaps even more important, today than ever before. But they are not new, and they are not special to globalizing cities.

So if globalization is not adequate to explain the shape of contemporary cities, what is? The details of an alternative answer should be part of a further research agenda. But I would suggest, at least as a working hypothesis, that the character of contemporary cities can be traced to one crucial fundamental aspect that they share: they are all phenomena of the Capitalist City. The Capitalist City goes under many different names. It is often spoken of as the industrial city, and indeed the shift from an agricultural or mercantile form of production to an industrial one, and certainly the rise of manufacturing, has had an indelible imprint on today's cities. Sometimes it goes by the name of the modern city, not to designate an architectural style but to differentiate it culturally from its predecessors, with their more traditional values and customs. And sometimes the defining characteristic is found in the rise of new economic classes and their changed relationship to each other as in the work of Karl Marx or Karl Polanyi, in which economic aspects are given priority. Perhaps, indeed, one might build on the analysis of globalization presented at the outset of this reading, suggesting it is a combination of specific advances in productive technology and social relationships, to argue that the Capitalist City is itself, and has been for centuries, always the result of changes in technology and social relations, in evolving combinations, with the globalizing city only the latest permutation of those changes, perhaps more extreme and more widespread than before, but nevertheless cut by the same tailors of the same cloth.

The general formulation, the fourth of the key research questions, given the above hypothesis, might then be twofold:

- As to any given characteristic of globalization (technological change, concentration of control, polarization, cultural homogenization, mobility, etc.), did that characteristic pre-exist the period of globalization, and, if so, when did it first appear, and does it differ today only in extent, rather than nature?
- As to any given characteristic of globalization, does it appear also as a characteristic of national systems (as opposed to global), and if so does it differ globally from its national manifestation only in extent, rather than nature?

Again, to give some examples for empirical research:

- *The shift from manufacturing to services.* Certainly the global assembly line, the new global division of labor, is a characteristic of globalization. But it is not a result of the declining importance of manufacturing; on the contrary, both the volume of goods produced and the number of workers engaged in manufacturing have increased since the mid-1960s. What has happened is a relocation of manufacturing activities from higher-wage to lower-wage countries. The driving force is one well known in industrial relations everywhere: profits can be increased if labor costs can be reduced. The quest for higher profits exerts a basic and ongoing pressure to hold wages down. That is part of capitalism. It was seen most recently in the move of textile plants from the United States north to its south, where labor legislation was less protective, unionization more limited, and wages lower. The threat, and its actualization, to move a plant from a higher

to a lower wage region has been a standard part of labor negotiations for at least a century. Moving overseas is only the latest manifestation of the underlying dynamics of capitalist production, facilitated (as in earlier times) by improvements in the technology of transportation and communications, but not determined by them. Whether this is so or not, and, if so, to what extent and how such moves differ today from earlier periods are matters capable of empirical investigation.

- *Brownfield sites.* Similarly the relocation of manufacturing activities to new sites, and the abandonment of (polluted) old sites, has to do with the relationship of land prices to the costs of production, in the context of other costs of location, specifically, accessibility. New technology has increased the options for business firms, and awareness of the impact of pollution has increased, but the dynamics are centuries old. The disconnect between the private profit-determined impact of relocation and its social costs and benefits is a characteristic of capitalism, not of globalization. Its extent can be empirically reviewed.
- *The mobility of capital.* Indeed, the mobility of capital has increased, but it has been increasing steadily over the centuries and may even not be, proportionately, higher today than at certain earlier times in history. Whether its impact on the spatial form of cities is different today than before is an empirical question. Cursory examination of how foreign capital invests in real estate development in foreign cities suggests that its patterns are no different from those of domestic capital; foreign capital uses domestic real estate appraisers, consultants, market analysts, brokers, even architects (if in consultation with "world-class" ones). Often, in fact, foreign capital co-ventures with local. Does the product then differ (except perhaps in size) from what it would be with purely domestic capital?

Other examples, more briefly:

- *Gentrification.* Is gentrification simply a continuation of the urban restructuring, the uneven spatial development of capitalism, that has been going on for centuries, with variations on extent and in the relative use of public versus private means?

■ *Edge Cities.* Is the search for the most efficient locations to accommodate the demands of expanding profit-oriented production, and its shifting spatial needs, largely a continuation of efforts that in earlier periods resulted in the exponential growth of some towns (from Gary to Wolfsburg) and the founding of new towns, and the pattern of private market-driven suburbanization almost everywhere?

■ *Centralization of control functions.* The impact of the centralization of control functions, resulting in larger and more ornate/elaborate structures in cities (e.g. downtown Manhattan, I.G. Farben in Frankfurt, Avenida in São Paulo, etc.), has its predecessors in the tendencies to monopolies and, internationally, cartels, over a previous century, tendencies long associated by economists with laws of capitalist societies. Is the rate of development different today than what it has been?

In general, the researchable question might be posed: is the spatial pattern that seems today to be the product of globalization any different in its tendency than similar patterns of change on an intra-national level in previous years (e.g. locational changes, capital investment, centralization of control, etc.)?

The agenda for research – and for the action that should follow from research – is thus extensive. It can encompass, at the scale of individual cities, the initial questions raised above: to what extent have the distinguishing characteristics of the contemporary city, as outlined in the first section of this reading, been the product of the developments and the actors involved in really existing globalization; if they are products of globalization, are they the result of technological changes or of changes in the balance of social, political and economic power; do they taken together create a new model of a global city applicable to all, or some, or only to global, cities? And then such an agenda should go further and explore what other models of the contemporary city might equally or better reflect its reality. If research is not to be merely empiricist, but to deal with the large questions of policy that confront us, it cannot shy away from addressing the large questions that detailed examination of the smaller ones raises. Research at the scale of the city and the time scale of contemporary globalization must be accompanied by research at the scale of the economic structures and the time scale of the post-feudal period. Foremost among the research questions at both scales, I suggest, is the problem of tracing the intertwined but separate impact of technological development and social relations, and to see what changes in either, or their relationship, might lead us to the better city that we would wish to see.

NOTES

1 We use the term "soft" by analogy to its use in zoning practice, where a "soft" site is spoken of as one not developed to the limits its legal zoning permits, i.e. one viewed as ripe for change and new development.
2 I owe the example to Susan Fainstein's (forthcoming) study of the Port Authority of New York and New Jersey.

REFERENCES FROM THE READING

Abu-Lughod, J. (1999) *New York, Chicago, Los Angeles: America's Global Cities.* Minneapolis, MN: University of Minnesota Press.

Fainstein, S. (forthcoming) Ground Zero's landlord: the role of the Port Authority of New York and New Jersey in the reconstruction of the World Trade Center site. In J. Mollenkopf (ed.) *The Politics of Rebuilding Downtown.* New York: Russell Sage Foundation.

Friedmann, J. and Wolff, G. (1982) World city formation: an agenda for research and action, *International Journal of Urban and Regional Research*, 6, 309–344.

Marcuse, P. (1989) "Dual City": a muddy metaphor for a quartered city, *International Journal of Urban and Regional Research*, 13, 4, 697–708.

Marcuse, P. (1991) Housing markets and labour markets in the quartered city. In J. Allen and C. Hamnett (eds) *Housing and Labour Markets: Building the Connections.* London: Unwin Hyman, 118–135.

Marcuse, P. (1997) Glossy globalization. In P. Droege (ed.) *Intelligent Environments*. Amsterdam: Elsevier Science.

Marcuse, P. (2002) Depoliticizing globalization: from neo-Marxism to the network society of Manuel Castells. In J. Eade and C. Mele (eds) *Understanding the City*. Oxford: Blackwell, 131–158.

Marcuse, P. and van Kempen, R. (eds) (1999) *Globalizing Cities: A New Spatial Order?* Oxford: Blackwell.

Marcuse, P. and van Kempen, R. (eds) (2002) *Of States and Cities: The Partitioning of Urban Space*. Oxford: Oxford University Press.

Smith, N. (2002) New globalism, new urbanism: gentrification as global urban strategy, *Antipode*, 34, 3, 427–450.

SEVEN

45

"Globalization and the Rise of City-regions"

from *European Planning Studies* (2001)

Allen J. Scott

Editors' introduction

Allen J. Scott is Distinguished Professor of Public Policy and Geography at UCLA. Scott is one of the most prominent figures in the "LA School" of urban studies and has written a number of seminal books on processes of industrialization, urbanization and regional development (see, for instance, Scott 1988, 1993). His work has explored the interplay between technology, inter-firm organization, labor markets and agglomeration economies. More recently, Scott has contributed to debates on cities and globalization through a major edited volume (Scott 2001) and a short monograph exploring the role of urban regions as motors of economic development under globalizing capitalism (Scott 1998). This reading is derived from Scott's work on the latter project. Building upon his earlier studies of urban agglomeration, Scott joins other global cities researchers in arguing for the enhanced significance of geographical proximity, and therefore of large-scale urban regions, under conditions of intensifying economic globalization. On this basis, Scott explores some of the new political spaces that have been emerging in conjunction with the consolidation of globalizing city-regions (see also Reading 20 by Hill and Kim, Reading 30 by Brenner and Reading 48 by Olds and Yeung). According to Scott, national state sovereignty has been partially undermined due to the accelerated "debordering" of national space-economies; meanwhile, new forms of political organization and regulatory experimentation are emerging on the subnational scale of urban regions. Scott's goal is to decipher some of the emergent contours of this "new regionalism," at once with reference to local economic development initiatives, new forms of interfirm coordination, land-use planning strategies, new forms of labor market regulation and, more generally, intensifying struggles over the meaning of urban citizenship. Scott suggests that new forms of collective order and institutional organization are emerging in city-regions throughout the world economy, and that the latter serve increasingly important regulatory functions within the global political system. Due to the ongoing clash of neoliberal and social democratic models of capitalism, Scott argues, the institutional shape and political form of regional governance remain intensely contested. While some scholars have questioned certain aspects of Scott's analysis – for instance, his apparent contention that national state power is declining – his analysis represents a provocative, lucid and empirically rich account of how global city-regions have become sites for the construction of new regulatory strategies and new models of political life.

INTRODUCTION

Contrary to many recent predictions (e.g. O'Brien 1992), geography is not about to disappear. Even in a globalizing world, geography does not become less important; it becomes more important because globalization enhances the possibilities of heightened geographic differentiation and locational specialization. Indeed, as globalization proceeds, an extended archipelago or mosaic of large city-regions is evidently coming into being, and these peculiar agglomerations now increasingly function as the spatial foundations of the new world system that has been taking shape since the end of the 1970s (Scott 1998). The internal and external relations of these city-regions and their complex growth dynamics present a number of extraordinarily perplexing challenges to researchers and policy-makers alike as we enter the twenty-first century.

There is an extensive literature on "world cities" and "global cities" that focuses above all on a concept of the cosmopolitan metropolis as a command post for the operations of multinational corporations, as a center of advanced services and information-processing activities, and as a deeply segmented social space marked by extremes of poverty and wealth. I seek to extend this concept so as to incorporate the notion of the wider region as an emerging political-economic unit with increasing autonomy of action on the national and world stages. I refer to this type of region by the term "global city-region."

Global city-regions constitute dense polarized masses of capital, labor, and social life that are bound up in intricate ways in intensifying and far-flung extra-national relationships. As such, they represent an outgrowth of large metropolitan areas – or contiguous sets of metropolitan areas – together with surrounding hinterlands of variable extent which may themselves be sites of scattered urban settlements. In parallel with these developments, embryonic consolidation of global city-regions into definite political entities is also occurring in many cases, as contiguous local government areas (counties, metropolitan areas, municipalities, etc.) club together to form spatial coalitions in search of effective bases from which to deal with both the threats and the opportunities of globalization. So far from being dissolved away as geographic entities by processes of globalization, city-regions are by and large actually thriving at the present time, and they are becoming increasingly central to the conduct and coordination of modern life.

GLOBALIZATION AND THE NEW REGIONALISM

In the immediate post-World War II decades almost all of the major capitalist countries were marked by strong central governments and relatively tightly bordered national economies. These countries constituted a political bloc within the framework of a *Pax Americana*, itself overlain by a rudimentary network of international arrangements (the Bretton Woods monetary system, the World Bank, the IMF, GATT, and so on) through which they sought to regulate their relatively limited – but rapidly expanding – economic interrelations. Over much of the post-war period, the most prosperous of these countries could be said to constitute a core zone of the world economy, surrounded in its turn by a peripheral zone of Third World nations, with a complex set of interdependences running between the two.

Today, after much economic restructuring and technological change, significant transformations of this older order of things have occurred across the world, bringing in their train the outlines of a new social grammar of space, or a new world system. One of the outstanding features of this emerging condition is the apparent though still quite inchoate formation of a multilevel hierarchy of economic and political relationships ranging from the global to the local. Four main aspects of this state of affairs call for immediate attention:

1. Huge and ever-increasing amounts of economic activity now occur in the form of long-distance, cross-border relationships. Such activity is in important ways what I mean by globalization as such, even though it remains far indeed from any ultimate point of fulfilment. Further, as globalization in this sense moves forward, it creates numerous conflicts and predicaments that in turn activate a variety of political responses and institution-building efforts. Practical expressions of such efforts

include a complete reorganization of international financial arrangements as compared with the post-war Bretton Woods system, together with the restructuring and reinforcement of international forums of collective decision-making and action such as the G7/G8 group, the OECD, the World Bank, the IMF, and a newly streamlined GATT, now known as the World Trade Organization. While these political responses to the pressures of globalization remain limited in scope and severely lacking in real authority, they are liable to expansion and consolidation as world capitalism continues its predictable expansion.

2. In part as a corollary of these same pressures, there has been a proliferation over the last few decades of multination blocs such the EU, NAFTA, MERCOSUR, ASEAN, APEC, CARICOM, and many others. These blocs, too, can be seen as institutional efforts to capture the benefits and control the negative externalities created by the steady spilling over of national capitalisms beyond their traditional political boundaries. They remain in various stages of formation at the present time, with the EU being obviously in the vanguard.

3. Sovereign States and national economies remain prominent, indeed dominant elements of the contemporary global landscape, though they are clearly undergoing many sea-changes. On the one hand, individual States no longer enjoy quite the same degree of sovereign political autonomy that they once possessed, and under conditions of intensifying globalization they find themselves less and less able or willing to safeguard all the regional and sectional interests within their jurisdictions. On the other hand, national economies have been subject to massive debordering over the last few decades so that it is increasingly difficult, if not impossible, to say precisely where, say, the American economy ends and the German or Japanese economies begin. As a result, some of the regulatory functions that were formerly carried out under the aegis of the central State have been drifting to higher levels of spatial resolution; at the same time, other functions have been drifting downward.

4. Accordingly, and most importantly for present purposes, there has of late been a resurgence of region-based forms of economic and political organization, with the most overt expression of this tendency being manifest in the formation of large global city-regions. These city-regions form a global mosaic that now seems to be over-riding in important ways the spatial structure of core–periphery relationships that has hitherto characterized much of the macro-geography of capitalist development.

Point (4) calls for some amplification. The propensity of many types of economic activity – manufacturing and service sectors alike – to gather together in dense regional clusters or agglomerations appears to have been intensifying in recent decades. This renewed quest for collective propinquity on the part of all manner of economic agents can in part be interpreted as a strategic response to heightened (global) economic competition in the context of a turn to post-Fordism in modern capitalism. Propinquity is especially important in this context because it is a source of enhanced competitive advantage for many types of firms (Scott 1988; Storper 1997), and, as a corollary, large regional production complexes are coming increasingly to function as territorial platforms for contesting global markets. At the same time, the diminishing capacity of central governments to deal with all the nuanced policy needs of each of the individual regions contained within their borders means that many regions are now faced with the choice of either passive subjection to external cross-border pressures, or active institution-building, policy-making, and outreach in an effort to turn globalization as far as possible to their advantage. Regions that take the latter course are by the same token faced with many new and unfamiliar tasks of political coordination and representation. Special urgency attaches to these tasks not only because of their economic import, but also because large city-regions function more and more as poles of attraction for low-wage migrants from all over the world, so that their populations are almost everywhere heavily interspersed with polyglot and often disinherited social groups. As a consequence of this, many city-regions today are being confronted with pressing issues related to political participation and the reconstruction of local political identity and citizenship.

THE POLITICAL ORDER OF GLOBAL CITY-REGIONS

The world system is thus currently in a state of rapid economic flux, leading in turn to many significant adjustments in patterns of political geography. On the one side, the profound changes that have been occurring on the economic front are giving rise to diverse responses and experiments in regulatory coordination at different geographic levels from the global to the local. On the other side, the new regulatory institutions that are now beginning to assume clearer outline on the world map, simultaneously reinforce the channeling of economic development into spatial structures that run parallel to the quadripartite political hierarchy described earlier. While the political shifts going on at each level in this hierarchy pose many perplexing problems, the level that is represented by the new global mosaic of city-regions is perhaps one of the least well understood. Moreover, precisely because the individual regional units at this level constitute the basic motors of a rapidly globalizing production system, much is at stake as they steadily sharpen their political identities and institutional foundations.

We may well ask, at the outset, how these regions are to be defined (in political-geographic terms) as territorial units with greater or lesser powers of coordinated action. In many instances, of course, the boundaries of given city-regions will tend to coincide with some pre-existing metropolitan area. But how will these boundaries be drawn when several different metropolitan areas lie in juxtaposition to one another, as, for example, in the case of the north-east seaboard of the US? And how far out into its hinterland will the political mandate of any city-region extend? The final geographic shape of any given global city-region must remain largely indeterminate in a priori terms. Even so, we can already perhaps see some of the outlines of things to come in the new regional government systems that have been put into place in a number of different European countries over the last couple of decades (Keating 2001), and in the maneuvering, some of which may bear fruit, some of which will certainly lead nowhere, that is currently gathering steam around prospective municipal alliances such as San Diego-Tijuana, Cascadia, the Trans-Manche region, Padania, Copenhagen–Malmo, Singapore–Johore–Batam, or Hong Kong–Shenzen. Note that a number of these alliances involve trans-border arrangements.

To an important extent, much of the political change going on in the world's large city-regions today represents a search for structures of governance capable of securing and enhancing their competitive advantages in a rapidly globalizing economic order. Agglomerated production systems are the arenas of both actual and potential region-wide synergies, but these synergies will always exist in some sub-optimal configuration so long as individual decision-making and action alone prevail in the economic sphere. These synergies have enormous relevance to the destinies of all the firms and workers in the immediate locality, and by the same token, they assume dramatic importance in a world where the continued spatial extension of markets brings each city-region into a position of vastly expanded economic opportunities, but also of greatly heightened economic threats from outside. The economies of large city-regions are thus intrinsically overlain by a field of collective order defined by these synergies, and this constitutes a crucial domain of social management. No matter what specific institutional form such management may assume (e.g. agencies of local government, private–public partnerships, civil associations, and so on), it derives its force and legitimacy from the positive role that coordinating agencies can play in regional economic systems by promoting and shaping critical increasing returns effects that would otherwise fail to materialize or that would be susceptible to severe misallocation. The possible shape and character of agencies such as these can be suggested by reference to strategies such as the fostering of agglomeration-specific technological research activities, the provision of high-risk capital to small start-up firms, the protection of certain kinds of infant industry, investments in upgrading workers' competencies, the cultivation of collaborative inter-firm relations, the promotion of distant markets for local products, and so on (Scott 1998). There is also, of course, a continued urgent need for more traditional types of urban planning to ensure that the negative effects of periodic land use and transportation breakdowns do not cut too deeply into local economic performance and social life.

SEVEN

The prospect of a mosaic of global city-regions, each of them characterized by an activist collectivity resolutely seeking to reinforce local competitive advantages, however, raises a further series of questions and problems. Rising levels of concerted regional activism can be expected to lead to specific kinds of destabilization and politicization of interregional relations, both within and across national boundaries. Consider, for example, the formation of regional alliances (such as the Four Motors for Europe Programme, or the recent (failed) linking of the London and Frankfurt stock exchanges) giving rise to complaints about unfair competition on the part of those excluded. Another example can be found in the currently prevalent attempts by the representatives of some regions to lure selected assets of other regions into their own geographic orbit, often at heavy social cost. Another can be deciphered in the development races that occur from time to time when different regions push to secure a decisive lead as the dominant center of some budding industry. Still another is evident in the expanding opportunities for multinational corporations to play one region off against another in competitive bidding wars for new direct investments. In view of the likelihood that stresses and strains of these types will be magnified as the new regionalism takes deeper hold, a need for action at the national, plurinational, and even eventually the global levels of political coordination is foreseeable in order to establish a framework of ground rules for the conduct of interregional relations (including aid to failing regions) and to provide appropriate forums for interregional problem-solving. The European Committee of the Regions, established under the terms of the Maastricht Treaty, may conceivably represent an early even if still quite fragile expression in the transnational sphere of this dawning imperative.

As these trends and tendencies come more resolutely to the surface, a further question arises as to what macro-political or ideological formations will be liable to assert a role in defining the calibrating frameworks for the institution-building and policy-making projects that can now be ever more strongly envisioned at various spatial levels. Giddens (1998) has forcefully argued that two main contending sets of political principles appear now to be moving toward a war of position with one another in relation to recent events on the world stage, certainly in the more economically advanced parts of the globe. One of these is a currently dominant neo-liberal view – a view that prescribes minimum government interference in and maximum market organization of economic activity (and that is sometimes but erroneously taken to be a virtually inescapable counterpart of globalization). In light of the above remarks regarding the urge to collective action in global capitalism and its various appendages, neo-liberalism strikes me as offering a seriously deficient political vision. The other is a renascent social democracy or social market approach. On the economic front, social democracy is prepared to acknowledge and to work with the efficiency-seeking properties of markets where these are consistent with standards of social fairness and long-term economic wellbeing, but to intervene selectively where they are not. As such, a social democratic politics would seem to be well armed to face up to the tasks of building the social infrastructures and enabling conditions (at every geographic level) that are each day becoming more critical to high levels of economic performance as the new world system comes increasingly into focus. At the city-region level, in particular, these tasks can be centrally identified with the compelling social need to promote those local levels of efficiency, productivity and competitiveness that markets alone can never fully secure.

There is a further forceful argument in favour of a social democratic approach to the governance of global city-regions, one that is associated with, yet that also goes well beyond, the need for remedial collective action in local economic affairs. Quite simply put, issues of representativeness and distributional impact are always in play in any political community, whether or not social management of the local economy is in some sense under way. In brief, the question of local democratic practice and how to establish effective forums of popular participation is inescapably joined to the more technocratic issues raised by the challenges of economic governance in global city-regions. This question takes on special urgency in view of the role of large global city-regions as magnets for low-wage migrants – many of them undocumented – from all over the world, so that often enough significant segments of their populations are made up of socially marginalized and politically dispossessed individuals. At the same time, and above and beyond any considerations of equity and social justice, enlargement of the sphere of democratic practice is an important practical means of registering and dealing with many of the social tensions that are especially prone to occur in dense social communities; and this remark in turn is based on the observation that the mobilization of voice in such communities is typically an important first step in the constructive treatment of their internal dysfunctionalities. Large city-regions, with their rising levels of social distress as a result of globalization, are confronted with a series of particularly urgent political challenges in this regard, not only because their internal conviviality is in jeopardy, but also because any failure to act is likely, too, to undermine the effectiveness of more purely economic strategies.

From all of this, it follows that some reconsideration of the everyday notion of citizenship is itself long overdue. An alternative definition of citizenship, one that is more fully in harmony with the unfolding new world system, would presumably assign basic political entitlements and obligations to individuals not so much as an absolute birthright, but as some function of their changing involvement and practical allegiances in given geographic contexts. In fact, traditional conceptions of the citizen and citizenship are vigorously and increasingly in question at every geographic level of the world system – for we are all rapidly coming to be, at one and the same time, participants in local, national, plurinational, and global communities – but nowhere as immediately or urgently as in the large global city-regions of the new world system (Holston 2001). Even though only a few tentative and pioneering instances of pertinent reforms in such regions are as yet in evidence (as in certain countries of the EU), more forceful experiments in local political enfranchisement will no doubt come to be initiated in the near future as city-regions start to deal seriously with the new economic and political realities that they face. In a world where mobility is continually increasing, it may not be entirely beyond the bounds of the conceivable that individuals will one day freely acquire title of citizenship in large city-regions many times over in conjunction with their movements from place to place throughout their lifetimes.

CODA

Globalization has potentially both a dark, regressive side and a more hopeful, progressive side. If the analysis presented here turns out to be in principle broadly correct, then those views that have been expressed of late in some quarters to the effect that any deepening trend to globalization must constitute a retrograde step for the masses of humanity can be taken as a salutary warning about a possible future world, but by no means as a representation of all possible future worlds. Insistent globalization under the aegis of a triumphant neo-liberalism would no doubt constitute something close to a worst-case scenario, leading to greatly increased social inequalities and tensions within city-regions and exacerbating the discrepancies in growth rates and developmental potentials between them. Alternative and realistic possibilities can be plausibly advanced, however, and I have tried to sketch out some of these in the preceding pages. Globalization, indeed, is the potential bearer of many significant social benefits. At this stage in history, its future course is still quite open-ended, and it will certainly be subject with the passage of time to many different kinds of political contestation, some of which will mold it in decisive ways. In particular, and as I have tried to indicate, globalization raises important new questions about economic governance or regulation at all spatial levels, and some form of social market politics seems to offer a viable, fair, and persuasive way of facing up to these questions.

Finally, while I have said little or nothing in this account about large cities in the less-developed countries of the world, it seems to me on the basis of both current trends and theoretical speculation – and with due acknowledgement of the enormous difficulties posed by the vicious circles in which they are often caught – that at least some of them might well be able to capitalize on the processes of urbanization and economic growth described above. These processes suggest that selected urbanized areas in a number of less-developed countries are likely eventually to accede as dynamic nodes to the expanding mosaic of global city-regions, just as places like Seoul, Taipei, Hong Kong, Singapore, Mexico City, São Paulo, and others, have done, and are doing, before them.

REFERENCES FROM THE READING

Giddens, A. (1998) *The Third Way*. Cambridge: Polity.

Holston, J. (2001) Urban citizenship and globalization. In A.J. Scott (ed.) *Global City-regions*. Oxford: Oxford University Press, 325–348.

Keating, M. (2001) Governing cities and regions: territorial reconstruction in a global age. In A.J. Scott (ed.) *Global City-regions*. Oxford: Oxford University Press, 371–390.

O'Brien, R. (1992) *Global Financial Integration: The End of Geography*. London: Pinter.

Scott, A.J. (1988) *New Industrial Spaces*. London: Pion.

Scott, A.J. (1993) *Technopolis*. Berkeley, CA: University of California Press.

Scott, A.J. (1998) *Regions and the World Economy*. Oxford: Oxford University Press.

Scott, A.J. (ed.) (2001) *Global City-regions*. Oxford: Oxford University Press.

Storper, M. (1997) *The Regional World*. New York: Guilford.

46

"The Global Cities Discourse: A Return to the Master Narrative?"

from *Transnational Urbanism* (2001)

Michael Peter Smith

Editors' introduction

Michael Peter Smith is currently Chair of the Community Studies and Development unit at the University of California, Davis. Smith has written, edited or co-edited 20 books on urban development and globalization, including *City, State and Market* (1988), *The Capitalist City* (with Joe R. Feagin, 1987) and, most recently, *Transnational Urbanism* (2001). Smith's earlier work examined the political economy of US cities during the post-1970s period and the changing role of state policy in the restructuring of urban space. While Smith's writings of the 1980s contributed significantly to the advancement of global cities research, he has more recently developed a provocatively critical relationship to this literature, in part through his ongoing ethnographic research on transnational migration flows, transnational social movements and the rise of "extra-territorial" citizenship. Smith's "agent-centered" critique of global cities scholarship is developed at some length in his 2001 book, *Transnational Urbanism*, from which this reading is excerpted.

For Smith, global cities research has focused excessively on the construction of objectivist, structural typologies that bracket the everyday experience of life and struggle within globalizing city-regions. According to Smith, it is essential to reconceptualize global cities as spaces of intense sociocultural interaction in which transnational social networks – composed of diverse, interconnected people and institutions – converge and interact. Smith argues that this perspective can transcend the methodological limitations of earlier approaches to global cities research by underscoring the role of social mobilization "from below" in the production of globalized urban spaces. In Reading 46, Smith concretizes his methodological agenda through a brief case study of grassroots urban protests against the Suharto regime in Jakarta, Indonesia following the Asian financial crisis. Smith reinterprets the process of "globalization" in this context as a multifaceted localization of diverse transnational social networks to produce contextually specific conditions of sociopolitical contestation. Smith's critique of global cities research raises a number of key questions regarding the limitations of political-economic approaches to the study of urbanization. While Smith's theoretical position remains controversial, it is evident that his alternative, agent-centered approach opens up an illuminating and highly original perspective on the social life of globalizing urban spaces.

Debates about "the global city" have taken on a recognizable, if not formulaic, character, poised somewhere on a conceptual and epistemological borderland where positivism, structuralism, and essentialism meet. The tendency to focus this debate around positivist taxonomies, urban hierarchies constructed on the basis of these taxonomies, and empirical efforts to map or even to formally model the "real" causes and consequences of global cities, leads participants in the debate to overlook the fact that the global cities discourse takes place within a wider public discourse on "globalization," which is itself a contested political project advanced by powerful social forces, not some "thing" to be observed by scientific tools. The global cities discourse constitutes an effort to define the global city as an objective reality operating outside the social construction of meaning. The participants in this debate argue about which set of material conditions are attributes of global cities and which cities possess these attributes. The debate generates alternative positivist taxonomies said to be occurring entirely outside our processes of meaning-making.

My own position on this debate is framed within the wider epistemological and ontological debate about social constructionism and the critique of ideology. The basic starting point of my argument is this: there is no solid object known as the "global city" appropriate for grounding urban research, only an endless interplay of differently articulated *networks, practices*, and *power relations* best deciphered by studying the agency of local, regional, national, and transnational actors that discursively and historically construct understandings of "locality," "transnationality," and "globalization" in different urban settings. The global city is best thought of as a historical construct, not a place or "object" consisting of essential properties that can be readily measured outside the process of meaning-making.

The global cities literature varies in its specification of the financial, informational, and migratory flows that intersect to constitute a global city. Representations of global cities nonetheless share a common conceptual strategy in which these global flows are envisaged as "coming together" within the jurisdictional boundaries of single cities like New York, London, Paris, Tokyo, or Los Angeles. This strategy thus localizes within the boundaries of particular cities highly mobile, transnational processes of capital investment, manufacturing, commodity circulation, labor migration, and cultural production. In so doing, the strategy sharply demarcates an "inside" from an "outside" and largely highlights what goes on inside global cities. When the global cities framework addresses relations among global cities it imposes a hierarchical ordering on the economic functions and production complexes assumed to be "integrating" the global cities across space. By hierarchically nesting criss-crossing transnational connections and imposing an economic ordering mechanism on global cities, the effort to construct a global urban hierarchy belies the often marked disarticulation among the financial flows, political alliances, mediascapes, and everyday sociocultural networks now transgressing borders and constituting what I term transnational urbanism.

RECONSIDERING THE GLOBAL CITY THESIS

Viewed from our current vantage point, global city assumptions about the systemic coherence of the urban hierarchy, the transterritorial economic convergence of global command and control functions, and the declining significance of the nation-state, are more difficult to maintain than they once were. For better or worse, cities have different histories, cultural mixes, national experiences, and modes of political regulation of urban space. These must be taken into account in any nuanced analysis of the localization of global processes. Their absence weakens the usefulness of the global cities thesis.

The quest for a fixed urban hierarchy should be abandoned, in my view, not only because hierarchical economic taxonomies are too static a formulation in the face of the volatility *of* capital investment flows, but also because of the multiple and often contradictory compositions of these flows (e.g. speculative vs. direct fixed investment). More importantly, this quest for a hierarchical ordering mechanism is fruitless because the local cultural spaces that are sites of transnational urbanism are also far less static than global city theorists assume. Far from reflecting a static ontology of "being" or "community," localities are dynamic constructions "in the making." They are multiply inflected sites of cultural and political as well as economic flows, projects, and practices. The cultural and political processes that go into their constant making and remaking are simply too dynamic to predict the course of urban development in advance of the actual political struggles through which contending spatial practices and cultural conflicts that constitute transnational urbanism are played out.

The global cities thesis centrally depends on the assumption that global economic restructuring precedes and determines urban spatial and socio-cultural restructuring, inexorably transforming localities by disconnecting them from their ties to nation-states, national legal systems, local political cultures, and everyday place-making practices. In the past decade extensive research questioning this core assumption on empirical grounds has opened up the discourse on global cities and globalization more generally (e.g., Cox 1997). I will extend the argument by advancing a social constructionism analysis which exposes the entire discourse of globalization as a "tightly scripted narrative of differential power" (Gibson-Graham 1996/1997: 1) that actually creates the powerlessness that it projects by contributing to the hegemony of prevailing globalization metaphors of capitalism's global reach, local penetration, and placeless logic.

Attention to the analysis of transnational social networks is one way out of this impasse. The practices of transnational political networks exemplify the ways in which situated actors socially construct historically specific projects that become localized within particular cities throughout the world, thereby shaping their urban politics and social life. This type of urban politics is constituted as people connected across borders in transnational networks interact with more local institutional structures and actors, as well as with such putatively more global actors as multinational corporations and international agencies, to produce urban change. The local and the global are thus mutually constitutive in particular places at particular times [. . .]

THE GLOBAL GOVERNANCE AGENDA

Today's grand narrative of economic globalization has been advanced most forcefully not by academics but by an emergent international monetarist regime, a set of institutional actors who have instituted a political offensive against developmental states and institutions. The globally oriented institutions spearheading this offensive were established under the auspices of the 1980s debt crisis to advance the monetarist agenda of global efficiency and financial credibility, against the nationally oriented institutions of developing countries. This globalist political project has, in turn, produced a series of struggles over the meaning of "the local," as cities and other localities, via their political, economic, and cultural actors and institutions, seek either to find a niche within the new global public philosophy, or resist pressures to "globalize," i.e. to practice fiscal austerity and conform to monetarist principles and policies.

The origins of the contemporary ideology of globalization are thus historically specific. They constitute efforts by powerful social forces to replace the developmentalist institutional framework of the 1960s to 1980s, premised on modernization theory, with a new mode of economic integration of cities and states to world market principles. Abandoning the argument that a convergent path of development will follow from the spread of Western-oriented "modernization," it is now argued that locational specialization or "niche formation" is an inevitable by-product of globalization. The principles legitimating economic globalization have been posited by their advocates as inevitable byproducts, or ruling logics, of the material condition of globalization "on the ground," rather than, as they are, the social constructs of historically specific social interests, including transnational corporate and financial elites, heads of international agencies, state managers that have embraced neoliberalism, various academic ideologues, and the managers of the IMF, the World Bank, and the World Trade Organization. This regime seeks to transform the discourse of development to one in which shrinking the state is a social virtue and structural adjustment an unavoidable imperative. This neoliberal regime of "global governance" is viewed as an incipient global ruling class whose efforts to achieve global economic management can only be thwarted when globalization itself is recognized as a historically specific and hotly contested project of social actors and agents rather than an inevitable condition of contemporary existence.

Many academics overestimate the coherence of this global project and underestimate the potential effectiveness of the oppositional forces they acknowledge. As grassroots social-movement scholars have shown, forceful national, local, and transnational political identities have sprung up to resist the hegemonic ideology and austerity policies imposed by the global neoliberal regime. Thus it is not possible to agree entirely with this pessimistic assessment of the global future. It is surprising, nonetheless, that the global cities scholars seem not to have given much thought to the question of whether their research agenda and its "objective" findings implicitly naturalize that very project by legitimating the "reality" of global cities as part and parcel of an unstoppable process of economic globalization. Unintentionally, their epistemology thus becomes the ontology of global cities.

Once globalization is unmasked as the most recent historical version of the free-market ideology, a more fruitful way forward is suggested by the trenchant critique of the globalization discourse offered by two leading post-Marxist feminist theorists writing under the pen name J.K. Gibson-Graham (1996/1997). Gibson-Graham argue that the globalization thesis, both in terms of the economy of global production and finance and in terms of the culture of global consumerism fueled by the assumed capitalist domination of global telecommunications, suffers from theoretical and empirical overreach, but has telling political consequences. These consequences – a politics of fear and subjection by workers, communities, and other potentially oppositional forces – are most apparent when the two threads of the globalization thesis are combined into a single masculinist grand narrative of the penetration of capitalist social relations not only of production and consumption but also of meaning-making and the constitution of subjectivity.

The global penetration metaphor may be challenged on theoretical and empirical grounds. Theoretically, Gibson-Graham argue that by relying on the sexual metaphor of penetration to invoke the power of capital, but failing to acknowledge that sexuality is a process of inter-penetration, not a one-way flow, the globalization discourse refuses a contingent counterversion of the capitalist economy "as penetrable by non-capitalist economic forms" and social relations (Gibson-Graham 1996/1997: 17–18). Relatedly, the very logic of the metaphor of the global reach of capitalist command and control arrangements can be reinterpreted as a project inherently subject to over-extension, a set of tasks often frustrated by global overreach [. . .]

The potential efficacy of active political resistance to global neoliberalism is insightfully illustrated in Gibson-Graham's narration of a transnational political campaign conducted during the early 1990s against a TNC, the Ravenwood Aluminum Company (RAC). The campaign was led by the United Steel-workers of America (USWA), working through the channels of several international labor organizations. It combined cross-border political lobbying by labor unions in 28 countries, on five continents, with a US-based consumer boycott. By going global, the unions and transnational grassroots activists involved in this campaign succeeded in obtaining a favorable labor contract in the USA and imposed restrictions on the key global trader who had organized the purchase of RAC in a leveraged buyout. Gibson-Graham (ibid.: 9–10) conclude their narrative with a metaphor of dogged determination, noting that "terrier-like, the USWA pursued the company relentlessly around the globe yanking and pulling at it until it capitulated. Using their own globalized networks, workers met internationalism with internationalism and eventually won."

The moral of the story is clear. Far from being the exclusive preserve of global capital, global space is a discursive arena and very much a contested terrain. One key dimension of globalization, the telecommunications revolution, is often assumed to be a straightforward tool of global capital in organizing production, directing financial flows, and orchestrating consumer desire. While it may facilitate these processes to some extent, it also has become a viable channel for spatially extending the contested terrain of globalization, enabling oppositional forces to jump scale and go global. New technologies of communication such as e-mail, fax, and the internet have become viable mechanisms facilitating the transnationalization of culture and politics, enabling cross-border information exchange, transnational political networking, and the sociopolitical organization of a wide variety of new forms of "transnational grassroots politics."

Such modes of transnational political networking include, but are not limited to, transnational labor organizing, international human rights campaigns, indigenous peoples' movements, multinational feminist projects, and global environmental activism. For better or worse, the transnationalization of communications provides access as well to such transnational networks of social action "from below" as Islamic fundamentalism, neo-Nazism, and the militia movement, a development that progressive scholars and social activists doubtless find less salutary.

TRANSNATIONAL URBANISM: BEYOND REIFICATION

In light of these dynamic political and cultural developments, it is time to move beyond the boundaries of the global cities discourse. Instead of pursuing the quest for a hierarchy of nested cities arranged neatly in terms of their internal functions to do the bidding of international capital, it is more fruitful to assume a less easily ordered urban world of localized articulations, where sociocultural as well as political-economic relations crisscross and obliterate sharp distinctions between inside and outside, local and global. These partially overlapping and often contested networks of meaning are relations of power that link people, places, and processes to each other transnationally in overlapping and often contested social relations rather than in hierarchical patterns of interaction.

Numerous examples of these criss-crossing linkages that literally "re-place" the urban, reconfiguring "the city" from a global epiphenomenon to a fluid site of contested social relations of meaning and power, could be investigated. To bring this chapter to a close, however, a historically specific example of what I have in mind may be helpful. I offer the case of the political crisis in Indonesia in which grassroots student protests and urban unrest in Jakarta and other Indonesian cities brought down the Suharto regime. These street-level forms of political practice have been represented as "local" responses to a perceived national crisis. Nevertheless, the protests, occurring in various cities throughout Indonesia, were occasioned by the transnational IMF austerity policies imposed on Indonesia from a putative

"outside" as a condition for an IMF bailout of the Suharto regime in the face of the Asian financial crisis. Moreover, global media representations of the interplay between Suharto and the IMF reinforced the outside–inside duality by featuring a widely circulated photograph of an IMF official standing behind Suharto with arms folded as he signed an IMF austerity agreement. While the bodily gesture was meaningless to the IMF official, it was pregnant with meaning within Indonesian cultural practice, symbolizing personal domination of one body by another and creating an impression of Suharto's declining strength as a symbol of national development and modernization. This impression was reinforced when the austerity policies were at first avoided and then implemented selectively "on the inside" by Suharto, who imposed sharp price increases on daily necessities such as food and fuel for ordinary citizens, while striving to protect his family's vast fortune in the face of a declining currency market and maintain the collusive business practices with ethnic Chinese business elites that had been the source of the family fortune.

A key target of the popular urban protests that eventually led to Suharto's ouster were the wealthy Chinese business elites of the Chinese diaspora, a transnational entrepreneurial class tied to networks of Chinese business elites operating in other cities in Asia and throughout the world. This group, comprising less than 3 per cent of the Indonesian population, dominates the nation's economy and has nearly 70 per cent of the nation's wealth, but has been absent from the military and the public political arena. Thus, a "Chinese" ethnic background became an easy initial target of nationalist and racist discourses and practices during the early stages of the financial crisis. Indeed, as criticism of Suharto and his inner circle mounted, the government tried to deflect the blame onto a more generalized target, the entire ethnic Chinese population.

This effort was initially partially successful, but when the student component of the urban protests emerged in Jakarta and other Indonesian university cities, the various student groups focused like a laser beam on political democratization as well as economic grievances. Their frequent demonstrations charging corruption and calling for Suharto's ouster captured as much, if not more, global media attention than the initial round of anti-Chinese violence. The growing student protests in various cities throughout Indonesia also further emboldened the discontented urban masses, who lashed out in a new round of urban rioting, looting, and burning directed largely against banks, shops, supermarkets, electronics stores, and the home of a prominent Chinese billionaire. These developments, in turn, refocused discontent on the vast wealth of Suharto and his family. In this changed and highly charged political context, Suharto was unable to defend his interests by commanding the army to repress the students or the urban masses. Instead the army largely stood by during the demonstrations and street rioting, as various military leaders maneuvered behind the scenes to oust Suharto and jockeyed for position in the new regime.

Suharto's fall from grace and power in Indonesia was accompanied elsewhere in Asia by the emergence of anti-IMF protests in cities like Seoul, Korea, where in late May 1998 tens of thousands of workers walked off their jobs and participated in street demonstrations in central Seoul directed against a state law making it easier to lay off workers. The law was passed by the South Korean government under pressure from the IMF, which had made increased "labor flexibility" a condition of granting Korea a $58 billion bailout package. While the workers railed against their government and the Korean corporate conglomerates, or *chaebol*, whose elites have escaped the hardships experienced by ordinary Koreans during the financial crisis, they left little doubt that they perceived the IMF as responsible for their plight. Standing in front of Seoul Station, the city's transportation hub, the workers raised their fists and chanted slogans like: "Let's destroy restructuring! Let's fight to secure our jobs!" The Korean Confederation of Trade Unions representing over 500,000 members followed up the demonstration by demanding that the state renegotiate the "labor flexibility" issue with the IMF.

In the face of increasingly visible popular opposition against their policies, both the IMF and the World Bank have been forced to reconsider the rigidity with which their agents have pursued the neoliberal agenda. For example, in July 1998 the World Bank resumed aid to Indonesia by granting a $1 billion loan under far less stringent terms than

had previously been expected of the Suharto regime. The new accord allowed for more social spending than previous agreements and permitted the state to run a sizable budget deficit, an item explicitly prohibited in previous agreements.

In the Indonesian case, a complex web of social relations and conflicting discourses of "globalization" came together and were localized in confrontations in Jakarta and other Indonesian cities. In the historically specific context of a general economic crisis in Asia, accompanied by urban protests in Indonesia and transnationally, a state-centered power structure that had colluded for three decades with a transnational economic network to regulate and appropriate the lion's share of Indonesia's economic wealth and expansion was significantly challenged. This alliance was caught in a collision course. They were squeezed between transnational regulatory bodies such as the IMF and the World Bank seeking to legitimate neoliberalism by ending "crony capitalism," and a series of geographically "local" irruptions of discontent in Indonesian cities by students who were connected to transnational communication circuits by internet information flows, and perceived themselves as central agents of democratization operating on a "global" stage. The student protests were accompanied by other episodes of urban violence by popular classes venting frustration at Chinese ethnic scapegoats or simply seeking survival in the face of austerity-induced economic hardships. These local and global crises of confidence were mediated by national cultural understandings, played out in the global media, and had effects beyond the boundaries of the nation-state. Though differently inflected, a related global–national–local discourse on economic "globalization" was being played out in Seoul. This is the stuff of the new urban politics of transnational urbanism.

The future course of transnational urbanism in Jakarta and Seoul are open-ended. Yet even a brief examination of these cases clearly illustrates the fruitfulness of viewing cities as sites where national and transnational practices become localized; local social actions reverberate transnationally, if not globally; and the networks connecting these social relations and practices intersect in time and space and can be discerned, studied, and understood.

A further advantage of this approach to urban studies is that a wide variety of cities, rather than a handful of sites of producer service functions, or a score of interwoven, mainly Western, centers of global command and control, become appropriate sites for comparative research. Viewing cities as contested meeting grounds of transnational urbanism invites their comparative analysis as sites of (a) the localization of transnational economic, sociocultural, and political flows; (b) the transnationalization of local socioeconomic, political, and cultural forces; and (c) the practices of the networks of social action connecting these flows and forces in time and space. These emergent transnational cities are human creations best understood as sites of multicentered, if not decentered, agency, in all of their overlapping untidiness. This is a project in which studying mediated differences in the patterns of intersecting global, transnational, national, and local flows and practices, in particular cities, becomes more important than cataloging the economic similarities of hierarchically organized financial, economic, or ideological command and control centers viewed as constructions of a single agent – multinational capital.

The "new urban politics" uncovered by this move is a disjointed terrain of global media flows, transnational migrant networks, state-centered actors that side with and oppose global actors, local and global growth machines and green movements, multilocational entrepreneurs, and multilateral political institutions, all colluding and colliding with each other, ad infinitum. The urban future following from this contested process of "place-making" is far less predictable but far more interesting than the grand narrative of global capital steam-rolling and swallowing local political elites and pushing powerless people around that inevitably seems to follow from the global cities model. Rather than viewing global cities as central expressions of the global accumulation of capital, all cities can then be viewed in the fullness of their particular linkages with the worlds outside their boundaries.

REFERENCES FROM THE READING

Cox, K. (ed.) (1997) *Spaces of Globalization.* New York: Guilford.

Gibson-Graham, J.K. (1996/1997) Querying globalization, *Rethinking Marxism*, 9, 1, 1–27.

Smith, M.P. (1988) *City, State and Market.* Cambridge, MA: Blackwell.

Smith, M.P. (2001) *Transnational Urbanism.* Cambridge, MA: Blackwell.

Smith, M.P. and Feagin, J.R. (eds) (1987) *The Capitalist City.* Cambridge, MA: Blackwell.

47

"Immigration and the Global City Hypothesis: Towards an Alternative Research Agenda"

from *International Journal of Urban and Regional Research* (2002)

Michael Samers

Editors' introduction

Michael Samers is Senior Lecturer in Geography at the University of Nottingham, UK. His research has focused on various aspects of critical economic geography, including the study of labor markets, informal work and immigration. Many of Samers' publications focus upon French cities, but he has also examined broader, European trends and written a number of articles dealing with US–European comparisons. In this reading, Samers takes up one of the key themes of Saskia Sassen's early research on global cities (see Readings 9 and 10), namely the relationship between immigration flows and the globalization of urban space. As Samers indicates, Sassen's arguments regarding immigration flows have received less attention in the global cities literature than her claims regarding the industrial and technological foundations of global city economies. Nonetheless, Samers suggests that this aspect of Sassen's initial research agenda retains considerable relevance for contemporary urban studies. Accordingly, in this reading, Samers reviews, critically evaluates, and attempts to "renew" some of Sassen's main lines of argument. After concisely summarizing Sassen's key claims regarding the interplay between global city formation and immigration, Samers surveys various problems of documentation, evidence and explanation that have emerged from her analysis, and which have subsequently been debated at some length in the literature. However, rather than abandoning Sassen's intellectual framework, Samers proposes to renew it through the strategic reformulation of key propositions. To this end, he elaborates five strategies for the further development of Sassen's approach with reference to various themes at the intersection between migration studies and urban research. Samers' research agenda on migration and global city formation also resonates closely with Michael Peter Smith's emphasis (see Reading 46) on the role of transnational social networks in global cities. Samers' chapter can be understood in part as an attempt to develop a theoretically grounded yet empirically testable methodological framework through which the emergence and evolution of such networks might be explored more systematically.

INTRODUCTION

During the 1990s, urban studies from a critical perspective seemed to experience a strange polarization. At the same time that the analytical and normative strands of neo-Weberian and Marxist urban political economy were slowly eclipsed by cultural studies and a related gamut of "post-ist" critiques (postmodern, post-Marxist, poststructural, postcolonial and transnational), the more economistic "Global City" hypothesis (henceforth GCH) began to set a certain agenda for research. But if this economistic discourse focused initially on the links between capital and labor, and especially migrant labor (Sassen 1988), then it soon turned more conspicuously to the business and technological dimensions of the theory and less to the relationship between cities and low income/"low-skilled" immigration. For example, a now leading group of researchers in the United Kingdom (the Globalization and World Cities Network, or GaWC, to which Sassen has been a regular contributor) has been preoccupied with identifying how putatively global cities are networked in terms of law, accountancy, consultancy and other financial firms.

If the original theoretical and empirical focus of the GCH has been overshadowed by "post-ist" urban research and a new propensity to concentrate on business and technological processes, then what value does Sassen's initial research (as well as similar studies in the "global city vein") still hold? In responding to this general question, this chapter seeks to redress the lack of focus on the relationship between immigration and the GCH. It aims to evaluate critically the "global city hypothesis" in relation to immigration in primarily (but not exclusively) Europe's large metropolitan regions. I do this initially by discussing Sassen's thesis, and then follow with an exploration of the subsequent literature that has sprouted from her arguments. Indeed, much of the research inspired by the GCH during the 1990s has either deployed her theoretical and conceptual arguments for the purposes of empirical verification or has questioned her framework altogether. In any case, I maintain that such a critical analysis of Sassen's ongoing research project and the parallel work of others is useful insofar as the GCH can be modified to address issues of urban inequality. I call this a "renewal" of the

"global city hypothesis." Finally, I offer an alternative research path (what I call a "reformulation" of the GCH) that combines the global city hypothesis with more recent debates around one element of the "post-ist" urban research – that is, transnationalism.

THE GLOBAL CITY HYPOTHESIS IN QUESTION

Saskia Sassen's (1988, 1991) arguments have felt the brunt of seemingly endless scrutiny. While it is not my intent to cover the broad spectrum of these criticisms, below I outline four reservations that have considerable gravity in terms of immigration.

Global cities and the expansion of labor migration?

The first reservation regarding Sassen's work is its apparent assumption that because of "globalization" (or at least a growth in FDI flows to third-world countries), global labor migration is expanding and therefore accompanying, if not necessarily driving, the growth of global cities. Yet, it should be pointed out that there is unreliable evidence regarding the massive expansion of "low-skilled" labor migration on a *global scale* since the 1970s (and by labor migration I mean migrating under the rubric of a formal job contract). This indeed may be happening, but again the international statistical evidence is difficult to assess (Ziotnick 1998). In Europe, low-skilled labor recruitment has certainly been reduced to a trickle with a few notable exceptions such as construction workers in Berlin, or temporary labor migration into French and Spanish agriculture, for example. Migration under the category of family reunification continues, and spouses and dependants may certainly search for work in these large metropolitan areas, but it too is limited in European countries. Thus, given the restrictions on labor recruitment, asylum and even family reunification, the linking of labor migration to the growth of global cities implies that much of the migration is undocumented. The anecdotal and patchy statistical evidence suggests that a significant proportion of migration to, or immigration within, Europe's largest cities is, in fact, undocumented (see, e.g., Samers 2001). But, again, comparative data, especially across European countries, is problematic given the legal dynamics of immigration policy. In short, if social scientists draw a link between the expansion of migration and the growth of global cities, then it rests on a certain amount of "guesswork" concerning the former.

Increasing informal employment?

There seems to be an implicit assumption that globalization means a dangerous cocktail of deregulation and increasing global competition, and hence the growth of informal employment. Thus, there is a considerable literature (mostly in the US and *not* in the European Union) which argues that global cities, together with their respective immigrant/"ethnic minority" concentrations, account disproportionately for the dynamism of these activities. And so emerges a second question about the GCH. As Williams and Windebank (1998: 83) point out: "The inevitable result is that they [researchers in the global city vein] identify what they seek: that informal employment is closely associated with such groups [immigrants/ethnic minorities] and that these groups engage in organised forms of exploitative, low-paid, informal employment." Regardless of the merit of these studies, Williams and Windebank insist that informal employment is not necessarily increasing in the advanced economies (note that their arguments are not discussed uniquely in the context of global cities). They insist further that immigrants do not dominate this work, even though the majority of immigrants are concentrated in informal activity. If one is to believe in the generality of the anecdotal evidence, it would seem that in many cases, immigrant employment in informal economic activity has been falling rather than growing throughout the 1990s in Europe. Nonetheless, other studies point to widespread sexual trafficking/prostitution and a growth in the number of unrecorded (female) domestic workers. In this sense, the contribution of immigrants to informal economic activity should not be seen as exclusive to "global cities." Furthermore, there is little evidence to suggest that informal economic activity is confined to global cities. For example, building and maintenance

work in the city of Liverpool (hardly, in anyone's imagination, a global city) functions with a veritable army of informally employed builders, plumbers, electricians and decorators. And El Paso, Texas, with its concentration of undocumented Mexican workers, constitutes the sixth largest garment district in the US (Spener and Capps 2001).

The growth of sweatshops?

A third and related reservation stems from the debate which surrounds the alleged growth since the 1970s of "sweatshops"/"downgraded manufacturing" in global cities. On the one hand, Waldinger and Lapp (1993) use an innovative methodology to show that there is little evidence of the growth of such "sweatshops" in the New York apparel industry, even if they acknowledge the presence of numerous garment contractors who violate a range of labor laws. Thus, the study of New York (by Waldinger and Lapp) and of the Randstad, Holland (by Kloosterman et al. 1998) conclude that in these cities, there is little change in their "post-Fordist economic structure." Rather, immigrants find employment in sectors where there are "vacancy chains" (as in bakeries in Amsterdam or Rotterdam) or in garment manufacturing (in New York or Los Angeles). In other words, in some sectors, high rates of forced shop closure and/or voluntary exit among natives are combined with lower rates of "start-up," thus offering a space for "third-world" migrants to literally set up shop. In short, this would imply out-migration of natives from certain economic activities and migrant in-migration without a necessary growth of sweatshop activity. On the other hand, Robert Ross' (2001) historical and contemporary study of sweatshops in the United States demonstrates that there has, in fact, been a steady increase in garment-related sweatshops since the 1970s in New York, Los Angeles/Southern California and San Francisco. This is the result, he argues, of a combination of processes associated with a new phase of global capitalism. These processes include "the declining capacity of the state to enforce labor laws, the increasing market power of retailers through concentration of sales; the competitive pressure brought about by massive imports from low-wage export platforms; and finally, the availability of a large pool of vulnerable immigrant labor" (ibid.: 28). Ross' thoughtful diachronic analysis seems more convincing than that of Waldinger and Lapp, but there is a glaring paucity of similar cross-national and interurban studies in the context of the European Union over the last decade. Thus, claims for the EU-wide growth of sweatshops since the 1970s require considerably more empirical demonstration.

The explanatory bias of the GCH

A fourth reservation is the exogenous explanatory bias within the GCH debate. That is, there is too much focus on economic globalization as *happening to* cities, and not enough on what might be considered endogenous processes – or, for example, how immigrants themselves structure the labor markets of large metropolises. In this sense, there is an enormous literature on "immigrant entrepreneurship," ethnic enclaves, immigrant/ethnic economies, migrant day laborers and so forth, which runs in parallel to the GCH, but which has not been the subject of any sustained synthesis. And notwithstanding some major world cities that do not have comparatively high levels of immigration, like Tokyo, it may in fact be the presence of such large-scale immigrant economic "communities" (with their attendant global financial remittances and their ability to incubate small business growth, rather than simply their complementarity to producer services employment) which partially distinguishes mega-cities from other more nationally oriented urban centers.

THE WEAKNESSES OF THE POLARIZATION DEBATE

While I do not intend to evaluate empirically the evidence for social polarization in large European cities, it may be useful to offer some critical comments on this literature. To this end, there are at least three glaring weaknesses that emerge from the polarization debate with respect to immigration. The first weakness is that the relevant studies do not or cannot capture statistically those immigrants who are deemed to be undocumented. In no subsequent discussion of Sassen's work have I noted any comprehensive recognition of the problems of statistical evaluation of income or occupational polarization in this regard. It is in fact Sassen herself who must be one of the few to point out this issue, especially in relation to informal labor markets (1991: 245). In short, if we can concede that many of the immigrants living in these putative global cities are undocumented, then using the available statistical data is not likely to be a very robust measure of inequality, whether occupational or income-based. Furthermore, the argument that European welfare states soften the hard edge of their market-oriented societies (at least relative to the United States) makes little sense for undocumented immigrants. In fact, evidence from London shows that undocumented immigrants are quite critical of the benefits system, and will do their utmost to avoid detection by not relying on social entitlements (Jordan 1999).

The second weakness, which follows from Peter Taylor's (1999) critique of the "evidential structure" of the world cities literature, is whether the evidence for, especially, income polarization is really sufficient given the transnational nature of economic activity among migrants. In other words, Taylor claims that the global city literature is based on nationally gathered statistics which are inadequate for assessing how cities are networked transnationally in the global economy. For example, in Chinese communities, many individuals and families rely on the *hui* (or the inter/intra-family pooling of money in the form of a savings bank) in order to facilitate business expansion, or just quotidian survival (White et al. 1987). As the *hui* and certainly other potentially transnational sources of income beyond employment exist, this is not easily captured by existing urban or national-level data. Needless to say, the implications here for urban policy are legion.

A third conceptual and methodological limitation is the lack of specificity about the timing (not to mention the spacing) of these "causal" (supply/demand) relationships as suggested by Sassen's original thesis. For example, Wacquant's (1999) otherwise wide-ranging overview of urban inequality seems to reference work from different moments of capital accumulation, job expansion and so on, without adequately addressing how the timing of these various studies might shape the relationship between cities, inequality and immigration. Indeed, as Gordon and Richardson note about urban income data: "The most important insight to be sought from income distribution data is this: What are the odds that any individual will remain at the bottom (or at the top or anywhere in between) and for how long? Social mobility is the real news" (1999: 577).

A RENEWAL OF THE GLOBAL CITIES LITERATURE THROUGH FIVE PROPOSITIONS

By marshalling together the criticisms of Sassen's GCH, as well as those of the polarization debate, we can move towards five propositions for a modified GCH. The theoretical and empirical results might translate into different and more appropriate urban policies. Indeed, a major difficulty with the GCH is its policy relevance. In other words, following Smith's (1998) intervention, we can ask why "global cities" should be an object of research? I would argue that the longevity of Sassen's approach rests on its ability to provide insights either into exposing inequalities, or ways of addressing these inequalities. Otherwise, I do not see the merit in pursuing this line of enquiry. We might ask more specifically, then, how Sassen's research and similar work contributes to urban policy? If Sassen is correct and the nature of immigration is related to just another round in the urbanization of global capitalist accumulation, what can or should be done about it? There are a number of axes in which urban governments can intervene. These include immigration policy, so-called "integration" policy and employment policy.

Reassess the relationship between international labor migration and urban labor markets

Comparative national-level data on migration is too incongruent and lacks sufficient disaggregation to draw a relationship between a growth in the labor supply and the expansion of global cities. And it is the ambiguity and hyperbole of statements about global labor migration that require us to relate the status of immigrants (undocumented, refugee/

asylum-seeker, family member with or without the right to work, student, entrepreneur etc.), and their background (skills, education, financial and commercial resources, and so forth) to labor market entry and the structure of urban labor markets. And this in turn must be combined with sensitivity to the timing of these migrations. In other words, one must be attentive to the relationship between changes in immigration policy and the growth of global cities.

Reassess the relationship between informal employment and global cities

By its very nature, informal employment is exceedingly difficult to capture in a quantitative sense, not least because businesses and employment are so ephemeral. And here we need a more precise notion of what constitutes informal economic activity (is it the "drug economy," prostitution, domestic labor, or simply the illegal production of textiles?) (Samers 2001). For example, it may be female prostitution and domestic work in the EU, rather than an illegal garment industry, that marks the informal character of economic activity in European global cities at the beginning of the twenty-first century. Nonetheless, if there is a strong relationship between mega-cities, immigration and the growth of informal employment, then this must be demonstrated rather than simply asserted. And it must be shown why informal economic activity is relatively unique to these cities. Otherwise, informal economic activity as a defining characteristic of global cities cannot be assumed. Thus, anecdotal qualitative evidence (and not quantitative estimations) is likely to provide valuable insights into the processes at work, and therefore help to sustain dynamic and appropriate urban policies with respect to potential employment regulations, job training, language training and housing.

Evaluate critically the growth of sweatshops and downgraded manufacturing

Like many forms of informal employment, statistical evidence across large metropolitan regions with regard to sweatshops/downgraded manufacturing is likely to prove elusive. Yet, again, qualitative case studies may offer a useful service in terms of public policy. They will not, on the other hand, help to assess whether these kinds of employment units are growing or declining, unless such studies are carried out *en masse*. And, as suggested in the previous proposition, there has been a tendency to focus disproportionately on garment manufacturing, and this may obscure the importance of other kinds of downgraded manufacturing. Again, the specific links between a growth in sweatshops/downgraded manufacturing (including how one might delineate such economic units) and how this is specific to global cities should be articulated.

Match exogenous and endogenous processes in global cities

There is a need to construct a sustained synthesis between processes of economic globalization (what I referred to earlier as exogenous processes) and the way in which these processes both provide the space for and constrain the economic activity of immigrants in urban labor markets (so-called endogenous processes). To this end, a dialogue between the GCH and the entire literature on "immigrant entrepreneurship," ethnic enclaves, immigrant/ethnic economies, day laboring and so forth, could prove enormously fruitful. Perhaps a combination of Kloosterman et al.'s (1999) concept of mixed embeddedness, an alternative conception of entrepreneurship based on day laborers as low-skilled entrepreneurs (Valenzuela 2001), and a sophisticated labor segmentation theory for understanding wage labor (see, e.g., Peck 1996) might prove to be one valuable path for research? Yet the point is not to substitute one discourse for another (i.e. endogenous processes for exogenous), but to show specifically how the world's largest cities may be distinguishable from other urban centers. In other words, how do the disproportionately large numbers of immigrants help to shape the urban labor markets of these largest cities, and what implications does this have for urban policy?

Rethink polarization and inequality

I have argued that the polarization debate suffers from a number of conceptual and empirical weak-nesses that have significant implications for both the GCH debate and urban policy. These include the "fuzziness" of the terms polarization and inequality, the national bias of statistics, the inattention to undocumented immigration, and the lack of temporal specificity. Thus, with regard to addressing these issues simultaneously, a starting point is an estimation of undocumented immigration. Notwithstanding the potentially insidious political and ethical implications of gathering or applying such statistics, and regardless of how paradoxical such measurements may seem (because they appear to officialize the unofficial), they are likely to provide a more robust measure of at least occupational, if not income, polarization/inequality. Yet this assumes that social scientists become more creative and look elsewhere for the relevant statistical sources. In fact, some complex statistical procedures have been adopted by various countries to measure the number of undocumented immigrants and/or undocumented immigrants working informally (SOPEMI 1999). This data would have to be combined with more conventional occupational, income, employment and micro-census statistics, and less conventional (*transnational*) data such as measurements of remittances, capital flows, the intra-ethnic pooling of capital, and other financial activities between the largest cities and the countries of emigration. Finally, this would have to be analysed through the lens of changes in immigration policy over at least a census period (i.e. roughly 10 years) in order to tease out the temporal dynamics of inequality/polarization.

In sum, I have pointed to five ways in which we can "renew" the GCH and parallel studies. These propositions together represent a considerable challenge to social scientists working in this vein, but it is my argument that if these are not addressed, then future studies into global cities are likely to be inadequate assessments of urban inequalities in the world's largest cities.

CONCLUSIONS

The GCH debate, and particularly the research agenda of GaWC, has been preoccupied recently with the technological dimensions of so-called "global cities." This chapter has sought to recover the role of immigration in large urban economies. Using mainly observations from European metropolises, I argued first that the GCH requires significant revision insofar as it can be used as a tool for addressing issues of inequality, and offered five propositions for a renewal of the existing contours of the GCH. Second, beyond these revisions, I suggested a complete reformulation of the debate, and sought to link it with ideas emanating from the literature on transnationalism.

Despite the move by researchers (such as the GaWC) to transcend the emphasis on "hierarchies" and explore the question of global city networks, I argue that this emphasis is at best incomplete, at worst misplaced. Instead, I have a very different idea of the possibilities of a "network paradigm" for global cities – one that focuses rather on cities as the locus for a "transnational political mobilization from below" (rather than above). The "Urban League" in the United States or the ATMF (with its headquarters in Paris) are examples of the sorts of interurban networks that I have in mind. However, the possibilities for increased political mobilization within a single global city, or indeed a network of such cities, should not blind us to falling into the trap of what Drainville (1998) calls the "fetishism of global civil society," nor what Mahler (1998) identifies as the weaknesses of the "transnationalism from below" idea. Nonetheless, such a theoretical move seems a necessary task, lest our understanding of "global cities" be confined to simply ranking them, outlining business networks, or categorizing their characteristics.

REFERENCES FROM THE READING

Drainville, A.C. (1998) The fetishism of global civil society. In M.P. Smith and L.E. Guarnizo (eds) *Transnationalism from Below*. New Brunswick, NJ: Transaction.

Gordon, P. and Richardson, H.W. (1999) Review essay: Los Angeles, city of angels? No, city of angles, *Urban Studies*, 36, 3, 555–591.

Jordan, W.J.O. (1999) End of Award Report to ESRC on Project 000236838: Undocumented immigrant workers in London, September 21.

Kloosterman, R., van der Leun, J. and Rath, J. (1998) Across the border: immigrants, economic opportunities, social capital, and informal business activities, *Journal of Ethnic and Migration Studies*, 24, 2, 249–268.

Kloosterman, R., van der Leun, J. and Rath, J. (1999) Mixed embeddedness: (in)formal economic activities and immigrant businesses in the Netherlands, *International Journal of Urban and Regional Research*, 23, 2, 252–266.

Mahler, S. (1998) Theoretical and empirical contributions: toward a research agenda for transnationalism. In M.P. Smith and L.E. Guarnizo (eds) *Transnationalism from Below*. New Brunswick, NJ: Transaction.

Peck, J. (1996) *Workplace*. New York: Guilford.

Ross, R.J.S. (2001) The new sweatshops in the United States: how new, how real, how many, and why? In G. Gereffi, D. Spener and J. Bair (eds) *Globalization and Regionalism*. Philadelphia, PA: Temple University Press.

Samers, M. (2001) Here to work: undocumented immigration and informal employment in Europe and the United States, *SAIS Review*, Winter–Spring, 21, 1, 131–145.

Sassen, S. (1988) *The Mobility of Labor and Capital*. New York: Cambridge University Press.

Sassen, S. (1991) *The Global City: New York, London, Tokyo*. Princeton, NJ: Princeton University Press.

Smith, M.P. (1998) The global city – whose social construct is it anyway?, *Urban Affairs Review*, 33, 4, 482–488.

SOPEMI (1999) *Trends in International Migration*. Paris: OECD.

Spener, D. and Capps, R. (2001) North American free trade and changes in the nativity of the garment industry workforce in the United States, *International Journal of Urban and Regional Research*, 25, 2, 301–326.

Taylor, P. (1999) So-called "world cities": the evidential structure within a literature, *Environment and Planning A*, 31, 1901–1904.

Valenzuela, A. (2001) Day laborers as entrepreneurs?, *Journal of Ethnic and Migration Studies*, 27, 2, 335–352.

Wacquant, L. (1999) Urban marginality in the coming millennium, *Urban Studies*, 36, 10, 1639–1647.

Waldinger, R. and Lapp, M. (1993) Back to the sweatshop or ahead to the informal sector?,

International Journal of Urban and Regional Research, 17, 1, 6–29.

White, P., Winchester, H. and Guillon, M. (1987) South-east Asian refugees in Paris: the evolution of a minority community, *Ethnic and Racial Studies*, 10, 1, 48–61.

Williams, C. and Windebank, J. (1998) *Informal Employment in the Advanced Economies*. London: Routledge.

Ziotnick, H. (1998) International migration 1965–96: an overview, *Population and Development Review*, 24, 3, 429–468.

SEVEN

48

"Pathways to Global City Formation: A View from the Developmental City-state of Singapore"

from *Review of International Political Economy* (2004)

Kris Olds and Henry Wai-Chung Yeung

Editors' introduction

Kris Olds and Henry Wai-Chung Yeung are both experts on urbanization and globalization in Southeast Asia. Olds formerly worked in the Department of Geography at the National University of Singapore and is currently Associate Professor of Geography at the University of Wisconsin, Madison. Olds has published widely on global city formation and urban restructuring in Pacific Rim cities such as Vancouver, Singapore and Shanghai, including a major book on the construction of urban mega-projects (Olds 2001) and several edited volumes dealing with various aspects of spatial change in contemporary Southeast Asia. Yeung is Associate Professor of Geography at the National University of Singapore. He has written a large number of books and articles on international business networks in Singapore, Hong Kong and the ASEAN region and is also an influential sociospatial theorist and economic geographer.

In this reading, which is excerpted from an article in a major journal of international political economy, Olds and Yeung extend the analytical scope and empirical focus of global cities research to explore a wider range of cases and developmental trajectories. Like the contributors to Part Four of this Reader, Olds and Yeung contend that many of the key arguments of world city theory have been derived from case studies of a relatively limited number of large cities located in the United States (New York), the United Kingdom (London) and Japan (Tokyo). By contrast, in an argument that also resonates closely with that of Reading 20 by Hill and Kim, Olds and Yeung suggest that the globalization of urbanization is occurring along differential, contextually specific pathways. Olds and Yeung's detailed review and critique of the world cities literature could not be included in the reading here, which focuses instead upon their constructive research agenda for the study of global cities. The heart of Olds and Yeung's argument is an attempt to differentiate three types of global cities – "hyper global cities," which correspond to the classic global cities emphasized in the literature, such as New York, London and Tokyo, "emerging global cities," in which national and/or local social forces and institutions are struggling to position a city strategically in the global economic system, and "global city-states," which correspond to the unique institutional and sociospatial arrangements associated with Hong Kong and Singapore. Additionally, like many of the contributors to Part Five of this Reader, Olds and Yeung emphasize the extensive role of "developmental" state institutions in the promotion and management of global city formation. While much of Olds and Yeung's analysis is inspired by their efforts to interpret the case of Singapore, their work suggests a number of more generally applicable methodological insights that can continue to inform research on global cities. In particular, Olds and Yeung urge researchers to analyze the contextually specific *processes* of global city formation rather than continuing to construct static, synchronic models of global city hierarchies.

The global/world city literature is, to a significant degree, grounded on empirical studies of two to three major global cities within large countries – London (in England), New York (in the United States) and, occasionally, Tokyo (in Japan) and Seoul (in South Korea). Moreover, many of its major contributors are based in the Anglo-American world. In such an intellectual context, we argue for the need to extend our existing global city research agenda such that it further recognizes the enormous varieties of global cities. There is also a need to support more research that investigates, in historically- and geographically-specific ways, the processes through which these "other" global cities are formed, transformed, and extended beyond their immediate urban territoriality. In short, we need stronger consideration of the differential pathways associated with global city formation processes.

This chapter aims to complement in a modest way attempts (by Saskia Sassen, John Friedmann and others) at better understanding the implications of globalization and transnationalism for cities, but in a way that emphasizes the *differential* paths that cities follow as they globalize or are globalized. We do this by moving out of the empirical terrain of North America and Europe to Pacific Asia, to develop an exploration of the interrelationships

Plate 34 Singapore (Henry Yeung)

between global city formation processes, the developmental state, and the unique characteristics of the contemporary city-state, with particular reference to the case of Singapore. Specifically, we develop a typology of global cities, not to encourage static thinking about forms of global cities, but instead to reinforce the dynamic pathways associated with varieties of global cities. In developing this typology, we pay particular attention to the third type of global city – the *global city-state* – and examine some key features of its governance process. In doing so, we hope to lay the ground for richer and more nuanced studies of global/world city formation; i.e. the process of becoming, versus the characteristics of being.

A TYPOLOGY OF GLOBAL CITIES

In our view the differences amongst established and emerging global cities can be attributed to different historical and geographical contexts of emergence and path dependency, and different configurations of internal institutional capacities and discursive practices (by strategic actors). Figure 1 presents three types of global cities: (1) hyper global cities; (2) emerging global cities; and (3) global city-states.

(1) *Hyper global cities*

The characteristics of *hyper global cities* as New York and London are very well known today as a result of several decades of global city research. These global cities are comprehensively integrated into a nested hierarchy of regional, national, and global economies. As portrayed in Figure 1, hyper global cities have strong embedded relationships with their immediate hinterland, the so-called "global city-region." These global city-regions have emerged as the fundamental spatial units of the contemporary global economy though they have no formal and particularly coherent political presence on the national or international level. The rapid emergence of hyper global cities is explained by the fact that globalization has accentuated the importance of spatial proximity and agglomeration in enhancing economic productivity and performance advantages. Large global city-regions function as

1. Hyper global cities

◄► Very well integrated into the global economy through both inward and outward flows

☼ Global city-regions

• Cities

2. Emerging global cities

→ More reliance on inward flows from the global economy

☼ Global city-regions

A/B Emerging global cities

• Cities

3. Global city-states

◄► Very well integrated into the global economy and experience direct influence

☼ Global city-regions

○ Global city-state

• Cities

Figure 1 A typology of global cities

territorial platforms for firms to compete in global markets. In Figure 1, these intense networks of flows are illustrated in the inner circle in which the global city is located.

Hyper global cities are not only embedded in their immediate global city-regions, but they are usually engaged in competitive relationships with other (global) city-regions in the same home country. This dimension of inter-regional interaction is very important to our understanding of why few dominant global city-regions can co-exist within one country unless there is significant degree of functional specialization. Indeed Figure 1 shows only one dominant global city-region within a country, although it has significant interaction with other regions in the same country (represented by various two-way arrows).

To a certain extent, the competitiveness of a particular global city-region is determined by its role and functions within global city networks that transcend specific regions and/or countries. For example, London is the only viable global city-region in the UK because it has developed such a strong momentum in serving as a strategic node in the global economy. Given this situation it is virtually impossible for another city-region in the UK to compete with London. The same logic may be extended in a sectoral sense and applied to understand the dominance of New York as a global city-region in the financial world of the US, San Francisco as a global city-region in the high-tech world, and Los Angeles as a global city-region in terms of cultural industries. All three global city-regions operate at different levels, and are embedded in disparate (albeit overlapping) geoeconomic networks that reach out across space. All three global city-regions owe their successes in becoming dominant global city-regions less to interaction and flows *within* their home countries, than to their articulation into the global (cultural) economy; an economy embedded in networked global city-regions that are dispersed across the globe. That said, all hyper-global cities/global city-regions are critically dependent upon much larger national markets for their survival.

To sum up, hyper global cities and their regions are deeply integrated into the contemporary global economy. They are relational city-regions of the highest order, attracting and distributing unprecedented volumes of material and non-material flows at a variety of scales. The transformation of these gateway cities is also related to historical context (e.g. imperialism and transnational migration) and geographical context (e.g. national capitals and agglomeration advantages). Once set in motion, these hyper global cities and their hinterland regions often gain momentum and the logic of path dependency becomes increasingly important.

(2) *Emerging global cities*

While some cities in today's world economy are hyper global cities, there are other cities that *strive* to become global cities. We designate these as emerging global cities. In the parlance of the Globalization and World Cities Study Group and Network (GaWC – see http://www.lboro.ac.uk/gawc/) these cities would be classified as Beta (second brightest) and Gamma (third brightest) world cities. In seeking to do so, these cities draw in significant resources and inputs from their home countries, as well as from multilateral institutions (in the case of developing country cities). As shown graphically in Figure 1, an emerging global city (A or B) has only limited relational linkages with the global economy (in a relative sense compared to hyper global cities). It is also much more dependent upon inward flows of development capital, people, goods and services and information from the global economy. Instead of acting as an interactive strategic node in the coordination of the global economy, emerging global cities act as coordination/channeling centers responsible for receiving or channeling inward flows. Such global cities do not (or cannot) facilitate the export of significant outward flows of development capital (or information) to service the global economy, as often expected in dominant definitions of global cities (such as London). That said they do facilitate outward flows of surplus capital in the forms of profits generated by TNCs. The flow arrows in Figure 1 are primarily unidirectional, representing inflows from the global cultural economy into a particular emerging global city (A) before these inflows are further redirected and/or distributed further down the urban hierarchy in that country. Because these cities are emerging and striving to become global cities, there is more potential for competition from other urban centers in the same host country. City (B) in Figure 1, for example, may pose as a challenger to the aspirations of (A) to become a global city. This competitive condition, of course, does not apply to those developing countries dominated by a primate city (e.g., Bangkok in Thailand or Kuala Lumpur in Malaysia).

The emergence of aspiring "global" cities depends significantly upon preconditions in terms of endowments of institutional resources, economic linkages at different spatial scales, and political fabric. For example, nation-states often deploy substantial resources (and initiate regulatory changes) in the goal of transforming these cities into global cities. The intention, of the nation-state and increasingly multilateral organizations such as the World Bank, is to use these cities to enable the nation to "plug" into the global economy. Such cities play a critical role for they act as the specific locales within a country where key actors and institutions analyze, represent, and associate with the global space of flows. This role is not a new one in the history of capitalism. In today's globalizing and post-colonial era, however, what is significant is that many nation-states in developing countries are engaging in novel discursive practices, and mobilizing disproportionate material resources to "construct" representations of entrepreneurial global cities. Malaysia's Multimedia Super Corridor project and Shanghai's Pudong mega urban development are two obvious examples of global city formation projects from the Asia Pacific region. In both cases, massive public resources have been poured into developing and promoting "show-case" projects that theoretically qualify both Kuala Lumpur and Shanghai as global cities. Both cities compete for hosting the tallest buildings in the world. Of course, the specific national contexts in the emergence of both Kuala Lumpur and Shanghai are quite different. But one common process in both emerging global cities is that there is a very strong political and institutional will to construct them as "national projects." Whether these emerging global cities will eventually converge in their characteristics and developmental pathways towards those hyper global cities may be a moot point. The critical condition is the sustainability of national efforts in developing particular cities to become global cities.

(3) *Global city-states*

Third, the above national-urban convergence in the process of global city formation is even more apparent in the third type of global city – the global city-state. To a large extent, city-states (e.g. Hong Kong and Singapore) are unique historical and geographical realities because the state is contained within a fully urbanized and spatially constrained territorial unit. The national and the urban/local scales are effectively juxtaposed under what we call (somewhat awkwardly) the *UrbaNational* scale. Global city-states are clearly different from hyper global cities and emerging global cities because they do not have an immediate hinterland within the same national territorial boundaries. To a significant degree, broader regions (e.g., Southeast Asia) and more

distant parts of the globe become their hinterland. The development of a terrain of extraterritorial influence emerges when the global city-state functions like (1) hyper global cities, both attracting in material and non-material flows, and in functioning as a command and control center for the flows and networks that reach out at regional (for the most part) and sometimes global scales. Not surprisingly, global city-states play key roles as international financial centers, acting as the basing sites for the intermediaries of global finance. The strong financial role of the global city-state is a key factor in the Alpha ("the brightest") status accorded to both Hong Kong and Singapore by GaWC researchers.

Referring to Figure 1, it is clear that in global city-states, the (national) state has a virtually direct access to the global economy. State policies can be shaped to develop the city-state into a global city-state. This process implies that the city-state must be not only an attractive location for material inflows from the global economy, but also an origin of development flows (versus mainly repatriated profits) to participate in the global economy. As understood here, the term "global reach" best captures the dynamics of the global city-state formation pathway. It illustrates how a specific territorial organization (e.g. the city-state) is able to extend its influence and relations in the global economy through encouraging both inward and outward flows of people, capital, goods and services, and information.

Global city-states differ from hyper global cities in at least three important ways. First, global city-states have the political capacity and legitimacy to mobilize strategic resources to achieve (national) objectives that are otherwise unimaginable in non-city-state global cities. This is the case because they are city-*states*; they are represented and governed by the state in all of its roles. When one recognizes that a unified state in city-states plays multiple regulatory roles, and that the territoriality of governance is minuscule in comparison to most nations (e.g., it takes 45 minutes to drive a car from one side of Singapore to the other), the unique nature and capacities of the global city-state becomes all the more evident (see Sim et al. 2003). This is an issue that development planners in larger nations (e.g., Sydney, Australia) are well aware of since both hyper global cities and emerging global cities are governed in a relatively more complex, less coherent, and less strategic fashion.

Second, and on a related note, global city-states are not constrained by the tensions inherent in national-versus-urban politics (or regional development politics) confronting so many developing, and, for that matter, developed, countries that aspire to "construct" their global cities. In other words there are no intra-national regions or cities competing for material and non-material resources. The politics of city/nation-building tends to be focused on the strengths and weaknesses of policy options rather than which intra-national territorial unit is deserving of attention and resources.

Third, the most prominent of these global city-states – Singapore and Hong Kong (until 1997) – are the products of colonialism, and then post-colonial political dynamics. Colonial origins helped to shape urban destinies that were (and still are) tightly intertwined with the evolving global economy. This colonial history has helped to engender an openness to constant change, and an outward-oriented and relatively cosmopolitan sensibility. Colonialism also helped to lay the legal, linguistic, and technological (esp., transport) foundations for integration into the contemporary global economy. Finally, postcolonial political dynamics (esp., the 1965 ejection of Singapore from Malaysia) concentrated the minds of politicians on the necessity of pursuing the global city pathway years before academics and planners were speaking of the "global city" or the "world city" (see Mauzy and Milne 2002).

How do global city-states acquire the capacity to spur on global reach? While there are a variety of factors to consider in such a discussion, it is clear that we must turn our attention to the building of institutional capacities in the global city-state. In particular, we consider two interrelated aspects of this process of building institutional capacities: developmentalism and political control.

To some ultra-globalists, globalization leads to the end of the nation-state (Ohmae 1990). This view fundamentally distorts the transformational nature of the nation-state in today's global economy. The reconstituted role of the nation-state in today's global economy does not necessarily diminish its role in governing its national space; the role of the state is simply being reshaped (Weiss 1998).

In the case of global city-states, the dialectical contest between the nation-state and global forces is becoming even more apparent. Whereas a global city-state may serve the global economy well through its role as a command and control node, the nation-state may have certain developmental objectives that run against the call for putting the global logic of capital above the local/national interests of citizens. To accomplish these sometimes contradictory objectives (of caring for citizens and serving the global economy) the nation-state in global city-states often takes on a developmental role. Developmentalism and the developmental state may sometimes be a historical legacy (e.g. in Japan and South Korea). They may also be a consequence of intense political struggles that ended with the dominance of one political power/coalition. Their emergence is therefore highly specific within particular historical and geographical contexts.

In Johnson's (1982) original study of post-war development in Japan, the "developmental state" is characterized by several attributes:

1. The top priority of state action, consistently maintained, is economic development, defined for policy purposes in terms of growth, productivity, and competitiveness rather than in terms of welfare. The substance of growth and competitiveness goals is derived from comparisons with external reference economies which provide the state managers with models for emulation.
2. The state is committed to private property and the market, and it limits its interventions to conform with this commitment.
3. The state guides the market with instruments formulated by an élite economic bureaucracy, led by a pilot agency or "economic general staff."
4. The state is engaged in numerous institutions for consultation and coordination with the private sector, and these consultations are an essential part of the process of policy formulation and implementation.
5. While state bureaucrats "rule," politicians "reign." Their function is not to make policy but to create economic and political space for the bureaucracy to maneuver in while also acting as a "safety valve" by forcing the bureaucrats to respond to the needs of groups upon which the stability of the system rests (i.e. to maintain the

relative autonomy of the state while preserving political stability). This separation of "ruling" and "reigning" goes with a "soft authoritarianism" when it comes to maintaining the needs of economic development *vis-à-vis* other claims and with a virtual monopoly of political power in a single political party or institution over a long period of time.

At a national scale, a developmental state that satisfies these conditions has much greater capacity to effect global reach in the building of an exploitable extraterritorial terrain in the aim of benefiting the city-state, while simultaneously enhancing the formation of global linkages via the attraction of FDI (and foreign firms). For example, in Singapore a plethora of state-directed institutions, policies, programs, and projects have emerged to spur on the outward investment process: this is in part because the historical underdevelopment of indigenous entrepreneurship in the private sector has convinced the state that regionalization drives cannot be effectively taken up by private sector initiatives only.

Focusing inwards, the political power and control of a developmental city-state distinguishes it from municipal governments in most global cities because it is able to bypass national-state/provincial-city politics typical in many global cities. In Singapore, for example, immigration policies and borders can be tightly managed to facilitate labor market restructuring – a capacity that no other global city has. One ministry (the Ministry of Education) develops and implements education policy from the pre-school to the tertiary levels. And on land use planning matters, the statutory board responsible for urban planning (the Urban Redevelopment Authority) answers directly to the Ministry of National Development. In turn, one key agent of national development (the Singapore Economic Development Board (EDB)) has near monopoly power at determining the strategic direction of the economy. Given that the EDB formulates and implements national economic development policy, and the URA then falls in line to ensure that land use planning supports EDB directives, the politics of urban change is highly charged, hierarchical in nature, and it rarely becomes complicated by citizen involvement procedures (compared to most Anglo-American global cities). Furthermore,

the consequence of intertwining the national and the urban is that *all* urban planning policies, programs and projects are suffused with the politics of nation-building in the post-colonial era. More pragmatically, 100 per cent of the country/city is planned by one authority, with every square centimeter of the city/island being managed in a fine-grained manner (Koolhaas and Mau 1995).

Given the role of the state *vis-à-vis* the limited size of the territory being governed, the global/world city formation process has been both rapid and unique. State guided urban restructuring, in the context of the rapid development of Pacific Asia over the last three decades, has facilitated the formation of deep and complex global economic linkages and interdependencies. The juxtaposition of both national and city governance in the hands of the developmental city-state necessarily implies that it is also able to extend its control over most aspects of social and political life of its citizens. The net outcome of this control is that the state is able to mobilize social actors and tremendous resources to meet its national objectives (e.g. global reach). It is also able to eliminate major opposition to its developmental policies through social control and discursive practices. Under these circumstances, the (nation) state becomes the city and the city becomes the (nation) state. The global reach of the city-state becomes an institutional extension of the influence and relations of the nation-state on a global scale.

It should be noted that not all actors in a city-state, and not all city-states, are willing and/or able to initiate and complete such processes of global city formation. Much depends upon existing political-economic and social-organizational processes, the capability of key actors (firms, state, and institutions) in exercising power to implement certain strategies that situate the city-state in a beneficial manner to the global spaces of flows, and the complex and intertwined influences of history and path dependency.

CONCLUSION

Analysts need to become more cognizant of the sheer variety of global cities, and the differential pathways to global (or world) city formation. Concurrently, analyses of urban transformation and global city formation must be situated within historically and geographically specific contexts: this is because the existing literature on global cities has (to date) focused somewhat too narrowly upon a few "champion examples." For us, however, the largest lacuna of knowledge in global city studies has been, and continues to be, in relationship to *global city formation processes*. A large volume of work has been focused on corporate service networks, and global service center functions but in the process the complex relationship and interdependency between globally active firms (and other institutions such as NGOs) and the state has become somewhat sidelined. Moreover, this complex relationship must be spatialized itself, with a deeper sense of the different state formations associated with different types of global cities at different historic periods. The complexity associated with global city formation processes is particularly apparent for us from the vantage point of Singapore, and our interests in the context of the evolution and development of the global city-state. While hyper global cities as London and New York continue to be reinvented by a rich assortment of agents and forces operating at a variety of scales, city-states such as Singapore and Hong Kong vigorously pursue relatively singular and focused UrbaNational developmental strategies.

The implication of recognizing the diversity of global cities is clear: there is no "cook book" approach to global city formation just as there is no "model city." In a policy sense, then, *the* global city model is in serious doubt. Calls for the attainment of global city status in many countries may also be unfounded or unrealistic, thereby shutting down alternative development scenarios that have the potential to be more appropriate and achievable given the continued diversity of conditions across space and time. Global cities, be they Alpha, Beta or Gamma global cities, or hyper, emerging or global city-state global cities, should not be viewed as an idealized end-state phenomenon. Instead all of these cities are the outcome of a wide range of processes, all of which are shaped by state institutions and spaces. Given the diversity of state roles and capacities around the world, we should therefore expect equally diverse global city formation and transformation processes.

REFERENCES FROM THE READING

Johnson, C. (1982) *MITI and the Japanese Economic Miracle*. Stanford, CA: Stanford University Press.

Koolhaas, R. and Mau, B. (1995) *Small, Medium, Large, Extra-Large*. Rotterdam: 010 Publishers.

Mauzy, D.K. and Milne, R.S. (2002) *Singapore Politics under the People's Action Party*. London: Routledge.

Ohmae, K. (1990) *The Borderless World*. London: Collins.

Olds, K. (2001) *Globalization and Urban Change*. Oxford: Oxford University Press.

Sim, L.L., Ong, S.E. and Agarwal, A. (2003) Singapore's competitiveness as a global city, *Cities*, 20, 2, 115–127.

Weiss, L. (1998) *The Myth of the Powerless State*. Ithaca, NY: Cornell University Press.

49

"World City Topologies"

from *Progress in Human Geography* (2003)

Richard G. Smith

Editors' introduction

Richard G. Smith is a social theorist and urban geographer who teaches in the Geography Department at Leicester University in the UK. While Smith's initial writings on global cities were co-authored with several scholars in the GaWC research group in Loughborough, he has recently moved away from the original propositions of this group in order to develop an innovative research agenda of his own. In linking poststructuralist theory to the urban political economy literature, Smith has developed a topological, actor-network approach to the study of global cities. As he explains in Reading 49, this theoretical integration of structuralist and poststructuralist, political-economic and discursive approaches has created a rather new way of looking at world cities: instead of viewing them as being locked into vertically scaled hierarchies, Smith suggests that global cities are sites within a tightly interconnected, horizontal network of places, actors and "actants."

Something exciting is happening in urban studies. A great experiment is afoot as slowly but surely old urbanism (the familiar or conventional urban studies we have become accustomed to) is being rejected and a new urbanism is coming into focus. The threat of an impasse in urban theory that Thrift identified some years ago seems a distant memory as a new urbanism based around globalisation and an ontology of movement, networks, flows, fluids, folds, mobilities, nonhumans, practices, and complexity is becoming more and more recognisable (e.g. see Thrift 1998; Amin and Thrift 2002; Smith 2003). In fact, the emergence of actor-network theory (ANT) out of some strands of poststructuralist thought (Deleuze in particular), and the emergence of non-representational theory

and complexity theory, is leading to a rethink of not only urban studies but also social sciences such as sociology and human geography.

Developing ideas from poststructuralism, ANT, non-representational theory, and complexity theory the paper strives to challenge dominant thinking about space, time, and scale in the literature on globalisation and world cities. The paper contributes to a topographical understanding of globalisation and world cities that is both non-scalar and non-linear as it defies both geographical borders and temporal frames. The paper takes the view that with globalisation space and time are pliable and moveable constructions that might or might not correspond to the spaces and times of nation-states.

SPACING WORLD CITIES

Before ANT and non-representational theory there was poststructuralism and the genius of Deleuze (often with Guattari), but his ideas have received scant attention in urban studies. Deleuze can help us conceptualise world city topologies because he gave much thought to the idea of space-time as folded. Deleuze's poststructuralist philosophy is one where boundaries, scales, and territories vanish through deterritorialisation as the world is conceptualised as a living dermis with an infinite bundle of (un)folds and surfaces that make space and time. For Deleuze and Guattari philosophy is akin to the Japanese art of origami – folding, unfolding, refolding (Deleuze 1993). Their world is one of lines that intersect and weave; "We think lines are the basic components of things and events. So everything has its geography, its cartography, its diagram" (Deleuze 1995: 33). There are no points in their philosophy (they hate pointillism), rather what would interest them is the lines that make up globalisation and the city; "I tend to think of things as sets of lines to be unravelled but also to be made to intersect. I don't like points . . . Lines aren't things running between two points; points are where several lines intersect. Lines never run uniformly, and points are nothing but inflections of lines. More generally, it's not beginnings and ends that count, but middles" (Deleuze 1995: 160–161). What interests Deleuze are (un)folds, the infinite labyrinth of fold to fold, that produces the world's topology as one of process that overwhelms the fictions of boundaries, limits, fixity, permanence, embedment. What is important to realise, as I come onto ANT, is that all folds are equally important, there are no masters and no servants, ". . . folds cannot be distinguished in terms of the essential and the inessential, the necessary and the contingent, or the structural and the ornamental. Every fold plays its part: every fold splays 'it' apart. The event of origami is in the (un)folding, just as the gift is in the wrapping: not as content, but as process" (Doel 1999: 18).

What the folds of Deleuze mean for the issue of propinquity has been taken up by some ANT theorists (e.g. Latour 1993). They observe that the near and far become less meaningful when a metric theory of space (and time) is rejected in favour of a topological theory where space-time is viewed as folded, crumpled, and multi-dimensional:

> If you take a handkerchief and spread it out in order to iron it, you can see in it certain fixed distances and proximities. If you sketch a circle in one area, you can mark out nearby points and measure far-off distances. Then take the same handkerchief and crumple it, by putting it in your pocket. Two distant points suddenly are close, even superimposed. If, further, you tear it in certain places, two points that were close can become very distant. This science of nearness and rifts is called topology, while the science of stable and well-defined distances is called metrical geometry (Serres and Latour 1995: 60).

The coordinates of distance and proximity are transformed by a folding, refolding, and unfolding that eschew ideas such as linearity. A topological rather than a geometrical spatial configuration emerges. But if that seems like a rather abstract, remote, and detached portrayal of space then a more accessible rendition can be found in Latour (1997a: 3):

> I can be one metre away from someone in the next telephone booth, and be nevertheless more closely connected to my mother 6,000 miles away; an Alaskan reindeer might be ten metres away from another one and they might be nevertheless cut off by a pipeline of 800 miles that make their mating for ever impossible; my son may sit at a school with a young Arab of his age but in spite of this close proximity in first grade they might drift apart in worlds that become at later grades incommensurable; a gas pipe may lie in the ground close to a cable television glass fiber and nearby a sewage pipe, and each of them will nevertheless continuously ignore the parallel worlds lying around them.

Here we learn not only that space does not add up (Thrift 1998), but that space is also rather messy, complex, juxtaposed, or perhaps that there are many kinds of space. In fact, for ANT theorists and any study of world city actor-networks (Smith 2003), the details are crucially important for getting at just how networks lengthen and endure.

TIMING WORLD CITIES

The notion of a linear time that is continuous and cumulative has come under considerable scrutiny by many writers. For such writers the "row of clocks on a wall" cliché that is so common in a global or world city would be melting and warped like Dali's famous painting "The Persistence of Memory" because, "...time is a polymorphous history of 'folds' which wrap around or melt into each other, producing a world of cross-talk, cross-tabulations, cross-infection and criss-crossing, a world of broad temporal sweeps but also of scattered pockets, tatters and emergencies" (Thrift 1998: 55). In other words, international standard time – the invention of serial time – is a fiction that hides the simultaneity of the world and consequently, "Time should be decartelized, and everyone should set his or her own" (Ballard 1992: 279). But even that view of globalisation, of global simultaneous multiple time, is too unified. Time is too complex to be absolutely defined (cf. Newton's absolute time; see Serres and Latour 1995). Time is not simply simultaneous and multiple it is with globalisation increasingly non-linear and complex. With globalisation any event can have unexpected, disproportionate, and emergent effects that are often distant in time and space from when and where they occurred. What is needed then is a more chaotic theory of time if we are to better appreciate that the networks in which global and world cities are involved are not stable precisely because when networks are immersed in time they, "...fluctuate, become very unstable, and bifurcate endlessly" (Serres and Latour 1995: 109).

With the temporalisation of networks there is constant movement, vibration, and everything is fluid; "...one must concede that everything is not solid and fixed and that the hardest solids are only fluids that are slightly more viscous than others. And that edges and boundaries are fluctuating. Fluctuating fluid" (Serres and Latour 1995: 107). A world of transitory hardenings and fluids lead one to think of mixtures and so let me turn now to think about what this rethinking of space and time means for scales, boundaries and the limits of global and world cities.

SCALING WORLD CITIES?

A rethinking of the "limits" of the city is needed. If "the city is everywhere and in everything" (Amin and Thrift 2002: 1) then a shift away from thinking about boundaries and limits altogether is necessary. The recently deceased Stephen Jay Gould recounts a story about boundaries in one of his books. Gould (1991: 200) tells the story of the Siamese twin girls "Ritta-Christina" from Sardinia who died in 1829 (see Figure 1). The commentators or pundits of that time went to some effort to try to draw a boundary either around the girls (to define as one person) or between the girls (to define as two persons). However, despite all the argument the issue could not be resolved precisely because it had

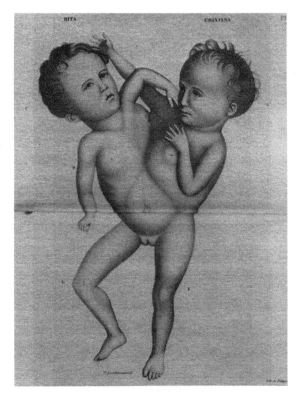

Figure 1 Challenging boundaries: Ritta-Christina

. . . no answer expressed in terms these pundits sought. Their categories were wrong or limited. The boundaries between oneness and twoness are human impositions, not nature's taxonomy. Ritta-Christina, formed from a single egg that failed to divide completely in twinning, born with two heads and two brains but only one lower half, was in part one, and in part two – not a blend, not one-and-a-half, but an object embodying the essential definitions of both oneness and twoness, depending upon the question asked or the perspective assumed.

In other words, "Ritta-Christina" defied the categories of the time (based upon a system of whole numbers and fractions), and pundits at the time were unable to make the jump to thinking of the girls as a continuum (a plane of consistency). Indeed, Gould's story suggests how we might now think about boundaries and so space and time in the contemporary city. What would cities be like if we were to think of them as no more than the undefined middle of a continuum? For example, how might we think of London or New York if we think of them as a continuum? If we were to change our mind set to accept this idea would we be more likely to agree with journalists (McGuire and Chan 2000) that New York and London are losing their specificity by becoming one bicontinental megalopolis (which perhaps could be called NY-LON)? Let me think a little bit more about this continuum and rejection of boundaries that Gould points to because as Foucault (1977: 34) himself observed, ". . . a limit could not exist if it were absolutely uncrossable . . ."

The theoretical vocabulary and language of political-economy is that of nests of scales that draw boundaries and define or categorise territories as local, regional, national, or global. But such a discourse of scalar and territorial relativisation stems flow and so is antithetical to a topology of circulation and network folding. To visualise the shift I am proposing see Figure 2. The diagram visualises the interaction of two cities: (a) shows two cities as separate and bounded entities. They are quite distinct and are reminiscent of how cities were portrayed as hermetically sealed or bounded entities by urbanists such as Mumford and Wirth; (b) portrays how relations between cities have been conceptualised and investigated by Beaverstock, Smith and Taylor. We used the idea of in-out

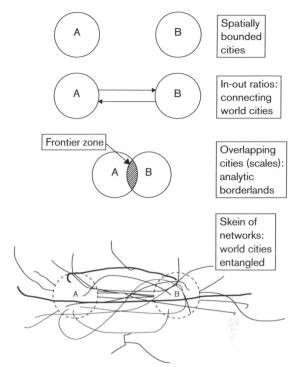

Figure 2 Conceptualising inter-city relations (two world cities: A and B)

ratios to build large data sets of the connections between world cities as part of a network; (c) visualises the idea of "frontier zones" that Sassen has been developing in recent years. It clearly shows Sassen's attempt to develop a new kind of "area studies" in globalisation research as she encourages us to investigate the spaces and times of zones of commonality that are produced when and where geographical scales overlap; (d) is a diagram that I have developed from Latour (1993) to try to begin to visualise world cities as "spatial nonconformities" (Law and Hetherington 2000), ". . . as having a fibrous, thread-like, wiry, stringy, ropy, capillary character that is never captured by the notions of levels, layers, territories, spheres, categories, structure, systems" (Latour 1998: 2). Cities here are hybrid and porous trans-local sites that are criss-crossed by the multiple lines of networks that are more or less long and more or less durable. Here a geography such as that produced by Sassen has melted precisely because there is no inside and outside if cities are open intensities where space and time are "out of joint."

WORLD CITIES AS "BODIES WITHOUT ORGANS"

One of the themes of Deleuze's philosophy (and borrowed from Antonin Artaud) is that of a "Body without Organs" (BwO) which forms a part of his (and Guattari's) schizoanalysis (these ideas resonate with the deconstruction of Derrida). The purpose of their philosophy is to counter, destabilise, short-circuit any force, power, or desire that strives to restrict, capture, fix, manage, redefine, specify or limit the flows that make the world a hotbed of flux and fluidity. In other words, the BwO is best thought of as a way of visualising the city as unformed, unorganised, and non-stratified, as always in the process of formation and deformation and so eluding fixed categories, a transient nomad space-time that does not dissect the city into either segments and "things" (a reductive Cartesianism) or structures and processes (a reductive political-economy). The BwO is a process of continuous coupling, chains of machines that facilitate endless flow and flux: "A flow of milk between a breast machine and a mouth machine, or a flow of words between a mouth machine and an ear machine . . ." (Bogue 1989: 91). In other words, a pathological reductive approach to the city that breaks it apart to produce ever more "units" for analysis (as in body → chest → sternum → heart → ventricle, etc.) is not for these poststructuralists (see Deleuze and Guattari (1987) on the differences between smooth deterritorialised space and striated territorialised space). Deleuze would prefer a horizontal ontology where objects are understood by their powers, surfaces and effects (Shields 1996), and are not explained away by vertical philosophies – such as Realism or Althusserian Marxism – that claim explanatory power by reducing surfaces, appearances, epiphenomena, superstructures, effects to little more than the product of deep underlying structures, systems, processes, forces, socio-economic relations. The BwO is not like Frankenstein's monster (a composite of stitched together parts) but is rather an undissected topology of intensities and relations. The point of thinking the city as a BwO (as an unconstrained flow, a set of relations, an actor-network) is to remind us to think through what it means to stop flows through an organising principle (e.g. Sassen's scales). By not stopping the flows (every-thing is made of flows) a new urbanism comes into view that bypasses vertical thinking by refusing to ossify or freeze the flow of the world into unities.

Researching global and world cities as BwOs (a continuum or plane of consistency) breaks with the philosophies, methods and styles of all previous urbanisms and consequently is presented here as an image of thought for a new urbanism that is concerned with a broader analysis, with connection rather than dissection, with the multiple rather than the singular, with the non-linear rather than the linear, with the disproportionate as well as the proportionate, and with the distant as well as the near. In short, the BwO is a way of not losing a city's rhizomatic hybridity, the range of its networks, and so a space-time that, ". . . is always more than 'mine, here and now', it is 'theirs, then and there'" (Shields 1996: 243).

ENDING

This chapter has argued that cities are heterogeneous assemblages of differential relations which v'b,a'e making, remaking, and unmaking the multiple spaces and times of urban networks. Or to quote Latour (1997b: 174) ". . . time and space are the consequences of the way in which bodies relate to one another." In this paper I have attempted to shift our thinking about world and global cities away from nouns, introversion, the singular, and exclusivity to verbs, extroversion, the multiple, and connectivity. I have argued that city networks are best understood not through points, lines, and boundaries (or the language of clustering, agglomeration, and localisation (see Amin and Thrift 2002)), but as BwOs, continuous circulations of flux and chaosmosis where there can be no summation and so no integrity. Analogous to Deleuze's rhizome, city networks are in constant movement, undergoing a series of transformations, translations, and traductions that defy capture by the exclusionary dualistic thinking of non-poststructuralist social theory.

In short, this chapter contributes to the development of a "new urbanism" in that it is attempting to transform the theory that drives the global and world cities literatures. The dominant theoretical ideas in the field of global/world cities research have remained unchanged for decades. Those following world-systems analysis (e.g. writers such as

Friedmann and Taylor etc.) or political-economy (e.g. writers such as Castells and Sassen etc.) are keen to talk about, or simply measure (if in fact they are doing that at all), global/world city networks without questioning the fundamental contradiction that the theories they follow were never designed to take the idea of networks seriously. Quite simply this paper is an attempt to do both something new and simultaneously to demonstrate that these dominant theories are at best not helping us make progress in understanding city networks, and at worst propelled just by inertia, like the *grande machine* in Alfred Jarry's novel *Le Surmâle* (1945) motivated by the cadaveric rigour of its cyclists, or, as Hegel put it, by "the life moving within itself, of that which is dead."

ALTERNATE ENDING

With this paper, and a previous paper (Smith 2003), I am beginning to work out a new research agenda (an alternative tradition) for globalisation and global/world cities research. I am starting to argue that after coming to a virtual "dead stop" (Thrift 1998) this research field can be restarted through an engagement with, and a development of, poststructuralist theory, actor-network theory, non-representational theory, and complexity theory. As an alternate ending I will briefly point to some of the beginnings or possibilities that these theories offer for further progress:

Poststructuralism: Deleuze's philosophy is above all one of connections and the idea that we know the world because we are connected with it (rather than the more conventional idea that we know "the real world" through objective knowledge). The BwO provides us with a way to visualise the connectivity of global and world cities in a less spatially and temporally regular world. The BwO does not have form or function, it is not a singularity, not a subject or even an object. Rather the BwO is an actor-network, a set of all kinds of relations, a spatial formation that embodies those topologies of movement that are concerning so many social sciences these days: fluids, flows, folds, and networks. In short, the BwO is how I visualise global and world cities because the focus is on relations and a refusal to ossify or freeze the flow of the world into unities.

Actor-Network Theory: Global networks, space and time, are not static, fixed, given, but are made, remade, and unmade. With ANT global and world cities can be viewed and researched as "switchers" (as middles, intermediaries, *in* networks), rather than centers of "command and control" as Sassen supposes. The idea that global cities somehow "command and control" the global economy is a highly problematic idea (see Thrift 1998), and it seems far more likely that global and world cities follow a networking logic than a command logic. From a Latourian perspective a city is comprised of networks of relations that are made up of humans, nonhumans, and representations that are all important in the functioning and maintenance of those networks. The capacity of a global or world city to "command and control" is governed by its participation as a "switcher" in networks because power is only exercised through any actants' ability to enrol and mobilise others to perform in "their" network. Consequently a successful network is an arrangement that enrols actants to produce apparently stable patterns of purpose and action. In short, I think that global and world cities are "switches," intermediaries, "middles" in a continuum (think of "Ritta-Christina").

Non-Representational Theory: Deleuze's philosophy and ANT point towards non-representational thought because they emphasise knowing through connection and participation. Non-representational theory is important because it reminds us to take the "doings" (the performances and practices) of actants in networks seriously. What is more it reminds us that what matters might not be written down, but may be non-cognitive, improvised or learned behaviour that is as yet unrecorded. Remembering the *as yet unrecorded* is important when thinking about how one should go about conducting empirical research into global and world city networks. In short, I think that what many people have considered to be unimportant is potentially very important. And that brings me on to complexity theory and the ideas of disproportionality and emergence.

Complexity Theory: Global and world cities are caught up in highly complex open-ended networks that are replete with space-time paths. Consequently it is obvious that theories need to be developed that can somehow "deal" with hyper-complexity. Poststructuralism and ANT do handle complexity, for example, Deleuze's philosophy of the fold enables one to appreciate that the world is complicated; "The multiple is not only what has many parts but also what is folded in many ways" (1993: 3). Folds are indeed everywhere, but it also seems to me that Deleuze and ANT are not enough and some of the ideas coming out of complexity theory might be tremendously useful for global urban studies. For example, complexity theory offers both a new way to conceptualise the *disproportionality* of global networks, and introduces the idea of *emergence* (rather than ideas of "shifts" in scale) which might well provide us with a way to avoid not seeing "the wood for the trees" when we focus on the details of this or that network in our empirical research.

In short, I think that through these theories progress can again be made in the field of global and world city research.

REFERENCES FROM THE READING

Amin, A. and Thrift, N. (2002) *Cities*. Cambridge: Polity.

Ballard, J.G. (1992) Project for a glossary of the twentieth century. In J. Crary and S. Kwinter (eds) *Zone 6: Incorporations*. New York: Zone Books, 268–279.

Bogue, R. (1989) *Deleuze and Guattari*. London: Routledge.

Deleuze, G. (1993) *The Fold: Leibniz and the Baroque*. Minneapolis, MN: University of Minnesota Press.

Deleuze, G. (1995) *Negotiations*. New York: Columbia University Press.

Deleuze, G. and Guattari, F. (1987) *A Thousand Plateaus*. Minneapolis, MN: University of Minnesota Press.

Doel, M. (1999) *Poststructuralist Geographies*. Edinburgh: Edinburgh University Press.

Foucault, M. (1977) *Language, Counter-Memory, Practice*. New York: Cornell University Press.

Gould, S.J. (1991) *Time's Arrow, Time's Cycle*. London: Penguin.

Jarry, A. (1945) *Le Surmâle*. Paris: Fasquelle.

Latour, B. (1993) *We Have Never Been Modern*. Hemel Hempstead: Harvester Wheatsheaf.

Latour, B. (1997a) The trouble with Actor-Network Theory. Available at: http://www.ensmp.fr/~latour/artpop/p67.html

Latour, B. (1997b) Trains of thought: Piaget, formalism, and the fifth dimension, *Common Knowledge*, 6, 170–191.

Latour, B. (1998) On actor-network theory: a few clarifications, 1–9. Available at: http://amsterdam.nettime.org/Lists-Archives/nettime-1-9801/msg00019.html (accessed February 15, 2005).

Law, J. and Hetherington, K. (2000) Materialities, spatialities, globalities. In J. Bryson, P. Daniels, N. Henry and J. Pollard (eds) *Knowledge, Space, Economy*. London: Routledge, 34–49.

McGuire, S. and Chan, M. (2000) The NY-LON life, *Newsweek*, November 13, 40–47.

Serres, M. and Latour, B. (1995) *Conversations on Science, Culture and Time*. Ann Arbor, MI: Michigan University Press.

Shields, R. (1996) A guide to urban representation and what to do about it: alternative traditions of urban theory. In A.D. King (ed.) *Representing the City*. London: Macmillan, 227–252.

Smith, R.G. (2003) World city actor-networks, *Progress in Human Geography*, 27, 1, 25–44.

Thrift, N. (1998) Distance is not a safety zone but a field of tension: mobile geographies and world cities. In S.G.E. Gravesteijn, S. Griensven and M. Smidt (eds) *Timing Global Cities*. Utrecht: Netherlands Geographical Studies, 54–66.

50

"The Urban Revolution"

from *The Urban Revolution*
(2003 [originally published 1968])

Henri Lefebvre

Editors' introduction

The French theorist Henri Lefebvre (1901–1991) has perhaps been the most important critical urban thinker since the early 1960s. His work on space (1991a), everyday life (1991b) and urbanism (1996, 2003) has influenced generations of critical urbanist scholars and activists. Although Lefebvre never wrote about global cities in the specific sense in which they are understood by most contributors to this volume, we would argue that his approach to urbanization still has considerable salience for the contemporary study of globalized urbanization. Lefebvre's masterful book, *The Urban Revolution*, was completed in the midst of the May 1968 revolt in Paris, when students and workers were engaged in a massive uprising against the French state and the capitalist order. Beyond the national political crisis that framed the uprising, colonial and postcolonial revolutions were never far afield: for it was partly the historic defeat of French colonialism in Algeria a few years previously that had set the stage for this metropolitan revolt. It is against this background that Lefebvre developed his famous claim that "society is completely urbanized." Rather than adopting then-fashionable notions of "industrial" and "postindustrial" society, Lefebvre argued that a new phase of human development was unfolding that was based on the worldwide (spatial) extension and (temporal) acceleration of urbanization processes. We would argue that the multifaceted sociospatial transformations Lefebvre described as the "urban revolution" must be viewed as an essential precondition for the emergence of a specialized world city network, as analyzed by the contributors to this Reader. Moreover, the notion of "complexification," as developed by Lefebvre below, usefully underscores the multilayered sociospatial fabric of contemporary global cities.

Lefebvre's contribution is also relevant to attempts to define the global city concept itself. While French students and workers revolted in Paris, Mao's Cultural Revolution was taking place in China. Influenced by leading intellectual Lin Biao, China's leaders maintained that the global system was now being divided into a "world city" and a "world countryside" – a rather crude version of the core–periphery model that was being developed during this same period by western theorists of the capitalist world-system.[1] Lefebvre, however, was unconvinced that the "world countryside" could succeed in resisting the urbanized core, particularly in light of the dramatic forms of capitalist (and state socialist) urbanization that were unfolding during the second half of the twentieth century; consequently, he takes issue with Mao's concept of a "world city." In Reading 50, Lefebvre argues forcefully against the notion that large-scale urbanization – the consolidation of the "world city" – could somehow be halted through revolutionary praxis based in the rural peripheries. Interestingly, versions of this notion continue to reappear, for instance in debates on postcolonialism and, most recently, in Michael Hardt and Antonio Negri's influential musings on "multitude" (2004: 123–124). In both cases, the showdown of the revolutionary rural masses with the urbanized core of the world-system is viewed as a key aspect of resistance to imperialism and empire. Other writers, however, have rejected such arguments and supported Lefebvre's viewpoint. John Friedmann (2002: 1–2), for example, suggests that "the urban transition will not be reversed" and that "[w]illful attempts at ruralization, such as in Mao Ze Dong's China or Pol Pot's Kampuchea, were thus never more than temporary reversals." Lefebvre's insistence on the impossibility of suppressing the complete urbanization of human existence thus provides a provocative, if very much open-ended, conclusion to the debates and analyses surveyed in this Reader.

FROM THE CITY TO URBAN SOCIETY

I'll begin with the following hypothesis: Society has been completely urbanized. This hypothesis implies a definition: An *urban society* is a society that results from a process of complete urbanization. This urbanization is virtual today, but will become real in the future.

Here, I use the term "urban society" to refer to the society that results from industrialization, which is a process of domination that absorbs agricultural production. This urban society cannot take shape conceptually until the end of a process during which the old urban forms, the end result of a series of *discontinuous* transformations, burst apart. An important aspect of the theoretical problem is the ability to situate the discontinuities and continuities with respect to one another. How could any absolute discontinuities exist without an underlying continuity, without support, without some inherent process? Conversely, how can we have continuity without crises, without the appearance of new elements or relationships?

Instead of the term "postindustrial society" – the society that is born of industrialization and succeeds it – I will use "urban society," a term that refers to tendencies, orientations, and virtualities, rather than any preordained reality. Such usage in no way precludes a critical examination of contemporary reality, such as the analysis of the "bureaucratic society of controlled consumption."

Economic growth and industrialization have become self-legitimating, extending their effects to entire territories, regions, nations, and continents. As a result, the traditional unit typical of peasant life, namely the village, has been transformed. Absorbed or obliterated by larger units, it has become an integral part of industrial production and consumption. The concentration of the population goes hand in hand with that of the mode of production. The *urban fabric* grows, extends its borders, corrodes the residue of agrarian life. This expression, "urban fabric," does not narrowly define the built world of cities but all manifestations of the dominance of the city over the country. In this sense, a vacation home, a highway, a supermarket in the countryside are all part of the urban fabric. Of varying density, thickness, and activity, the only regions untouched by it are those that are stagnant or dying, those that are given over to "nature." As this global process of industrialization and urbanization was taking place, the large cities exploded, giving rise to growths of dubious value: suburbs, residential conglomerations and industrial complexes, satellite cities that differed little from urbanized towns. Small and midsize cities became dependencies, partial colonies of the metropolis. In this way my hypothesis serves both as a point of arrival for existing knowledge and a point of departure for a new study and new projects: complete urbanization. The hypothesis is anticipatory. It prolongs the fundamental tendency of the present. Urban society is gestating in and through the "bureaucratic society of controlled consumption." The expression "urban society" meets a theoretical need. It is more than simply a literary or pedagogical device, or even the expression of some form of acquired knowledge; it is an elaboration, a search, a conceptual formulation. A movement of thought toward a certain *concrete,* and perhaps toward *the* concrete, assumes shape and detail. This movement, if it proves to be true, will lead to a practice, *urban practice*, that is finally or newly comprehended.

Similarly, by "urban revolution" I refer to the transformations that affect contemporary society, ranging from the period when questions of growth and industrialization predominate (models, plans, programs) to the period when the urban problematic becomes predominant, when the search for solutions and modalities unique to urban society are foremost. Some of these transformations are sudden; others are gradual, planned, determined. But which ones? This is a legitimate question. It is by no means certain in advance that the answer will be clear, intellectually satisfying, or unambiguous. The words "urban revolution" do not in themselves refer to actions that are violent. Nor do they exclude them.

We can draw an axis as follows:

$$0 \text{——————————} 100\%$$

The axis runs from the complete absence of urbanization ("pure nature," the earth abandoned to the elements) on the left to the completion of the process on the right. A signifier for this signified – the *urban* (the urban reality) – this axis is both spatial and temporal: spatial because the process extends through space, which it modifies; temporal because it develops over time. Temporality, initially of secondary importance, eventually becomes the predominant aspect of practice and history. This schema presents no more than an aspect of this history, a division of time that is both abstract and arbitrary and gives rise to operations (periodizations) that have no absolute privilege but are as necessary (relative) as other divisions.

The rise of the mercantile city, which was grafted onto the political city but promoted its own ascendancy, was soon followed by the appearance of industrial capital and, consequently, the *industrial city*. This requires further explanation. Was industry associated with the city? One would assume it to be associated with the *non-city*, the absence or rupture of urban reality. We know that industry initially developed near the sources of energy (coal and water), raw materials (metals, textiles), and manpower reserves. Industry gradually made its way into the city in search of capital and capitalists, markets, and an abundant supply of low-cost labor. It could locate itself anywhere, therefore, but sooner or later made its way into existing cities or created new cities, although it was prepared to move elsewhere if there was an economic advantage in doing so. Just as the political city resisted the conquest

– half-pacific, half-violent – of the merchants, exchange, and money, similarly the political and mercantile city defended itself from being taken over by a nascent industry, industrial capital, and capital itself. But how did it do this? Through corporatism, by establishing relationships. Historical continuity and evolution mask the effects and ruptures associated with such transitions. Yet something strange and wonderful was also taking place, which helped renew dialectical thought: the non-city and the anti-city would conquer the city, penetrate it, break it apart, and in so doing extend it immeasurably, bringing about the urbanization of society and the growth of the urban fabric that covered what was left of the city prior to the arrival of industry. This extraordinary movement has escaped our attention and has been described in piecemeal fashion because ideologues have tried to eliminate dialectical thought and the analysis of contradictions in favor of logical thought – that is, the identification of coherence and nothing but coherence. Urban reality, simultaneously amplified and exploded, thus loses the features it inherited from the previous period: organic totality, belonging, an uplifting image, a sense of space that was measured and dominated by monumental splendor. It was populated with signs of the urban within the dissolution of urbanity; it became stipulative, repressive, marked by signals, summary codes for circulation (routes), and signage. It was sometimes read as a rough draft, sometimes as an authoritarian message. It was imperious. But none of these descriptive terms completely describes the historical process of implosion–explosion (a metaphor borrowed from nuclear physics) that occurred: the tremendous concentration (of people, activities, wealth, goods, objects, instruments, means, and thought) of urban reality and the immense explosion, the projection of numerous, disjunct fragments (peripheries, suburbs, vacation homes, satellite towns) into space.

The *industrial city* (often a shapeless town, a barely urban agglomeration, a conglomerate, or conurbation like the Ruhr Valley) serves as a prelude to a *critical zone*. At this moment, the effects of implosion–explosion are most fully felt. The increase in industrial production is superimposed on the growth of commercial exchange and multiplies the number of such exchanges. This growth extends from simple barter to the global market, from the simple exchange between two individuals all the way to the exchange of products, works of art, ideas, and human beings. Buying and selling, merchandise and market, money and capital appear to sweep away all obstacles. During this period of generalization, the effect of the process – namely the urban reality – becomes both cause and reason. Induced factors become dominant (inductors). The *urban problematic* becomes a global phenomenon. Can urban reality be defined as a "superstructure" on the surface of the economic structure, whether capitalist or socialist? The simple result of growth and productive forces? Simply a modest marginal reality compared with production? Not at all. Urban reality modifies the relations of production without being sufficient to transform them. It becomes a productive force, like science. Space and the politics of space "express" social relationships but react against them. Obviously, if an urban reality manifests itself and becomes dominant, it does so only through the urban problematic. What can be done to change this? How can we build cities or "something" that replaces what was formerly the City? How can we reconceptualize the urban phenomenon? How can we formulate, classify, and order the innumerable questions that arise, questions that move, although not without considerable resistance, to the forefront of our awareness? Can we achieve significant progress in theory and practice so that our consciousness can comprehend a reality that overflows it and a possible that flees before its grasp? We can represent this process as follows: *implosion–explosion* (urban concentration, rural exodus, extension of the urban fabric, complete subordination of the agrarian to the urban).

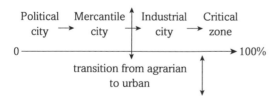

The onset of urban society and the modalities of urbanization depend on the characteristics of society as it existed during the course of industrialization (neocapitalist or socialist, full economic growth or intense automation). The onset of urban society at different times, the implications and consequences of these initial differences, are part of the problematic associated with the urban phenomenon, or simply the "urban." These terms are preferable to the word "city," which appears to designate a clearly defined, definitive *object*, a scientific object and the immediate goal of action, whereas the theoretical approach requires a critique of this "object" and a more complex notion of the virtual or possible object. Within this perspective there is no science of the city (such as urban sociology or urban economy), but an emerging understanding of the overall process, as well as its term (goal and direction).

The urban (an abbreviated form of urban society) can therefore be defined not as an accomplished reality, situated behind the actual in time, but, on the contrary, as a horizon, an illuminating virtuality. The *virtual object* is nothing but planetary society and the "global city," and it stands outside the global and planetary crisis of reality and thought, outside the old borders that had been drawn when agriculture was dominant and that were maintained during the growth of exchange and industrial production. Nevertheless, the urban problematic can't absorb every problem. There are problems that are unique to agriculture and industry, even though the urban reality modifies them. Moreover, the urban problematic requires that we exercise considerable caution when exploring the realm of the possible. It is the analyst's responsibility to identify and describe the various forms of urbanization and explain what happens to the forms, functions, and urban structures that are transformed by the breakup of the ancient city and the process of generalized urbanization. Until now the critical phase was perceived as a kind of black box. We know what enters the box, and sometimes we see what comes out, but we don't know what goes on inside. This makes conventional procedures of forecasting and projection useless, since they extrapolate from the actual, from a set of facts. Projections and forecasts have a determined basis only in the fragmentary sciences: demography, for example, or political economy. But what is at stake here, "objectively," is a totality.

URBAN SOCIETY

During this exploration, the urban phenomenon appears as something other than, as something more than, a superstructure (of the mode of production). I say this in response to a form of Marxist dogmatism that manifests itself in a variety of ways. The urban problematic is worldwide. The same problems are found in socialism and in capitalism – along with the failure to respond. Urban society can only be defined as global. Virtually, it covers the planet by recreating nature, which has been wiped out by the industrial exploitation of natural resources (material and "human"), by the destruction of so-called natural particularities.

Moreover, the urban phenomenon has had a profound effect on the methods of production: productive forces, relationships of production, and the contradictions between them. It both extends and accentuates, on a new plane, the social character of productive labor and its conflict with the ownership (private) of the means of production. It continues the "socialization of society," which is another way of saying that the urban does not eliminate industrial contradictions. It does not resolve them for the sole reason that it has become dominant. What's more, the conflicts inherent in production (in the relationships of production and capitalist ownership as well as in "socialist" society) hinder the urban phenomenon, prevent urban development, reducing it to growth. This is particularly true of the action of the state under capitalism and state socialism.

To summarize then: Society becomes increasingly complex with the transition from the rural to the industrial and from the industrial to the urban. This multifaceted complexification affects space as well as time, for the complexification of space and the objects that occupy space cannot occur without a complexification of time and the activities that occur over time. This space is occupied by interrelated networks, relationships that are defined by interference. Its homogeneity corresponds to intentions, unified strategies, and systematized logics, on the one hand, and reductive, and consequently simplifying, representations, on the other. At the same time, differences become more pronounced in populating this space, which tends, like any abstract space, toward homogeneity (quantitative, geometric, and logical space). This, in turn, results in conflict and a strange sense of unease. For this space tends toward a unique code, an absolute system, that of exchange and

exchange value, of the logical thing and the logic of things. At the same time, it is filled with subsystems, partial codes, messages, and signifiers that do not become part of the unitary procedure that the space stipulates, prescribes, and inscribes in various ways.

The theory of complexification anticipates the revenge of development over growth. The same is true for the theory of urban society. This revenge is only just beginning. The basic proposition, that growth cannot continue indefinitely and that the means can remain an end without a catastrophe occurring, still seems paradoxical.

These considerations evoke the prodigious extension of the urban to the entire planet, that is, urban society, its virtualities and potential. It goes without saying that this extension-expansion is not going to be problem-free. Indeed, it has been shown that the urban phenomenon tends to overflow borders, while commercial exchange and industrial and financial organizations, which once seemed to abolish those territorial limits (through the global market, through multinationals), now appear to reaffirm them. In any event, the effects of a possible rupture in industry and finance (a crisis of overproduction, a monetary crisis) would be accentuated by an extension of the urban phenomenon and the formation of urban society.

I have already introduced the idea of the "global city," generally attributed to Maoism, if not Mao Tse-tung himself. I would now like to develop this idea. The global city extends the traditional concept and image of the city to a global scale: a political center for the administration, protection, and operation of a vast territory. This is appropriate for the oriental city within the framework of an Asian mode of production. However, urban society cannot be constructed on the ruins of the classical city alone. In the West, this city has already begun to fragment. This fragmentation (explosion–implosion) may appear to be a precursor of urban society. It is part of its problematic and the critical phase that precedes it. However, a known strategy, which specifically makes use of urbanism, tends to view the political city as a decision-making center. Such a center is obviously not limited to collecting information upstream and distributing it downstream. It is not just a center of abstract decision making but a center of power. Yet power requires wealth, and vice versa. That is, the decision-making center, in the strategy being ana-lyzed here, will serve as a point of attachment to the soil for a hyperorganized and rigidly system-atized state. Formerly, the entire metropolitan land area played a central role with respect to the colonies and semicolonies, sucking up wealth, imposing its own order. Today, domination is consolidated in a physical locale, a capital (or a decision-making center that does not necessarily coincide with the capital). As a result, control is exer-cised throughout the national territory, which is transformed into a semicolony.

Part of my analysis may appear at first glance to correspond to the so-called Maoist interpreta-tion of the "global city," but this interpretation raises a number of objections. There is nothing that prevents emerging centers of power from encoun-tering obstacles and failing. What's more, any contradictions that occur no longer take place between city and country. The principal contra-diction is shifted to the urban phenomenon itself: between the centrality of power and other forms of centrality, between the "wealth-power" center and the periphery, between integration and segregation.

Is the urban phenomenon the *total social phe-nomenon* long sought for by sociologists? Yes and no. Yes, in the sense that it tends toward totality without ever achieving it, that it is essentially totalizing (centrality) but that this totality is never effected. Yes, in the sense that no partial deter-minism, no fragmentary knowledge can exhaust it; it is simultaneously historical, demographic, geo-graphic, economic, sociologic, psychologic, semi-ologic, and so on. It "is" that and more (thing or non-thing) besides: *form, for* example. In other words, a void, but one that demands or calls forth a content. If the urban is total, it is not total in the way a thing can be, as content that has been amassed, but in the way that thought is, which con-tinues its activity of concentration endlessly but can never hold or maintain that state of concentration, which assembles elements continuously and dis-covers what it has assembled through a new and different form of concentration. Centrality defines the u-topic (that which has no place and searches for it). The u-topic defines centrality.

But neither the separation of fragment and content nor their confused union can define (and therefore express) the urban phenomenon. For it incorporates a *total reading*, combining the vocabularies (partial readings) of geographers, demographers, economists, sociologists, semiolo-gists, and others. These readings take place on dif-ferent levels. The phenomenon cannot be defined by their sum or synthesis or superposition. In this sense, it is not a totality. Similarly, it overcomes the separation between accident and necessity, but their synthesis doesn't determine it, assuming such synthesis can be determined. This is simply a repetition of the paradox of the urban phenom-enon, a paradox that in no way gives it preced-ence over the fundamental paradox of thought and awareness. For it is undoubtedly the same. The urban is specific: it is localized and focused. It is locally intensified and doesn't exist without that localization, or center. Thought and thinking don't take place unless they are themselves localized. The specificity of the fact, the event, is a given. And, consequently, a requirement. Near order occurs around a point, taken as a (momentary) center, which is produced by practice and can be grasped through analysis. This defines an isotopy. At the same time, the urban phenomenon is colossal; its prodigious extension-expansion cannot be con-strained. While encompassing near order, a *distant order* groups distinct specificities, assembles them according to their differences (heterotopies). But isotopy and heterotopy clash everywhere and always, engendering an *elsewhere*. Although ini-tially indispensable, the *transformed* centrality that results will be reabsorbed into the fabric of space-time. In this way the dialectical movement of the specific and the colossal, of place and non-place (elsewhere), of urban order and urban disorder assumes form (reveals itself as form).

The urban is not produced like agriculture or industry. Yet, as an act that assembles and dis-tributes, it does create. Similarly, manufacturing at one time became a productive force and eco-nomic category simply because it brought together labor and tools (technology), which were formerly dispersed. In this sense, the urban phe-nomenon contains a praxis (urban practice). Its form, as such, cannot be reduced to other forms (it is not isomorphous with other forms and structures), but it absorbs and transforms them.

The procedure for accessing urban reality as a form is reversed once the process is complete. In this way we can use linguistics to define isotopy and heterotopy. Once they have been identified in the urban text, these concepts assume a different meaning. Isn't it because human habitations assume the form that they do that they can be recognized in discourse? The urban is associated with a discourse and a route, or pathway. And it is for this reason, or formal cause, that there are different discourses and pathways in language. One cannot be separated from the other. Although different, language and dwelling are indissolubly combined. Is it surprising then that there is a *paradigm of* the urban (high and low, private and public), just as there is for habiting (open and closed, intimate and public), although neither the urban nor habiting can be defined by a simple discourse or by a system? If there is any logic inherent in the urban and the habiting it implies, it is not the logic of a system (or a subject or an object). It is the logic of thought (subject) that looks for a content (object). It is for this reason that our understanding of the urban requires that we simultaneously abandon our illusions of subjectivity (representation, ideology) and objectivity (causality, partial determinism).

The urban consolidates. As a form, the urban transforms what it brings together (concentrates). It consciously creates difference where no awareness of difference existed: what was only distinct, what was once attached to particularities in the field. It consolidates everything, including determinisms, heterogeneous materials and contents, prior order and disorder, conflict, preexisting communications and forms of communication. As a transforming form, the urban destructures and restructures its elements: the messages and codes that arise in the industrial and agrarian domains.

The urban also contains a negative power, which can easily appear harmful. Nature, a desire, and what we call culture (and what the industrial era dissociated from nature, while during predominately agrarian periods, nature and culture were indissoluble) are reworked and combined in urban society. Heterogeneous, if not heteroclite, these contents are put to the test. Thus, by way of analogy, agricultural exploitation (the farm) and the enterprise (which came into existence with the rise of manufacturing) are put to the test, are transformed, and are incorporated in new forms within the urban fabric. We could consider this a form of second-order creativity (*poiesis*), agricultural and industrial production being forms of first-order creativity. This does not mean that the urban phenomenon can be equated with second-order discourse, metalanguage, exegesis, or commentary on industrial production. No, second-order creation and the secondary naturality of the urban serve to *multiply* rather than reduce or reflect creative activity. This raises the issue of an activity that produces (creates) meanings from elements that already possess signification (rather than units similar to phonemes, sounds or signs devoid of signification). From this point of view, the urban would create situations and acts just as it does objects.

NOTE

1 We are grateful to Klaus Ronneberger and Horst Müller for helping us clarify this connection in the work of Lefebvre.

REFERENCES FROM THE READING

Friedmann, J. (2002) *The Prospect of Cities.* Minneapolis, MN: University of Minnesota Press.

Hardt, M. and Negri, A. (2004) *Multitude.* New York: Penguin.

Lefebvre, H. (1991a [1974]) *The Production of Space*, trans. D.N. Smith. Cambridge, MA: Blackwell.

Lefebvre, H. (1991b [1947]) *Critique of Everyday Life, Volume 1*, trans. J. Moore. London: Verso.

Lefebvre, H. (1996) *Writings on Cities*, trans. E. Kofman and E. Lebas. Cambridge, MA: Blackwell.

Lefebvre, H. (2003 [1968]) *The Urban Revolution*, trans. R. Bononno. Minneapolis, MN: University of Minnesota Press.

ILLUSTRATION CREDITS

PLATES

1 Toronto © Roger Keil
2 La Défense, Paris © Roger Keil
3 Amsterdam © Roger Keil
4 London © Ute Lehrer
5 Manchester © Roger Keil
6 New York City © Roger Keil
7 New York City © Roger Keil
8 Chicago © Roger Keil
9 Los Angeles © Roger Keil
10 World Trade Center, New York, 1980 © Roger Keil
11 DGTAL HOUSE, London © Roger Keil
12 Tokyo © Takashi Machimura
13 New York City © Roger Keil
14 London © Roger Keil
15 Tokyo © Takashi Machimura
16 Renaissance Center, Detroit © Roger Keil
17 Spui Straat, Amsterdam © Roger Keil
18 Homeless camp, Los Angeles, 1987 © Roger Keil
19 Los Angeles © Roger Keil
20 Cape Town © Roger Keil
21 Walvis Bay, Namibia © Roger Keil
22 Frankfurt © Roger Keil
23 Green London © Ute Lehrer
24 Chongqing: the latest city in the world seeks the global image © Anne Haila
25 Conservative campaign advertising for the World City 1989 © Roger Keil
26 Frankfurt: activist map © Roger Keil
27 Frankfurt: northern suburb © Roger Keil
28 Amsterdam: squatter symbol © Roger Keil
29 Potsdamer Platz, Berlin © Roger Keil
30 Construction of Potsdamer Platz © Ute Lehrer
31 Info Box and Berlin Wall © Ute Lehrer
32 Shanghai © Henry Yeung
33 Frankfurt MesseTurm © Roger Keil
34 Singapore © Henry Yeung

FIGURES

Except as noted below, for figure source information, see copyright information for the reading in which the figure appears.

Figure 2 on page 349: World City Lego © 2005 The LEGO Group. LEGO and the LEGO logo are trademarks of the LEGO Group. The LEGO Group does not sponsor or endorse this textbook.

COPYRIGHT INFORMATION

PART ONE

1 Hall, Peter, "The metropolitan explosion." In *The World Cities*, 3rd edition. London: Heinemann, 1–5 © 1984 Orion Publishing, London. Reproduced from *The World Cities* by Peter Hall (© 1966, 1977 Peter Hall) by permission of PFD (www.pfd.co.uk) on behalf of Professor Sir Peter Hall.

2 Braudel, Fernand, "Divisions of space and time in Europe," In *The Perspective of the World,* Volume 3 of *Civilization and Capitalism, 15th–18th Century*, trans. Sian Reynolds. New York: Perennial Library, 21–45 © 1984 William Collins Sons & Co. Ltd. and Harper & Row, Publishers, Inc. First published in France as *Le Temps du Monde* © 1979 Librarie Armand Colin, Paris. Reprinted by permission of HarperCollins Publishers, Inc.

3 Rodriguez, Nestor and Feagin, Joe R., "Urban specialization in the world system: an investigation of historical cases," *Urban Affairs Quarterly*, 22, 2, 187–220 © 1986 Sage. Reprinted by permission of Sage Publications, Inc.

4 Abu-Lughod, Janet, "Global city formation in New York, Chicago and Los Angeles: an historical respective," In *New York, Chicago, Los Angeles: America's Global Cities*. Minneapolis, MN and London: Minnesota University Press, 2–5 (excerpts from introduction and concluding chapter) © 1999 University of Minnesota Press. Reprinted by permission.

5 Cohen, Robert B., "The new international division of labor, multinational corporations and urban hierarchy." In Michael Dear and Allen J. Scott (eds) *Urbanization and Urban Planning in Capitalist Societies*. London and New York: Methuen, 287–315 © 1981 Routledge/Taylor & Francis. Reprinted by permission.

6 Friedmann, John and Wolff, Goetz, "World city formation: an agenda for research and action," *International Journal of Urban and Regional Research*, 6, 3, 309–344 © 1982 Blackwell. Reprinted by permission of Blackwell Publishing.

7 Friedmann, John, "The world city hypothesis," *Development and Change*, 17, 69–83 © Reprinted with the permission of Blackwell Publishing.

PART TWO

8 Sudjic, Deyan, "100-mile cities." In *The 100-mile City*. San Diego, CA: Harcourt Brace © 1992 Deyan Sudjic. Reprinted by permission of Harcourt, Inc.

PART THREE

PART FOUR

PART FIVE

PART SIX

Index

Figures and tables in *italic*
"n" after page number indicates material in notes